$$R_T = \frac{1}{\Sigma G} \quad \text{(ohms)} \tag{13-8}$$

$$R_T = \frac{1}{\dfrac{1}{R_1} + \dfrac{1}{R_2} + \dfrac{1}{R_3}} \quad \text{(ohms)} \tag{13-9}$$

$$I_T = I_1 + I_2 + I_3 \quad \text{(amperes)} \tag{13-10}$$

$$I_T = \Sigma I \quad \text{(amperes)} \tag{13-11}$$

$$I_n = \frac{I_T G_n}{\Sigma G} \quad \text{(amperes)} \tag{13-12}$$

$$I_n = \frac{E}{R_n} \quad \text{(amperes)} \tag{13-13}$$

$$I_n = E G_n \quad \text{(amperes)} \tag{13-14}$$

$$R_{Th} \equiv R_N \tag{13-15}$$

$$R_N = R_{Th} = \frac{1}{\dfrac{1}{R_1} + \dfrac{1}{R_2}} = \frac{1}{\dfrac{1}{10} + \dfrac{1}{30}} = 7.5\ \Omega$$

$$I_N = E_{Th}/R_{Th} \quad \text{where } R_{Th} = R_N \tag{13-16}$$

$$E_{Th} = I_N R_N \quad \text{where } R_N = R_{Th} \tag{13-17}$$

$$y = mx + b \tag{15-1}$$

$$\text{Slope} = m = \frac{y_1 - y_2}{x_1 - x_2} \tag{15-2}$$

$$\begin{vmatrix} a_1 & b_1 \\ a_2 & b_2 \end{vmatrix} = a_1 b_2 - a_2 b_1 \tag{16-1}$$

$$\begin{cases} a_1 x + b_1 y = k_1 \\ a_2 x + b_2 y = k_2 \end{cases} \tag{16-2}$$

$$\Delta = \begin{vmatrix} a_1 & b_1 \\ a_2 & b_2 \end{vmatrix} = a_1 b_2 - a_2 b_1 \tag{16-3}$$

$$x = \frac{\begin{vmatrix} k_1 & b_1 \\ k_2 & b_2 \end{vmatrix}}{\Delta} = \frac{k_1 b_2 - k_2 b_1}{\Delta} \tag{16-4}$$

$$y = \frac{\begin{vmatrix} a_1 & k_1 \\ a_2 & k_2 \end{vmatrix}}{\Delta} = \frac{a_1 k_2 - a_2 k_1}{\Delta} \tag{16-5}$$

$$\begin{vmatrix} a_1 & b_1 & c_1 \\ a_2 & b_2 & c_2 \\ a_3 & b_3 & c_3 \end{vmatrix} = \begin{aligned} a_1 b_2 c_3 + b_1 c_2 a_3 + c_1 a_2 b_3 \\ - a_3 b_2 c_1 - b_3 c_2 a_1 - c_3 a_2 b_1 \end{aligned} \tag{16-6}$$

$$\begin{cases} a_1 x + b_1 y + c_1 z = k_1 \\ a_2 x + b_2 y + c_2 z = k_2 \\ a_3 x + b_3 y + c_3 z = k_3 \end{cases} \tag{16-7}$$

$$\Delta = \begin{vmatrix} a_1 & b_1 & c_1 \\ a_2 & b_2 & c_2 \\ a_3 & b_3 & c_3 \end{vmatrix} \tag{16-8}$$

$$x = \frac{\begin{vmatrix} k_1 & b_1 & c_1 \\ k_2 & b_2 & c_2 \\ k_3 & b_3 & c_3 \end{vmatrix}}{\Delta} \tag{16-9}$$

$$y = \frac{\begin{vmatrix} a_1 & k_1 & c_1 \\ a_2 & k_2 & c_2 \\ a_3 & k_3 & c_3 \end{vmatrix}}{\Delta} \tag{16-10}$$

$$z = \frac{\begin{vmatrix} a_1 & b_1 & k_1 \\ a_2 & b_2 & k_2 \\ a_3 & b_3 & k_3 \end{vmatrix}}{\Delta} \tag{16-11}$$

$$h_{fe} = \frac{\Delta i_c}{\Delta i_b} \bigg|_{v_{ce}} \tag{17-1}$$

$$h_{oe} = \frac{\Delta i_c}{\Delta v_{ce}} \bigg|_{i_b} \quad \text{(siemens)} \tag{17-2}$$

$$y_{fs} = \frac{\Delta i_d}{\Delta v_{gs}} \bigg|_{v_{ds}} \quad \text{(siemens)} \tag{17-3}$$

THIRD EDITION

MATHEMATICS
APPLIED
TO ELECTRONICS

James H. Harter
Mesa Community College
Mesa, Arizona

Wallace D. Beitzel
TRW Space and Technology Group
Redondo Beach, California

A RESTON BOOK
PRENTICE HALL, Englewood Cliffs, New Jersey 07632

Library of Congress Cataloging-in-Publication Data

HARTER, JAMES H.
    Mathematics applied to electronics.

    "A Reston book."
    Includes index.
    1. Electronics—Mathematics.   I. Beitzel, Wallace D.
II. Title.
TK7835.H28 1988   510'.246213   87–29273
ISBN 0-13-562299-9

Editorial/production supervision: *Kathryn Pavelec*
Cover photo: Courtesy of *Garry Gay/The Image Bank*
Cover design: *Meryl Poweski*
Manufacturing buyer: *Margaret Rizzi/Peter Havens*
Illustrator: *Marolyn Young*

Printed in the United States of America

10 9 8 7 6 5 4 3 2 1

ISBN   0-13-562299-9

PRENTICE-HALL INTERNATIONAL (UK) LIMITED, *London*
PRENTICE-HALL OF AUSTRALIA PTY. LIMITED, *Sydney*
PRENTICE-HALL CANADA INC., *Toronto*
PRENTICE-HALL HISPANOAMERICANA, S.A., *Mexico*
PRENTICE-HALL OF INDIA PRIVATE LIMITED, *New Delhi*
PRENTICE-HALL OF JAPAN, INC., *Tokyo*
SIMON & SCHUSTER ASIA PTE. LTD., *Singapore*
EDITORA PRENTICE-HALL DO BRASIL, LTDA., *Rio de Janeiro*

CONTENTS

4   ALGEBRA FUNDAMENTALS I   72

5   ALGEBRA FUNDAMENTALS II   86

6   SOLVING EQUATIONS   105

7   QUANTITIES AND UNITS OF MEASUREMENT   129

8   APPLYING MATHEMATICS TO ELECTRICAL CIRCUITS   150

CONTENTS

PREFACE

This book provides an understanding of mathematics as it is applied to electronics. It may be used in a formal setting such as a community college or in a self-study program. *Mathematics Applied to Electronics* is for those studying electronic technology, computer technology, or electromechanical automation technology.

Since the electronics curriculum needs the support of a large and diverse amount of mathematics, the content of this text is based on a "trade-off" between a detailed, formal, proof orientation and the need for expediency in developing a broad, general mathematics ability.

The sequence of the chapters and of the topics within each chapter has been planned to be as vital as possible and to be compatible with the circuits books currently in use. The book provides the reader with a wide exposure to mathematics while still expediting the learning process. This process is enhanced through the use of the calculator, which is an integral part of the text. The calculator is used whenever calculations are performed.

The early chapters include selected topics from prealgebra. The purpose of these initial chapters is twofold: first, to introduce the use of the calculator at the onset; and second, to focus attention on specific electronics-related mathematical topics at a level easily comprehended—thus providing the reader with positive reinforcement.

These early chapters are followed by several chapters dealing with the fundamentals of algebra, including the evaluation of formulas. This series of chapters culminates with a chapter devoted to linear equations.

Each section of theoretical chapters is followed by one or more application chapters. The application chapters serve to reinforce what was presented previously and to provide the reader with the all-important opportunity to transfer mathematical skills to electronic concepts.

Interspersed throughout the book are chapters and topics dealing with analytical geometry. These chapters are essential because so much valuable information is presented in graphical form in handbooks and data sheets.

The transcendental functions are covered after the algebraic functions. Included in these chapters are the logarithmic, exponential, and trigonometric functions. These topics are followed by a series of chapters covering the mathematics of alternating current. The text concludes with chapters dealing with math analysis, number systems, and computer logic. These final chapters are directed to those who need preparation in the study of calculus and computer technology.

This text is designed to guide the reader through the learning process by providing a means of coordinating the instruction in the classroom with outside assignments. The reader is helped by hundreds of detailed examples, figures, and problems. The use of SI units throughout enables the reader to make an easy transition to any one of the circuits books in use today.

The authors wish to thank the users of this text for their helpful comments and suggestions. Many of these suggestions have been incorporated into this edition. We would also like to acknowledge and thank Marjorie Streeter for her imaginative design and expert coordination of the production of the first edition, Patricia Rayner for her talent and experience in the production of the second edition, and Marolyn Young for her extraordinary effort in the preparation of the illustrations. Finally, we acknowledge Faybeth Harter for her typing of the initial and final draft of the manuscript and for her help in keeping the vocabulary and reading level simplified.

JAMES HARTER
Flagstaff, Arizona

WALLACE BEITZEL
Redondo Beach, California

1

SURVIVING
MATHEMATICS

This chapter is intended to help you to *"survive"* the educational process. More specifically, it is to help you survive mathematics applied to electronics. An overview of the scope and structure is presented. Assistance is offered in helping you to select a calculator. Then, we let you in on the assumptions that we have made about you.

1-1 YOU

"You are the best you there is." Be kind to yourself and read through this chapter. It is written to give you an understanding of the text and how to survive the stress-producing process of becoming educated.

You are from varied backgrounds, of different ages, and have had different experiences, but all of you have a common goal—to learn more about *"mathematics applied to electronics."* Give yourself a gift and listen to what we have to say to you.

1-2 THE SCOPE AND STRUCTURE OF THE TEXT

Scope

We have selected from the field of mathematics those topics that support your career goal in electronics. We have tailored the topics and have streamlined the presentation so that both the special needs of the technology are met and a reasonable level of instruction is maintained.

The text material starts at the prealgebra level, covers topics in algebra, number notation, units of measurement, and graphing; moves into systems of simultaneous equations, logarithmic functions, trigonometric functions, circular functions, mathematics of phasors, and dc and ac electric circuits; and concludes with math analysis, number systems, and computer logic.

Structure

A gradual progression from the known to the unknown, from the simple to the complex is achieved in the sequencing of the chapters and the topics within each chapter. The rigorous, meticulous, in-depth pursuit of particular topics common to formal courses in mathematics is not found in this book. A particular topic may be introduced in the early section of the text and then reintroduced at a later time for a more detailed or in-depth application. By using this technique, a reinforcing of the concept is possible.

We have purposefully structured the book so that you will receive a positive feeling about the course you are taking. We have provided you with hundreds of detailed examples to make it possible for you to "learn on your own" and become responsible for educating yourself. If the class you are in is taught in a traditional lecture demonstration manner, then the self-educating aspects of the text may be used to prepare for the next day's lesson and to assist in doing the assigned out-of-class work. In addition, if you have been absent, the material missed may be studied and mastered on an individual basis.

1-3 THE LEARNING PROCESS

The process of becoming educated is usually accompanied by a feeling of uncertainty. Learning a new idea for the first time, coupled with the experience of attending college, is very stressful. Many students face these situations and far too many fail in the first weeks of the semester because they were unable to "adjust" to the learning experience.

We as educators recognize these feelings and have made a conscious effort to keep the stress level down by:

1. Selecting only the topics necessary for a rounded exposure to mathematics as it applies to electronics.
2. Integrating the scientific calculator into the learning process.
3. Designing the presentation so that you may have sufficient time to become familiar with the ideas and learn the concepts.
4. Providing detailed examples, tables, illustrations, and explanations; thus sufficient information is presented in a variety of ways so that you will be interested and motivated.
5. Presenting the concepts in a clear, uncluttered manner reinforced with examples and graded problem sets.
6. Setting the reading level to communicate with you rather than to impress you or your instructor.

1-4 STUDY AND SURVIVAL

How, where, and when you study this text will greatly influence your survival of the educational process. If survival is in your plans, then prepare for class, do your homework, and study for your tests. You can "set yourself up to fail." You can make excuses and be preoccupied with "more important activities." You need to recognize that your education is your responsibility, not the responsibility of the instructor. Your survival in a course is dependent solely upon your choices of how, where, and when you study.

How

Study by reading the material in each section. Work through each example with pencil, paper, and calculator. If you are to become educated, you must be an active participant in the process.

Where

Study where you won't be distracted by friends, noise, or activities. Study where you will have all the materials needed for study. Remember, you must work through and write down the examples and problems. Success in mathematics comes by you *doing,* not by watching someone else.

When

Study each day. As a student you have a very limited time with which to work. Most of you have full- or part-time jobs, families, and other commitments, which when coupled with going to school leaves very little time for study. You must understand that much of your education takes place outside the classroom. When you do required assignments that use pencil, paper, and a calculator, you are educating yourself by becoming an active participant in the educational process. As a rule, one to three hours of active study are needed to fully learn a new idea. To corrupt an old saying, "an hour a day, keeps drop-out away."

The object of education is to survive the process of becoming educated. The means of survival is through active study, participation, and preparation. Be kind to yourself by choosing to study daily.

1-5 SELECTING A CALCULATOR

First and foremost, purchase a *full-function* scientific calculator with display and rounding capability. Use the information in Table 1-1 to guide you in your purchase, making sure that the calculator has *all* of the functions listed. Double check to be sure that the →P and →R are among the special function keys and that SCI is also on the keyboard. An inadequate calculator will cost you time and energy that would better be spent elsewhere.

Calculators use one of two data entry systems—either the *reverse Polish notation* (RPN) or the *algebraic entry system* (AES). The RPN system, commonly called the *Polish system,* is based on the work of the Polish mathematician Jan Lukasiewicz. Unlike the algebraic entry system, the Polish system does not use parentheses when entering numerical expressions into the calculator.

In calculators with the algebraic entry system (commonly called the *algebraic system*), the operators (+ − ÷ × etc.) come between the two num-

TABLE 1-1

SUMMARY OF CALCULATOR FUNCTIONS OR OPERATIONS

Typical Key Symbol	Function/Operation	Comments	Introduced in Chapter	Alternative Key Stroke
$+$	Add	Simple arithmetic operations	2	
$-$	Subtract		2	
$\times$	Multiply		2	
$\div$	Divide		2	
CHS	Change sign		2	$+/-$
y^x	Exponential	Raise number to power	2	x^y
$1/x$	Reciprocal	Calculate the reciprocal	3	
$\sqrt{x}$	Square root	Calculate square root	3	
x^2	Square x	Square a number	3	
FIX	Fix point notation	Display and rounding	3	
SCI	Scientific notation		3	
ENG	Engineering notation		3	
STO	Store	Memory		
RCL	Recall			
$x \blacktriangleright y$	Exchange x and y			EXC
EE	Enter exponent		3	EEX
log	Common logarithm	Logarithmic and exponential functions	22	
10^x	Common antilog		22	INV LOG
ln x	Natural logarithm		22	
e^x	Natural antilogarithm		22	INV LN
SIN	Sine	Trigonometric functions	24	
COS	Cosine		24	
TAN	Tangent		24	
SIN^{-1}	Arc sine	Inverse trigonometric functions	24	INV SIN
COS^{-1}	Arc cosine		24	INV COS
TAN^{-1}	Arc tangent		24	INV TAN
DEG	Degree	Angular mode selection	24	DRG
RAD	Radian		24	
→P	Rectangular to polar	Polar/rectangular coordinate conversion	26	
→R	Polar to rectangular		26	

bers, as in 7 × 3. This is not the case with reverse Polish notation. Before you purchase either an *"algebraic"* or a *"Polish"* calculator, try out various calculators, talk to other students in your department about their calculators, listen to and talk with your instructor, and look over Table 1-1 to become familiar with the functions and operations needed in a calculator.

Today's full-function scientific calculators are usually programmable. You will find that once the program feature is learned and understood, you will have added flexibility in your calculations. Being able to program your calculator to repeatedly solve a function for different values of the variables can be both convenient and timesaving. From time to time you will see in the margin the note ( ), which will alert you to topics and problems that are appropriate for the programmable calculator. Although these noted topics may be learned without the programmable calculator, we would encourage you to learn to use your calculator to its fullest.

1-6 ASSUMPTIONS MADE BY US ABOUT YOU

We have made several assumptions about you that we would like to share with you:

- We assume that you have mastered the skills of arithmetic, including adding, subtracting, multiplying, dividing, fractions (both decimal fractions and built-up fractions), and percentage.
- We assume that you will have a scientific calculator with you while studying this text. We have made provisions in the text for a lesser calculator, but if you are to get the most out of your education, the scientific calculator will be necessary.
- We assume that you have an active interest in the field of electronics.
- We assume that you will work through each example. A great deal of information has been included in the examples.
- We assume that you will have your owner's guide for your calculator available with your calculator.

1-7 GENERAL INFORMATION

In addition to the previous assumptions, you need to be aware of the following:

- The leading zero in decimal numbers, as in 0.357, is used to set off the decimal point.
- Numbers are set off in groups of three without commas, as in 5 282 621 rather than 5,282,621.
- Whole numbers such as 50, 18, and 143 are written with an *implied* decimal point to the right of the last digit in the number. Numbers contain-

ing decimal fractions such as 1.3, 0.59, and 10.15 are written with an *explicit* decimal point.

- Selected answers to the exercises are given in Appendix B.
- No trigonometric or logarithmic tables have been included in this text because the scientific calculator is being used exclusively.

We wish you well in your study of this text and know that the field you have selected will bring challenge and reward to you.

> *The world belongs to the dissatisfied. . . . The great underlying principle of all human progress is that divine discontent which makes men strive for better conditions and improved methods.*
>
> *Charles Proteus Steinmetz (1865–1923)*

2

SELECTED PREALGEBRA TOPICS

In this chapter you will prepare for the study of algebra. Whole numbers are used to introduce you to several concepts used in algebra, including signed numbers, symbols of grouping, and operators.

Some abbreviations have been used to add clarity to the examples. The symbols within *boxes* are used to indicate calculator operations. The other symbols are used to indicate manual operations. The following have been used in this chapter and are used throughout the book:

$+$	and **A** add	$	\ \	$	take absolute value
$-$	and **S** subtract	CHS	change sign
$\times$	and **M** multiply	y^x	raise to an exponent
$\div$	and **D** divide	$\therefore$	therefore

It is assumed that you have read Chapter 1 and that you will be using a calculator to aid in your calculations. If you have not read Chapter 1, go back and read it before continuing Chapter 2.

2-1 NATURAL NUMBERS AND NUMBER SYSTEMS

Our first introduction to mathematics was counting. The numbers used were one, two, three, and so forth, and are called the counting numbers or *natural numbers*. Since zero is not used in counting, zero is not a natural number.

Number Systems

The system of numbers used by most people in the world is the decimal system. This system is a base-ten system. Ten symbols, 0, 1, 2, 3, 4, 5, 6, 7, 8, 9, are used in this number system. These ten symbols are called *digits.*

There are several other number systems in use today. Since you are studying electronics, you will probably learn about the number systems in use by computers. Table 2-1 is a summary of some of these number systems.

TABLE 2-1

NUMBER SYSTEMS

System Name	Number of Symbols	Symbols Used	Name of Symbol
Binary	Two	01	Binary digits or bits
Octal	Eight	01234567	Octal digits
Decimal	Ten	0123456789	Digits
Hexadecimal	Sixteen	0123456789 ABCDEF	Hex digits

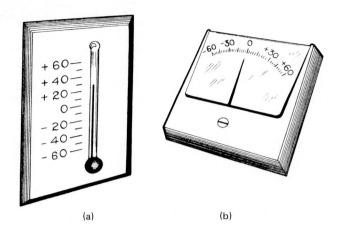

(a)   (b)

FIGURE 2–1
**Use of signed numbers: (a) temperature measured above (+) and below (−) zero; (b)
voltage measured above (+) and below (−) zero reference voltage.**

Notice that the number of symbols used in the number system is the same as the base of the system: base two (binary), 2 symbols; base eight (octal), 8 symbols; base ten (decimal), 10 symbols; base sixteen (hexadecimal), 16 symbols. Notice also that each system starts with zero.

2-2 SIGNED NUMBERS

To solve the wide variety of problems found in electronics, more is needed than the natural numbers. Signed numbers are needed. Numbers preceded by a + sign are called *positive numbers;* those proceded by a − sign are called *negative numbers.* A plus sign (+) *may* be written before a number to show that it is a positive number, or the plus sign *may* be left off. A negative sign (−) is **always** written before a number to show that it is a negative number. If a number has no sign before it, it is assumed to be a positive number.

Look at the thermometer pictured in Figure 2-1(a). Notice that two kinds of numbers are shown. The positive numbers are measuring the distance above the zero. The negative numbers are measuring the distance below the zero. Thus, zero is a reference point from which positive numbers are greater and negative numbers are less. Zero is an invented symbol that is *neither positive nor negative.*

In the following exercise many of the problems can be done mentally, but some require calculations. Use the calculator when you need it.

EXERCISE 2-1

1. If +3 stands for a gain of $3, what number would be used to stand for a loss of $7?

2. If an increase of 15% in cost of resistors is shown by +15, how would a decrease of 3% be shown?

3. If −8 means 8 meters below sea level, how can 20 meters above sea level be represented?

4. In the diagram below, distance to the right of the starting point is positive and distance to the left is negative. Complete the following chart for points B, C, E, F, and G.

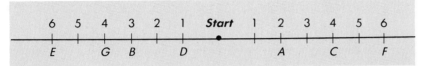

Points	A	B	C	D	E	F	G
Distance	+2			−1			

5. How would you keep score in a game if you were "down 8 points"? Would you use 8, +8, or −8 to stand for "down 8"?

6. Which is colder, −13° or −18°?

7. Below are 12 test scores for a class of students in an electronics math course. The average grade on the test is 65. Show by signed numbers the amount each score is above and below the average score of 65. The first grade is 55, which is 10 below the average. So a −10 is written below the 55. The next grade is 65. The difference between it and 65 is zero, so 0 is written below the 65. Complete the table for the remaining scores.

Test Score	55	65	90	40	80	50	75	95	30	65	100	45
Difference	−10	0										

8. A man operating an electronics store kept a record of his monthly sales. He compared them with the monthly sales of the year before. He used a plus sign to show an increase in sales and a minus sign to show a loss in sales. Complete the following table:

Monthly Sales	Jan.	Feb.	Mar.	Apr.	May
This year	4250	7840	5628	10112	6339
Last year	3020	8118	3782	9286	7002
Change	+1230				

TABLE 2-2

OPERATORS

Operator	Expression	Equation
+	2 + 6	2 + 6 = 8
−	6 − 2	6 − 2 = 4
×	2 × 6	2 × 6 = 12
÷	6 ÷ 2	6 ÷ 2 = 3

2-3 NUMERICAL EXPRESSIONS AND EQUATIONS

In arithmetic, symbols or operators are used to indicate addition (+), subtraction (−), multiplication (×), and division (÷). These same symbols are also used in algebra.

To show that two numbers are to be added, such as 2 and 5, write the *numerical expression:* 2 + 5. This expression is read "the sum of 2 and 5" or simply "2 + 5." The result of adding 2 and 5 is 7. Seven is the sum of 2 and 5. To state that "2 + 5" and "7" are the same number, write the numerical equation: 2 + 5 = 7. This equation is read "the sum of 2 and 5 *is equal to* 7" or simply "2 plus 5 *equals* 7." Table 2-2 shows the forming of numerical expressions and equations for the arithmetic operators of +, −, ×, ÷.

EXAMPLE 2-1 Use the pair of numbers 8 and 4 and the arithmetic operators of Table 2-2 to form numerical expressions and equations for each of the operators.

Solution Given 8, 4, and +, −, ×, ÷.
Form expressions:

$$8 + 4, \quad 8 - 4, \quad 8 \times 4, \quad 8 \div 4$$

Form equations:

$$8 + 4 = 12$$
$$8 - 4 = 4$$
$$8 \times 4 = 32$$
$$8 \div 4 = 2$$

This solution is summarized in Table 2-3.

TABLE 2-3

SUMMARY OF EXAMPLE 2-1		
Operator	**Expression**	**Equation**
+	8 + 4	8 + 4 = 12
−	8 − 4	8 − 4 = 4
×	8 × 4	8 × 4 = 32
÷	8 ÷ 4	8 ÷ 4 = 2

EXERCISE 2-2

1. Use the four operators of Table 2-2 and the following pairs of numbers to form numerical expressions for each operator.
   (a) 10, 5 (b) 16, 8 (c) 20, 4 (d) 60, 12
2. Use the four operators of Table 2-2 and the following pairs of numbers to form numerical expressions for each operator.
   (a) 15, 3 (b) 21, 7 (c) 54, 6 (d) 64, 4
3. Use the pairs of numbers in problem 1 to form numerical equations for each of the operators of Table 2-2.
4. Use the pairs of numbers in problem 2 to form numerical equations for each of the operators of Table 2-2.
5. Write a numerical expression that indicates 3 is subtracted from 5.
6. Write a numerical expression that indicates 4 is multiplied by 6.
7. Write a numerical equation that indicates the value of 6 is divided by 3.
8. Write a numerical equation that indicates the value of 18 is subtracted from 20.

2-4 ORDER OF OPERATIONS

So far we have worked with numerical expressions that used only one operator. We have performed addition, subtraction, multiplication, and division. Now expressions with more than one operator will be considered. For example, the expression $9 + 6 \div 3$ has both addition and division. This expression represents a number, but what number? If 9 and 6 are added first, the answer is 15, which is then divided by 3 to result in 5. On the other hand, if 6 is divided by 3 first, the answer is 2, which is then added to 9 to result in 11. Since there is only one expression, it can represent only one number.

To avoid having two answers, the order in which the operations are done must be decided. Otherwise, you would never know which answer was correct. Fortunately, mathematicians have come to an agreement on the order in which the operations are to be done. This agreement, called a *convention,* is shown in Convention 2-1.

CONVENTION 2-1. ORDER OF OPERATIONS

In a series of operations:

1. Multiplication and division ($\times$ and $\div$) are done first in the order they occur in the expression left to right.
2. Addition and subtraction ($+$ and $-$) are done next in the order they occur in the expression left to right.

By using the conventional order of operations to solve the previous example of $9 + 6 \div 3$, first divide and then add. So the expression $9 + 6 \div 3 = 9 + 2 = 11$.

To add clarity to the examples, the letters A, S, M, and D will be used to indicate addition, subtraction, multiplication, and division. When the example focuses on using a calculator, calculator key symbols will be used to indicate the operation. Thus, $\boxed{+}$ means addition, $\boxed{-}$ subtraction, $\boxed{\times}$ multiplication, and $\boxed{\div}$ division on your calculator.

EXAMPLE 2-2 Using the conventional order of operation, compute the number represented by the numerical expression

$$2 + 3 \times 6 - 4$$

Solution By Convention 2-1, multiplication is done before addition and subtraction.

$$\text{M: } 3 \times 6 \qquad 2 + 18 - 4$$

By Convention 2-1, the addition and subtraction are performed from left to right.

$$\text{A: } 2 + 18 \qquad 20 - 4$$
$$\text{S: } 20 - 4 \qquad 16$$

Many calculators have aids for working *chain calculations.* Check your calculator and owner's guide for these features: memory, roll stack, and parentheses. These features can each in its own way *"save"* an intermediate answer for later use. This eliminates having to write out all the steps of a calculation. Refer to your owner's guide on how to use your particular calculator.

EXAMPLE 2-3   Use your calculator to compute the value of

$$5 \times 3 + 6 \times 2$$

Solution   Convention 2-1 states that the two multiplications are to be performed first and then the products are to be added.

$\boxed{\times}$ $5 \times 3 = 15$

Store this result.

$\boxed{\times}$ $6 \times 2 = 12$

Recall and add:

$\boxed{+}$ $15 + 12 = 27$

∴ $5 \times 3 + 6 \times 2 = 27$

EXERCISE 2-3

In each of the following numerical equations, state whether the answer (right-hand number) is correct or not. If the answer is correct, write *true*. If the answer is not correct, write *false* and then state the correct answer.

1. $7 + 3 - 2 = 8$ 2. $14 - 3 \times 2 = 22$
3. $8 \times 2 - 3 = 14$ 4. $12 - 12 \div 3 = 0$
5. $2 + 6 \div 2 = 5$ 6. $-2 + 4 \times 5 = 18$
7. $12 \div 3 + 8 = 12$ 8. $6 \div 2 \times 3 = 1$
9. $-5 - 2 \div 2 = -4$ 10. $-9 \times 2 \div 3 = -6$

Calculator Drill

Use your calculator with Convention 2-1 to perform the following chain calculations. NOTE: You will find all the answers to the calculator drills in Appendix B.

11. $2 + 4 \times 7$ 12. $3 - 5 \times 3$
13. $8 + 9 \div 3$ 14. $-6 + 12 \div 4$
15. $6 \div 2 \times 3$ 16. $15 \times 2 \div 5$
17. $-8 \div 2 \times 6$ 18. $-16 \div 4 \times 3$
19. $2 \times 3 + 4 \times 2$ 20. $5 \times 3 + 7 \times 2$

21. $3 \times 3 - 4 \times 5$ 22. $8 \times 4 - 9 \times 3$
23. $15 \div 3 - 8 \times 2$ 24. $18 \div 6 + 4 \times 5$
25. $9 \div 3 - 14 \div 7$ 26. $-10 \div 2 - 35 \div 7$
27. $39 \div 3 + 64 \div 4$ 28. $27 \div 9 + 20 \div 10$
29. $6 + 12 - 4 \times 2$ 30. $5 - 2 - 18 \div 6$

2-5 SYMBOLS OF GROUPING

As we have just learned, there is an agreed-upon order in solving a numerical expression. Suppose, however, that one needs to add before multiplying. How can we say "add first, then multiply"? By using symbols of grouping, the normal order of operation can be overridden. When a particular operation needs to be performed first, or if a particular operation needs to be emphasized, then parentheses ( ) are used.

EXAMPLE 2-4   Compute the number represented by the expression

$$4 \times (3 + 2)$$

Solution   Since $3 + 2$ is enclosed in parentheses, it is done first.

A: $(3 + 2)$ 4×5
M: 4×5 20

Besides parentheses ( ), several other symbols are used to show grouping. These include brackets [ ] and braces { }. When these signs of grouping are used together, they are used in the following order: parentheses ( ) first, then brackets [ ], and finally braces { }. When one pair of grouping symbols is enclosed in another, the operation in the inner symbol is performed first.

EXAMPLE 2-5   Compute the number represented by the expression

$$[2 + (6 - 3)] \times 5$$

Solution   Work within the parentheses first, and then within the brackets:

S: $(6 - 3)$ $[2 + 3] \times 5$
A: $[2 + 3]$ 5×5
M: 5×5 25

16

EXAMPLE 2-6   Compute the number represented by the numerical expression

$$3 + [(6 + 1) + (6 + 8 \div 4 \times 2)]$$

Solution   Work within each set of parentheses first, and then within the brackets:

A: $(6 + 1)$ $3 + [7 + (6 + 8 \div 4 \times 2)]$
D: $8 \div 4$ $3 + [7 + (6 + 2 \times 2)]$
M: 2×2 $3 + [7 + (6 + 4)]$
A: $(6 + 4)$ $3 + [7 + 10]$
A: $[7 + 10]$ $3 + 17$
A: $3 + 17$ 20

EXERCISE 2-4

1. In working with the expression $[(6 + 3) - (2 \times 2)] + 1$, either of two operations may be performed first. What are they?
2. In working with the expression $[(9 - 2) \times (5 + 4)] - (3 + 2)$, any of three operations may be performed first. What are they?

Calculator Drill
Compute the number represented by the following numerical expressions. Use your calculator in the solution.

3. $(6 + 3) \div 9$
4. $(17 - 2) \times 2 \div 3$
5. $5 \times (12 - 10) \div 5$
6. $80 \div (4 + 12)$
7. $(8 + 9 - 5) \div 3$
8. $(2 + 3) + (5 + 4)$
9. $55 - 6 \times (3 + 4)$
10. $(7 + 10 + 8) \div 5$
11. $12 + 5 \times (6 + 7)$
12. $30 - 2 \times (11 - 7)$
13. $6 - [(4 + 1) - (3 + 2)]$
14. $12 + [6 \div 3 + (7 - 4)]$
15. $21 \times (16 \div 2) + 26$
16. $1040 \div (47 - 39) - 123$
17. $[17 + (14 \times 3)] - (3 \times 19)$
18. $[(5 + 4) \times (10 \div 5)] \div (54 \div 27)$
19. $[(9 + 3) \div 3 - (18 \div 9)] + 1$
20. $[(6 - 1) \div 5 + (4 \times 4)] - [(3 \times 3) + 1]$
21. $\{[(12 \times 2) \div (8 \times 2 \div 4)] + 3\} - (3 + 3) \div 2$
22. $\{[(9 + 7) \times 2 \div 2 + (8 \div 2)] - 4\} \div 16$

2-6 DOUBLE MEANING OF + AND −

Before learning to add algebraically, one must first have additional understanding of the meaning of the two symbols plus (+) and minus (−). In arithmetic, + is used as the sign (operator) for addition, while − is used as the sign (operator) for subtraction. In algebra, each of these symbols has two uses. The + may be used as the operator for addition or as the sign for a positive number. The − may be used as the operator for subtraction or as the sign for a negative number. In the expression −3, the minus indicates a negative number, while the minus in 5 − 3 indicates subtraction. In the expression (−3) + (−7), the plus means to add and the minuses indicate negative numbers. When double signs occur in an expression, they may be simplified to a single sign. Table 2-4 shows how to combine double signs into a single sign.

EXAMPLE 2-7   Combine the double sign in (−3) + (−7) to a single sign.

Solution   Remove the parentheses:

$$-3 + -7$$

Replace + − with −:

$$-3 - 7$$

EXAMPLE 2-8   Simplify 6 − (−8).

Solution   Remove the parentheses:

$$6 - -8$$

Replace − − with +:

$$6 + 8$$

TABLE 2-4

COMBINING DOUBLE SIGNS			
Double Sign	*Single Sign*	*Double Sign*	*Single Sign*
+ +	+	− +	−
+ −	−	− −	+

Rewrite each of the following expressions with all the double signs combined into single signs. Use Table 2-4.

1. $3 + (-2)$
2. $-16 + (+3)$
3. $-8 - (-2)$
4. $(+4) - (+3)$
5. $2 - 3 + (-7)$
6. $(-9) - (+3) - (-5)$
7. $1 + (+2) - (+6)$
8. $2 + 3 - +8$
9. $7 - -1 + (-4)$
10. $-5 - +7 - -3$

2-7 ABSOLUTE VALUE OF A SIGNED NUMBER

When you work with signed numbers, it is sometimes easier to manipulate the number without the sign. The *absolute value* of a signed number is just the number without the sign. The absolute value of a number is indicated by placing a vertical bar on either side of the number. The following example will show you how to indicate that the absolute value is being taken.

EXAMPLE 2-9 Indicate the absolute value of -5 and $+7$.

Solution Absolute value of -5 is indicated by $|-5|$.

Absolute value of $+7$ is indicated by $|+7|$.

As you have seen, the absolute value of a number is indicated by placing a vertical bar on either side of the number. The solution to the operation of taking the absolute value is the number without any sign. Therefore, the absolute value of a number is understood to be positive.

EXAMPLE 2-10 Take the absolute value of -5 and $+7$.

Solution Indicate the absolute value of -5:

$$|-5| \Rightarrow 5$$

Indicate the absolute value of $+7$:

$$|+7| \Rightarrow 7$$

Take the absolute value of each of the following signed numbers, combining double signs first.

1. $+3$	2. -4	3. -9
4. $+2$	5. -8	6. $+(+3)$
7. $-(+7)$	8. $-(-6)$	9. $-(+1)$

2-8 COMBINING SIGNED NUMBERS

When we add two numbers together, a *sum* is formed. The numbers used to form the sum are called the *terms* of the sum. For example, in the numerical equation $2 + 5 = 7$, the terms of the sum are 2 and 5. The addition and subtraction of signed numbers is done by applying three rules. Each of these rules will be presented with examples to illustrate their application.

RULE 2-1. ADDING LIKE SIGNED NUMBERS

To add two numbers with the *same* sign:

1. Add their absolute values.
2. Give the sum the same sign as the terms.

EXAMPLE 2-11 Add -5 and -7.
   Solution Use Rule 2-1 to evaluate $-5 + (-7)$:

$$| \quad |: -5, -7 \quad\quad 5, 7$$
$$A: 5 + 7 \quad\quad 12$$

Assign $(-)$ to the sum: -12

$$\therefore \quad -5 + (-7) = -12$$

EXAMPLE 2-12 Add 3 and 7.
   Solution Use Rule 2-1 to evaluate $3 + 7$:

$$| \quad |: +3, +7 \quad\quad 3, 7$$
$$A: 3 + 7 \quad\quad 10$$

Assign + sign: +10

$$\therefore \quad 3 + 7 = +10$$

RULE 2-2. ADDING OPPOSITE SIGNED NUMBERS

To add two signed numbers with *different* signs:

1. Subtract the smaller absolute value from the larger.
2. Give the difference the sign of the larger term.

EXAMPLE 2-13   Add 8 and -3.
  Solution   Use Rule 2-2 to evaluate $8 + (-3)$:

$$| \quad |: +8, -3 \quad 8, 3$$

Subtract the smaller term from the larger:

$$S: 8 - 3 \quad 5$$

Assign the sign of the larger term (+):   +5

$$\therefore \quad 8 + (-3) = +5$$

EXAMPLE 2-14   Add -9 and 2.
  Solution   Use Rule 2-2 to evaluate $-9 + 2$.

$$| \quad |: -9, +2 \quad 9, 2$$

Subtract the smaller term from the larger:

$$S: 9 - 2 \quad 7$$

Assign the sign of the larger term ($-$): -7

$$\therefore \quad -9 + 2 = -7$$

EXAMPLE 2-15 Subtract -3 from 7.

Solution Use Rule 2-3 to evaluate $7 - (-3)$.

Change sign of -3 and restate problem using Rule 2-1:

$$
\begin{array}{ll}
 & 7 + 3 \\
|\quad|: +7, +3 & 7, 3 \\
\quad\text{A: } 7 + 3 & 10
\end{array}
$$

Assign $(+)$ to sum: $+10$

$$\therefore \quad 7 - (-3) = 10$$

Observation When you see the word *subtract,* think **sign change.** In Example 2-15 you may see that the statement "subtract -3 from 7" is translated into the numerical expression $7 - (-3)$. By applying the methods for simplifying the double sign found in Table 2-4, this expression becomes $7 + 3$. However, rather than concern yourself with condensing the double sign, it is easier to think, *sign change and add.*

EXAMPLE 2-16 Subtract 6 from 2.

Solution To subtract 6 from 2, change the sign of 6 and add. Thus, 2 subtract 6 becomes 2 add -6. Use Rule 2-2 to evaluate

$$
\begin{array}{ll}
 & 2 + (-6) \\
|\quad|: 2, -6 & 2, 6
\end{array}
$$

Subtract the smaller term from the larger:

$$\text{S: } 6 - 2 \quad 4$$

Assign the sign of the larger $(-)$: -4

$$\therefore \quad 2 - 6 = -4$$

As in Example 2-16, subtraction can result in a negative number. What would such a result mean? We have seen that negative numbers are used to indicate amounts below a reference value, for instance, the temperature below freezing, the distance below sea level, and the amount by which a checking account is overdrawn.

EXAMPLE 2-17 You have $42.00 in your checking account. You write checks for the following amounts: $6.00, $18.00, $7.00, and $13.00. How much do you have in your account?

Solution

$$42.00 - (6.00 + 18.00 + 7.00 + 13.00)$$
$$42.00 - 44.00$$
$$-2.00$$

$\therefore$ Your account is overdrawn by $2.00.

From this example we see that $-$$2.00 means that you have insufficient funds in your account and will no doubt be charged a service charge for the overdraft. This certainly has real meaning for you.

EXERCISE 2-7

Combine the following signed numbers by adding or subtracting as indicated. Use the rules of addition and subtraction, and solve without the use of your calculator.

Add the following numbers.

1. 7 and 3	2. -3 and -5	3. 3 and $+2$
4. -6 and -1	5. 8 and 4	6. -5 and -7
7. -17 and -6	8. 21 and $+14$	9. -13 and -39

Subtract the first number from the second in the following problems.

10. -5 from 3	11. 6 from 2	12. 8 from -3
13. 12 from -4	14. -18 from -3	15. 9 from -4
16. 32 from 14	17. 12 from -7	18. 8 from $+16$

Using signed numbers, rewrite the following phrases into numerical expressions and then compute the number represented by the expression.

19. Earning $21 and spending $16
20. Climbing up 14 stairs and then climbing up 17 more
21. A temperature rise of 18° and a fall of 26°
22. A gain of $3.00 and a loss of $9.00
23. Going 12 steps backward and then going 16 steps forward
24. A raise of $10 in pay and an increase of $3 in taxes

Using a Calculator with Signed Numbers

Your calculator is designed to work with signed numbers. It uses the operator keys $+$ and $-$ for addition and subtraction. It uses the change sign key $+/-$ or $\boxed{\text{CHS}}$ to enter negative numbers. It has been programmed to perform arithmetic with signed numbers. The following example will help you to become familiar with these features. It is recommended that you consult your owner's guide for additional information on working with signed numbers.

EXAMPLE 2-18   Evaluate $-7 + 8 - 6$.

Solution   Use your calculator. Enter 7, then change the sign.

$\boxed{\text{CHS}}$	7	-7
$\boxed{+}$	8	1
$\boxed{-}$	6	-5

$$\therefore \quad -7 + 8 - 6 = -5$$

EXERCISE 2-8

Calculator Drill
Evaluate each of the following expressions.

1. $-18 + (-2)$
2. $47 + (-17)$
3. $12 - (-4)$
4. $-18 - (-14)$
5. $-27 + 8$
6. $17 - 13$
7. $-19 - 17$
8. $22 + (-17)$
9. $-13 + 9$
10. $-25 - (-12)$
11. $6 + (-3) + 5$
12. $120 - 30 - (-10)$
13. $75 - 15 + (-12)$
14. $-13 + (-15) + 39$
15. $-37 + 14 - (23)$

TABLE 2-5

RELATIONAL OPERATORS

Symbol	Name	Use to:	Example	Read:
$=$	Is equal to	Show equality	$3 + 5 = 8$	3 plus 5 equals 8
$\neq$	Is not equal to	Show inequality	$2 \neq 9$	2 is not equal to 9
$>$	Is greater than	Show inequality and relative size	$6 > 1$	6 is greater than 1
$<$	Is less than	Show inequality and relative size	$4 < 5$	4 is less than 5
$\geq$	Is greater than or equal to	Show relative size	$7 \geq 1 + 5$	7 is greater than or equal to 1 plus 5
$\leq$	Is less than or equal to	Show relative size	$0 \leq 3$	0 is less than or equal to 3

2-9 RELATIONAL OPERATORS

We have learned that a numerical equation uses the equal sign ($=$) to say that the numbers or expressions on either side of the equal sign are exactly the same. The equal sign is one of the relational operators.

The relationship of inequality may be shown in several ways. Suppose that we wanted to state the relationship between 5 and 7. We cannot say that 5 and 7 are equal, but we can say that they are not equal. This is stated as the mathematical sentence $5 \neq 7$. The not-equal symbol ($\neq$) is a relational operator. We also know that 7 is greater than 5. Using the greater-than relational operator ($>$), we can write $7 > 5$. The statement "$7 > 5$" is read "seven is greater than five." Finally, by using the less-than relational operation ($<$), we can write $5 < 7$. The inequality "$5 < 7$" is read "five is less than seven." Table 2-5 is a summary of these relational operators.

EXERCISE 2-9

Use the appropriate relational operator in place of the words to express each statement.

1. 7 is less than 10.
2. 5 is smaller than 8.
3. 23 is more than 13.
4. -4 is greater than -7.
5. 3 is the same as -10 plus 13.
6. 3 plus 8 is not the same as -2.
7. The sum of 2 and 7 is greater than 3.
8. The sum of 8 and -3 is less than the difference between 25 and 7.

Use $=$, $>$, or $<$ in place of "?" to make each statement true.

9. $7 ? -12 + 5$
10. $-8 ? 3 + 4$
11. $|+6| ? |-3|$
12. $-5 ? |-9|$

13. $-3 ? 13 - (10 + 3)$
14. $36 + 4 ? 48 - 12$
15. $-3 + 4 ? -7 + (-2)$
16. $8 + 2 - 4 ? -8 + 4$
17. $102 - 8 ? 91 - 89$
18. $-68 + 14 ? -59 + 2$

Compound Inequalities

In electronic applications of relational operators we often write compound inequality statements. These statements use two inequalities that have a number in common. For example, $5 < 8$ and $8 < 10$. These statements may be combined into a single compound statement of $5 < 8 < 10$. This statement is read "5 is less than 8 which is less than 10." Here is another compound inequality: $63 > 32 > 25$. This statement is read "63 is greater than 32 which is greater than 25."

In a compound inequality, both inequalities must be true in order for the compound inequality to be true. If either inequality is false, the entire statement is false.

EXAMPLE 2-19   Determine if $(15 - 3) < 20 < (14 - 8)$ is a true statement.

Solution   Combine and remove parentheses:

$$\text{S: } (15 - 3) \qquad 12 < 20 < (14 - 8)$$
$$\text{S: } (14 - 8) \qquad 12 < 20 < 6$$

Determine if the first inequality is true:
  Yes, 12 is less than 20.
Determine if the second inequality is true:
  No, 20 is not less than 6.
  $\therefore$ The compound inequality is false.

EXERCISE 2-10

Determine which of the following compound statements are true and which are false.

1. $3 < 6 < 10$
2. $2 > 5 > 3$
3. $17 < 28 < 7$
4. $11 > 9 > 7$
5. $-7 < 3 < 4$
6. $-4 > -3 > -6$
7. $13 > 18 > -21$
8. $-5 < -4 < -2$
9. $(3 - 7) > 3 > (-9 + 3)$

2-10 MULTIPLYING WITH SIGNED NUMBERS

When we multiply two numbers together, we form a *product*. The numbers used to form the product are called the *factors* of the product. For example, in the numerical equation $2 \times 3 = 6$, 2 and 3 are the factors of the product 6. Six is the product of the factors 2 and 3.

The multiplication of signed numbers is done by applying two rules. Each of these rules will be presented with examples to illustrate its application.

RULE 2-4. MULTIPLYING LIKE SIGNED NUMBERS

To multiply two numbers with the *same* sign:

1. Multiply the absolute values.
2. Give the product a plus sign (+).

EXAMPLE 2-20   Multiply -3 times -4.
      Solution   Use Rule 2-4 to evaluate $(-3) \times (-4)$:

$$| \; |: -3, -4 \quad\quad 3, 4$$
$$\mathrm{M}: 3 \times 4 \quad\quad 12$$

Assign plus sign (+) to product: $+12$

$$\therefore \quad (-3) \times (-4) = 12$$

EXAMPLE 2-21   Multiply 3 times 4.
      Solution   Use Rule 2-4 to evaluate $(+3) \times (+4)$:

$$| \; |: +3, +4 \quad\quad 3, 4$$
$$\mathrm{M}: 3 \times 4 \quad\quad 12$$

Assign plus sign (+) to product: $+12$

$$\therefore \quad (+3) \times (+4) = 12$$

RULE 2-5. MULTIPLYING UNLIKE SIGNED NUMBERS

To multiply two numbers with *different* signs:

1. Multiply the absolute values.
2. Give the product a minus sign $(-)$.

EXAMPLE 2-22   Multiply -8 times 3.
   Solution   Use Rule 2-5 to evaluate -8×3:

$$| \ \ |: -8, 3 \qquad 8, 3$$
$$\text{M: } 8 \times 3 \qquad 24$$

Assign minus sign $(-)$ to the product: -24

$$\therefore \quad -8 \times 3 = -24$$

EXAMPLE 2-23   Multiply 5 times -3.
   Solution   Use Rule 2-5 to evaluate $5 \times (-3)$:

$$| \ \ |: 5, -3 \qquad 5, 3$$
$$\text{M: } 5 \times 3 \qquad 15$$

Assign minus sign $(-)$ to product: -15

$$\therefore \quad 5 \times (-3) = -15$$

From the previous examples, observe that the product of two numbers with like signs is positive, while the product of two numbers with unlike signs is negative. Thus, in multiplication, *like signs give a positive result* and *unlike signs give a negative result.* This concept is graphically presented in Table 2-6.

When using your calculator, remember that it is designed to work with signed numbers. It has been programmed to perform multiplication with signed numbers according to the rules you have learned.

EXAMPLE 2-24   Use your calculator to multiply $15 \times (-6)$.

Solution

$$\begin{array}{ll} \text{Enter 15} & 15 \\ \boxed{\times}\ 6\ \boxed{\text{CHS}} & -90 \end{array}$$

$$\therefore \quad 15 \times (-6) = -90$$

Observation   The indicated keystrokes are to remind you of the operations used in the calculation, but they may not be the same strokes and they may not be in the same order that you will use with your calculator.

EXERCISE 2-11
Use Rules 2-4 and 2-5 to multiply the following numbers.

1. 2 and -3	2. $+8$ and 4	3. -5 and -2
4. -1 and -2	5. -5 and 7	6. -8 and 3
7. -6 and -7	8. $+4$ and $+3$	9. $+2$ and -2
10. -8 and $+5$	11. 9 and -10	12. -7 and -1

Calculator Drill
Using your calculator, compute the number represented by each of the following numerical expressions.

13. -31×40	14. $28 \times (-15)$	15. $+16 \times (+12)$
16. $34 \times (-27)$	17. -92×102	18. $-72 \times (-13)$
19. $+18 \times 13$	20. -47×52	21. $-37 \times (-53)$
22. $-49 \times (-25)$	23. -13×94	24. $37 \times (-14)$

TABLE 2-6

MULTIPLYING SIGNED NUMBERS			
Signs of Factors	**Sign of Product**	**Signs of Factors**	**Sign of Product**
+ times +	+	+ times −	−
− times −	+	− times +	−

$$\frac{\text{Dividend}}{\text{Divisor}} = \text{Quotient} \qquad \frac{\text{Numerator}}{\text{Denominator}} = \text{Numerator/Denominator}$$

(a)   *(b)*

FIGURE 2–2

Parts of a fraction. (a) The result of dividing the dividend by the divisor is the quotient; (b) Another name for dividend is numerator. Another name for divisor is denominator. The bar may be horizontal or diagonal.

2-11 DIVIDING WITH SIGNED NUMBERS

Division is indicated in arithmetic by the division symbol ($\div$), as in $6 \div 2 = 3$. In algebra, division is usually indicated as a fraction. A fraction is made up of three parts, as shown in Figure 2-2: (1) the numerator (the dividend or the number to be divided), (2) the denominator (the divisor or the number to do the dividing), and (3) the bar between the numerator and the denominator. The bar may appear horizontally or diagonally ($-$ or $/$), as pictured in Figure 2-2(b). Like parentheses, the bar is a symbol of grouping. Thus, the bar controls the order of operation. The following example will demonstrate this important property of the bar.

EXAMPLE 2-25   Discuss how to evaluate the following expression and then show the solution: $\dfrac{2 + 13}{5}$.

Discussion   Since $2 + 13$ is grouped together by the bar, we evaluate $2 + 13$ first. We then divide by 5. Notice that the usual order of division before addition has been overridden by the bar, a symbol of grouping.

Solution

$$\frac{2 + 13}{5}$$

A: $2 + 13$ $15/5$
D: $15/5$ 3

The following rules are used when dividing signed numbers. Each of these rules will be presented with examples to illustrate its application.

RULE 2-6. DIVIDING LIKE SIGNED NUMBERS

To divide two numbers with the *same* sign:

1. Divide the absolute values.
2. Give the result a plus sign ($+$).

EXAMPLE 2-26   Divide -8 by -4.

   Solution   Use Rule 2-6 to evaluate $-8/-4$:

$$| \quad |: -8, -4 \qquad 8, 4$$
$$\text{D}: 8/4 \qquad 2$$

Assign plus sign $(+)$: $+2$

$$\therefore \qquad -8/-4 = 2$$

RULE 2-7. DIVIDING UNLIKE SIGNED NUMBERS

To divide two numbers with *different* signs:

1. Divide the absolute values.
2. Give the result a minus sign $(-)$.

EXAMPLE 2-27   Divide 6 by -2.

   Solution   Use Rule 2-7 to evaluate $6/-2$:

$$| \quad |: 6, -2 \qquad 6, 2$$
$$\text{D}: 6/2 \qquad 3$$

Assign minus sign $(-)$: -3

$$\therefore \qquad 6/-2 = -3$$

EXAMPLE 2-28   Use your calculator to evaluate $\dfrac{50-11}{-13}$.

   Solution   The bar is a sign of grouping, so combine the numbers in the numerator.

	Enter 50	50
	$\boxed{-}$ 11	39
$\boxed{\div}$ 13	$\boxed{\text{CHS}}$	-3

Use Rules 2-6 and 2-7 to evaluate the following expressions.

1. 15/3
2. −8/2
3. 6/−2
4. 12/−6
5. −14/7
6. 10/5
7. −18/−6
8. −15/−5
9. 16/−4
10. −20/−10
11. 25/5
12. 24/8

Calculator Drill
In the following, remember that the bar is a sign of grouping.

13. −27/9
14. −36/4
15. −34/−17
16. −64/16
17. 42/−14
18. 75/−15

19. $\dfrac{7 + 2}{3}$
20. $\dfrac{8 - 6}{2}$
21. $\dfrac{24 + 36}{-12}$

22. $\dfrac{-13 - 7}{-4}$
23. $\dfrac{19 - 11}{5 + 3}$
24. $\dfrac{-8 + 2}{-3}$

25. $\dfrac{5 + 15}{-15 + 5}$
26. $\dfrac{-12}{25 - 13}$
27. $\dfrac{16 - 6}{-2 + 7}$

28. $\dfrac{3 + 17}{-5 + 15}$
29. $\dfrac{-15 + 15}{20 + 15}$
30. $\dfrac{-19 + (-89)}{17 - 29}$

2-12 INTRODUCTION TO EXPONENTS

In the equation $4 \times 4 = 16$, recall that the number 16 is the *product* and the two 4's are called the *factors* of the product. This equation could be written in a shorter way by using exponent notation. The equation $4 \times 4 = 16$ becomes $4^2 = 16$. The small 2 written above and to the right of the 4 is called an *exponent*. The expression 4^2 is read "four squared" or "four to the second power." The exponent tells how many times a number is used as a factor in a product.

EXAMPLE 2-29   Read the following expression and indicate its meaning; then compute the number it represents: 5^4.

Discussion 5^4 is read "five to the fourth power" and means $5 \times 5 \times 5 \times 5$.

Solution
$$5^4 = 5 \times 5 \times 5 \times 5$$
$$= 625$$

$$\therefore \quad 5^4 = 625$$

EXAMPLE 2-30   Use exponents to simplify the following expression:

$$8 \times 8 \times 8 \times 8 \times 8 \times 8 \times 8 \times 8$$
$$\times 8 \times 8 \times 8 \times 8 \times 8 \times 8$$

Solution   Count the number of times 8 is used as a factor:

14 times

Rewrite using an exponent: 8^{14}

$\therefore$ $8 \times 8 \times 8 \times 8 \times 8 \times 8 \times 8 \times 8$
$\times 8 \times 8 \times 8 \times 8 \times 8 \times 8 = 8^{14}$

EXERCISE 2-13

Read the following expressions and explain what each means.

1. 5^2	2. 6^3	3. 2^4	4. 3^5
5. 4^4	6. 7^2	7. 1^3	8. 8^7

Simplify each of the following expressions by rewriting the expression using exponents.

9. $3 \times 3 \times 3$ 10. 4×4

11. $8 \times 8 \times 8 \times 8$ 12. $2 \times 2 \times 2 \times 2 \times 2$

13. $5 \times 5 \times 5$ 14. $10 \times 10 \times 10 \times 10$

15. $9 \times 9 \times 9 \times 9$ 16. $6 \times 6 \times 6 \times 6$

17. $7 \times 7 \times 7 \times 7 \times 7 \times 7 \times 7$ 18. $13 \times 13 \times 13$

19. 14×14 20. $3 \times 3 \times 3 \times 3 \times 3$

Using a Calculator with Exponents

Many calculators can evaluate expressions like 2^5 with a special function key such as $\boxed{y^x}$ or $\boxed{x^y}$. Check your calculator and your owner's guide to see how your calculator performs this operation. Pay special attention to the order in which the numbers are to be entered. If your calculator does not have this feature, you will have to perform the repeated multiplications.

EXAMPLE 2-31   Evaluate 5^4 using your calculator.
  Solution

$$\boxed{y^x}\ 5, 4 \qquad 625$$

$$\therefore \qquad 5^4 = 625$$

EXERCISE 2-14

Calculator Drill
Compute the number represented by each of the following expressions.

1. 2^5	2. 4^3	3. 9^3	4. 5^3
5. 8^4	6. 7^3	7. 6^5	8. 3^5
9. 10^3	10. 4^8	11. 2^7	12. 9^5

SELECTED TERMS

absolute value   The value of a number without regard to its sign.

chain calculation   The technique of solving arithmetic problems using a calculator without storing intermediate results.

exponent   In a power, the number of times the base is used as a factor in a product; e.g. 2^3, the base is 2, the exponent is 3; thus, 2 is used three times in the product ($2 \times 2 \times 2$).

factors   A product is formed when two or more numbers are multiplied; each of the numbers in the product is called a factor of the product.

inequality   A statement that two numerical expressions do not have the same value.

natural numbers   The set of numbers (1, 2, 3, etc.) used in counting; the counting numbers.

numerical equation   A statement that declares two numerical expressions have the same value.

numerical expression   A number or a list of numbers joined by arithmetic operators; e.g., 9, or $25 - 8$.

terms   A sum is formed when two or more numbers are added; each of the numbers in the sum is called a term of the sum.

3

NUMBER NOTATION
AND OPERATION

In this chapter, ways to write answers to calculations are presented. After completing this chapter, you will have a knowledge of scientific notation, engineering notation, significant figures, and rounding. Rules of exponents are also presented as they apply to powers of ten. Additionally, you will have had an opportunity to use your calculator to multiply, divide, square, take square roots, and find reciprocals.

In addition to the abbreviations used in Chapter 2, the following abbreviations are new to this chapter:

DN	decimal notation	$\boxed{x^2}$	square a number
R	round	$\boxed{EE}$	exponent entry
SF	significant figures	$\boxed{\sqrt{}}$	square root
SN	scientific notation	$\boxed{ENG}$	engineering notation

The following symbols are new to this chapter and are listed here for your reference:

$\equiv$	identically equal to	$\approx$	approximately equal to
$\pm$	plus or minus	$\Rightarrow$	yields

Special attention is given in this chapter to some features your calculator might have. Therefore, it would be helpful to have the owner's guide for your calculator handy as you study.

3-1 POWERS OF TEN NOTATION

Powers of ten notation provides a way to express any number as a decimal number times a power of ten. This will enable us to conveniently work with both very large and very small numbers. In electronics, it is not uncommon to work with extremely small decimal fractions like 0.000 0073 and 0.000 000 000 435. By applying the concepts of powers of ten notation, even these numbers may be expressed in a simple manner.

In Chapter 2 you learned how to use exponents. You know that $10 \times 10 \times 10$ can be expressed as 10^3. Since 1000 is $10 \times 10 \times 10$, we may conclude that $1000 = 10^3$. Suppose that we wanted to write 5000 using powers of ten notation; 5000 is 5×1000 or 5×10^3. Therefore, 5000 can be written as 5×10^3 in *powers of ten notation*. Figure 3-1 shows the pattern of a number written in powers of ten notation. Table 3-1 shows several examples of numbers written in powers of ten notation.

Notice in Table 3-1 that the decimal fractions 0.0234, 0.0087, and 0.345 have negative exponents. Also notice that the placement of the decimal point in the powers of ten notation is not consistent. The decimal point may be

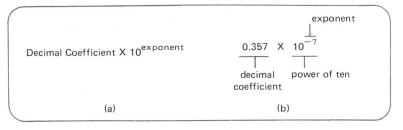

FIGURE 3-1
Powers of ten notation. (a) Pattern for powers of ten notation is a decimal coefficient times 10 to an exponent; (b) example of powers of ten notation.

TABLE 3-1

POWERS OF TEN NOTATION EXAMPLES	
Decimal Notation	*Powers of Ten Notation*
12 430	124.30×10^2
8350	83.5×10^2
0.0235	235×10^{-4}
104 200	10.42×10^4
0.0087	8.7×10^{-3}
192	0.192×10^3
0.345	34.5×10^{-2}

placed anywhere you desire with one restriction: there must be at least one digit to the left of the decimal point. This allows a great deal of flexibility when forming decimal numbers into powers of ten notation.

The steps used to form decimal numbers into powers of ten notation are summarized in Rule 3-1. They are also pictured in Figure 3-2.

RULE 3-1. EXPRESSING DECIMAL NUMBERS IN POWERS OF TEN
            NOTATION

To express a decimal number in powers of ten notation:

1. Form the decimal coefficient by moving the decimal point to the desired position.
2. Determine the exponent by counting the number of places the decimal point was moved.

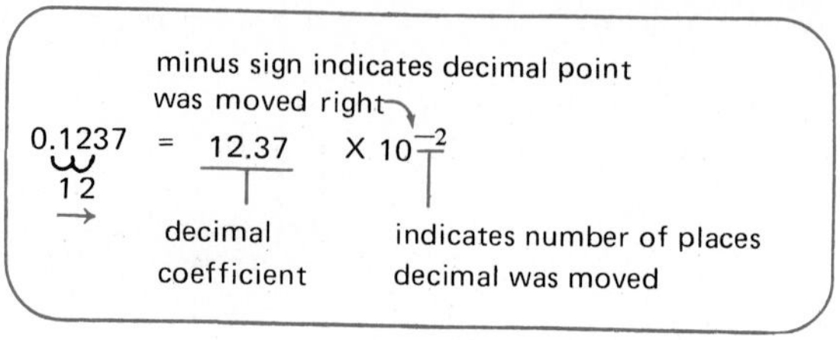

FIGURE 3–2
Forming powers of ten notation.

3. The proper sign for the exponent is selected as:
   (a) A minus sign if the decimal point is moved to the right.
   (b) A plus sign if the decimal point is moved to the left.
4. Write the number in powers of ten notation as a product of the decimal coefficient and the power of ten.

EXAMPLE 3-1   Write 0.003 725 in powers of ten notation. Place the decimal between the seven and the two.

Solution   Use Rule 3-1.

Step 1:   Move the decimal point:

$$0.003\ 725 \Rightarrow 37.25$$

Step 2:   Count places moved:

$$0.003\ 725 \qquad \text{moved four places}$$
$$1\ 2\ 3\ 4$$

Step 3:   Determine the sign of the exponent. Since the decimal point was moved right, a minus sign is assigned to the exponent: (-4).

Step 4:   Form powers of ten notation:

$$37.25 \times 10^{-4}$$
$$\therefore \quad 0.003\ 725 = 37.25 \times 10^{-4}$$

EXAMPLE 3-2   Write $1_{\wedge}320$ in powers of ten notation. Move the decimal to the place indicated by the caret ($\wedge$).

Solution Use Rule 3-1.
Step 1: Move the decimal point:

$$1320 \Rightarrow 1.320$$

Step 2: Count places:

$$1320 \quad \text{moved three places}$$

Step 3: Determine the sign of exponent. Since the decimal point was moved left, the exponent is positive: $(+3)$.

Step 4: Form the powers of ten notation:

$$1.320 \times 10^3$$
$$\therefore \quad 1320 = 1.320 \times 10^3$$

EXERCISE 3-1

Express the following decimal numbers in powers of ten notation. Since the decimal point may be placed anywhere, a caret ($\wedge$) has been placed under the number to show you where to place the decimal point.

1. 75$_\wedge$4 000 000
2. 0.000 8$_\wedge$83 36
3. 0.003$_\wedge$72
4. 9$_\wedge$976
5. 4$_\wedge$01
6. 0.000 006 0$_\wedge$9
7. 0.5531$_\wedge$
8. 176$_\wedge$2
9. 0.002$_\wedge$49
10. 3$_\wedge$1 982
11. 82$_\wedge$92
12. 0.3$_\wedge$12
13. 28$_\wedge$6 514
14. 7$_\wedge$21.12
15. 1.788$_\wedge$2
16. 42.06$_\wedge$8

3-2 SCIENTIFIC NOTATION AND DECIMAL NOTATION

Scientific Notation

In the preceding section you learned that there is a great deal of flexibility in forming powers of ten notation. The decimal point could be placed anywhere desired. Scientific notation is a special form of powers of ten notation. In scientific notation the decimal point is always placed to the right of the leftmost nonzero digit. Table 3-2 shows several examples of scientific notation. You will notice that in every instance the decimal point is located in the same place—to the right of the first nonzero digit.

As can be seen from Table 3-2, scientific notation has a very definite form. Rule 3-2 outlines the procedure for writing a decimal number in scientific notation.

RULE 3-2. EXPRESSING DECIMAL NUMBERS IN SCIENTIFIC NOTATION

To express a decimal number in scientific notation:

1. Form the decimal coefficient by placing the decimal point to the right of the leftmost nonzero digit.
2. Determine the exponent by counting the number of places the decimal point was moved.
3. The proper sign for the exponent is selected as:
   (a) A minus sign if the decimal point is moved to the right.
   (b) A plus sign if the decimal point is moved to the left.
4. Write the number in scientific notation as a product of the decimal coefficient and the power of ten.

EXAMPLE 3-3   Express 0.008 72 in scientific notation.
Solution   Use Rule 3-2.
Step 1:   Form the decimal coefficient.

$$0.008\,72 \Rightarrow 8.72$$

TABLE 3-2

COMPARISON OF NOTATION

Decimal Notation	Powers of Ten Notation	Scientific Notation
5350	53.50×10^2	5.350×10^3
0.002 87	287×10^{-5}	2.87×10^{-3}
0.033 33	333.3×10^{-4}	3.333×10^{-2}
26 220	26.220×10^3	2.6220×10^4
0.0805	80.5×10^{-3}	8.05×10^{-2}

Steps 2–3: Count places and determine the sign of the exponent.

$$0.008\,72 \qquad \text{moved three places to the right } (-3)$$

Step 4: Form the number.

$$8.72 \times 10^{-3}$$
$$\therefore \quad 0.008\,72 = 8.72 \times 10^{-3}$$

EXERCISE 3-2

Express the following numbers in scientific notation.

1. 300	2. 2000	3. 280
4. 36	5. 0.50	6. 0.009
7. 0.36	8. 0.85	9. 7 052
10. 0.008 83	11. 0.0063	12. 992
13. 0.001	14. 409	15. 578 000
16. 0.004 93	17. 5550	18. 43 500
19. 0.68	20. 0.007 04	21. 0.013 13
22. 18.52	23. −282.3	24. 0.151
25. −0.008 25	26. 1316.2	27. −47.8

Decimal Notation

Numbers that are written without powers of ten are written in decimal notation. For example, 75 824, 0.005 13, and 19.3 are in decimal notation. In electronics, it is sometimes necessary to change numbers in powers of ten notation to decimal notation. Rule 3-3 gives a general procedure for doing that.

RULE 3-3. CONVERTING FROM POWERS OF TEN NOTATION
         TO DECIMAL NOTATION

To change from powers of ten notation to decimal notation, form the decimal number from the decimal coefficient and:

1. If the exponent is positive, then move the decimal point to the right the number of places equal to the exponent. If necessary, add trailing zeros to hold the position of the decimal point.

2. If the exponent is negative, then move the decimal point to the left the number of places equal to the exponent. If necessary, add leading zeros to hold the position of the decimal point. The number will be in decimal notation.

EXAMPLE 3-4   Write (a) 6.28×10^3 and (b) 5.31×10^{-4} in decimal notation. Use Rule 3-3.

Solution   (a) Move the decimal point three places to the right:

$$6.28 \times 10^3 = 6280. = 6280.$$

$$1\ 2\ 3$$
$$\longrightarrow$$

Notice that the zero was added to fill out the number.

$$\therefore \quad 6.28 \times 10^3 = 6280$$

(b) Move the decimal point four places to the left:

$$5.31 \times 10^{-4} = 0.000\ 531 = 0.000\ 531$$

$$4\ 3\ 2\ 1$$
$$\longleftarrow$$

Notice that zeros were added to fill out the number.

$$\therefore \quad 5.31 \times 10^{-4} = 0.000\ 531$$

Table 3-3 contains more examples that illustrate the use of Rule 3-3.

TABLE 3-3

APPLICATION OF RULE 3-3

Powers of Ten Notation	Movement of Decimal Point		Decimal Notation
	Direction	No. of Places	
6.38×10^{-3}	Left	3	0.006 38
0.904×10^5	Right	5	90 400
34.16×10^{-4}	Left	4	0.003 416
7.24×10^6	Right	6	7 240 000
5.00×10^3	Right	3	5000

EXERCISE 3-3
Write the following numbers in decimal notation.

1. 6.35×10^{-2}
2. 1.058×10^{6}
3. 382.8×10^{-6}
4. 52.42×10^{3}
5. 9.935×10^{-3}
6. 7015×10^{2}
7. 462×10^{-2}
8. 14.86×10^{-5}
9. 0.8831×10^{2}
10. $0.003\,62 \times 10^{4}$
11. -1.83×10^{5}
12. 0.252×10^{-1}
13. 288.2×10^{-3}
14. -192.76×10^{2}
15. -0.0715×10^{-2}
16. 3652.5×10^{-5}

Expressing Powers of Ten Notation in Scientific Notation

In the sections to follow you will be asked to express answers in scientific notation. Many times the numbers will be expressed in powers of ten notation. Rule 3-4 details the procedures for writing powers of ten numbers in scientific notation.

RULE 3-4. EXPRESSING POWERS OF TEN NOTATION IN SCIENTIFIC NOTATION

To change from powers of ten notation to scientific notation:

1. Form the decimal coefficient by placing the decimal point to the right of the leftmost nonzero digit.
2. Determine the *exponent increment* by counting the number of places the decimal point is moved and assign the sign:
   (a) A minus sign if the decimal point is moved to the right.
   (b) A plus sign if the decimal point is moved to the left.
3. Combine the exponent of the powers of ten notation with the exponent increment of step 2.
4. Write the number in scientific notation as a product of the decimal coefficient and the power of ten.

EXAMPLE 3-5 Express 49.1×10^{3} in scientific notation.
Solution Use Rule 3-4.

Step 1: The decimal is moved left one place:

$$49.1 \Rightarrow 4.91$$
$$\underset{\underset{\leftarrow}{1}}{\smile}$$

Step 2: The exponent increment is $+1$.

Step 3: Combine the exponent of the powers of ten notation (3) with the exponent increment (1):

$$3 + 1 = 4$$

Step 4: 4.91×10^4.

$$\therefore \quad 49.1 \times 10^3 = 4.91 \times 10^4$$

EXAMPLE 3-6 Express $0.005\ 28 \times 10^5$ in scientific notation.

Solution Use Rule 3-4.

Step 1: The decimal is moved right three places:

$$0.005\ 28 \Rightarrow 5.28$$
$$\underset{\underset{\longrightarrow}{1\ 2\ 3}}{\overset{}{\smile\smile\smile}}$$

Step 2: The exponent increment is -3.

Step 3: Combine the exponent of the powers of ten notation (5) with the exponent increment (-3):

$$5 + (-3) = 2$$

Step 4: 5.28×10^2.

$$\therefore \quad 0.005\ 28 \times 10^5 = 5.28 \times 10^2$$

EXERCISE 3-4

Write each of the following numbers in scientific notation.

1. 0.0273×10^{-2} 2. 820×10^4
3. 604×10^{-5} 4. 32.8×10^4
5. 0.019×10^7 6. 59.8×10^{-5}

7. 13.13×10^4	8. $0.003\ 52 \times 10^6$
9. 42.6×10^{-1}	10. 0.081×10^2
11. 20.7×10^3	12. 1073×10^4
13. 78.9×10^{-5}	14. -0.428×10^2
15. -493×10^{-4}	16. 45.78×10^{-2}

3-3 SIGNIFICANT FIGURES AND ROUNDING

Significant Figures

When using some calculators to solve a problem, the number of digits or figures displayed by the calculator varies. There may be as many as ten or as few as one digit displayed as the answer. The question of how many figures the answer should contain depends upon the number of *significant figures* in each of the numbers used in the calculation. Suppose that 4.58 is divided by 2.13 and the calculator indicates the answer to be 2.150 234 742. Should all ten figures be written as the answer? Certainly not! Since each of the original numbers had only three significant figures, then the answer should have only three figures. Therefore, we would *round* the answer to 2.15.

The following guidelines will help you determine the number of significant figures in a decimal number and how many to include in your answer.

GUIDELINES 3-1. DETERMINING NUMBER OF SIGNIFICANT
FIGURES

1. Usually, all figures in a decimal number are significant.
2. If zero is used to hold the decimal point in its place, then it is not significant. For example, the zeros in 0.024 and 0.0038 *are not* significant. However, the zeros in 30.7, 108.9, and 8.30 *are* all significant.
3. The placement of the decimal point does not affect the number of significant figures in a number. Thus, each of the numbers 9.360, 0.009 360, and 936.0 has four significant figures.
4. Use judgment in forming the answer to a calculation. Most of the numbers used in electronics are significant only to two or three figures.
5. Use powers of ten or scientific notation to indicate the significant digits of very large and very small quantities. For example, writing 470 000 ohms to indicate the resistance of a resistor might indicate six-figure significance, but the manufacturer's tolerance of the component is only significant to three figures. By expressing the number

in scientific notation as 4.70×10^5 or as 470×10^3 in powers of ten notation, only three significant figures are indicated.

6. If the number lacks sufficient digits to indicate the desired significance, then add zeros to the number to fill it out. For example, it is desired to express 80 to three significant figures; then one zero is added to 80 to make it 80.0.

Rounding

Rounding is used to shorten the calculator answer to the desired number of significant figures. The following very simple rounding procedure is the one programmed into most calculators with display and rounding options.

RULE 3-5. ROUNDING

To round to the desired number of significant figures:

1. Decide on the number of significant figures in the answer.
2. Examine the first digit to the right of the last significant figure:
   (a) If it is 5 or more, then add 1 to the last significant figure.
   (b) If it is 4 or less, then leave all figures alone.
3. Drop everything to the right of the last significant figure.

EXAMPLE 3-7   Express 740 800 to three significant figures using scientific notation.

Solution   Use Rule 3-5:

$$\text{SN:} \qquad 7.408\ 00 \times 10^5$$

R to 3 SF: The fourth digit is 8, which is greater than 5; so add 1:

$$7.41 \times 10^5$$

∴ $740\ 800 \Rightarrow 7.41 \times 10^5$ to three significant figures.

EXAMPLE 3-8   Express 0.008 895 to three significant figures using scientific notation.

Solution   Use Rule 3-5:

$$\text{SN:} \qquad 8.895 \times 10^{-3}$$

R to 3 SF: The fourth digit is 5, so 1 is added to 9, which becomes 10; so 89 becomes 90:

$$8.90 \times 10^{-3}$$
$$\therefore \qquad 0.008\ 895 \Rightarrow 8.90 \times 10^{-3}$$

EXERCISE 3-5

Express the following numbers in scientific notation to three significant figures. NOTE: Use the $\boxed{\text{SCI}}$ calculator stroke, if available. Consult your owner's guide for the use of this stroke.

1. 756.83	2. 99.05	3. 72
4. 0.008 109	5. 14.14	6. 1776
7. 0.1066	8. 149 200	9. 186.4
10. 3 481 500	11. 0.0593	12. 0.04
13. 0.000 877 5	14. 55.55	15. 99 999

3-4 MULTIPLICATION WITH POWERS OF TEN

In this section you will use your calculator to multiply decimal numbers and numbers expressed in both scientific notation and powers of ten notation. The ease of carrying out these operations is governed to some degree by the features found on your calculator. Some calculators have display and rounding options that round the answer to the desired number of significant figures in either decimal or scientific notation. Consult your owner's guide to see if your calculator has this feature.

EXAMPLE 3-9   Evaluate 0.187×4530 with your calculator. Write the answer in both scientific and decimal notation with three significant figures.

Solution

$$\boxed{\times} \qquad 0.187 \times 4530 = 847.11$$
$$\text{SN:} \qquad 847.11 = 8.4711 \times 10^{2}$$

R to 3 SF: 8.47×10^2
DN: $8.47 \times 10^2 = 847$
$\therefore$ $0.187 \times 4530 = 8.47 \times 10^2 = 847$

EXAMPLE 3-10   Evaluate 83×9967 with your calculator. Express the answer in both scientific and decimal notation with three significant figures.

Solution

$\boxed{\times}$ $83 \times 9967 = 827\ 261$
SN: $8.27\ 261 \times 10^5$
R to 3 SF: 8.27×10^5
DN: $827\ 261$ to six significant figures

NOTE: This number cannot be clearly written in decimal notation to three significant figures because $827\ 000$ can also indicate six-place significance.

From the preceding example you can see that decimal notation cannot always be written to indicate the desired number of significant figures. Only if we use scientific notation can we consistently express a number to the desired significance.

The following rule summarizes the steps used to multiply numbers that are expressed in scientific and powers of ten notation.

RULE 3-6. MULTIPLYING WITH POWERS OF TEN

To multiply manually numbers expressed in scientific and powers of ten notation:

1. Multiply the decimal coefficients.
2. Add the exponents.
3. Form the product.

EXAMPLE 3-11   Solve $10^{-7} \times 10^3 \times 10^{-2}$ manually.
Solution   Use Rule 3-6.

Step 1: Not needed.
Step 2: Add exponents:

$$-7 + 3 + (-2) = -6$$

Step 3: Form the product:

$$10^{-6}$$
$$\therefore \quad 10^{-7} \times 10^3 \times 10^{-2} = 10^{-6}$$

EXAMPLE 3-12 Solve $6.37 \times 10^4 \times 2.40 \times 10^{-2}$, expressing the answer in scientific notation to three significant figures.

Solution Use Rule 3-6.
Step 1: Multiply decimal coefficients:

$$6.37 \times 2.40 = 15.288$$

Step 2: Add exponents:

$$4 + (-2) = 2$$

Step 3: Form the product:

$$15.288 \times 10^2$$

R to 3 SF: $\quad 15.3 \times 10^2$

SN: $\quad 1.53 \times 10^3$

$\therefore \quad 6.37 \times 10^4 \times 2.40 \times 10^{-2}$
$= 1.53 \times 10^3$

When working with powers of ten, we sometimes encounter the situation where 10 is raised to the zero power (10^0). By definition, *any number* (except zero) *raised to the zero power is 1.* So $10^0 = 1$. You should remember this important definition.

Before you start the next exercise, we call your attention to a very useful feature that your calculator might have. It is the exponent entry key $\boxed{\text{EE}}$. If your calculator has this feature, learn to use it, as it can greatly simplify your work with scientific and powers of ten notation.

EXAMPLE 3-13   Use your calculator to multiply $3.15 \times 10^5 \times 8.29 \times 10^{-8}$.

Solution   Enter 3.15 $\boxed{\text{EE}}$ 5.

$\boxed{\times}$ 8.29 $\boxed{\text{EE}}$ $\boxed{\text{CHS}}$ 8 = 0.026 113 5

R to 3 SF: 0.0261

SN: 2.61×10^{-2}

∴ $3.15 \times 10^5 \times 8.29 \times 10^{-8}$
$= 2.61 \times 10^{-2}$

Observation   The last two steps may be replaced with the formatting function on your calculator. See your owner's guide.

EXERCISE 3-6
Combine the following exponents.

1. $10^2 \times 10^3$
2. $10^8 \times 10^{-9}$
3. $10^{-5} \times 10^{-2}$
4. $10^{-3} \times 10^7$
5. $10^1 \times 10^0 \times 10^{-1}$
6. $10^8 \times 10^{-3} \times 10^{-5}$
7. $10^{-9} \times 10^{-3} \times 10^{-6}$
8. $10^2 \times 10^4 \times 10^5$
9. $10^{-4} \times 10^7 \times 10^3$
10. $10^7 \times 10^0 \times 10^{-9}$

Calculator Drill
Perform the following multiplications and express the answers in scientific notation to three significant figures.

11. 1084×326
12. 63.0×847
13. 0.0054×0.0400
14. $0.9035 \times 463\ 528$
15. 26.78×4020
16. 0.0332×4007
17. 0.0178×0.111
18. $203 \times 0.000\ 919$
19. 998×1995
20. 14.38×0.770
21. $4.38 \times 10^7 \times 2.00 \times 10^{-2}$
22. $9.900 \times 10^{-2} \times 3.28 \times 10^1$
23. $7.03 \times 10^2 \times 3.00 \times 10^{-2}$
24. $6.8375 \times 10^{-7} \times 2.046 \times 10^5$
25. $992 \times 4.35 \times 10^{-6}$
26. $0.0828 \times 4.336 \times 10^{-4}$
27. $2.024 \times 10^3 \times 52.0$
28. $92.02 \times 1.096 \times 10^{-6}$
29. $7.40 \times 10^{-9} \times 0.483$
30. $0.003\ 94 \times 7.083 \times 10^4$

3-5 DIVISION WITH POWERS OF TEN

In this section you will learn to divide numbers written in both decimal and powers of ten notation. Again you will have the opportunity to use your calculator, with special emphasis on entering exponents and formatting answers.

EXAMPLE 3-14 Divide 83.7 by 5.82 with your calculator. Express the answer in scientific notation to two significant figures.

Solution

$$\boxed{\div} \quad 83.7 \div 5.82 = 14.3814$$
$$\text{SN:} \quad 1.43814 \times 10^1$$
$$\text{R to 2 SF:} \quad 1.4 \times 10^1$$
$$\therefore \quad 83.7 \div 5.82 = 1.4 \times 10^1$$

RULE 3-7. DIVIDING WITH POWERS OF TEN

To divide numbers that are expressed in either scientific notation or in powers of ten notation:

1. Divide the decimal coefficients.
2. Determine the exponent of 10 by subtracting the exponent of the 10 in the denominator from the exponent of 10 in the numerator.
3. Form the quotient.

EXAMPLE 3-15 Divide 8×10^{-4} by 4×10^{-8}.

Solution Use Rule 3-7.

Step 1: Divide the decimal coefficients:

$$8/4 = 2$$

Step 2: Subtract exponents:

$$-4 - (-8) = 4$$

Step 3: Form the quotient:

$$2 \times 10^4$$

$$\therefore \quad \frac{8 \times 10^{-4}}{4 \times 10^{-8}} = 2 \times 10^4$$

EXAMPLE 3-16   Divide 10^{-6} by 10^3.
    Solution   Use Rule 3-7.
    Step 1:   Not used.
    Step 2:   Subtract exponents:

$$-6 - (3) = -9$$

    Step 3:   Form quotient:

$$\therefore \quad 10^{-6}/10^3 = 10^{-9}$$

EXAMPLE 3-17   Divide 43.82×10^4 by 51.32×10^{-3}, expressing the answer in scientific notation to three significant figures.
    Solution   Use your calculator.
              Enter 43.82 $\boxed{\text{EE}}$ 4.

$$\boxed{\div} \quad 51.32 \ \boxed{\text{EE}} \ 3 \ \boxed{\text{CHS}} \ = 8.5386 \times 10^6$$

R to 3 SF: 8.54×10^6

$$\therefore \quad \frac{43.82 \times 10^4}{51.32 \times 10^{-3}} = 8.54 \times 10^6$$

EXERCISE 3-7

Mentally divide the following powers of ten.

1. $10^3/10^2$ 2. $10^5/10^7$ 3. $10^{-3}/10^{-4}$

4. $10^8/10^{-3}$ 5. $10^9/10^4$ 6. $(5 \times 10^6)/10^2$

7. $\dfrac{5 \times 10^3}{10^2}$ 8. $\dfrac{6 \times 10^{-5}}{2 \times 10^{-9}}$ 9. $\dfrac{9 \times 10^{12}}{3 \times 10^3}$

10. 7×10^6 11. $\dfrac{8 \times 10^0}{4 \times 10^7}$ 12. $\dfrac{2 \times 10^2}{10^2}$

Calculator Drill

Using your calculator, divide the following. Express the answer in scientific notation to three significant figures.

13. $\dfrac{702 \times 10^2}{18.6}$ 14. $\dfrac{32.2 \times 10^{-6}}{0.83}$ 15. $\dfrac{1.93 \times 10^2}{22.4 \times 10^6}$

16. $\dfrac{2.00 \times 10^4}{823 \times 10^{-3}}$ 17. $\dfrac{0.008\ 70 \times 10^{-3}}{226}$ 18. $\dfrac{3800 \times 10^{-5}}{0.0282}$

19. $\dfrac{0.762 \times 10^4}{320 \times 10^3}$ 20. $\dfrac{80.3 \times 10^{-6}}{0.0323}$ 21. $\dfrac{561}{0.013 \times 10^{-4}}$

3-6 POWERS, PRODUCTS, AND FRACTIONS RAISED TO POWERS

In Chapter 2 you learned about exponents; in this section, more about exponents is explored as they apply to powers of ten.

Power of a Power

What does the expression $(10^3)^3$ mean? It means to use 10^3 as a factor in a product three times: $10^3 \times 10^3 \times 10^3 = 10^9$. This example shows that a power of ten raised to a power is equal to 10 raised to the product of the two exponents. Thus, $(10^3)^3 = 10^{3 \times 3} = 10^9$.

RULE 3-8. POWERS AND POWERS OF TEN

To raise a power of ten to a power, multiply the exponents.

EXAMPLE 3-18 Solve $(10^{-5})^2$.

Solution Use Rule 3-8.

$$(10^{-5})^2 = 10^{-5 \times 2} = 10^{-10}$$
$$\therefore \quad (10^{-5})^2 = 10^{-10}$$

EXERCISE 3-8
Mentally solve each of the following.

1. $(10^2)^3$ 2. $(10^5)^3$ 3. $(10^4)^2$ 4. $(10^2)^2$
5. $(10^6)^3$ 6. $(10^{-5})^6$ 7. $(10^{-7})^2$ 8. $(10^{-3})^{-3}$
9. $(10^{-2})^6$ 10. $(10^3)^{-8}$ 11. $(10^5)^{-2}$ 12. $(10^{-9})^3$

Power of a Product

A product raised to a power is called a *power of a product*. This expression is a power of a product: $(3 \times 10^3)^2$. From our understanding of exponents, we know that $(3 \times 10^3)^2 = (3 \times 10^3) \times (3 \times 10^3)$ or $3 \times 10^3 \times 3 \times 10^3$, which may be written $3 \times 3 \times 10^3 \times 10^3$. Multiplying, we get 9×10^6. We arrive at this same answer by applying Rule 3-9.

RULE 3-9. POWERS AND PRODUCTS

To simplify the power of a product:

1. Apply the power to each factor.
2. Form the product.

EXAMPLE 3-19 Simplify $(3 \times 10^3)^2$.
Solution Use Rule 3-9.
Step 1: Apply power to each factor:

$$3^2 = 9$$
$$(10^3)^2 = 10^6$$

Step 2: Form the product:

$$9 \times 10^6$$
$$\therefore \quad (3 \times 10^3)^2 = 9 \times 10^6$$

EXAMPLE 3-20 Simplify $(2 \times 10^{-5})^4$.
Solution Use Rule 3-9.
Step 1: Apply power to each factor:

$$2^4 = 16$$
$$(10^{-5})^4 = 10^{-20}$$

Step 2: Form the product:

$$16 \times 10^{-20}$$
$$\therefore \quad (2 \times 10^{-5})^4 = 16 \times 10^{-20}$$

EXERCISE 3-9
Simplify each of the following.

1. $(2 \times 10^3)^2$ 2. $(3 \times 10^2)^2$ 3. $(5 \times 10^{-4})^2$

4. $(6 \times 10^{-5})^2$ 5. $(2 \times 10^4)^3$ 6. $(3 \times 10^6)^3$
7. $(8 \times 10^{-6})^2$ 8. $(4 \times 10^{-7})^2$ 9. $(9 \times 10^1)^2$
10. $(2 \times 10^3)^5$ 11. $(7 \times 10^{-5})^2$ 12. $(10 \times 10^{-4})^3$

Power of a Fraction

A fraction raised to a power is called a *power of a fraction*. The expression $(\frac{3}{10})^3$ is a power of a fraction. From our understanding of exponents, we know that $(\frac{3}{10})^3 = \frac{3}{10} \times \frac{3}{10} \times \frac{3}{10}$, which may be written $(3 \times 3 \times 3)/(10 \times 10 \times 10)$. Simplifying, we get $\frac{27}{1000} = 0.027$. We will arrive at this same answer by applying Rule 3-10.

RULE 3-10. POWERS AND FRACTIONS

To simplify the power of a fraction:

1. Work within the parentheses first.
2. Raise the resulting quotient to the power.

EXAMPLE 3-21

Simplify $\left(\dfrac{10^3}{10^6} \right)^2$.

Solution   Use Rule 3-10.
Step 1:   Work within parentheses:

$$\left(\frac{10^3}{10^6} \right) = 10^{3-6} = 10^{-3}$$

Step 2:   Raise to the power:

$$(10^{-3})^2 = 10^{-3 \times 2} = 10^{-6}$$

$$\therefore \quad \left(\frac{10^3}{10^6} \right)^2 = 10^{-6}$$

EXAMPLE 3-22

Simplify $\left(\dfrac{62.2 \times 10^4}{2.81 \times 10^{-5}} \right)^2$, expressing the answer in scientific notation to three significant figures.

Solution   Use Rule 3-10 and your calculator.

$$\boxed{\div} \qquad \frac{62.2 \times 10^4}{2.81 \times 10^{-5}} = (22.1352 \times 10^9)$$

$$\boxed{x^2} \qquad (22.1352 \times 10^9)^2 = 489.968 \times 10^{18}$$

R to 3 SF: $\qquad 489.968 \times 10^{18} \Rightarrow 490 \times 10^{18}$

SN: $\qquad 490 \times 10^{18} = 4.90 \times 10^{20}$

$$\therefore \qquad \left(\frac{62.2 \times 10^4}{2.81 \times 10^{-5}}\right)^2 = 4.90 \times 10^{20}$$

There are several ways to square a number on a calculator depending upon the kind you have. Some calculators use the times sign $\boxed{\times}$ for squaring, while others have a square key $\boxed{x^2}$. Yet others have both these features and a key to raise a positive number to any power $\boxed{y^x}$. Consult your owner's guide to determine your option in squaring (x^2) and cubing (x^3) numbers.

EXERCISE 3-10
Mentally solve the following.

1. $\left(\dfrac{10^3}{10^5}\right)^2$ 2. $\left(\dfrac{10^4}{10^2}\right)^3$ 3. $\left(\dfrac{10^{-2}}{10^3}\right)^3$

4. $\left(\dfrac{10^{-3}}{10^5}\right)^4$ 5. $\left(\dfrac{10^6}{10^2}\right)^2$ 6. $\left(\dfrac{10^{-4}}{10^{-3}}\right)^5$

7. $\left(\dfrac{10^{-4}}{10^{-4}}\right)^3$ 8. $\left(\dfrac{10^1}{10^7}\right)^4$ 9. $\left(\dfrac{10^{-2}}{10^4}\right)^3$

10. $\left(\dfrac{10^8}{10^5}\right)^2$ 11. $\left(\dfrac{10^6}{10^5}\right)^3$ 12. $\left(\dfrac{10^7}{10^{-4}}\right)^2$

Calculator Drill
Using your calculator, compute the following, expressing the answers in scientific notation to three significant figures.

13. $\left(\dfrac{6.00 \times 10^3}{3.00 \times 10^{-6}}\right)^2$ 14. $\left(\dfrac{27.0 \times 10^{-2}}{3.00 \times 10^4}\right)^2$

15. $\left(\dfrac{532 \times 10^2}{72.8 \times 10^{-3}}\right)^2$ 16. $\left(\dfrac{413 \times 10^4}{825}\right)^2$

17. $\left(\dfrac{0.282 \times 10^2}{4.07 \times 10^{-3}}\right)^2$ 18. $\left(\dfrac{0.008\,73 \times 10^6}{0.323 \times 10^{-2}}\right)^2$

19. $\left(\dfrac{8 \times 10^2}{4 \times 10^{-3}}\right)^3$ 20. $\left(\dfrac{27 \times 10^3}{9 \times 10^4}\right)^3$

21. $\left(\dfrac{360 \times 10^{-5}}{25 \times 10^{-3}}\right)^3$ 22. $\left(\dfrac{0.547}{9.31 \times 10^3}\right)^3$

3-7 SQUARE ROOTS AND RADICALS

To take the square root of a number, we simply enter the number into our calculator and press the square-root key $\boxed{\sqrt{}}$. What was once a very difficult and time-consuming task is now as simple as adding when we are aided by a calculator. There is, however, more to be known about taking a square root than the mechanics of pressing the square-root key.

Square Root

You are now very familiar with the meaning of 4^2. It, of course, means to use 4 twice as a factor in a product. A *square root* of a product is one of the two equal factors that were used to form the product. So the square root of 16 ($\sqrt{16}$) is 4. This is because the two equal factors of 16 are 4. If $4 \times 4 = 16$, then what does $(-4) \times (-4)$ equal? 16! Then -4 is also a square root of 16.

RULE 3-11. EVALUATING SQUARE ROOTS

Every positive number has two square roots that have equal absolute values, but are opposite in sign.

EXAMPLE 3-23   What are the square roots of 196?
    Solution   Use your calculator and Rule 3-11.

$$\boxed{\sqrt{}} \qquad \sqrt{196} = 14$$

So the square roots of 196 are $+14$ and -14.
    Check

$$\boxed{\times} \qquad (+14) \times (+14) = 196$$

$$\boxed{\times} \qquad (-14) \times (-14) = 196$$
$$\therefore \qquad \text{The square roots of 196 are } \pm 14.$$

NOTE: The symbol $\pm$ is read "plus or minus."

If you want only the positive square root of a number, say 25, called the *principal square root,* then you indicate this as $\sqrt{25}$ without a sign. The negative square root is indicated as $-\sqrt{25}$. If both positive and negative roots are desired, then the $\pm$ symbol is used, as in $\pm\sqrt{25}$. In solving electrical problems, generally we use only the positive or principal square root. In this book, unless otherwise noted, we shall work with the principal root of the number.

EXAMPLE 3-24   Evaluate $-\sqrt{240}$, expressing the answer in decimal notation to four significant figures.

Solution

$$\boxed{\sqrt{}} \qquad -\sqrt{240} = -15.49$$
$$\therefore \qquad -\sqrt{240} = -15.49 \text{ to four significant}$$
$$\text{figures.}$$

EXERCISE 3-11

Calculator Drill
Evaluate the following, expressing the answers to four significant figures:

1. $\sqrt{372.0}$ 2. $\sqrt{0.008\,700}$ 3. $\pm\sqrt{0.7850}$

4. $\sqrt{99.92}$ 5. $\sqrt{1020}$ 6. $-\sqrt{65.92}$

7. $\sqrt{0.008\,825}$ 8. $\pm\sqrt{7.285 \times 10^{-1}}$ 9. $-\sqrt{192.3 \times 10^4}$

10. $\sqrt{20\,970 \times 10^{-4}}$ 11. $-\sqrt{0.0587 \times 10^3}$ 12. $\pm\sqrt{409.5 \times 10^{-3}}$

Radical

The symbol $\sqrt{}$ is the *radical sign.* The radical sign combined with the bar, a sign of grouping, is written over the number. This combination of number and symbol is called a *radical.* The number under the radical sign is called the *radicand.* Figure 3-3 summarizes these concepts.

Because the bar is a sign of grouping, a radicand that is an arithmetic expression is computed first before the root is taken.

EXAMPLE 3-25 Solve $\sqrt{16 + 9}$.

Solution Enter 16 and add 9:

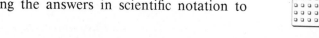

$$\boxed{+} \quad 25$$
$$\boxed{\sqrt{}} \quad 5$$
$$\therefore \quad \sqrt{16 + 9} = 5$$

EXERCISE 3-12

Calculator Drill
Evaluate the following, expressing the answers in scientific notation to three significant figures.

1. $\sqrt{82.0 - 18.0}$
2. $\sqrt{-108 + 133}$
3. $\sqrt{8.00 \times 32.0}$
4. $\sqrt{64.0/4.00}$
5. $\sqrt{7.35 + 12.4}$
6. $\sqrt{802 \times 2.39}$
7. $\sqrt{0.950/0.003\ 80}$
8. $\sqrt{1.22 + 56.3}$
9. $\sqrt{72.1 \times 406}$
10. $-\sqrt{0.0580 - 0.248 + 2.03}$
11. $\sqrt{3.56 \times 0.923 \times 10^4}$
12. $\sqrt{15 + 8 - 7}$

Fractional Exponents

Since fractional exponents follow all the rules for powers, it is generally easier to work with fractional exponents rather than with radical signs. The square root of a number may be indicated by a fractional exponent of $\frac{1}{2}$. Thus,

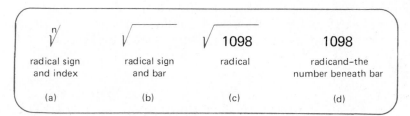

radical sign and index	radical sign and bar	radical	radicand–the number beneath bar
(a)	(b)	(c)	(d)

FIGURE 3–3
Summary of the parts of a radical.

$\sqrt{100} = 100^{1/2} = 100^{0.5}$. By raising a number to the $\frac{1}{2}$ or 0.5 power, we mean *"compute the principal square root of the number."*

EXAMPLE 3-26   Use your calculator to evaluate $256^{1/2}$.
   Solution   Enter 256:

$$\boxed{\sqrt{}} \qquad 16$$
$$\therefore \qquad 256^{1/2} = 16$$

If your calculator does not have provision for entering numbers in powers of ten notation, then you must change the numbers to decimal notation before entering them into the calculator. The following example assumes that this is the case.

EXAMPLE 3-27   Evaluate $(732 \times 10^3)^{0.5}$, expressing the answer in scientific notation to three significant figures.

   Solution   Convert to decimal notation:

$$(732 \times 10^3) = 732\,000$$

Enter 732 000:

$$\boxed{\sqrt{}} \qquad (732\,000)^{0.5} = 855.57$$
R to 3 SF : $855.57 \Rightarrow 856$
SN: $856 = 8.56 \times 10^2$
$\therefore \qquad (732 \times 10^3)^{0.5} = 8.56 \times 10^2$

EXERCISE 3-13

Calculator Drill
With the aid of your calculator, evaluate the following, expressing each of your answers in scientific notation to three significant figures.

1. $242^{1/2}$
2. $0.0116^{1/2}$
3. $52.4^{0.5}$
4. $1380^{0.5}$
5. $9.93^{1/2}$
6. $(672 \times 10^5)^{1/2}$
7. $(14.3 \times 10^{-3})^{0.5}$
8. $(7.28 \times 10^4)^{0.5}$
9. $(203 \times 10^5)^{1/2}$
10. $(0.775 \times 10^3)^{1/2}$
11. $(805.3/0.385)^{1/2}$
12. $(43.8 \times 10^2 + 2.01)^{0.5}$

Square Root of a Negative Number

Although we will not be working with the square root of a negative number until Chapter 26, we mention it here because of its curious nature. Suppose that we want to determine the square root of -49. Place this in your calculator and press the square-root key. You get a strange response, but not an answer. Your calculator is indicating an improper operation.

From this we conclude that the calculator is no help in our investigation. Exploring, we try $7 \times 7 = 49$, not -49, so 7 is not the square root of -49. We next try $-7 \times -7 = 49$, not -49, so -7 is not the square root. If neither 7 nor -7 is the square root, then what is? Since it is not a number that we know about, it must be some other kind of number that we do not know about. This new kind of number is called an *imaginary number*. This very important kind of number is used to analyze ac circuits.

3-8 RECIPROCALS

Numbers like $\frac{2}{3}$ and $\frac{3}{2}$ are *reciprocals* of each other because their product equals 1. Thus, $\frac{2}{3} \times \frac{3}{2} = (2 \times 3)/(3 \times 2) = \frac{6}{6} = 1$. For example:

- 2 and $\frac{1}{2}$ are reciprocals because $2 \times \frac{1}{2} = \frac{2}{2} = 1$.
- -1 and -1 are reciprocals because $-1 \times -1 = 1$.
- 0 has no reciprocal because 0 times any number is 0, *not* 1.

To form the reciprocal of a number, simply place one (1) over that number.

EXAMPLE 3-28 Write the reciprocal of 8 and then compute its value. Express the answer in decimal notation rounded to three significant figures.

Solution Write the reciprocal:

$$\frac{1}{8}$$

$$\boxed{\div} \quad \frac{1}{8} = 0.125$$

$$\therefore \quad \text{The reciprocal of 8 is } \frac{1}{8} = 0.125.$$

EXAMPLE 3-29 Write the reciprocal of 23.82. Compute its value, expressing the answer in scientific notation to four significant figures.

Solution   Write the reciprocal:

$$1/23.82$$

Use the reciprocal key on your calculator.
Enter 23.82.

$\boxed{1/x}$ 4.198×10^{-2}
$\therefore$ The reciprocal of 23.82 is $1/23.82$
$= 4.198 \times 10^{-2}$.

Most calculators have a reciprocal key $\boxed{1/x}$. You should use this feature of your calculator to aid in the solution of the following set of problems.

┌ EXERCISE 3-14

Calculator Drill
Compute the reciprocal of the following numbers. Express the answers in scientific notation to three significant figures.

1. 5.00	2. 4.00	3. 6.00
4. 9.00	5. 3.00	6. 92.6
7. 0.0835	8. 776	9. 503
10. 0.001 38	11. 6.28×10^3	12. -920
13. 4.09×10^{-2}	14. 1.86×10^{-3}	15. 70.7×10^2
16. -47.3		

Division by Multiplying

Division can be accomplished by taking the reciprocal of the divisor and then multiplying. For example, $\frac{16}{8} = 16 \times \frac{1}{8} = 2$. This is useful when performing combined operations on your calculator.

EXAMPLE 3-30   Solve $\frac{20}{4}$ by multiplying.

Solution   Use your calculator.
Enter 4.

$\boxed{1/x}$ 0.25
$\boxed{\times} \, 20$ 5.00

$\therefore$ $\frac{20}{4} = 0.25 \times 20 = 5.00$

Division with a fractional divisor is done by reciprocating the divisor and then multiplying. Rule 3-12 summarizes the steps in solving this type of problem.

RULE 3-12. DIVIDING BY FRACTIONS

To divide a number by a fraction:

1. Reciprocate the divisor (invert the fraction).
2. Multiply the numerators.
3. Form a new fraction of the product over the denominator.
4. Take the quotient of the resulting fraction.

EXAMPLE 3-31 Find the value of 16/(4/3) by multiplying.
Solution Use Rule 3-12.
Step 1: Reciprocate the denominator:

$$\frac{3}{4}$$

Step 2: Multiply numerators:

$$16 \times 3 = 48$$

Step 3: Form a new fraction:

$$\frac{48}{4}$$

Step 4: Simplify:

$$\frac{48}{4} = 12$$
$$\therefore \quad 16/(4/3) = (16 \times 3)/4 = 12$$

EXERCISE 3-15
Use your calculator to aid in the solution of the following problems. First reciprocate the denominator and then multiply. Express the answers in decimal notation to four significant figures.

1. $\dfrac{48.00}{16.00}$ 2. $\dfrac{84.00}{6.00}$ 3. $\dfrac{96.00}{480.0}$

4. $\dfrac{720.0}{432.0}$	5. $\dfrac{20.00}{5/4}$	6. $\dfrac{18.00}{9/2}$
7. $\dfrac{7.09}{1/2}$	8. $\dfrac{15.4}{3/12}$	9. $\dfrac{17.36}{0.0820}$
10. $\dfrac{0.0385}{3.23 \times 10}$	11. $\dfrac{723.6}{0.6923}$	12. $\dfrac{0.004\ 356}{3/125}$

3-9 COMBINED OPERATIONS

The solution of electronic problems may require that several operations be performed. For example, the power dissipated by a resistance can be computed by squaring the current and multiplying by the resistance. To avoid confusion when solving problems involving powers or roots, the following convention for combined operations has been agreed upon.

CONVENTION 3-1. COMBINED OPERATIONS

It is conventional in problems with several operations to solve the problem from left to right and to:

1. Raise to powers and extract roots first.
2. Multiply and divide next.
3. Add and subtract next.
4. Set priority of operation by interrupting the usual order with parentheses ( ), the bar ($-$), or other signs of grouping.

EXAMPLE 3-32 Solve $\dfrac{5^2 \times 8}{3}$, expressing the answer in decimal notation to three significant figures.

Solution

Square 5: $\qquad \dfrac{5^2 \times 8}{3} = \dfrac{25 \times 8}{3}$

M: $25 \times 8 \qquad 200/3$

D: $200/3 \qquad 66.7$

$\qquad \therefore \quad \dfrac{5^2 \times 8}{3} = 66.7$

EXAMPLE 3-33
Solve $\dfrac{3 \times \sqrt{18 \times 32} + 18}{9}$.

Solution The solution begins with the radicand.

$\boxed{\times}$	$18 \times 32 \Rightarrow 576$	$\dfrac{3 \times \sqrt{576} + 18}{9}$
$\boxed{\sqrt{}}$	$\sqrt{576} \Rightarrow 24$	$\dfrac{3 \times 24 + 18}{9}$
$\boxed{\times}$	$3 \times 24 \Rightarrow 72$	$\dfrac{72 + 18}{9}$
$\boxed{+}$	$72 + 18 \Rightarrow 90$	$90/9$
$\boxed{\div}$	$90/9 \Rightarrow 10$	10

$$\therefore \quad \frac{3 \times \sqrt{18 \times 32} + 18}{9} = 10$$

EXERCISE 3-16

Evaluate each of the following expressions. Express the answers in scientific notation to three significant figures.

1. $6 - 3 \times 4$

2. $6.82 - 3.11 \times 4.32$

3. $5.25 \times 2.03 - 4.42$

4. $\dfrac{7.09}{2.82 \times 6.55}$

5. $0.835 \div 2.81 \times 52.6$

6. $\dfrac{12.2 + 9.51}{2.03}$

7. $7 \times 5 \div 3 - 6 + 2$

8. $(1.28 + 6.92)3.5$

9. $\dfrac{(75.4)(0.952)}{(-4.05)(12.8)}$

10. $\dfrac{(-8.38)(-5.51)}{-4.22}$

11. $\dfrac{3 \times 8 + 5^2}{12}$

12. $(44 \times 16/4)^2$

13. $15 - 18 + 4^2 \times 3$

14. $\sqrt{12/3 + 9 \times 4}$

15. $\dfrac{8.25 \times 10^4 \times 6.02 + 5}{\sqrt{6}}$

16. $\dfrac{11.5 - (18.4)^2 + (22.1)^{0.5}}{0.0821 \times 10^4}$

17. $\dfrac{193^{0.5} + 0.0391^2 \times 16.0}{13.9}$

18. $\dfrac{0.707^2 + 844/12.3}{3.00^{0.5}}$

$$19. \quad \frac{\sqrt{18.2} \times 4.17^2 - 18}{8/12}$$

$$20. \quad \frac{(12.8^2 \times 14.3/8.03)^2}{4.81 \times 23.2^2}$$

3-10 APPROXIMATIONS

We are including this section on approximations so that you will not be a *slave to your calculator*. By approximating the answer to an arithmetic expression before solving with your calculator, you will have a check on your work. Also, the techniques of approximations may be applied for quick electronic circuit calculations. The emphasis in approximations is on speed, not accuracy. The approximation is carried out mentally or with pencil and paper—but not with a calculator!

The following are general guidelines to aid in approximating answers. Note that ≈ means *approximately equal to*.

GUIDELINES 3-2. APPROXIMATING

1. Express each number in scientific notation to one significant figure. For example, 285 becomes 3×10^2; 0.0835 becomes 8×10^{-2}.

2. When multiplying, express the product to the nearest 10. For example, $6 \times 3 \approx 20$, $6 \times 8 \approx 50$, and $3 \times 4 \approx 10$.

3. Express improper fractions as integers. For example, $\frac{9}{4} \approx 2$, $\frac{7}{5} \approx 1$, and $\frac{8}{3} \approx 3$.

4. When solving proper fractions, add a zero to the numerator and decrease the power of ten by 1. Then, treat as an improper fraction. For example, $\frac{2}{7}$ becomes $20 \times 10^{-1}/7 \approx 3 \times 10^{-1}$, and $\frac{3}{5}$ becomes $30 \times 10^{-1}/5 \approx 6 \times 10^{-1}$.

5. When squaring, express the answer as simply as possible. For example, $4^2 \approx 20$, $3^2 \approx 10$, and $8^2 \approx 60$.

6. When taking the square root of a number, first express the number in powers of ten notation with one or two digits in the decimal coefficient and an exponent of ten divisible by 2. Then divide the exponent by 2 and approximate the square root of the decimal coefficient. For example, $\sqrt{3285} \approx (30 \times 10^2)^{0.5} \approx 30^{0.5} \times 10^{2 \times 0.5} \approx 5 \times 10^1$, and $\sqrt{328} \approx (3^{0.5} \times 10^{2 \times 0.5}) \approx 2 \times 10^1$.

EXAMPLE 3-34 Approximate $\dfrac{0.043 \times 2820}{\sqrt{5}}$.

Solution Express in scientific notation:

$$\frac{4 \times 10^{-2} \times 3 \times 10^3}{\sqrt{5}}$$

Simplify numerator by multiplying:

$$(10 \times 10^1)/\sqrt{5}$$

Approximate denominator:

$$(10 \times 10^1)/2$$

Simplify:

$$5 \times 10^1$$

$$\therefore \quad \frac{0.043 \times 2820}{\sqrt{5}} \approx 5 \times 10^1$$

EXAMPLE 3-35 Approximate $\dfrac{0.0327 \times (542)^{0.5}}{77.4}$.

Solution Express in scientific notation:

$$\frac{3 \times 10^{-2} \times (5 \times 10^2)^{0.5}}{8 \times 10^1}$$

Approximate the square root:

$$\frac{3 \times 10^{-2} \times (2 \times 10^1)}{8 \times 10^1}$$

Multiply numerator:

$$\frac{6 \times 10^{-1}}{8 \times 10^1}$$

Divide; use step 4 of Guidelines 3-2:

$$\frac{60 \times 10^{-1}}{8} \times \frac{10^{-1}}{10^{1}} \approx 8 \times 10^{-3}$$

$$\therefore \quad \frac{0.0327 \times (542)^{0.5}}{77.4} \approx 8 \times 10^{-3}$$

EXERCISE 3-17

Approximate each of the following expressions.

1. $\frac{2}{7}$

2. $\frac{2}{9}$

3. $\frac{83}{230}$

4. $0.572/67.4$

5. 528×41

6. 0.37×132

7. 0.752×0.0904

8. $292 \times 0.41 \times 1.43$

9. $\frac{0.083 \times 272}{4.32}$

10. $\frac{682}{0.047 \times 6.12}$

11. $\frac{35^2 \times 8.07}{216}$

12. $\frac{12.2 \times \sqrt{92}}{0.0162}$

13. $\frac{1776}{582 \times 820}$

14. $\frac{\sqrt{0.82}}{4.9 \times 0.76}$

15. $\frac{0.043 \times 2820}{5}$

16. $\frac{10.6 \times 311 \times 5.92}{\sqrt{14.5} \times 32.4}$

17. $\frac{4.13^2 + 378 + 15.6^2}{27.6/3.5^2}$

18. $\frac{0.083 \times \sqrt{1.97}}{75.2 \times 10^{-3} + 0.0982}$

19. $\frac{3.75 \times 18.25 \times 10^{-1}}{\sqrt{16.9}/2.91^2}$

3-11 ENGINEERING NOTATION

Numbers written in powers of ten notation which have exponents that are multiples of 3 are written in *engineering notation*. Table 3-4 shows several examples of numbers written in engineering notation. You will notice that each number written in engineering notation has an exponent of ten that is a multiple of 3. Decimal numbers, written in engineering notation, are used to form multiples and submultiples of *metric units*.

As you see in Table 3-4, engineering notation is a special form of powers of ten notation. Rule 3-13 summarizes the procedure for writing a decimal number in engineering notation. Example 3-36 demonstrates the formation of engineering notation as outlined in Rule 3-13.

TABLE 3-4

COMPARISON OF NOTATION

Decimal Notation	Engineering Notation*
3200	3.20×10^3
62 400 000	62.4×10^6
9 880 000 000	9.88×10^9
0.0255	25.5×10^{-3}
0.000 082	82.0×10^{-6}†

*Expressed to three significant figures.
†A zero was added to *fill* out the number.

RULE 3–13. EXPRESSING DECIMAL NUMBERS IN ENGINEERING NOTATION

To express a decimal number in engineering notation:

1. Express the number in scientific notation.
2. If necessary, adjust the exponent in the scientific notation by subtracting 1 or 2 from the exponent to make the exponent a multiple of 3.
3. Move the decimal point in the decimal coefficient to the right one or two places to correspond to the change in the exponent.
4. Write the number in engineering notation as a product of the decimal coefficient and the power of ten.

EXAMPLE 3-36   Write 0.000 72 in engineering notation.
Solution   Use Rule 3–13.
Step 1:   Express in scientific notation:

$$0.000\ 72 = 7.2 \times 10^{-4}$$

Step 2:   Subtract 2 from the exponent:

$$10^{-4-2} = 10^{-6}$$

Step 3:   Move the decimal point two places to the right:

$$7.2 \rightarrow 720$$

Step 4: Form the number:

$$720 \times 10^{-6}$$

The steps of Rule 3-13 may be carried out quickly, simply, and automatically with the ENG stroke of the calculator. If your calculator has this stroke, use it to do the following exercise. Consult your calculator owner's guide for the procedure for using this stroke.

EXERCISE 3-18

Write each of the following decimal numbers in engineering notation with three significant figures.

1. 1320	2. 564 000	3. 0.000 085
4. 8648	5. 0.0047	6. 4 280 000
7. 0.008	8. 47 300	9. 0.000 572

Change the following numbers written in engineering notation to decimal notation. If needed, use Rule 3-3 for converting from powers of ten notation to decimal notation.

10. 27×10^{-3}	11. 0.284×10^3	12. 7.32×10^6
13. 420×10^3	14. 8.20×10^{-9}	15. 0.57×10^6
16. 1.25×10^{-6}	17. 94.5×10^6	18. 0.39×10^{-3}

3-12 EXPONENTIAL NOTATION

When working with computer programs, very large and very small numbers may be handled by means of *exponential notation*. Exponential notation as used in computer programming is a variation of powers of ten notation. The "×10" of the powers of ten notation is replaced with an "E."

Suppose that you wish to enter 62 000 into the computer. This number could be entered in exponential notation as 6.2E4, which means 6.2×10^4. Table 3-5 shows several examples of exponential notation as it is used with such programming languages as FORTRAN, BASIC, and APL.

EXERCISE 3-19

Write each of the following numbers in exponential notation with three significant figures. Form the answer so the decimal point is located to the right of the first nonzero digit, as in 9.52E04.

TABLE 3-5

COMPARISON OF NOTATION		
Decimal Notation	Powers of Ten Notation	Exponential Notation
0.0028	28×10^{-4}	28.E-04*
745	7.45×10^2	7.45E02
92 800	92.8×10^3	92.8E03
0.0843	84.3×10^{-3}	84.3E-03
5 128 000	512.8×10^4	512.8E04

*The leading zero in the exponent and the decimal point directly in front of the E may be omitted. Thus 28.E-04 may also be written 28E-4. Consult your programming book for specific language applications and requirements.

1. 62.8×10^4 2. 4.72×10^{-6} 3. 19 200.
4. 0.003 24   5. 873.5×10^4 6. 3.300×10^{-10}
7. 9 282 746   8. 0.448 12   9. 0.001 987
10. 3 128   11. 902.7   12. 0.012 48

Write the following numbers in decimal notation.

13. 92E-01   14. 194.2E03   15. 3.81E-04
16. 54.9E05   17. 203.8E-03   18. 829.5E-07
19. 4.482E02   20. 13.06E04   21. 17.76E02
22. 537E-05   23. 8.034E03   24. 12.89E-03

SELECTED TERMS

engineering notation   Numbers written in powers of ten notation, having exponents that are multiples of 3 (e.g., 83.7×10^{-3}).

exponential notation   Numbers written in powers of ten notation having an E (in place of $\times$ 10) in the notation (e.g., 23.6E09).

principal square root   The positive square root.

radical   An expression of the form $\sqrt[n]{a}$.

scientific notation   Numbers written in powers of ten notation having the decimal point placed after the leftmost nonzero digit. (e.g., 5.73×10^2).

4

ALGEBRA FUNDAMENTALS I

In this chapter and the one that follows, the *language of algebra* is introduced. Like any written language, algebra uses many special symbols and technical words. It is very important that you learn the names of these symbols and become familiar with the vocabulary of algebra. Once this is done, you will then be ready to solve equations and work with the *techniques of algebra.*

4-1 VARIABLES, SUBSCRIPTS, AND PRIMES

Variables

As your understanding of algebra grows, so will your appreciation of how algebraic methods are used to complement arithmetic methods for solving problems. The effectiveness of algebra comes from its use of letters to stand in for numbers. These letters are called *variables.* Variable is the technical term for a letter which is used in place of a number. The number represented by a letter may be known or unknown, it may be constant or changing. The important idea is that the letter represents a number and obeys the laws of numbers.

The letters most commonly used as variables in electronic problems are taken from the letters of the English and Greek alphabets. Table 4-1 is a par-

TABLE 4-1

GREEK LETTERS COMMONLY ASSOCIATED WITH ELECTRONIC APPLICATIONS			
Name	*Capital*	*Lowercase*	*Used to Designate:*
Alpha		α	Angles, temperature coefficient, current transfer ratio
Beta		β	Angles, current transfer ratio
Delta	Δ		Increment
Eta		η	Efficiency
Theta		θ	Phase angle, thermal impedance
Lambda		λ	Wavelength
Mu		μ	Permeability, micro, amplification factor
Pi		π	3.1416 (ratio of circumference of circle to diameter)
Rho		ρ	Specific resistance
Sigma	Σ	σ	Summation (capital), conductivity (lowercase)
Phi	Φ	ϕ	Magnetic flux (capital), phase angles and angles (lowercase)
Psi		ψ	Electric flux
Omega	Ω	ω	Ohms (capital), angular velocity (lowercase)

tial listing of the Greek alphabet. It contains the letters commonly associated with electronic applications. Since these letters occur in electronics, you should learn their names and how to write them. Appendix A contains a complete listing of the Greek alphabet.

Subscripts

In naming variables for electronics use, sometimes there is a need for the same letter to be used more than once. In this case, a subscript or a prime is used. A *subscript* is a number, a letter, or a group of letters written to the right and below a variable.

EXAMPLE 4-1   Using the letter V for voltage, select variables to represent three different voltages.

Solution   Since each voltage is different, a subscript will be used with each variable. The numbers 1, 2, 3, or the letters a, b, c could be used as subscripts. Each is shown:

$$V_1 \text{ read ``} V \text{ one'' or ``} V \text{ sub one''}$$
$$V_2 \text{ read ``} V \text{ two'' or ``} V \text{ sub two''}$$
$$V_3 \text{ read ``} V \text{ three'' or ``} V \text{ sub three''}$$
$$V_a \text{ read ``} V \text{ } a \text{'' or ``} V \text{ sub } a \text{''}$$
$$V_b \text{ read ``} V \text{ } b \text{'' or ``} V \text{ sub } b \text{''}$$
$$V_c \text{ read ``} V \text{ } c \text{'' or ``} V \text{ sub } c \text{''}$$

Primes

A *prime* is an apostrophe mark made to the *right* and *above* a variable. It is placed in the same position as an exponent. It is used to distinguish variables with the same name.

EXAMPLE 4-2   Using the Greek letter β, select variables to represent two corresponding angles in two triangles.

Solution   Use β to represent the angle in the first triangle. Use β', "beta prime," for the angle in the second triangle. See Figure 4-1.

FIGURE 4–1

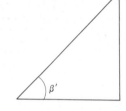

FIGURE 4–1
The two corresponding angles are
labeled β and β'.

EXERCISE 4-1
Read each of the following variables.

1. L_1	2. I_{max}	3. G_m
4. k'	5. R_p	6. ω_0
7. R'	8. β_2	9. μ
10. m'	11. R_{ac}	12. i_p
13. e'	14. C_{cb}	15. λ_t

4-2 INDICATING MULTIPLICATION

In arithmetic, multiplication was indicated by the multiplication sign ($\times$). When we indicate multiplication in algebra, we may use the multiplication sign ($A \times B$), a dot ($A \cdot B$), parentheses ($A)(B$), or no sign at all (AB). These four ways of indicating multiplication are summarized in Table 4-2. Note that we cannot omit the multiplication sign when multiplying numbers because they would be misunderstood. For example, we would understand 32 *(three two)* as the number thirty-two and not as three times two.

When a multiplication sign ($\times$) is used to indicate multiplication, care must be taken to make it look different from the letter x. When using the dot to indicate multiplication, be careful to place it above the writing line so that it does not appear as a decimal point. For example, $2 \cdot B$, not $2 . B$.

TABLE 4-2

EXAMPLES INDICATING MULTIPLICATION		
Symbol	**Possible Ways of Expression**	
$\times$	3×2 $\quad$ $2 \times b$	$\rho \times \omega$
$\cdot$	$3 \cdot 2$ $\quad$ $2 \cdot b$	$\rho \cdot \omega$
( )	$(3)(2)$ $\quad$ $(2)(b)$	$(\rho)(\omega)$
None	None $\quad$ $2b$	$\rho\omega$

4-3 GENERAL NUMBER

Numbers that are represented by letters are called *literal numbers* or *general numbers.* By using general numbers rather than words, electrical formulas can be easily stated. Ohm's law, an important formula of electronics, is easily expressed as $E = IR$. To express this same idea in words, the following statement would be needed. One volt of electromotive force (E) causes one ampere of current (I) to flow through a resistance (R) of one ohm. As you can see, written expressions are cumbersome and more easily misunderstood, whereas formulas written with general numbers are concise and less likely to be misunderstood.

4-4 ALGEBRAIC EXPRESSIONS

An *algebraic expression* is a collection of variables, numbers, and exponents connected by operators ($+$, $-$, $\times$, $/$, $\sqrt{}$) and signs of grouping. Table 4-3 shows several examples of algebraic expressions.

Algebraic expressions are classified by mathematicians in several ways. One way is to describe them as *rational expressions* or *irrational expressions.* A rational expression is an algebraic expression that contains no radical signs and has only whole numbers as exponents. An irrational expression is an algebraic expression that contains a fractional exponent or a radical sign. In Table 4-3, Examples (1), (4), and (7) are irrational expressions; the rest are rational expressions.

Rational expressions are further classified as either polynomial or fractional expressions. A rational expression that has no variable as a divisor (no division with a variable) is called a *rational integral expression,* or simply a *polynomial.* A rational expression that has a variable as a divisor (division with a variable) is called a *fractional expression.* In Table 4-3, examples (2), (3), (6), and (8)–(11) are polynomials. Examples (5) and (12) are fractional expressions. Figure 4-2 summarizes the classification of algebraic expressions.

TABLE 4-3

EXAMPLES OF ALGEBRAIC EXPRESSIONS	
(1) $\sqrt{8y}$	(2) $8y + 5$
(3) $(x + 2)(x - 3)$	(4) $x^{3.5}$
(5) $2/y$	(6) $\frac{1}{2}$
(7) $\sqrt{2x + y}$	(8) $3x$
(9) $\dfrac{R_1 + 2}{3}$	(10) $5x^2 - \frac{1}{3}x - 5$
(11) 3	(12) $\dfrac{R_b + R}{\alpha + \beta'}$

EXAMPLE 4-3   In the following algebraic expressions, list all the terms that apply to each. Are they irrational, rational, polynomial, or fractional expressions?

$$\text{(a) } 3x + y/2 \qquad \text{(b) } \frac{1}{x + y} \qquad \text{(c) } \sqrt{4x - 3}$$

Solution   (a) $3x + y/2$ rational, polynomial expression

(b) $\dfrac{1}{x + y}$ rational, fractional expression

(c) $\sqrt{4x - 3}$ irrational expression

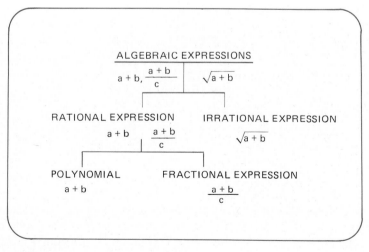

FIGURE 4-2
Classification of algebraic expressions.

Now that you have an understanding of algebraic expressions, the *general number* can be fully defined. Every number, variable, rational expression, and irrational expression is a *general number*. Because each of these (number, variable, rational expression, and irrational expression) is a numerical quantity, each follows all the rules for numbers.

EXAMPLE 4-4   List several general numbers.

Solution   Because every number, variable, rational expression, and irrational expression is a general number, all entries in Table 4-3 are general numbers. Examples would be 6.32, x, $5/x$, and $\sqrt{3/5}$.

EXERCISE 4-3

In the following algebraic expressions, list all the terms that apply to each. Describe whether each is a rational or an irrational expression. If rational, then further indicate whether each is a polynomial or a fractional expression.

1. 12.7

2. $\sqrt{5x}$

3. $\dfrac{a^2 + 3b}{5y}$

4. $5x + \sqrt{6y}$

5. $1 + \sqrt{y}$

6. $1/x$

7. $\eta_0 + 2$

8. $7\pi + (x/3)$

9. $\sqrt{3x}$

10. $\sqrt{x + y}$

11. $4x^{0.5}$

12. $(3/x) + \sqrt{5y}$

13. $5 + 3\mu$

14. 7.0

15. $5x^2 + \dfrac{1}{3x}$

4-5 PRODUCTS, FACTORS, AND COEFFICIENTS

A *product* is formed when two or more general numbers (numbers, variables, rational expressions) are multiplied together. Thus, $3y$ is a product; it is the result of multiplying 3 (a number) times y (a variable).

The general numbers that are used to form a product are called the *factors* of the product. Since $3 \times y = 3y$, three and y are factors of $3y$. Three is called the *numerical factor*; y is called the *literal factor*.

A *coefficient* of a product is any factor of a product. Each factor is the coefficient of the other factor. In the product $3y$, the 3 is the coefficient of y and y is the coefficient of 3. The numerical factor of a product is called the *numerical coefficient*. If a variable has no written numerical coefficient, then the numerical coefficient is *understood to be 1.* For example, T means $1T$ and *rs* means $1rs$.

EXAMPLE 4-5   In the product $5ab$, state the coefficient of:
(a) ab (b) 5   (c) a

Solution

(a) The coefficient of ab is 5.
(b) The coefficient of 5 is ab.
(c) The coefficient of a is $5b$.

EXERCISE 4-4
State the numerical coefficient of each product.

1. $3x$	2. $\frac{2}{5}\mu$	3. ab
4. $\frac{1}{2}zy$	5. b	6. $0.3c$
7. $2y\alpha$	8. 0.8ω	9. $\frac{1}{3}s$
10. 1.2λ	11. $\frac{7}{4}\beta$	12. 16η

State the coefficient of π in each of the following problems.

13. 2π	14. πb	15. πr^2
16. $2\pi fL$	17. $5b\pi$	18. $\sqrt{\pi S}$
19. $16\pi d^2$	20. π	21. $7.4\pi\omega$

4-6 COMBINING LIKE TERMS

An algebraic expression made up of a number, a variable, a product, or a quotient is called a *term.* The algebraic expression $5y$ is a term, as are 7, $-2cd$, $b(2 + c)$, x^2, β, $3a/5$, $1/x$, and $\sqrt{5x}$. Notice that both rational and irrational algebraic expressions may be terms.

Like Terms

Terms that have the same literal factors, but different numerical coefficients, are called *like terms.* Examples are $5a$ and $2a$, 8μ and 2μ, $2x^2y$ and $0.5x^2y$. Like terms can be added and subtracted. For example, $2R + 5R = 7R$.

RULE 4-1. COMBINING LIKE TERMS

To combine *like terms:*

1. Combine the numerical coefficients.
2. Form the new term from the combined numerical coefficients and the literal factor.

EXAMPLE 4-6 Simplify $2a + 5a$ by combining like terms.

Solution Use Rule 4-1.

Step 1: Combine numerical coefficients:

$$2 + 5 = 7$$

Step 2: Form new term:

$$7a$$
$$\therefore \quad 2a + 5a = 7a$$

EXAMPLE 4-7 Simplify $3a - 4a$ by combining like terms.

Solution Use Rule 4-1.

Step 1: Combine numerical coefficients:

$$3 - 4 = -1$$

Step 2: Form new term:

$$-1a$$
$$\therefore \quad 3a - 4a = -a$$

EXERCISE 4-5

Simplify each of the following algebraic expressions by combining like terms.

1. $2a + 3a$
2. $7\omega - 3\omega$
3. $9x + 3x$
4. $-ab - 6ab$
5. $-c - 6c$
6. $7y - (-8y)$
7. $5b - (+7b)$
8. $2\mu - 3\mu + 7\mu$

9. $4h + 3h - 2h$
10. $4k - 9k - k$
11. $4A + 3B - B + 6A$
12. $16x + 5y - 4x$
13. $c^2b - 4c^2b$
14. $x^2y^2 - 3x^2y^2$

15. $5x^2 - 3x - 7 - 2x^2 + 6x + 9$
16. $5y^2 + 3y - 8y^2 + 2y$
17. $2x - 3x^3 + 2x^2 - 4x - x^2 + x^3$
18. $9x^2y - 2xy^2 - 3x^2y$
19. $3xy^2 - 5x^2y - 3xy^2$
20. $2y - 3y^2 - 4y^3 - 2y + 4y^3 - y^2$

4-7 POLYNOMIALS

Polynomials are sums of products. Certain polynomials have special names. If the polynomial has only one term, it is called a *monomial*. If the polynomial has two terms, it is called a *binomial*. A three-term polynomial is called a *trinomial*. Special names are not used for polynomials with more than three terms. Table 4-4 is a list of sample polynomials.

EXERCISE 4-6

Of the following rational expressions, which can be classified as (a) monomials, (b) binomials, (c) trinomials, (d) fractional expressions?

1. $5h$
2. $2a^2 + 3a + 5$
3. $\pi + 2\lambda$
4. $27y^3$
5. $\dfrac{e + ir}{4x}$
6. 28
7. $E_1 + E_2 + IR$
8. $\frac{5}{8}$
9. $x + 5y$
10. $\dfrac{\mu + e_0}{e_i}$
11. $\dfrac{\eta_1 + \alpha_0}{\Sigma}$
12. $3\psi^3 + 2\psi^2 + \psi$

TABLE 4-4

SAMPLE POLYNOMIALS	
Special Name	**Example**
Monomial	$3y^2$
Binomial	$7y^2 + 4$
Trinomial	$m^2 - mn + n^2$

4-8 ADDING POLYNOMIALS

When adding polynomials, there are two important concepts of addition. First, two terms may be added in either order. Thus, the sum $2x + 3x$ is $5x$; also the sum $3x + 2x$ is $5x$. This concept is called the *commutative law of addition*. Second, three terms may be grouped in either manner to indicate which addition is to be performed first. Thus, $(5x + 3x) + 4x = 5x + (3x + 4x)$. This concept is called the *associative law of addition*. In summary, **addition is independent of order and of grouping.**

EXAMPLE 4-8 Simplify $6x^2 + 5 + 3x + 5x + 2x^2 + 4$.
Solution Apply the commutative and associative laws in order to group like terms:

$$(6x^2 + 2x^2) + (5 + 4) + (3x + 5x)$$

Combine like terms:

$$8x^2 + 9 + 8x$$

$$\therefore \quad 6x^2 + 5 + 3x + 5x + 2x^2 + 4$$

$$= 8x^2 + 9 + 8x$$

RULE 4-2. ADDING POLYNOMIALS

To add polynomials:

1. Arrange the terms in both polynomials in the same order.
2. Write like terms under each other.
3. Add each column, connecting the sums with their correct sign.

EXAMPLE 4-9 Add $6x^2 + 5 + 3x$ and $5x + 2x^2 + 4$.
Solution Use Rule 4-2.
Step 1: Arrange terms:

$$(6x^2 + 3x + 5) + (2x^2 + 5x + 4)$$

Step 2: Place like terms in columns, then add:

$$\begin{array}{r} 6x^2 + 3x + 5 \\ \underline{2x^2 + 5x + 4} \end{array} \cdot$$

Step 3: $\qquad 8x^2 + 8x + 9$

EXAMPLE 4-10 Add $3m + 6n - 4m^2$ and $2 + 6m^2 - 8n$.
      Solution Use Rule 4-2.
      Step 1: Arrange terms:

$$(-4m^2 + 3m + 6n) + (6m^2 - 8n + 2)$$

      Step 2: Place like terms in columns, then add:

$$
\begin{array}{l}
-4m^2 + 3m + 6n \\
\underline{6m^2 - 8n + 2} \\
\end{array}
$$

      Step 3: $\qquad\qquad 2m^2 + 3m - 2n + 2$

EXAMPLE 4-11 Add $-\theta^3 - 7 + \theta^2$ and $-\theta^2 + \theta^3 + 5$ and $-2 - \theta^3 + \theta^2$.
      Solution Use Rule 4-2.
      Step 1: Arrange terms:

$$(-\theta^3 + \theta^2 - 7) + (\theta^3 - \theta^2 + 5) + (-\theta^3 + \theta^2 - 2)$$

      Step 2: Place like terms in columns, then add:

$$
\begin{array}{l}
-\theta^3 + \theta^2 - 7 \\
\theta^3 - \theta^2 + 5 \\
\underline{-\theta^3 + \theta^2 - 2} \\
\end{array}
$$

      Step 3: $\qquad\qquad -\theta^3 + \theta^2 - 4$

EXERCISE 4-7

Read each of the following polynomials, and state the terms in descending powers of the variables.

1. $3x^2 + 2 + 2x$
2. $5y^2 + 4 + 8y$
3. $9 + 2x^2 + 4x$
4. $6 + 3\phi^2 + 13\phi$
5. $3n + 2n^2 - 11$
6. $5\eta - 2 + 2\eta^2$
7. $-8V + 2 - V^2 + 2V^3$
8. $-5 + 3\alpha^3 - \alpha^2 + 4\alpha$
9. $x^2 + 3y^2 + 2xy$
10. $\omega^3 + 2\omega^2\pi + 2\pi^2 + 3 + \omega\pi^2$

Add the following.

11. $2y + 3$
    $4y + 6$

12. $3x + 5$
    $2x - 4$

13. $x^2 + 3$
    $x^2 - 3$

14. $\mu - \beta^2$
    $\mu + \beta^2$

15. $2x + 4y + \ z$
    $\ x - 2y + 3z$

16. $3\psi^2 + 4\psi - 1$
    $\ \psi^2 \qquad + 1$

17. $V^3 + 2V^2 - V$
    $\qquad V^2 + V + 5$

18. $\ \ c^3 - 2c^2 + \ c$
    $-2c^3 \qquad\ + 3c - 9$

19. $\ \ \ 2x + \ y - 3z$
    $\qquad - 2y + 4z$
    $-3x \qquad\ - z$

20. $-\alpha^3 + \ \alpha^2 - 7\alpha$
    $\ 2\alpha^3 - \ \alpha^2 + 4\alpha$
    $-\alpha^3 + 2\alpha^2 - 2\alpha$

21. $n + 5$ and $2n + 3$
22. $2\omega - 7$ and $3\omega + 7$
23. $3x^3 - x^2$ and $x^3 - 4x$
24. $5\mu^4 + 3\mu^2$ and $\mu^3 - 8\mu^2$
25. $4m - 2n$ and $3m + 3n$
26. $-3t + S$ and $2t + 4S$
27. $3x^2 - 2x$ and $-x^2 - 3x$
28. $5y^2 - 2xy$ and $-10y^2 + 14xy$
29. $7.2\omega - 3.1\rho$ and $0.8\omega + 5.7\rho$
30. $-4.2V + 0.5W$ and $3.9W - 1.8V$

Simplify each expression by adding like terms.

31. $(x^3 - 4x^2 - 7x + 6) + (2x^3 - 5x + 2)$
32. $(m^2 - mn + n^2) + (4n^2 + m^2 - 3mn)$
33. $(6a^3 - 5a) + (7a - 11a^3 + 5b^2)$
34. $(4x^2 - 2x^3 + 3x + 1) + (x^3 + 3x^2 - 1 - 5x)$
35. $(-3t^2 + t^2 - 2 + 5t) + (-5t - t^2 + 2 + 3t^2)$
36. $(2x + 4y - 12) + (-5y + 7x - 2) + (-1 - 3y + 4x)$
37. $(7\beta - 9 + \theta) + (-5\alpha + \theta) + (3\alpha - 2\theta + 8\beta)$
38. $(-3y^2 - 3y + 2) + (-3y^2 + 7) + (-7y - 5y^2)$
39. $(x - y - z) + (2z - 5x + 3y) + (2y - 7x + 2z)$
40. $(7\lambda^2 - 6 + 7\lambda) + (7\lambda - 4 + \lambda^2) + (-1 + 2\lambda^2)$

Work the following problems, expressing the coefficients in the answers in decimal notation to three significant figures.

41. $(9.26x^2 - 4.25x + 8.25) + (-3.81x^2 + 5.01x + 9.88)$
42. $(30.1x^2 + 22.8x - 14.3) + (5.11x^2 - 24.3x - 13.4)$
43. $(10.9x^2 - 7.65x + 4.33) + (18.88x^2 - 7.52x + 7.99)$
44. $(52.4x^2 - 14.2x - 46.5) + (-12.3x^2 - 5.53x + 18.3)$

SELECTED TERMS

associative law of addition   Three terms may be grouped in either way to indicate which addition is performed first; thus, (a + b) + c = a + (b + c).

coefficient of a product   Any factor of a product; each factor is the coefficient of the other factor(s).

commutative law of addition   Two terms may be added in either order; thus, a + b = b + a.

fractional expression   Algebraic expression having a variable in the denominator of one or more terms.

irrational expression   Algebraic expression containing a fractional exponent or a radical sign.

like terms   Having the same literal factors but different numerical coefficients.

polynomial   Rational expression having no variable as a divisor.

rational expression   Algebraic expression having whole numbers as exponents and containing no radical signs.

variable   Symbol for a number that may be constant or varying in an algebraic expression or equation.

5

ALGEBRA FUNDAMENTALS II

This chapter continues our discussion of the concepts of algebra. By becoming familiar with the topics presented here, you will be ready to solve equations.

5-1 MULTIPLYING MONOMIALS

In multiplication, as in addition, we will use several important concepts. First, two factors may be multiplied together in either order. Thus, $A \times B = B \times A$. This concept is called the *commutative law of multiplication*. Second, three factors may be grouped in either manner to indicate which multiplication is to be performed first. Thus, $(5 \times A) \times D = 5 \times (A \times D)$. This concept is called the *associative law of multiplication*. In summary, **multiplication is independent of order and of grouping.**

In Section 2-12, the use of exponents was introduced. At that time we stated that the exponent tells us how many times a number is used as a factor in a product. Thus, $y^3 = y \cdot y \cdot y$. In Section 3-4 you were shown how to multiply with powers of ten. Recall that powers of ten can be multiplied by adding their exponents. Thus, $10^3 \times 10^4 = 10^7$. In like manner, powers of a single variable may be multiplied by adding their exponents.

RULE 5-1. MULTIPLYING POWERS OF A VARIABLE

To multiply powers of a variable, add the exponents:

$$a^m a^n = a^{m+n}$$

EXAMPLE 5-1   Solve $x^3 \cdot x^4$.
    Solution

$$x^3 \cdot x^4 = x^{3+4} = x^7$$
$$x^3 \cdot x^4 = 7$$

Check   We can check our answer by "long hand." Thus,

$$x^3 = x \cdot x \cdot x$$
$$x^4 = x \cdot x \cdot x \cdot x$$
$$x^3 \cdot x^4 = (x \cdot x \cdot x)(x \cdot x \cdot x \cdot x) = x^7.$$

When multiplying monomials we form the product by applying Rule 5-2.

To multiply monomials:

1. Multiply the numerical coefficients.
2. Multiply the literal factors.
3. Form the product by multiplying the numerical factor and the literal factor.

EXAMPLE 5-2 Multiply $5a$ by 3.
Solution Use Rule 5-2.
Step 1: Multiply numerical coefficients:

$$3 \times 5 = 15$$

Step 2: Not needed.
Step 3: Form the product:

$$15a$$
$$\therefore \quad 5a \times 3 = 15a$$

EXAMPLE 5-3 Multiply $3y^5$ by $2y^4$.
Solution Use Rule 5-2.
Step 1: Multiply numerical coefficients:

$$3 \times 2 = 6$$

Step 2: Multiply literal factors:

$$y^5 y^4 = y^{5+4} = y^9$$

Step 3: Form the product:

$$6y^9$$
$$\therefore \quad 3y^5(2y^4) = 6y^9$$

EXAMPLE 5-4 Multiply $-6c$ by $-3b$.
Solution Use Rule 5-2.

Step 1: Multiply numerical coefficients:

$$(-6)(-3) = 18$$

Step 2: Multiply literal factors:

$$c \cdot b = cb$$

Step 3: Form the product:

$$18bc$$
$$\therefore \quad (-6c)(-3b) = 18bc$$

NOTE: It is conventional to put variables in alphabetical order.

EXAMPLE 5-5 Multiply $-7IR$ by $6IZ$.
   Solution Use Rule 5-2.

Step 1: $(-7)(6) = -42$

Step 2: $(IR)(IZ) = IIRZ = I^2RZ$

Step 3: $-42I^2RZ$

$$\therefore \quad (-7IR)(6IZ) = -42I^2RZ$$

EXERCISE 5-1
Multiply each of the following.

1. $x^5 \cdot x^4$ 2. $y^2 \cdot y^6$
3. $\omega^3 \cdot \omega^{-3}$ 4. $a^5 \cdot a^{-2}$
5. $b \cdot b$ 6. $-4(3a)$
7. $-7y(-2)$ 8. $-6z(2)$
9. $7(-5ab)$ 10. $4\mu(-8\beta)$
11. $10^2 \cdot 10^{-4}$ 12. $-5c \times -3c$
13. $(-xy)(x^3y^2)$ 14. $(ab^2)(-7ab)$
15. $(-8\phi)(-2\theta\eta)$ 16. $(3a^5)(-7a)$
17. $(-8x^3)(-2x^{-3})$ 18. $(-2ab)(4cb)$

19. $(2y^{-3})(-4yx^2)$ 20. $(-2x^3)(5y^4)$
21. $2x^{-2} \cdot 3x \cdot x^3$ 22. $-5(a^2)(ab)$
23. $(-2y)(-2y)(-2y)$ 24. $(6ab)(-bc)(-3ac)$

5-2 MULTIPLYING A MONOMIAL AND A BINOMIAL

When multiplying a monomial and a binomial, we will use the *distributive law of multiplication over addition,* commonly called the *distributive law.* This law states that each term in the binomial is multiplied by the monomial, and the resulting products are added together as in Figure 5-1.

RULE 5-3. DISTRIBUTIVE LAW OF MULTIPLICATION

To multiply a polynomial by a monomial:

1. Multiply each term of the polynomial by the monomial.
2. Add the resulting products.

EXAMPLE 5-6   Multiply $(a - b)$ by 6.
Solution   Use Rule 5-3 with the expression $6(a - b)$.
Step 1:   Multiply each term by 6:

$$6(a) = 6a, \; 6(-b) = -6b$$

Step 2:   Add the products:

$$6a - 6b$$

$$\therefore \quad 6(a - b) = 6a - 6b$$

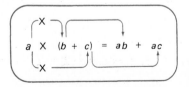

FIGURE 5-1
Distributive law: *a* is distributed over *b* and *c*.

EXAMPLE 5-7   Multiply $5x + 3$ by $2y$.

Solution   Use Rule 5-3 with the expression $2y(5x + 3)$.

Step 1: $(2y \cdot 5x) + (2y \cdot 3)$

Step 2: $10xy + 6y$

$\therefore \quad 2y(5x + 3) = 10xy + 6y$

EXAMPLE 5-8   Multiply $-3x^2 - 2x$ by $-x$.

Solution   Use Rule 5-3 with the expression $-x(-3x^2 - 2x)$.

$(-x \cdot -3x^2) + (-x \cdot -2x)$
$(3x^3) + (2x^2)$
$\therefore \quad -x(-3x^2 - 2x) = 3x^3 + 2x^2$

Before you begin work on Exercise 5-2, review Sections 2-6 and 2-10.

EXERCISE 5-2

Apply the distributive law to each of the following.

1. $3(7 + 2)$
2. $-2(5 - 3)$
3. $2(y - 3)$
4. $-4(x + 3)$
5. $3(\mu - 5)$
6. $-1(a - 2)$
7. $-Z(3 - Z)$
8. $a(-5 - 2a)$
9. $\beta(c - 4)$
10. $-d(a + 6)$
11. $-3(x - y)$
12. $-6(U - V)$
13. $-2a(a - 2b)$
14. $3x(x^2 - y)$
15. $-7\theta(2\omega + 5\phi)$
16. $-x^3(x^2 + xy)$
17. $y^4(-y^3 - 4y^2)$
18. $8ab(-a - b)$
19. $-3cd(-d^2 - c^2)$
20. $-5xy(x^2 - y^2)$
21. $3s(s^2 + 4s)$

5-3 MULTIPLYING A MONOMIAL AND A POLYNOMIAL

In this section we will build on a technique used in the preceding section. To multiply a monomial by a polynomial, we use the distributive law (Rule 5-3).

EXAMPLE 5-9   Multiply $3x^2 - x + 3$ by $-2x$.

Solution   Use Rule 5-3 with the expression $-2x(3x^2 - x + 3)$.

Distribute $-2x$ over each term in the trinomial:

$$(-2x \cdot 3x^2) + (-2x \cdot -x) + (-2x \cdot 3)$$
$$-6x^3 + 2x^2 - 6x$$
$$\therefore \quad -2x(3x^2 - x + 3) = -6x^3 + 2x^2 - 6x$$

EXAMPLE 5-10 Multiply $3xy^2 - 2x + 3y - 7$ by $-3xy$.

Solution Use Rule 5-3 with $-3xy(3xy^2 - 2x + 3y - 7)$.
Distribute $-3xy$ over each term in the polynomial:

$$(-3xy \cdot 3xy^2) + (-3xy \cdot -2x) +$$
$$(-3xy \cdot 3y) + (-3xy \cdot -7)$$

$$-9x^2y^3 + 6x^2y - 9xy^2 + 21xy$$

$$\therefore \quad -3xy(3xy^2 - 2x + 3y - 7)$$
$$= -9x^2y^3 + 6x^2y - 9xy^2 + 21xy$$

EXERCISE 5-3
Apply the distributive law to each of the following.

1. $4(3x + 2y + 5)$
2. $-3(2a + b - c)$
3. $-7(-3u + 2v - w)$
4. $-1(2\rho - 3\lambda - 6)$
5. $5(-8x + y - 3)$
6. $2(7 + 2x - 3y)$
7. $-6(-\mu + 2\theta - 3)$
8. $-8(-2a + b - 3c)$
9. $-x(2x + 4 - y)$
10. $2x(-3 + 4y - z)$
11. $-3h(-4j + 5\theta - 1)$
12. $4b(-2b^2 - b + 2)$
13. $-3ab(2a - 3b + 4c)$
14. $-a^2c(-a + bc - 1)$
15. $3x^2(2x^2 - 7x - 6)$
16. $-2y^2(-y - 2x - 3)$
17. $-2ab(a^2 - b^2 + c - 3)$
18. $-x^2yz(z^2 - 2x^2 + xy + 4)$
19. $-1(a^2 + 2b - 3c + 5)$
20. $-1(x^2 - 3x + y - 7)$

Calculator Drill
Work each of the following problems, expressing the coefficients in the answers in decimal notation to three significant figures.

21. $3.87y(1.95y^2 - 6.41x + 4.05)$
22. $12.8b(-5.32b + 3.55c - 8.17d)$
23. $-42.8\mu(2.93\mu^2 - 5.25\mu + 7.15)$
24. $-3.81a^2(16.5b^2 + 4.32a - 8.47)$

5-4 SUBTRACTING POLYNOMIALS

Recall from Section 2-8 that we subtract by changing the sign and adding. But a polynomial has more than one sign, so what do we do? Distribute the sign change over each term of the polynomial.

RULE 5-4. SUBTRACTING POLYNOMIALS

To subtract one polynomial from another:

1. Change the sign of **each and every** term in the polynomial to be subtracted.
2. Combine like terms as in addition.

EXAMPLE 5-11   Subtract $2x - 3$ from $5x + 4$.
Solution   Use Rule 5-4.
Step 1:   Change both signs of $2x - 3$:

$$2x - 3 \Rightarrow -2x + 3$$

Step 2:   Add like terms:

$$\begin{array}{r} 5x + 4 \\ -2x + 3 \\ \hline 3x + 7 \end{array}$$

$$\therefore \quad (5x + 4) - (2x - 3) = 3x + 7$$

Signs of grouping preceded by a plus sign may be removed without affecting the signs of the terms. Signs of grouping preceded by a minus sign may be removed by changing the sign of **each and every** term. These concepts are demonstrated in the following examples.

EXAMPLE 5-12   Simplify $(8a + 3) - (5a - 2)$ by removing parentheses and adding like terms.
Solution   Remove the parentheses.
Observation   There is an implied plus sign in front of $(8a + 3)$.

$$(8a + 3) - (5a - 2) \Rightarrow 8a + 3 - 5a + 2$$

Combine like terms:

$$3a + 5$$

$$\therefore \quad (8a + 3) - (5a - 2) = 3a + 5$$

EXAMPLE 5-13 Simplify $(3x^2 + 3x - 1) - (x^2 - 5x + 4)$.
Solution Remove the parentheses:

$$(3x^2 + 3x - 1) - (x^2 - 5x + 4) \Rightarrow$$
$$3x^2 + 3x - 1 - x^2 + 5x - 4$$

Combine like terms:

$$2x^2 + 8x - 5$$

$$\therefore \quad (3x^2 + 3x - 1) - (x^2 - 5x + 4)$$
$$= 2x^2 + 8x - 5$$

EXERCISE 5-4

1. Subtract $2a$ from $3a - 8$.
2. Subtract $-5x$ from $7x + 3$.
3. Subtract $2a - 3c$ from $6c + a$.
4. Subtract $4\eta + 2\omega$ from $-3\eta - 2\omega + 1$.
5. Subtract $5x^2 + 2x - 1$ from $-7x^2 - 4x + 3$.

Simplify each expression by removing parentheses and combining like terms.

6. $5y - (3y - 2)$ 7. $(3x + 2) - 3x$
8. $-7\mu - (2\mu + 6)$ 9. $7\theta - (4\theta + \beta)$
10. $(a + b) - (a + b)$ 11. $(x - y) - (x + y)$
12. $(2x + 3y) - (2y + 3x)$ 13. $(3x + 2y) - (x + y)$
14. $(2a + 3b) - (2a - b)$ 15. $(-4\mu + \pi) - (-5\mu - \pi)$
16. $(-3x - 7) - (-x + 5)$
17. $(V^2 - 3U + 2) - (-V^2 - 2U + 2)$

18. $(a^3 - 3a^2 + 4) - (2a^3 + 5a^2 - 5a)$
19. $(2x^4 - 7x^2 + 1) - (x^4 - 3x^2 - x + 3)$
20. $(3\theta^3 - \theta^2 + 2\theta) - (5\theta^3 - \theta^2 + \theta - 7)$
21. $(-3y^2 + 4y - 5) - (2y^2 - 7)$

5-5 ADDITIONAL WORK WITH POLYNOMIALS

This section is intended to allow you to improve your skills in working with addition and multiplication as it applies to simple polynomial expressions. Before working through the following examples, review the rules pertaining to signs of grouping in Section 2-5.

EXAMPLE 5-14   Simplify $2a[-3(5a - 4) - 5a]$.
  Solution

$$M: -3 \qquad 2a[-15a + 12 - 5a]$$
$$A: -15a - 5a \qquad 2a[-20a + 12]$$
$$M: 2a \qquad -40a^2 + 24a$$
$$\therefore \qquad 2a[-3(5a - 4) - 5a]$$
$$= -40a^2 + 24a$$

EXAMPLE 5-15   Simplify $2x - [4 + 2(x - 2)]$.
  Solution

$$M: 2 \qquad 2x - [4 + 2x - 4]$$
$$A: 4 - 4 \qquad 2x - [0 + 2x]$$

Remove brackets:

$$2x - 0 - 2x$$
$$A: 2x - 2x \qquad 2x - 2x = 0$$
$$.. \qquad 2x - [4 + 2(x - 2)] = 0$$

EXAMPLE 5-16   Simplify $-3[-2y(y^2 - 2) - 3y(-2y^2 + 1)]$.
   Solution

$$
\begin{aligned}
\text{M: } -2y \qquad & -3[-2y^3 + 4y - 3y(-2y^2 + 1)] \\
\text{M: } -3y \qquad & -3[-2y^3 + 4y + 6y^3 - 3y] \\
\text{A: } -2y^3 + 6y^3 \qquad & -3[4y^3 + 4y - 3y] \\
\text{A: } 4y - 3y \qquad & -3[4y^3 + y] \\
\text{M: } -3 \qquad & -12y^3 - 3y \\
\therefore \qquad & -3[-2y(y^2 - 2) - 3y(-2y^2 + 1)] \\
& = -12y^3 - 3y
\end{aligned}
$$

EXERCISE 5-5
Simplify the following.

1. $x \cdot x \cdot x - (2x^2 + 4)$
2. $-a(a^2 + a - 5) + a \cdot a$
3. $-3y - 4y(2 + y)$
4. $2x + 4x(-3x + 1)$
5. $-(\theta + \eta - 2) + 2(\theta - \eta)$
6. $5x + 3(2 - x)$
7. $3 - 7[2x(x - 1)]$
8. $y^2 - [y \cdot y + 3(-2y + 2)]$
9. $3c[-(2a + 3) - 5c]$
10. $-5a - [3 + 4a(4 - 3a)]$
11. $5\mu + [7 - 2\mu(6 - 3\mu)]$
12. $-3\pi[-2(\pi + 3) - 6\pi]$
13. $-\theta - [3\theta - 2(\theta - 1)]$
14. $-5[4c(c \cdot c - 3) - 2c(c^2 + 1)]$
15. $3[-x^3 + 2x(x^2 + x - 3) - 4]$
16. $2[2t(t - 4) - 2(t^2 + 4t)]$

5-6 DIVISION OF MONOMIALS

Dividing Powers of a Variable

In Section 5-1 we multiplied powers of a single variable by adding their exponents. Division of powers of a single variable is just as easy.

RULE 5-5. DIVIDING POWERS OF A VARIABLE

To divide powers of a variable, subtract the exponent in the denominator from the exponent in the numerator.

$$a^m/a^n = a^{m-n}$$

> **EXAMPLE 5-17**   Solve x^5/x^2.
>   *Solution*   Use Rule 5-5.
>
> $$x^5/x^2 = x^{5-2} = x^3$$
> $$\therefore \quad x^5/x^2 = x^3$$

Division by Zero

In this book we will not consider division by zero. *Division by zero is not a permissible operation.* You may wish to try dividing by zero on your calculator.

Implied Exponent

When a variable is written without any exponent, the variable occurs only once as a factor. This means that the exponent is 1 ($x = x^1$). When the exponent of 1 is omitted, it is an *implied exponent*.

Zero Exponent

Any number (other than zero) divided by itself is equal to 1. For example, $3/3 = 1$, $a/a = 1$, $x^3/x^3 = 1$. If we applied Rule 5-5 to x^3/x^3, the result would be x^0. Since there can only be one answer, $x^0 = 1$. *Any number (except zero) or any variable raised to the zero power is 1 by definition.*

Division of Monomials

In dividing monomials we apply the preceding concepts. In applying the following rule, we will treat each variable separately.

RULE 5-6. DIVIDING ONE MONOMIAL BY ANOTHER

To divide one monomial by another:

1. Divide the numerical coefficients.
2. Divide the literal factors.
3. Form the quotient.

EXAMPLE 5-18   Simplify $9y/3y$.

    Solution   Use Rule 5-6.

    Step 1:   Divide numerical coefficients:

$$9/3 = 3$$

    Step 2:   Divide literal factors:

$$y/y = 1$$

    Step 3:   Form the quotient:

$$(3)(1) = 3$$
$$\therefore \quad 9y/3y = 3$$

EXAMPLE 5-19   Divide $15x^2y$ by $5xy$.

    Solution

$$
\begin{aligned}
\text{D: } 5 \qquad & \frac{15x^2y}{5xy} = \frac{3x^2y}{xy} \\
\text{D: } x \qquad & \phantom{\frac{15x^2y}{5xy}} = 3xy/y \\
\text{D: } y \qquad & \phantom{\frac{15x^2y}{5xy}} = 3x \\
\therefore \qquad & 15x^2y/5xy = 3x
\end{aligned}
$$

EXERCISE 5-6

Find the quotients of the following.

1. $a^5 \div a$ 2. $x^8 \div x^2$ 3. $c^{10} \div c^8$

4. $-x^4 \div x^3$ 5. $\mu^{10} \div \mu^3$ 6. $\theta^{12} \div \theta^3$

7. $3y^8/y^8$ 8. $12t^4/3t^2$ 9. $27y^3z^6/-9yz$

10. $\dfrac{-32a^5b^4}{4a^5b^2}$ 11. $\dfrac{-8c^{10}d^3}{-4c^{12}d^3}$ 12. $\dfrac{48\eta^5\lambda^7}{16\lambda^3}$

13. $\dfrac{4^2a^3b^5}{-2ab}$ 14. $\dfrac{-27x^5y^2}{3^2x^3y^4}$ 15. $\dfrac{-20c^2t}{-5t}$

5-7 DIVIDING A POLYNOMIAL BY A MONOMIAL

Division is the inverse of multiplication. Remember, to multiply a polynomial by a monomial, each term in the polynomial is multiplied by the monomial. Similarly, to divide a polynomial by a monomial, each term in the polynomial is divided by the monomial.

RULE 5-7. DIVIDING A POLYNOMIAL BY A MONOMIAL

To divide a polynomial by a monomial, divide each term of the polynomial by the monomial.

EXAMPLE 5-20   Divide $12x^3 + 8x^2 + 4x$ by $4x$.
Solution   Use Rule 5-7.

$$\frac{12x^3 + 8x^2 + 4x}{4x} = \frac{12x^3}{4x} + \frac{8x^2}{4x} + \frac{4x}{4x}$$
$$= 3x^2 + 2x + 1$$
$$\therefore \quad (12x^3 + 8x^2 + 4x)/4x = 3x^2 + 2x + 1$$

EXAMPLE 5-21   Divide $5a^3b + 15a^2b^2 + 10ab^3$ by $5ab$.
Solution

$$\frac{5a^3b + 15a^2b^2 + 10ab^3}{5ab}$$
$$= \frac{5a^3b}{5ab} + \frac{15a^2b^2}{5ab} + \frac{10ab^3}{5ab}$$
$$= a^2 + 3ab + 2b^2$$
$$(5a^3b + 15a^2b^2 + 10ab^3)/5ab$$
$$= a^2 + 3ab + 2b^2$$

EXERCISE 5-7
Divide the following.

1. $\dfrac{9x + 12}{3}$ 2. $\dfrac{8c + 16}{8}$ 3. $\dfrac{21 + 12y}{3}$

4. $\dfrac{10a^2 + 15}{5}$ 5. $\dfrac{18U + 6V}{6}$ 6. $\dfrac{25\theta^2 + 20}{5}$

7. $\dfrac{6b^3 + 12a^3}{3}$

8. $\dfrac{-18\alpha + 6}{6}$

9. $\dfrac{24x^2 + 12xy}{4x}$

10. $\dfrac{y^3 + y^2}{y^2}$

11. $\dfrac{a^5 + a^3}{a^2}$

12. $\dfrac{27x^3 + 21x^2 + 15x}{3x}$

13. $\dfrac{13c^3 + 9c^2 + 11c^4}{c^2}$

14. $\dfrac{2ax^2 + 4ax - 6a^2x}{2ax}$

15. $\dfrac{5x^2y + 10x^2y^2 + 15xy^3}{5xy}$

16. $\dfrac{3\psi^3\phi + 6\psi\phi^2 + 9\psi^2\phi^2}{3\psi\phi}$

17. $\dfrac{40a^2b^2 + 30a^2b + 20a^2b^2}{10a^2b}$

18. $\dfrac{16U^5 + 32U^4 + 8U^3}{-8U^3}$

19. $\dfrac{28\mu^7 + 16\mu^5 + 20\mu^3}{4\mu^5}$

20. $\dfrac{16y^2z^3 + 8y^3z^4 + 4yz^3}{2y^2z^3}$

5-8 FACTORING POLYNOMIALS WITH A COMMON MONOMIAL FACTOR

Factoring

From the distributive law we learned that $a(b + c) = ab + ac$. We know that $ab + ac$ and $a(b + c)$ are two algebraic expressions for the same number. The number expressed as a product is $a(b + c)$. This form is called the *factored form*. Similarly, $3(x + 1)$ is the factored form of $3x + 3$. The process of finding the factors of a sum is called *factoring*. Table 5-1 shows several examples of polynomials that have been factored.

> **RULE 5-8. FACTORING POLYNOMIALS CONTAINING A COMMON MONOMIAL FACTOR**
>
> To factor a polynomial that contains a common monomial factor:
>
> 1. By inspection, determine the factors that are common to each term of the polynomial. Form the common factor.
> 2. Divide the polynomial by the common factor.
> 3. Write the factored expression as a product of the common factor and the quotient.

TABLE 5-1

EXAMPLES OF FACTORING

Polynomial	Factors	Factored Form
$5a - 5$	5 and $a - 1$	$5(a - 1)$
$2x^2 - x$	x and $2x - 1$	$x(2x - 1)$
$8y^2 + 4y + 12$	4 and $2y^2 + y + 3$	$4(2y^2 + y + 3)$
$4xy + 8x^2y$	$4xy$ and $1 + 2x$	$4xy(1 + 2x)$

EXAMPLE 5-22 Factor $6x + 12$.

Solution Each term of $6x + 12$ is divisible by both 3 and 6. In factoring, where there is more than one possible common factor, the largest common factor is selected. So 6 is the common monomial factor. The other factor is the quotient of $6x + 12$ and 6; that is, $(6x + 12)/6 = x + 2$. The factored form of $6x + 12$ is expressed as a product of 6 and $x + 2$, which is $6(x + 2)$.

$$\therefore \quad 6x + 12 = 6(x + 2)$$

EXAMPLE 5-23 Factor $2a^4 + 6a^3 + 8a^2$.

Solution Use Rule 5-8.

Step 1: By inspection we see that 2 is the largest numerical factor that exactly divides 2, 6, and 8; and a^2 is the highest power of a that divides a^4, a^3, and a^2 exactly. So $2a^2$ is the common monomial factor of $2a^4 + 6a^3 + 8a^2$. It is important that both the numerical and the literal factors be found.

Step 2: Divide by $2a^2$:

$$\frac{2a^4 + 6a^3 + 8a^2}{2a^2} = a^2 + 3a + 4$$

Step 3: Write the factored expression as a product:

$$2a^2(a^2 + 3a + 4)$$

$$\therefore \quad 2a^4 + 6a^3 + 8a^2 = 2a^2(a^2 + 3a + 4)$$

Write the following polynomials in factored form. Check your solutions by multiplying the factors.

1. $6a + 12$
2. $8\mu + 4$
3. $6 + 9b$
4. $5x + 25$
5. $3m + 3$
6. $ax + ay$
7. $\omega u + \omega v$
8. $6x + 6y$
9. $7\phi + 7\theta$
10. $3c^2 + 7c$
11. $\rho^2 + 5\rho$
12. $4x^2 + 5x$
13. $7xy + 21y$
14. $10h^2 + 5h^3$
15. $8a^3 + 16a^2$
16. $7st^3 + 5s^2t$
17. $ax + 3a$
18. $2\pi r^2 + 2\pi rh$
19. $a^2b^3c + a^4b^5c^3$
20. $4d + 6 + 10d^2$
21. $5a^2 + 15a + 20$
22. $6 + 3\lambda + 18\lambda^2$
23. $ab + a^2b^2 + a^3b^3$
24. $U^4V^2 + U^3V^3 + U^2V^4$
25. $16\beta^2\pi^2 + 8\beta^2\pi + 32\beta^2$
26. $42x^2y + 35x^2 + 14x^2y^2$
27. $30\eta^2\alpha + 15\eta\alpha - 25\eta\alpha^2$
28. $12a^3b^3 + 6a^2b^3 + 18ab^3$
29. $5y^3z^2 + 15y^2z + 10y^4z$
30. $18\theta^3\phi^3 + 12\theta^2\phi^4 + 27\theta\phi^5$
31. $21y^4x + 18y^3x^2 + 27xy^3$

5-9 EVALUATING ALGEBRAIC EXPRESSIONS

The process of *finding the value* of an algebraic expression is called *evaluating the expression*. This process is carried out by substituting numbers for the variables in the expression, and then simplifying the result to a single number.

EXAMPLE 5-24 Find the value of $2a^2b$ when $a = 5$ and $b = 3$.

Solution Substitute the numbers for the variables and simplify:

$$2a^2b = 2(5)^23 = 2(25)3 = 150$$
$$\therefore \quad 2a^2b = 150 \text{ when } a = 5 \text{ and } b = 3$$

EXAMPLE 5-25

Find the value of $\dfrac{x^2 - \sqrt{4y}}{8}$ when $x = 4$ and $y = 16$.

Solution Substitute:

$$\frac{x^2 - \sqrt{4y}}{8} \Rightarrow \frac{4^2 - \sqrt{4(16)}}{8}$$

$$\frac{16 - \sqrt{64}}{8} = \frac{16 - 8}{8} = \frac{8}{8} = 1$$

$$\therefore \quad \frac{x^2 - \sqrt{4y}}{8} = 1 \text{ when } x = 4 \text{ and } y = 16$$

EXAMPLE 5-26   Evaluate $\dfrac{3x^3 - (2xy + y^2)}{2x - 7}$ when $x = 2$ and $y = 3$.

   Solution   Substitute:

$$\frac{3x^3 - (2xy + y^2)}{2x - 7} \Rightarrow \frac{3(2)^3 - [(2)(2)(3) + (3)^2]}{2(2) - 7}$$

$$\frac{3(8) - (12 + 9)}{4 - 7} = \frac{24 - 21}{-3} = \frac{3}{-3} = -1$$

$$\therefore \quad \frac{3x^3 - (2xy + y^2)}{2x - 7} = -1 \text{ when } x = 2 \text{ and } y = 3$$

EXERCISE 5-9

Calculator Drill
Evaluate the following expressions for the given values. Express your
answers in decimal notation, and round the answer to three significant
figures.
   If $x = 3$ and $y = 4$, find the value of the following.

1. xy 2. $2x - y$ 3. $\dfrac{3y + x}{3}$

4. $x^2 - \sqrt{y}$ 5. $5y^2$ 6. $y^3 - 10$

7. $-2(y - x)^2$ 8. $3x^2 + 5y$ 9. $2x^2 - xy$

If $a = 2$, $b = 3$, and $c = 4$, evaluate the following.

10. $2a^2 + 5$ 11. $a^3 - b^2$

12. $a^2 + b^2 - c^2$ 13. $25 - abc$

14. $c^2 - ab$ 15. $(12 - 2c)(2c - b)$

16. $(c - 2)(bc - a)$ 17. $(14 - 2c)/b$

18. $c^2 - (ab) + 3$ 19. $\dfrac{\sqrt{2c - 2} - (b + c)}{a^2 + cb}$

20. $\dfrac{5(2c - b)}{b(c - b)} - \dfrac{7(b + 1)}{2c + b}$ 21. $b^2 - 5/a^2$

If $x = 3.72$ and $y = 1.15$, evaluate the following.

22. $(3xy^2 - 2xy) - (9xy - 3xy^2)$

23. $4x(3x^2 - 2y)$

24. $(6x^2 - 4\sqrt{y} + 8y) \div 2x$

25. $(10x^2 + y - \sqrt{5}) \div (2x - 3y)$

SELECTED TERMS

distribution law   when multiplying a monomial and a polynomial, each term in the polynomial is multiplied by the monomial.

evaluating an expression   Substituting numbers for variables in an algebraic expression.

factoring   The process of finding factors of a sum.

implied exponent   Any number or variable without an exponent has an *implied exponent* of 1.

zero exponent   Any non-zero number or variable raised to the zero power is equal to 1.

6

SOLVING EQUATIONS

This chapter will be limited to fairly elementary equations so that you may gain a solid understanding of how to solve algebraic equations by concentrating on the algebraic principles.

6-1 EQUATIONS

A sentence saying that one general number is equal to a second general number is called an *equation*. The expressions that are joined by the equal sign (=) to form an equation are called *members* of the equation. Thus, in the equation $x - 3 = 4$, the left member is "$x - 3$" and "4" is the right member. Equations may be read left to right or right to left. Figure 6-1 shows the structure of an equation. Equations are divided into two types: identical equations and conditional equations. An *identical equation* is true for all values of the variable contained in the equation. Thus, the identical equation $2(x + 1) = 2x + 2$ is true for any value of x. Identical equations are also called *identities*. A *conditional equation* is only true for particular values of the variable contained in the equation. Thus, the conditional equation $x + 2 = 5$ is true only when $x = 3$. That is $3 + 2 = 5$. We will be concerned with conditional equations in this chapter. We will use the word *equation* to mean conditional equation.

EXERCISE 6-1
Study each of the following sentences to determine if it is true or false.

1. $3 + 2 = 4 + 1$
2. $6 - 2 = 2 + 2$
3. $6(2 + 1) = 15$
4. $8 + 2 \neq 12 - 3$
5. $5^2 = 29 - 4$
6. $9 + 3 \neq 15 - 3$
7. $8(4) = 42 - 10$
8. $3 - 2 > 8 - 11$
9. $-5 - 4 < -7 + 6$
10. $7(2 - 4) = 7(2) - 4(7)$
11. $6(-3) + 4 > 7/(-7) - 14$
12. $(2 \times 3) \times 5 < 3 \times (5 \times 2)$

In each of the following equations, a value for the variable has been given to the right. Determine if the value given for the variable makes the equation true.

13. $x + 3 = 11,$ $\quad x = 8$
14. $I - 4 = 7,$ $\quad I = 3$
15. $-5\mu = -15,$ $\quad \mu = -3$
16. $a^2 = 6,$ $\quad a = 3$
17. $7 - \beta = 7,$ $\quad \beta = 0$
18. $-R + 5 = 3^2, R = -4$
19. $y^3 + 4 - 6 = 12,$ $\quad y = 2$
20. $x^{0.5} - 4(2) = -15 + 9, x = 4$

FIGURE 6–1
Structure of an equation.

6-2 FINDING THE ROOT OF AN EQUATION

The value of the variable that makes the equation true is called the *solution* of the equation or the *root* of the equation. The root is a number that makes the equation a true statement. The equation is said to be *satisfied* by the root of the equation.

EXERCISE 6-2
Select the root from the values at the right of each equation.

1. $x + 3 = 4$, $\{-2, 1, 5\}$
2. $2x + 1 = 5$, $\{1, 2, 3, 4\}$
3. $m + 2 = -6$, $\{-8, -6, -4, 4\}$
4. $2\lambda = -12$, $\{-8, -6, -4\}$
5. $x + 1 = 0$, $\{-1, 0, 1, 2\}$
6. $-5y = -13 + 3$, $\{-3, -2, 2, 3\}$
7. $\alpha - 3 = 6$, $\{5, 7, 9, 11\}$
8. $3(x + 2) = 15$, $\{1, 2, 3\}$
9. $-2(\phi - 4) = 0$, $\{3, 4, 5\}$
10. $-(x + 2) = 1$, $\{-1, -2, -3\}$

6-3 USING ADDITION TO TRANSFORM EQUATIONS

In solving an equation it is usual to isolate the variable in the left member of the equation. To do this, an *equivalent* equation is formed by *transforming* the original equation. Two equations are equivalent if they have the same roots.

Among the common operations used to transform an equation are:

- Addition (or subtraction) of the same number to (or from) both members of the equation.
- Multiplication (or division) of members of the equation by the same number.

EXAMPLE 6-1   Solve $x - 3 = 5$.

Solution   Add 3 to both members:

$$A: 3 \quad x - 3 + 3 = 5 + 3$$
$$x + 0 = 8$$
$$x = 8$$

Check   Is 8 the root of the original equation? Let's check by substituting 8 for x in $x - 3 = 5$.

$$x - 3 = 5$$
$$8 - 3 = 5$$
$$5 = 5 \text{ Yes!}$$

$$\therefore \quad x = 8 \text{ is the root of } x - 3 = 5.$$

In the preceding example, 3 was added to both members of the equation so that -3 would be eliminated from the left member. This, then, leaves the *unknown isolated* in the left member. The reason a 3 was added is that it nullifies (makes zero) -3. Because of this property of addition, the following rule is very useful in solving equations.

RULE 6-1. ADDITION AXIOM

The same number added to both members of an equation results in an equivalent equation.

EXAMPLE 6-2   Solve $y - 5 = 2$.
   Solution   Add 5 to both members:

$$A: 5 \quad y - 5 + 5 = 2 + 5$$
$$y + 0 = 7$$
$$y = 7$$

   Check   Does $y - 5 = 2$ when $y = 7$?

$$7 - 5 = 2$$
$$2 = 2 \text{ Yes!}$$

$$\therefore \quad y = 7 \text{ is the root.}$$

Checking the Solution

In the preceding examples, we checked the answer by substituting the computed root into the original equation. This is routinely done to *"catch"* numerical mistakes that might occur in transforming the equation. It is important that you develop a habit of checking the solution.

Subtraction Axiom

In solving equations, sometimes it is necessary to nullify a positive number. This can be achieved by applying the following rule to subtract the positive number from both members of the equation.

RULE 6-2. SUBTRACTION AXIOM

The same number subtracted from both members of an equation results in an equivalent equation.

EXAMPLE 6-3   Solve $x + 7 = 14$.
   Solution   Subtract 7 from both members:

$$S: 7 \quad x + 7 - 7 = 14 - 7$$
$$x + 0 = 7$$
$$x = 7$$

Check   Does $x + 7 = 14$ when $x = 7$?

$$7 + 7 = 14$$
$$14 = 14 \text{ Yes!}$$

$$\therefore \quad x = 7 \text{ is the root.}$$

EXERCISE 6-3

Solve each of the following equations using the addition or subtraction axiom. Check the root by substituting into the original equation.

1. $x + 6 = 8$	2. $x + 3 = 9$
3. $x - 5 = 5$	4. $\lambda - 7 = 8$
5. $y + 2 = 10$	6. $h + 3 = -15$
7. $\theta - 1 = 4$	8. $m - 7 = 18$
9. $y + 10 = 13$	10. $x + (-3) = 5$
11. $n + (-4) = -5$	12. $-13 + \beta = 24$
13. $-18 + C = -14$	14. $62 + K = 62$
15. $\tau + 17 = 0$	16. $K - 4 = -7$
17. $22 + m = 9$	18. $26 + \omega = -8$
19. $x + 11 = 11$	20. $y - 2 = -5$
21. $-45 + t = -25$	

6-4 USING MULTIPLICATION TO TRANSFORM EQUATIONS

To solve an equation, the variable must have a coefficient of 1. A fractional coefficient can be transformed into 1 by multiplying both members of the equation by the reciprocal of the fraction.

RULE 6-3. MULTIPLICATION AXIOM

Multiplying both members of an equation by the same nonzero number results in an equivalent equation.

EXAMPLE 6-4   Solve $\frac{1}{3}Y = 5$.

Solution   Multiply by 3, the reciprocal of $\frac{1}{3}$:

$$M: 3 \qquad 3(\tfrac{1}{3}Y) = 3(5)$$
$$1Y = 15$$
$$Y = 15$$

Check   Does $\frac{1}{3}Y = 5$ when $Y = 15$?

$$(\tfrac{1}{3})15 = 5$$
$$5 = 5 \text{ Yes!}$$

$$\therefore \qquad Y = 15 \text{ is the root.}$$

EXAMPLE 6-5   Solve $\dfrac{1}{-7}x = -4$.

Solution   Multiply by -7, the reciprocal of $1/-7$:

$$M: -7 \qquad -7\dfrac{1}{-7}x = -7(-4)$$
$$1x = 28$$
$$x = 28$$

Check   Does $\dfrac{1}{-7}x = -4$ when $x = 28$?

$$\dfrac{1}{-7}(28) = -4$$
$$-4 = -4 \text{ Yes!}$$

$$\therefore \qquad x = 28 \text{ is the root.}$$

Division Axiom

An equation with a coefficient of the unknown greater than 1 may be transformed to an equation with a coefficient equal to 1 by dividing both members by the coefficient of the unknown.

RULE 6-4. DIVISION AXIOM

Dividing both members of an equation by the same nonzero number results in an equivalent equation.

EXAMPLE 6-6   Solve $5x = 15$.

Solution   Divide by 5, the coefficient of x:

$$D: 5 \qquad 5x/5 = 15/5$$
$$x = 3$$

Check   Does $5x = 15$ when $x = 3$?

$$5(3) = 15$$
$$15 = 15 \text{ Yes!}$$

$$\therefore \qquad x = 3 \text{ is the root.}$$

EXAMPLE 6-7   Solve $-2x = 14$.

Solution   Divide by -2, the coefficient of x:

$$D: -2 \qquad -2x/-2 = 14/-2$$
$$x = -7$$

Check   Does $-2x = 14$ when $x = -7$?

$$-2(-7) = 14$$
$$14 = 14 \text{ Yes!}$$

$$\therefore \qquad x = -7 \text{ is the root.}$$

┌ EXERCISE 6-4

Apply the multiplication or division axiom to each of the following equations. Check the solutions.

1. $3y = 12$ 2. $9m = 72$ 3. $8\mu = 48$
4. $5x = 40$ 5. $x/3 = 5$ 6. $h/4 = 9$

7. $y/6 = 10$
8. $\psi/8 = 3$
9. $6K = -36$
10. $-5t = -15$
11. $-4P = 28$
12. $-3d = 0$
13. $y/2 = 48$
14. $18y = -72$
15. $-16\beta = 64$
16. $-2x = 32$
17. $x/-9 = 7$
18. $K/3 = -13$
19. $-7\phi = -98$
20. $13x = 52$
21. $5r = -10$
22. $-17x = -85$
23. $-y/13 = -4$
24. $19w = -57$

Apply the addition or subtraction axiom and then the multiplication or division axiom to each of the following equations. Check the solutions.

25. $x/2 + 6 = 8$
26. $3y - 5 = 10$
27. $3 - 4k = 19$
28. $28 - 9W = -8$
29. $3y - 6 = -15$
30. $h/5 - 12 = -14$

6-5 ADDITIONAL TECHNIQUES

In solving a given equation, several techniques may be needed to transform the equation into the solution. We will explore several additional techniques in this section.

Combining Like Terms in Equations

Complicated looking equations can often be made to look simpler by combining like terms. Like terms are always combined before isolating the unknown.

EXAMPLE 6-8   Solve $8x - 5 + 3x = 12 + 16$.
   Solution   Combine like terms:

$$8x - 5 + 3x = 12 + 16$$
$$11x - 5 = 28$$
A: 5 $11x - 5 + 5 = 28 + 5$
$$11x = 33$$
D: 11 $11x/11 = 33/11$
$$x = 3$$

Check   Does $8x - 5 + 3x = 12 + 16$ when $x = 3$?

$$8(3) - 5 + 3(3) = 12 + 16$$
$$24 - 5 + 9 = 12 + 16$$
$$28 = 28 \text{ Yes!}$$

$$\therefore \quad x = 3 \text{ is the solution.}$$

EXERCISE 6-5
Solve each equation.

1. $2x + 7x = 18$	2. $4x - 2x = 12$
3. $5x + 4x = 35 - 8$	4. $5I + 2I - 3I = 22 - 6$
5. $R + 2R - R = 1$	6. $7E - 5E + E = 45$
7. $9\phi - 2\phi + 6\phi = 39$	8. $3y + 4y + 7y = 70$
9. $0.5m + 0.5m = 1$	10. $3\omega + 4\omega = 7 + 35$

Solving Equations with the Unknown in the Right Member

Since the unknown is conventionally isolated in the left member, it then becomes time-consuming to manipulate an equation that has the unknown in the right member. To counter this situation, we may apply Rule 6-5, the *symmetric property of equality*.

RULE 6-5. SYMMETRIC PROPERTY OF EQUALITY

If the members of an equation are interchanged, the resulting equation is equivalent to the original equation. *Example:* If $A = B$, then $B = A$.

EXAMPLE 6-9   Solve $9 = 3y$.
     Solution   Use Rule 6-5 to *interchange the members:*

$$3y = 9$$
$$\text{D: } 3 \quad 3y/3 = 9/3$$
$$y = 3$$

Check Does $9 = 3y$ when $y = 3$?

$$9 = 3(3)$$
$$9 = 9 \text{ Yes!}$$

$\therefore \quad y = 3$ is the solution.

EXAMPLE 6-10 Solve $9 + 8 = 4x - x + 5$.

Solution Combine like terms:

$$17 = 3x + 5$$
$$\text{S: 5} \quad 17 - 5 = 3x + 5 - 5$$
$$12 = 3x$$

Interchange members:

$$3x = 12$$
$$\text{D: 3} \quad 3x/3 = 12/3$$
$$x = 4$$

Check Does $9 + 8 = 4x - x + 5$ when $x = 4$?

$$9 + 8 = 4(4) - (4) + 5$$
$$9 + 8 = 16 - 4 + 5$$
$$17 = 17 \quad \text{Yes!}$$

$\therefore \quad x = 4$ is the solution.

EXERCISE 6-6

Solve each equation.

1. $16 = 4y$
2. $-49 = 7x$
3. $3 = \phi - 3 + 5$
4. $-2 = -7 + \alpha - 2\alpha$
5. $4 = -5 - 3R$
6. $-6 = \frac{1}{2}I$
7. $-3 = \frac{1}{4}E$
8. $26 = -13h$
9. $-7 = 2\mu + 3$
10. $13 = 5y - 2 - 2y$
11. $3 + 12 = 6t - 2t - 1$
12. $25 - 13 = 4W + 2 + W$

13. $13 - 3 = 3 + 7u - 5 - 3u$
14. $29 - 6 = 6 - 2v + 2 + 7v$
15. $-42 + 35 = 4\eta - 16 + 3\eta + 9$

Solving Equations with the Unknown in Both Members

So far, the unknown has been in only one member of the equation. We will now work with equations in which the unknown appears in both members.

EXAMPLE 6-11 Solve $5x = 2 + 3x$.

Solution Collect all the terms containing x in the left member:

$$\text{S: } 3x \qquad 5x - 3x = 2 + 3x - 3x$$

Combine like terms:

$$2x = 2$$
$$\text{D: } 2 \quad 2x/2 = 2/2$$
$$x = 1$$

Check Does $5x = 2 + 3x$ when $x = 1$?

$$5(1) = 2 + 3(1)$$
$$5 = 2 + 3$$
$$5 = 5 \text{ Yes!}$$

$$\therefore \quad x = 1 \text{ is the solution.}$$

EXAMPLE 6-12 Solve $-7x = 24 - x$.

Solution Collect all terms containing x in the left member:

$$\text{A: } x \qquad -7x + x = 24 - x + x$$

Combine like terms:

$$-6x = 24$$
$$\text{D: } -6 \quad -6x/-6 = 24/-6$$
$$x = -4$$

Check   Does $-7x = 24 - x$ when $x = -4$?

$$-7(-4) = 24 - (-4)$$
$$28 = 24 + 4$$
$$28 = 28 \text{ Yes!}$$

$\therefore \quad x = -4$ is the solution.

EXERCISE 6-7
Solve each equation.

1. $3\Delta = \Delta + 4$ 2. $6N = N + 4$
3. $6x = 20 - 4x$ 4. $-10\beta = 24 + \beta$
5. $-3y = 15 - 2y$ 6. $2R - 10 = 12R$
7. $5\pi + 13 = -8\pi$ 8. $-20 - 6m = 4m$
9. $2\theta = -7\theta + 18$ 10. $-8 - 5I = 3I$
11. $5\eta - 1 = 2\eta + 14$ 12. $E - 6 = 10 - 3E$
13. $9y - 16 + 6y = 11 + 4y - 5$
14. $3a + 5 - 8a = 7 - 9a - 12$
15. $4z + 14 + 2z = 12 - 3z - 8$
16. $21 - 3C - 7 = 6C - 2 - C$

6-6 EQUATIONS CONTAINING PARENTHESES

To solve an equation containing parentheses, we must first remove the parentheses. This may require the application of the distributive law of multiplication.

EXAMPLE 6-13   Solve $5(x + 2) = 20$.
   Solution   Distribute 5 over $(x + 2)$:

$$5x + 10 = 20$$
S: 10 $5x + 10 - 10 = 20 - 10$
$$5x = 10$$
D: 5 $5x/5 = 10/5$
$$x = 2$$

Check   Does $5(x + 2) = 20$ when $x = 2$?

$$5(2 + 2) = 20$$
$$5(4) = 20$$
$$20 = 20 \text{ Yes!}$$

$\therefore \quad x = 2$ is the solution.

EXAMPLE 6-14   Solve $10i - 2(i - 8) = (5i + 4)$.
Solution   Distribute -2 over $(i - 8)$:

$$10i - 2i + 16 = (5i + 4)$$

Remove the ( ) in the right member:

$$10i - 2i + 16 = 5i + 4$$

Combine like terms:

$$8i + 16 = 5i + 4$$
S: $5i$ $\quad 8i + 16 - 5i = 5i + 4 - 5i$
$$3i + 16 = 4$$
S: 16 $\quad 3i + 16 - 16 = 4 - 16$
$$3i = -12$$
D: 3 $\quad 3i/3 = -12/3$
$$i = -4$$

Check   Does $10i - 2(i - 8) = (5i + 4)$ when $i = -4$?

$$10(-4) - 2(-4 - 8) = (5(-4) + 4)$$
$$-40 - 2(-12) = (-20 + 4)$$
$$-40 + 24 = -16$$
$$-16 = -16 \text{ Yes!}$$

$\therefore \quad i = -4$ is the solution.

EXAMPLE 6-15   Solve $7R - (3 - 2R) + 6 = 21 + 3(R - 4)$.
Solution   Distribute -1 over $(3 - 2R)$ in the left member and distribute 3 over $(R - 4)$ in the right member:

$$7R - 3 + 2R + 6 = 21 + 3R - 12$$

Combine like terms:

$$9R + 3 = 9 + 3R$$

S: 3 $\quad 9R + 3 - 3 = 9 + 3R - 3$

$$9R = 6 + 3R$$

S: 3R $\quad 9R - 3R = 6 + 3R - 3R$

$$6R = 6$$

D: 6 $\quad 6R/6 = 6/6$

$$R = 1$$

Check Does $7R - (3 - 2R) + 6 = 21 + 3(R - 4)$ when $R = 1$?

$$7(1) - (3 - 2 \cdot 1) + 6 = 21 + 3(1 - 4)$$
$$7 - (1) + 6 = 21 + 3(-3)$$
$$12 = 21 - 9$$
$$12 = 12 \text{ Yes!}$$

$$\therefore \quad R = 1 \text{ is the solution.}$$

EXERCISE 6-8

Solve each equation.

1. $2(i - 3) = 16$
2. $-3(5 + e) = 3$
3. $4 + 5(\beta - 1) = -6$
4. $3(Z + 6) = 21$
5. $15 - (2x + 10) = -1$
6. $4(-2 + 3\lambda) = -20$
7. $-6 = 3(8 - 2R_1)$
8. $E_1 = 5(E_1 - 4)$
9. $-(-2W + 6) = 3W - 18$
10. $4Q_1 + 6 = -2(Q_1 + 3)$
11. $-(-2L + 13) + 2L = 59$
12. $6\phi + 15 = 7(\phi - 2)$
13. $-4(3 + R_1) + 5(R_1 + 4) = 0$
14. $(7 - K)3 = (-8 + 3K)$
15. $(3\theta - 3) = (\theta + 5)$
16. $9I_1 - (6I_1 - 2) = 8$
17. $13N - (3 + 12N) = -3$
18. $7(C' - 5) = 6 - (C' + 1)$
19. $6V_t - (V_t - 7) = (V_t + 15)$
20. $5A_v - (1 - A_v) + 4 = 9$
21. $5(\alpha + 2) - 4(\alpha + 1) - 3 = 0$
22. $12g_m - (3 - g_m) = 7(5g_m - 1) - 18g_m$
23. $3(8X_C - 2) - 3(1 - X_C) + 8 = 8$
24. $11 + 4(\sigma - 17) = 21(\sigma - 2) - 5(3 - \sigma)$

6-7 SOLVING FORMULAS

In electronics, as in all sciences, there are many laws. These laws are often expressed as equations. This type of equation is called a *formula.* It is through formulas that mathematics is applied to electronics, because formulas express relationships between physical quantities in mathematical language.

Ohm's law is fundamental to all electronics. It describes the relationship between voltage, current, and resistance. As a formula, Ohm's law is written $E = IR$.

In applying formulas, we must sometimes transform the stated formula into an equivalent formula so that a particular quantity may be found. This concept is demonstrated in the following examples.

EXAMPLE 6-16 Solve for R in the formula $P = I^2R$. Express the answer in terms of P and I^2 as a new formula.

Solution

$$P = I^2R$$
$$\text{D: } I^2 \quad \frac{P}{I^2} = \frac{I^2R}{I^2}$$
$$P/I^2 = R$$

Interchange members:

$$\therefore \quad R = P/I^2$$

We see in Example 6-16 that the formula $P = I^2R$ was transformed into $R = P/I^2$ using the rules previously learned in this chapter.

EXAMPLE 6-17 Solve $E = IR_1 + IR_2$ for I.

Solution Factor I from the right member:

$$E = I(R_1 + R_2)$$
$$\text{D: } R_1 + R_2 \quad E/(R_1 + R_2) = I(R_1 + R_2)/(R_1 + R_2)$$
$$E/(R_1 + R_2) = I$$

Interchange members:

$$\therefore \quad I = \frac{E}{R_1 + R_2}$$

120

EXAMPLE 6-18   Solve $I = E/R$ for R.
    Solution

$$\text{M: } R \qquad IR = ER/R$$
$$IR = E$$
$$\text{D: } I \qquad IR/I = E/I$$
$$\therefore \qquad R = E/I$$

EXAMPLE 6-19   Solve $\alpha\beta + \alpha = \beta$ for β.
    Solution

$$\text{S: } \alpha \qquad \alpha\beta + \alpha - \alpha = \beta - \alpha$$
$$\alpha\beta = \beta - \alpha$$

$$\text{S: } \beta \qquad \alpha\beta - \beta = \beta - \alpha - \beta$$
$$\alpha\beta - \beta = -\alpha$$

Change the sign of each term by multiplying the equation by -1:

$$\text{M: } -1 \qquad -1(-\beta) - 1(\alpha\beta) = -1(-\alpha)$$
$$\beta - \alpha\beta = \alpha$$

Factor the left member:

$$\beta(1 - \alpha) = \alpha$$
$$\text{D: } (1 - \alpha) \qquad \beta(1 - \alpha)/(1 - \alpha) = \alpha/(1 - \alpha)$$
$$\therefore \qquad \beta = \frac{\alpha}{1 - \alpha}$$

EXAMPLE 6-20   Solve $I_C = \beta I_B + I_{CBO}(\beta + 1)$ for β.
    Observation   This is an actual formula used with transistors. There are three currents denoted by I_C, I_B, and I_{CBO}. Even though the third subscript is composed of three letters, it is only a subscript.
    Solution   Distribute I_{CBO} over $(\beta + 1)$:

$$I_C = \beta I_B + \beta I_{CBO} + I_{CBO}$$
$$\text{S: } I_{CBO} \qquad I_C - I_{CBO} = \beta I_B + \beta I_{CBO} + I_{CBO} - I_{CBO}$$
$$I_C - I_{CBO} = \beta I_B + \beta I_{CBO}$$

Factor right member:

$$I_C - I_{CBO} = \beta(I_B + I_{CBO})$$

$$D: I_B + I_{CBO} \qquad \frac{I_C - I_{CBO}}{I_B + I_{CBO}} = \beta \frac{I_B + I_{CBO}}{I_B + I_{CBO}}$$

$$\frac{I_C - I_{CBO}}{I_B + I_{CBO}} = \beta$$

Interchange members:

$$\therefore \quad \beta = \frac{I_C - I_{CBO}}{I_B + I_{CBO}}$$

EXERCISE 6-9

Solve each formula for the indicated variable in terms of the other variables:

Given:	Solve for:
1. $v = f\lambda$	λ
2. $C = Q/V$	V
3. $E = IR$	R
4. $R = E^2/P$	E^2
5. $r_p = \dfrac{\mu}{g_m}$	g_m
6. $I_C = \beta I_B$	β
7. $Q_0 = \dfrac{BW}{f_0}$	f_0
8. $V_{CC} = I_C R_1 + V_{CE}$	R_1
9. $E_{BB} - E_1 = I_B E_p$	E_1
10. $E_1(R_1 + R_2) = E_T R_1$	E_1
11. $R_0 I_L = E_N - E_L$	E_L
12. $X_C = \dfrac{1}{\omega C}$	C
13. $-x = t/RC$	R
14. $Q = \omega_1 L/R$	ω_1

15. $I_2R_1 = I_1G - I_2R_2$ I_2

16. $R = \rho L / A$ ρ

17. $T_J - T_A = P_D \theta_{JA}$ θ_{JA}

18. $cd = 0.2KA(N - 1)$ N

19. $I_1 = \dfrac{E_T - E_1}{R}$ E_1

20. $E_f C_1 + E_f C_2 = E_1 C_1 - E_2 C_2$ C_2

21. $V_1(Y_1 Y_2 - Y^2) = iY_2$ Y_2

22. $R_s Q^2 = R_p - R_s$ R_s

23. $I_T R_{th} = E_{th} - I_T R_L$ I_T

24. $I_E = I_B + \beta I_B$ I_B

25. $A_v R_L + A_v r_p = \mu R_L$ R_L

26. $R_2 = \dfrac{C_3 P_0}{D_2}$ D_2

6-8 FORMING EQUATIONS

In a word problem, written descriptive information is given about numerical quantities and how they relate to one another. To solve word problems, we must *first* form an equation from the written words, and *second* we must solve the equation.

EXAMPLE 6-21   Express each of the statements using algebraic symbols. Let x represent the unknown number.

STATEMENT A   A number increased by 4 equals 17.
Equation $x + 4 = 17$
Solution $x = 13$

STATEMENT B   A number decreased by 8 equals 20.
Equation $x - 8 = 20$
Solution $x = 28$

STATEMENT C   A number increased by twice the number equals 33.
Equation $x + 2x = 33$
Solution $x = 11$

STATEMENT D One-third of a number diminished by 2 is 3.

Equation
$$\tfrac{1}{3}x - 2 = 3$$

Solution
$$\tfrac{1}{3}x = 3 + 2$$
$$x = 3(5)$$
$$x = 15$$

STATEMENT E Twice a number plus 3 equals the number minus 6.

Equation
$$2x + 3 = x - 6$$

Solution
$$2x - x = -6 - 3$$
$$x = -9$$

EXERCISE 6-10

Use algebraic symbols to form an equation for each of the statements. Let x represent the unknown number. Solve each of the equations.

1. Eight times a number is 40.
2. Four times a number, increased by twice the number, equals 42.
3. Six times a number, diminished by 5, equals 25.
4. A number less 3 is 11.
5. Sixteen diminished by a number is 9.
6. Five less than one-half a number is 6.
7. Add 2 to a number, and double the sum; the result is 18.
8. Add 6 to a number, and decrease the sum by 7; the result is 15.
9. Six times a number diminished by twice the number is 52.
10. Five added to twice a number less 2 equals the number diminished by 3.

6-9 SOLVING WORD PROBLEMS

Solving word problems in mathematics and troubleshooting an electronic system have many similarities. In each case, the problem is verbalized; each may appear complicated, causing confusion in the solution; each requires a systematic approach that includes the use of symbols; a successful conclusion is reached after one or more trials.

Because word problems require that you have *interpretive skills,* you may find them very challenging. We have included the following procedures for solving word problems to aid in developing analytical and interpretive skills. There is no easy way to learn how to troubleshoot or to solve word problems; the skill only comes by a consistent, conscious effort over a relatively long period of time.

GUIDELINES 6-1. GENERAL PROCEDURE FOR SOLVING WORD
                PROBLEMS

1. Carefully read the problem.
2. Select a variable to represent the unknown number.
3. If there is more than one condition described, represent all conditions in terms of the selected variable.
4. Form an equation from the described condition.
5. Solve the equation for the variable.
6. Check the solution to see that all the described conditions of the problem are satisfied.

EXAMPLE 6-22   A test technician tested 84 electronic systems. Some of the systems tested were rejected. Five times as many systems were accepted as were rejected. How many systems were rejected?

Solution   Let x represent the number of systems rejected, and $5x$ represent the number of systems accepted. The number of systems rejected plus the number of systems accepted equals the number of systems tested.

$$x + 5x = 84$$
$$6x = 84$$
$$x = 14$$

∴   14 systems were rejected.

EXAMPLE 6-23   A leading manufacturer of solid-state memories sells three models of its memory system. The storage capacity of the largest model is four times the capacity of the intermediate model. The storage capacity of the intermediate model is twice that of the smallest model. If the total combined capacity of all three models is 45 056 bytes of storage, find the capacity of each memory system. (Computer memory storage is measured in units called *bytes*.)

Solution   Let N = the capacity of the smallest memory
 $2N$ = the capacity of the intermediate memory
 $4(2N)$ = the capacity of the largest memory

Observation   Each of the capacities is related to the smallest memory by using the variable N. No new variable has been introduced.

$$N + 2N + 4(2N) = 45\ 056$$
$$N + 2N + 8N = 45\ 056$$
$$11N = 45\ 056$$
$$N = 4096$$

∴   4096 bytes is the capacity of the smallest memory.
8192 bytes is the capacity of the intermediate memory.
32 768 bytes is the capacity of the largest memory.

EXERCISE 6-11

Solve each problem using Guidelines 6–1.

1. When five is added to an unknown number, the sum is twice the unknown number. Find the unknown number.
2. Two times the sum of twelve and an unknown number is equal to twenty. Determine the unknown number.
3. When twice the sum of six and an unknown number is added to the unknown number, the result is equal to ten added to the unknown. Find the unknown number.
4. Three times the result of subtracting an unknown number from nine is nine. Determine the unknown number.
5. Twice the sum of four and three times an unknown number is equal to five times the sum of four times the unknown number and ten. Find the unknown number.
6. Three transformers purchased from a local supply house cost $16 after a $2 discount. What is the cost of a single transformer without a discount?
7. A large supplier of electronic components installed a computerized inventory system. In checking the daily printout of the supply of a particular transistor, it was found that there were 1210 fewer transis-

tors listed than the amount recorded on the printout of the day before. If the daily printout shows 3126 transistors, what was the number recorded the day before?

8. A special rectangular screen room, which is used to make RF (radio-frequency) measurements, was measured and found to be three times as long as it is wide. If the sum of all the sides is 24 meters, what is the length and width of the room?

9. One resistor and two capacitors cost 82¢. If the capacitor costs 23¢ more than the resistor, find the cost of one capacitor.

10. A 30-m piece of *twin-lead* wire (TV lead-in wire) is cut into two pieces. One piece is 2 m shorter than three times the length of the other piece. Find the length of the shorter piece.

11. An 18-minute video tape was made to train electronic technicians on the use of the digital voltmeter (DVM). The part of the tape about voltage measurements lasted 3 minutes longer than twice the time about resistance measurement. How many minutes of instruction was spent on voltage measurements?

12. A service technician took 35 minutes to disassemble and reassemble a tape recorder. If it took 7 minutes less to reassemble than it took to disassemble, how long did it take to disassemble this unit?

13. A resistor costs R cents and a capacitor costs five times more than a resistor. If the combined cost of one resistor and one capacitor is 18 cents, what is the cost of a single resistor and a single capacitor?

14. Divide a 96-centimeter (cm) piece of hookup wire into three parts so that the first piece is three times as long as the second, and the third piece is as long as the sum of the other two pieces.

15. Two semiconductor diodes, one a rectifier, the other a zener diode, have an average operating temperature of 160°C. What is the operating temperature of each diode if the rectifier diode has a 20°C higher operating temperature than the zener diode? (The average of two values is found by adding the two quantities and dividing by 2.)

SELECTED TERMS

conditional equation   True only for particular values of the variable in the equation.

formula   An equation expressing the relationship between physical quantities.

identical equation   True for all values of the variable in the equation.

members of the equation   The expressions on either side of equal signs.

root of the equation   A value of the variable that makes the equation true. A solution.

symmetric property of equality   Allows the members of an equation to be interchanged.

7

QUANTITIES AND UNITS OF MEASUREMENT

The application of electronics involves measurements of many physical quantities. These measurements have numerical values that can be used in formulas. They also have units associated with them. A measurement without units is useless. If I tell you a building is 16 high, you still do not know how tall it is. It may be 16 feet, 16 meters, or 16 stories.

In this chapter you will learn the units commonly used in electronics and other areas of technology.

7-1 SYSTEMS OF MEASUREMENT

At present, two systems are used in the United States for measuring quantities. These are the English system and the metric system.

English System

The English system has developed over many centuries; it has its origin in England in the thirteenth century. It uses units and subunits that were not systematically chosen. Thus, if the name foot, inch, and yard were not known to you, you would have no clue that each was related to the other. Furthermore, without prior knowledge of the size of each of them, you would not know that 12 inches equal 1 foot, or 3 feet equal 1 yard. In the English system, each time a multiple or submultiple of a unit was needed, a new name was applied. Table 7-1 lists part of the English system.

Because of its complexity and other shortcomings, the English system has not been adopted as a standard for the various scientific and engineering

TABLE 7-1

ENGLISH SYSTEM FOR MECHANICS		
Physical Quantity	*Unit Name*	*Unit Symbol*
	BASE UNITS	
Length	Foot	ft
Force	Pound	lb
Time	Second	s
	DERIVED UNITS	
Mass	Slug	$lb \cdot s^2/ft$
Pressure	Pound per square foot	lb/ft^2
Work	Foot-pound	$ft \cdot lb$
Power	Horsepower	hp
Energy	British thermal unit	Btu

fields. Instead, the International System of Units (SI metric system) has been selected as the standard system.

SI Metric System

The metric system, like the English system, has undergone changes to get to its present form. The metric system in use today is called the *International System of Units,* abbreviated SI. The SI metric system has been adopted by the Institute of Electrical and Electronic Engineers (IEEE). This system uses unit prefixes to form multiples and submultiples of a unit. Thus, centi*meter,* *meter,* and kilo*meter* are all used to measure the same quantity, length. The SI metric system keeps the same unit name for both multiples and submultiples. Because of this simplicity, the SI metric system is *"the standard of the industry."*

7-2 THE SI SYSTEM OF MEASUREMENT

The International System of Units has seven base units, two supplementary units, several derived units with special names, and many derived units with compound names.

Base Units

The seven base units are the building blocks from which the derived units are constructed. Each base unit is defined by a very precise measurement standard that gives the exact value of the unit. Table 7-2 lists the seven SI base units.

TABLE 7-2

SI BASE UNITS			
Physical Quantity	**Physical Symbol***	**Unit Name**	**Unit Symbol†**
Length	l	meter	m
Mass	m	kilogram	kg
Time	t	second	s
Electric current	I	ampere	A
Temperature	T	kelvin	K
Luminous intensity	I	candela	cd
Amount of substance	M	mole	mol

*Standardized letters used in formulas or equations to represent the physical quantity in the calculation.

†An abbreviation used to indicate the units in the statement of physical measurement.

Great care must be used when writing the symbol (abbreviation) for a particular unit. The uppercase letters must be easily distinguished from the lowercase letters. In Table 7-2, you might have noticed that all the abbreviations (unit symbols) are lowercase letters except the A for ampere and the K for kelvin. Because these two units are named for scientists, uppercase letters are used.

SI Derived Units

The SI derived units are formed from the previously defined SI base and supplementary units. Many derived units have convenient names and symbols in place of the complicated ones that would result from derivation from the base unit. The watt (W), a measure of both mechanical and electrical power, is such a unit. Its base units include the meter, kilogram, and second. Without the special name watt and the symbol W, we would express power as $m^2 \cdot kg/s^3$, read "meter squared kilogram per second cubed." Table 7-3 lists some of the SI-derived units with special names.

7-3 SELECTED PHYSICAL QUANTITIES

Electric Charge

The coulomb (C) expressed in SI units is the *ampere second* (A · s). It is a measure of the quantity of charge (one coulomb is equal to the charge of 6.242×10^{18} electrons). The name coulomb was selected to honor the French physicist Charles A. de Coulomb (1736–1806). The symbol Q (for quantity) is used to represent charge in electrical formulas. See Figure 7-1.

TABLE 7-3

SI-DERIVED UNITS WITH SPECIAL NAMES

Physical Quantity	Physical Symbol	Unit Name	Unit Symbol
Capacitance	C	farad	F
Conductance	G	siemens	S
Electric charge	Q	coulomb	C
Electromotive force	E	volt	V
Energy, work	W	joule	J
Force	F	newton	N
Frequency	f	hertz	Hz
Inductance	L	henry	H
Power	P	watt	W
Resistance	R	ohm	Ω

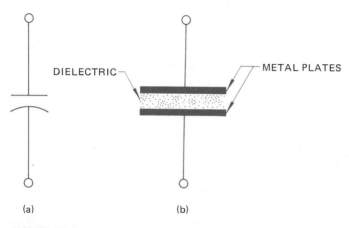

DIELECTRIC

METAL PLATES

(a)

(b)

FIGURE 7-1
The capacitor is a device for storing charge. (a) Schematic symbol used for a capacitor; (b) electric charge is stored in the dielectric between the metal plates.

Electric Current

The ampere (A) is one of the seven base SI units. The name ampere was selected to honor the French physicist André M. Ampère (1775–1836). The symbol I (for intensity) is used to represent current in electrical formulas. See Figure 7-2.

Electromotive Force

The volt (V) expressed in SI units is joules per coulomb (J/C). The name volt was selected to honor the Italian inventor of the battery, Alessandro Volta (1745–1827). Both the symbols E, for a voltage rise, and V, for a voltage drop, are used in electrical formulas. See Figure 7-3.

Resistance

The ohm (Ω) expressed in SI units is volts per ampere (V/A). It is a measure of the opposition to the movement of electrical current. One ohm of resistance limits the intensity of electrical current to one ampere when the applied voltage is one volt. The name ohm was selected to honor the German physicist Georg Simon Ohm (1787–1854). The symbol R is used to represent resistance in electrical formulas. See Figure 7-4.

Conductance

The siemens (S) expressed in SI units is amperes per volt (A/V). It is a measure of the ease of the movement of electrical current. One siemens of conductance permits the electrical current of one ampere to flow when the applied

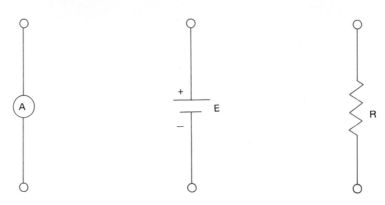

FIGURE 7-2
Schematic symbol used to represent an ammeter. The ammeter is used to measure electric current.

FIGURE 7-3
Schematic symbol for a dc voltage source. The longer line indicates the positive terminal and *E* represents electromotive force.

FIGURE 7-4
Schematic symbol for any kind of resistance.

voltage is one volt. The name siemens was selected to honor the German inventor Ernst Werner von Siemens (1816–1892). The symbol G is used to represent conductance in electrical formulas.

Energy, Work

The joule (J) expressed in SI units is the newton-meter (N · m). One joule of work is done when an applied force of one newton moves an object one meter. The joule is also used as a unit of energy. In an electrical system, work is done in moving an electric charge. The name joule was selected to honor the English physicist James P. Joule (1818–1889). The symbol W (for work) is used to represent work and energy in electrical formulas. See Figure 7-5.

FIGURE 7-5
The cape buffalo does work by moving the cart in the direction of the arrow by exerting a horizontal force on the cart. Work = force × distance.

Power

The watt (W) expressed in SI units is joules per second (J/s). It is a measure of the rate of using (or producing) energy or doing work. The watt is both a measure of electrical and mechanical power. When one joule of work is done in one second, the rate of doing work is one watt. The name watt was selected to honor the Scottish engineer James Watt (1736–1819). The symbol P (for power) is used to represent power in electrical formulas. See Figure 7-6.

Force

The newton (N) expressed in SI units is joules per meter (J/m). It is the amount of force needed to accelerate a mass of one kilogram to one meter per second in one second. The name newton was selected to honor the English scientist, astronomer, and mathematician Sir Isaac Newton (1642–1727). The symbol F is used to represent force in formulas. Newton's second law, $F = ma$, is possibly the most important equation in mechanics.

Temperature

Degree Celsius (°C) is commonly used for temperature measurement in electronics. Degree Celsius is the same size as the SI base unit of temperature, the kelvin (K). The two temperature scales start from different reference points. The same temperature measured in kelvin is 273.15 greater than that measured in degrees Celsius.

$$T(°C) = T(K) - 273.15$$

This concept is pictured in Figure 7-7. The name Celsius was selected to honor the Swedish astronomer Anders Celsius (1701–1744). See Figure 7-8.

Time

The SI base unit for time is the second (s). Other common units of time are:

- Minute (min)   1 min = 60 s

FIGURE 7–6
The electric motor converts electrical power to mechanical power. The electric power (P) supplied to the motor is computed by the formula $P = EI$.

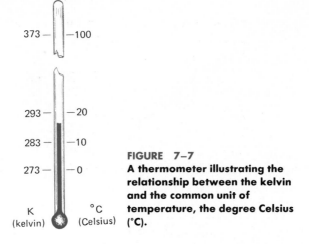

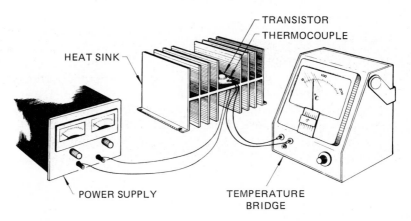

FIGURE 7-7
A thermometer illustrating the relationship between the kelvin and the common unit of temperature, the degree Celsius (°C).

FIGURE 7-8
A temperature bridge and thermocouple are being used to measure the case temperature of a power transistor.

- Hour (h)   1 h = 60 min
- Day (d)   1 d = 24 h

EXERCISE 7-1
Match one item from the list at the right to one item at the left. Some items in the right list may be used more than once.

1. electrical current	a. N	
2. voltage rise	b. S	
3. J/s	c. coulomb	

4. o─/\/\/\─o   d. $N \cdot m$
5. English unit   e. °C
6. SI base unit   f. I
7. V/A   g. IEEE
8. A/V   h. R
9. force   i. E
10. Q j. horsepower
11. joule   k. power
12. temperature   l. ampere

7-4 FORMING DECIMAL MULTIPLES AND SUBMULTIPLES OF THE SI UNITS

When working with the SI units, it often becomes necessary to use larger or smaller units of a particular quantity. Multiples and submultiples of the SI units are formed by using the prefixes in Table 7-4, which lists most of the prefixes used in science and technology. Those prefixes in boldface type are commonly used with electrical units.

In writing numbers with dimensions, it is usually suitable to select a prefix so that the numerical values may be expressed as numbers from 0.100 to 1000. An appropriate multiple or submultiple of the SI unit is chosen so the number value lies in the conventional range. Table 7-5 shows several exam-

TABLE 7-4

PREFIXES AND SYMBOLS FOR MULTIPLES AND SUBMULTIPLES OF THE SI UNITS

Prefix	Symbol	Phonic	Multiple or Submultiple	Power of Ten
kilo	k	ki′lō	1 000	10^3
mega	M	meg′ə	1 000 000	10^6
giga	G	ji′gə	1 000 000 000	10^9
tera	T	ter′ə	1 000 000 000 000	10^{12}
deci	d	des′ə	0.1	10^{-1}
centi	c	sen′tə	0.01	10^{-2}
milli	m	mil′ē	0.001	10^{-3}
micro	μ	mī′krō	0.000 001	10^{-6}
nano	n	nān′ō	0.000 000 001	10^{-9}
pico	p	pē′kō	0.000 000 000 001	10^{-12}
femto	f	fĕm′tō	0.000 000 000 000 001	10^{-15}

TABLE 7-5

FORMING MULTIPLES AND SUBMULTIPLES

Measured or Computed Value	Engineering Notation	Expressed as Multiple or Submultiple of SI Unit
0.002 80 A	2.80×10^{-3} A	2.80 mA
8 470 000 Hz	8.47×10^{6} Hz	8.47 MHz
47 000 Ω	47×10^{3} Ω	47 kΩ
0.009 20 V	9.20×10^{-3} V	9.20 mV
0.000 000 053 s	53×10^{-9} s	53 ns
1390 W	1.39×10^{3} W	1.39 kW

ples of how the prefixes of Table 7-4 are used to form multiples and submultiples of the SI units.

RULE 7-1. FORMING MULTIPLES AND SUBMULTIPLES

To form multiples and submultiples of SI units:

1. Write the number in engineering notation (if the quantity is electrical, then the decimal point is placed so the exponent is a multiple of 3).
2. The decimal coefficient is usually expressed as a number from 0.100 to 1000.
3. Form the multiple or submultiple by writing the decimal coefficient with the engineering notation replaced with the appropriate unit prefix and unit symbol.

EXAMPLE 7-1 Write 270 000 Ω using a prefix from Table 7-4.

Solution A Apply Rule 7-1 and use the prefix kilo:

$$270\ 000\ \Omega = 270 \times 10^{3}\ \Omega = 270\ k\Omega$$

Solution B Apply Rule 7-1 and use the prefix mega:

$$270\ 000\ \Omega = 0.270 \times 10^{6}\ \Omega = 0.270\ M\Omega$$

Either answer is correct as it meets the guidelines for expressing dimensioned numbers as numbers from 0.100 to 1000.

Solution C Use the ENG calculator stroke:

$$\text{ENG} \qquad 270\ 000 \Rightarrow 270\ k\Omega$$

$$\therefore \qquad 270\ 000\ \Omega = 270\ k\Omega \text{ or } 0.270\ M\Omega$$

EXAMPLE 7-2   Change 38.3 μS to the basic unit siemens.

Solution   Replace μ with 10^{-6}:

$$38.3\ \mu S = 38.3 \times 10^{-6}\ S$$

Express in decimal notation:

$$0.000\ 038\ 3\ S$$
$$\therefore \qquad 38.3\ \mu S = 0.000\ 038\ 3\ S$$

EXAMPLE 7-3   Convert 0.892 meter to centimeters.

Solution   Express 0.892 m in powers of ten notation with an exponent of -2:

$$0.892\ m = 89.2 \times 10^{-2}\ m$$

Replace 10^{-2} m with centimeters:

$$89.2\ cm$$
$$\therefore \qquad 0.892\ m = 89.2\ cm$$

When adding or subtracting dimensioned numbers, first convert each of the numbers to basic units without a prefix, then add or subtract. Write the answer with an appropriate unit prefix.

EXAMPLE 7-4   Add 1.308 kV and 72.60 V. Express the answer with an appropriate prefix to four significant figures.

Solution   Convert 1.308 kV to volts:

$$1.308\ kV = 1308\ V$$
$$A: \qquad 1308\ V + 72.60\ V = 1380.6\ V$$
$$1.381\ kV$$
$$\therefore \qquad 1.308\ kV + 72.60\ V = 1.381\ kV$$

EXERCISE 7-2
Express the following quantities in the named multiple or submultiple unit.

	Express in:		Express in:
1. 282 Ω	kΩ	2. 0.041 A	mA
3. 1.76×10^{-4} H	μH	4. 6940 V	kV
5. 19.0×10^5 W	MW	6. 0.560 m	cm
7. 1520 Hz	kHz	8. 72×10^{-11} F	pF
9. 180 000 Ω	MΩ	10. 47.2×10^{-10} s	ns

Express the following without unit prefixes.

11. 628 cm	12. 2.35 kV
13. 17.0 mJ	14. 3.30 kΩ
15. 902 μS	16. 487 kN
17. 0.854 kW	18. 250 mS
19. 47.0 μF	20. 702 mg

Calculator Drill
Express each of the following as a multiple or submultiple of the stated SI unit. Apply Rule 7-1 or use the ENG calculator stroke. State the answer to three significant figures.

21. 7834 V	22. 0.000 083 26 s
23. 0.065 378 H	24. 17 342 W
25. 0.152 S	26. 51 621.35 Hz
27. 680 800 Ω	28. 0.000 000 91 F
29. 0.020 39 A	30. 362.08 N

Add the following and state the answer to three significant figures (use an appropriate prefix with the SI unit).

31. 10.0 μA and 0.175 mA
32. 235 W and 2.87 kW
33. 330 kΩ and 1.20 MΩ
34. 75.0 μs and 822 ns
35. 1.02 V and 93.7 mV

7-5 EVALUATING FORMULAS

When substituting numerical quantities into a formula, the units representing the physical quantity *cannot* be separated from the number, nor can the unit size be neglected. When working with formulas, the usual procedure is to substitute the quantity in the basic unit size into the formula. For example, when substituting into the formula for electrical power, $P = EI$, substitute voltage (E) in volts and current (I) in amperes. This is accomplished by converting the prefix into a power of ten.

EXAMPLE 7-5   Compute the power dissipation of an industrial furnace that requires 60.0 A of current at a voltage of 0.440 kV.

Solution   Consider unit size and check to see that each unit is in its basic form:
Change 0.440 kV to 0.440×10^3 V.
Substitute into $P = EI$:

$$P = 0.440 \times 10^3 \times 60.0$$
$$= 26\ 400$$
$$\therefore \quad P = 26.4 \text{ kW}$$

The preceding example shows that unit size must be accounted for in solving a formula. As you become more familiar with the physical quantities used in electronic technology, you will be able to take shortcuts. For now, it is best to convert the unit prefixes into powers of 10 before substituting into the formula.

EXAMPLE 7-6   The equation for inductive reactance is $X_L = 2\pi fL$. Compute X_L in ohms when $f = 2.35$ MHz and $L = 100\ \mu$H.

Solution   Convert to basic units:

$$2.35 \text{ MHz} = 2.35 \times 10^6 \text{ Hz}$$
$$100\ \mu\text{H} = 100 \times 10^{-6} \text{ H}$$

Substitute:

$$X_L = 2\pi fL$$
$$= 2 \times 3.14 \times 2.35 \times 10^6 \times 100 \times 10^{-6}$$

Solve:

$$X_L = 1476.5$$
$$= 1.48 \times 10^3$$
$$\therefore \quad X_L = 1.48 \text{ k}\Omega$$

Although it is conventional to substitute basic units into a formula, some formulas have been derived for special cases. When this is done, a statement is made noting the change from basic units to prefix units. Example 7-7 explores such a situation.

EXAMPLE 7-7 A leading semiconductor manufacturer provided a data sheet for the integrated operational amplifier in Figure 7-9. This data sheet presents two special for-

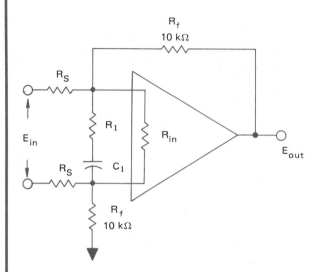

FIGURE 7-9
Schematic for the integrated operational amplifier of Example 7-7. The triangular symbol represents the amplifier.

mulas for determining satisfactory values for resistance R_1 (Ω) and for capacitance C_1 (μF).

$$R_1 = 5R_f \ \Omega \qquad C_1 = 0.04/R_f \ \mu F$$

In these special formulas, R_f is measured in kilohms (kΩ), not in ohms. Determine R_1 and C_1 when R_f is 10 kΩ.

Solution Summarize the information given:

R_f is in (kΩ). R_f has been selected as 10 kΩ.
R_1 is in ohms (Ω). $R_1 = 5R_f \ \Omega$.
C_1 is in microfarads (μF). $C_1 = 0.04/R_f \ \mu F$.

Determine the values for R_1 and C_1 when R_f is 10 kΩ:

$$R_1 = 5R_f \ \Omega \qquad\qquad C_1 = \frac{0.04}{R_f} \ \mu F$$

Substitute 10 for R_f (not 10 k) in each equation:

$$R_1 = 5 \times 10 \qquad\qquad C_1 = \frac{0.04}{10}$$

$$R_1 = 50\ \Omega \qquad\qquad C_1 = 0.004\ \mu F$$

$$\therefore \quad R_1 = 50\ \Omega \text{ and } C_1 = 4\ nF$$

EXERCISE 7-3

Evaluate each of the following formulas for the indicated quantity. Transform the given equation if needed. Substitute the given values into the formulas. Use your calculator to aid in the solution. State the answer to three significant figures. Use an appropriate prefix with the base unit.

Formula	Solve for:	Given:	Answer Basic Unit
1. $E = IR$	I	$E = 52.3$ V $R = 0.470$ kΩ	A
2. $P = I^2 R$	R	$P = 52.0$ W $I = 22.0$ mA	Ω
3. $X_C = \dfrac{1}{2\pi f C}$	X_C	$f = 10.0$ kHz $C = 0.100\ \mu F$	Ω
4. $C = \pi d$	d	$C = 3.82$ m	m
5. $T = K - 273.15$	T	$K = 872$ K	°C
6. $P = W/t$	t	$P = 15.8$ W $W = 0.273$ kJ	s
7. $R_T = R_1 + R_2$	R_2	$R_1 = 3.30$ kΩ $R_T = 10.1$ kΩ	Ω
8. $G_1 = I/V_1$	V_1	$G_1 = 500\ \mu S$ $I = 75$ mA	V
9. $\theta = \omega t$	ω	$t = 8.33$ ms $\theta = 3.14$ rad	rad/s
10. $r_p = \mu/g_m$	g_m	$\mu = 80$ (no units) $r_p = 15$ kΩ	S

The following are special formulas. Pay attention to the specified units.

	Formula	Solve for:	Given:	Answer Basic Unit
11.	$\lambda = \dfrac{300}{f}$ λ is in m f is in MHz	λ	$f = 372$ kHz	m
12.	$T = \dfrac{I_m R_m}{2}$ I_m is in μA R_m is in kΩ T is in $^\circ$C	R_m	$T = 50^\circ$ C $I_m = 0.152$ mA	Ω
13.	$F = ma$ m is in kg a is in m/s^2	F	$m = 7.20$ kg $a = 9.8$ m/s^2	N

7-6 UNIT ANALYSIS AND CONVERSION BETWEEN SYSTEMS

When converting from basic units to prefix units or converting between the English and SI systems, you may become confused in knowing when to divide and when to multiply. If you use a technique called *unit analysis,* you are less likely to have this problem. This technique uses the concept that a quantity may be multiplied by *1* without changing its value. Thus, the identity 100 centimeters = 1 meter may be expressed as a fraction:100 cm/1 m = 1 or, in its reciprocal form, 1 m/100 cm = 1. In either case, we see the ratio (fraction) is worth 1 and has no dimensions.

EXAMPLE 7-8 Convert 8230 cm to meters.
  Solution Use the identity 1 m/100 cm = 1.
  Multiply 8230 cm by 1:

$$8230 \text{ cm} = 8230 \text{ cm} \times 1$$

Substitute 1 m/100 cm for 1:

$$8230 \text{ cm} = 8230 \text{ c\cancel{m}} \times \frac{1 \text{ m}}{100 \text{ c\cancel{m}}}$$

$$\therefore \quad 8230 \text{ cm} = 82.3 \text{ m}$$

RULE 7-2. UNIT ANALYSIS TECHNIQUE

Unit analysis is generally applied by:

1. Expressing an identity (identical equation) that has the desired units.
2. Transforming the identity into a ratio having as the numerator the desired units.
3. Multiplying the unit to be acted on by the ratio.
4. Writing the resulting product with the desired units.
5. Repeating the preceding steps if more than one conversion is needed.

EXAMPLE 7-9 Convert $\frac{1}{2}$ hour (h) to seconds (s).

Solution

$$1 \text{ h} = 60 \text{ min and } 1 \text{ min} = 60 \text{ s}$$

Converting hours to minutes:

$$1/2\cancel{h} \times \frac{60 \text{ min}}{1 \cancel{h}} = \frac{60 \text{ min}}{2}$$

Converting minutes to seconds:

$$\frac{60 \cancel{\text{min}}}{2} \times \frac{60 \text{ s}}{1 \cancel{\text{min}}} = \frac{60 \times 60 \text{ s}}{2} = 1800 \text{ s}$$

$$\therefore \quad \tfrac{1}{2} \text{ h} = 1800 \text{ s}$$

Electrical energy is measured in joules; however, for some purposes this is a very small unit, so commercial power companies sell energy in a larger unit, the kilowatt-hour (kWh). Figure 7-10 pictures a kilowatt-hour meter.

FIGURE 7–10
Kilowatt-hour meter used to measure electrical energy.

EXAMPLE 7-10 Using unit analysis, show that the kilowatt-hour is an energy unit by converting 1 kWh to joules.

Solution We need an identity containing watts and joules:

$$1 \text{ W} = 1 \text{ J/s}$$

Written as a ratio:

$$\frac{1 \text{ J}}{1 \text{ W} \cdot \text{s}} = 1$$

We need a second identity containing seconds and hours:

$$\frac{3600 \text{ s}}{1 \text{ h}} = 1$$

Use these identities:

$$1 \text{ kWh} = 1 \times 10^3 \text{ Wh} \times \frac{3600 \text{ s}}{\text{h}} \times \frac{1 \text{ J}}{1 \text{ W} \cdot \text{s}}$$

Remove the common units (hour, watt, second):

$$1 \text{ kWh} = 1 \times 10^3 \text{ W}\cancel{\text{h}} \times \frac{3600 \cancel{\text{s}}}{\cancel{\text{h}}} \times \frac{1 \text{ J}}{1 \cancel{\text{W}} \cdot \cancel{\text{s}}}$$
$$= 3.6 \times 10^6 \text{ J}$$

$$1 \text{ kWh} = 3.6 \text{ MJ}$$
$$\therefore \quad \text{kilowatt-hour is an energy unit.}$$

Unit analysis may also be used when converting between systems. Table 7-6 gives several conversion factors (identities). There are many intersystem conversion factors listed in engineering, physics, and chemical handbooks.

EXAMPLE 7-11 Convert 10 lbm to kilograms.

Solution

$$1 \text{ kg} = 2.205 \text{ lbm}$$

$$10 \text{ lbm} \times \frac{1 \text{ kg}}{2.205 \text{ lbm}} = 4.535 \text{ kg}$$

$$\therefore \quad 10 \text{ lbm} = 4.535 \text{ kg}$$

EXAMPLE 7-12 Compute the area of a circle in square millimeters (mm²) if the radius is 1.00 inch. The formula for area is $A = \pi r^2$, where A is area and r is the radius.

Solution Compute the radius in centimeters:

$$1 \text{ in.} = 2.54 \text{ cm}$$

$$r = 1.0 \text{ in.} \times \frac{2.54 \text{ cm}}{1 \text{ in.}}$$

$$= 2.54 \text{ cm}$$

Convert centimeters to millimeters:

$$10 \text{ mm} = 1 \text{ cm}$$

$$r = 2.54 \text{ cm} \times \frac{10 \text{ mm}}{1 \text{ cm}}$$

$$= 25.4 \text{ mm}$$

TABLE 7-6

CONVERSION FACTORS			
1 inch	2.540 centimeters	1 gallon	3.785 liters
1 meter	39.37 inches	1 cubic yard	0.7646 cubic meter
1 millimeter	0.0394 inch	1 Btu	1055 joules
1 ounce	28.35 grams	1 horsepower	745.7 watts
1 kilogram	2.205 pounds mass (lbm)		

Substitute:

$$A = \pi r^2 = \pi(25.4)^2$$
$$= 3.14 \times 645.2$$
$$\therefore \quad A = 2026 \text{ mm}^2$$

In Example 7-12, the answer was given in square millimeters (mm²), which is the recommended way of indicating square measurement.

EXERCISE 7-4

Apply unit analysis to the solution of the following problems. Express the answers to three significant figures, and use powers of ten notation where needed to express this significance.

1. Express 17.2 kilometers (km) in meters (m).
2. Express 5.00 grams (g) in milligrams (mg).
3. Express $\frac{1}{50}$ second (s) in milliseconds (ms).
4. Express 0.023 square meter (m²) in square centimeters (cm²): 1 m² = 1 × 10⁴ cm².
5. Express 2.38 × 10⁴ square millimeters (mm²) in square centimeters (cm²).
6. Express 1.16 hours (h) in seconds (s).
7. Convert 9.76 British thermal units (Btu) to joules (J).
8. Convert 16 horsepower (hp) to kilowatts (kW).
9. Convert 0.738 watt-hours (Wh) to joules (J).
10. Convert 3.2 pounds mass (lbm) to grams (g).
11. The velocity (speed) of sound in dry air is 1087 feet per second (ft/s). Compute the velocity of sound in centimeters per second (cm/s).
12. The velocity (speed) of light is given as 186.7 × 10³ miles per second. Compute the velocity of light in centimeters per second (cm/s).
13. Derive the conversion factor for yards (yd) to meters (m).
14. Derive the conversion factor for quarts (qt) to liters (l).

SELECTED TERMS

base unit   The building blocks of the measurement system. A precise standard that gives an exact value of the unit.

derived unit Formed from base units. Usually given names other than the derived unit names--Watt, volt, ohm, etc., are examples of names of derived units.

SI International system of units. The metric system.

unit analysis Technique used in reduction within a system of measurement, or in conversion between systems of measurement, in which conversion factors are expressed as fractions having values equal to 1.

8

APPLYING MATHEMATICS TO ELECTRICAL CIRCUITS

150

In this chapter we will be concerned with using the mathematical concepts presented in the previous chapters to solve problems associated with electrical circuits.

The behavior of an electrical circuit is described by Ohm's law. This law relates the three parameters of an electrical circuit: the electromotive force, the electrical current, and the electrical resistance. Each of these three important parameters will be used along with Ohm's law to compute circuit characteristics and conditions.

8-1 CURRENT, VOLTAGE, AND RESISTANCE

Current

Electrical current may be thought of as a movement of electrons within the electrical circuit. It takes the charge possessed by a very large number of electrons moving in a circuit to have a measurable current. Electrical current is measured in amperes.

One ampere of current flows when the charge possessed by 6.24×10^{18} electrons passes a point in a circuit in one second. Mathematically, this is stated as Equation 8-1.

$$I = Q/t \quad \text{(amperes)} \qquad (8\text{-}1)$$

where I = current in amperes (A)
$\quad Q$ = quantity of charge in coulombs (C)
$\quad t$ = time in seconds (s)

From Equation 8-1 we learn that electrical current is the rate at which electrons move through a circuit, and that one ampere and one coulomb per second are the same (1 A = 1 C/s).

EXAMPLE 8-1 What is the current in a circuit where 325 C of charge passes a point in 25 s?

Solution Use Equation 8-1. Substitute 325 C for Q and 25 s for t:

$$I = Q/t = 325/25 = 13 \text{ A}$$
$$\therefore \quad I = 13 \text{ A}$$

Electric current is measured by an instrument called an ammeter. See Figure 8-1.

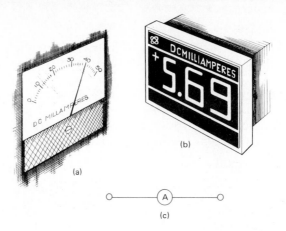

FIGURE 8–1
**Measuring current flow: (a) ammeter; (b) digital ammeter;
(c) schematic symbol for ammeter.**

Voltage

The word voltage is used to indicate both the *rise* in potential due to a generator or a battery, as pictured in Figure 8-2, and the *fall* in potential when current flows through the circuit resistance. Since the word voltage is so common in the language of electronics, we will make a conscious effort to indicate a difference between a voltage rise and a voltage drop.

The letter *E* is used to indicate a *voltage rise*. The rise in potential created by the battery is the electromotive force (emf) that causes current to flow through the resistance of the circuit. The letter *E* is used to indicate a voltage source.

The letter *V* is used to indicate a *voltage drop*. A voltage drop occurs in a circuit when current flows through an electrical load, such as the lamp in Figure 8-3(b).

From Figures 8-3(a) and (b) you may see that in each figure the voltmeter connected across the source voltage registers 9 V. However, the voltmeter across the lamp registers 9 V only when current is flowing in the circuit. **There can only be a voltage drop in an electrical circuit when there is a current flow.**

A voltmeter is used to make voltage measurements. When making voltage measurements, the voltmeter is placed across the load in the circuit. Two voltmeters are pictured in Figure 8-4.

Voltage is expressed as energy per unit charge (joule/coulomb) as stated in Equation 8-2.

$$E \text{ (or } V) = W/Q \qquad \text{(volts)} \qquad (8\text{-}2)$$

where E (or V) = potential difference in volts (V)
W = energy (work) in joules (J)
Q = amount of electric charge in coulombs (C)

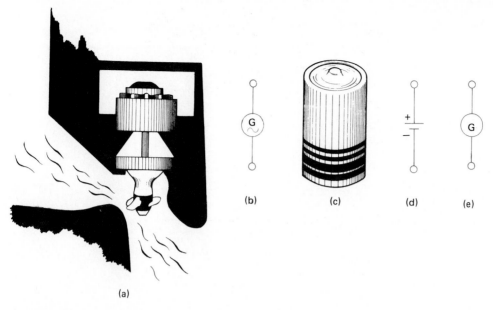

(b) (c) (d) (e)

FIGURE 8–2
Direct and alternating sources of electromotive force: (a) hydroelectric alternating-current generator; (b) schematic symbol for an alternating-current generator; (c) direct-current battery (cell); (d) schematic symbol for any direct-current source; (e) universal schematic symbol for either an ac or dc generator.

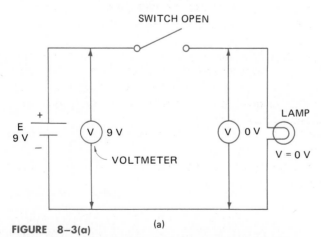

SWITCH OPEN

E
9 V

+

−

V 9 V

VOLTMETER

V 0 V

LAMP

V = 0 V

(a)

FIGURE 8–3(a)
The source is indicated by the letter E. With the switch open no current flows and the lamp is not lit. There is no potential drop (voltage) seen across the lamp. The meter reads 0 V.

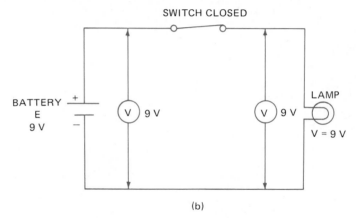

SWITCH CLOSED

BATTERY
E
9 V

LAMP
V = 9 V

(b)

FIGURE   8–3(b)
With the switch closed, current flows, the lamp lights, and there is potential drop (voltage) seen across the lamp. The meter indicates 9 V.

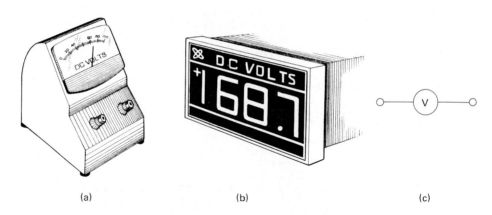

(a)   (b)   (c)

FIGURE   8–4
Measuring voltage: (a) dc voltmeter; (b) digital voltmeter (DVM); (c) schematic symbol for a voltmeter.

EXAMPLE 8-2   What potential difference (voltage) will be measured across the lamp of Figure 8-5 if 150 mC of charge flows through the lamp and releases 0.900 J of energy as heat and light?

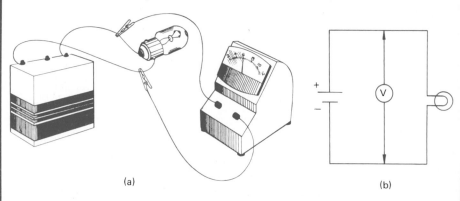

FIGURE 8–5
Circuit for Example 8–2: (a) pictorial diagram; (b) schematic diagram.

Solution   Use Equation 8-2. Substitute 0.900 J for W and 150 mC for Q:

$$V = W/Q = \frac{0.900}{0.150} = 6.00 \text{ V}$$

∴   The potential difference (voltage) across the lamp is 6.00 V.

EXAMPLE 8-3   The current of Figure 8-5 was measured at a constant 150 mA for 2.0 min. What is the voltage of the battery if 108 J of energy was used?

Solution   Solve Equation 8-1 for Q:

$$I = Q/t$$
$$Q = I(t)$$

Substitute 150 mA for I and 2.0 min for t:

$$Q = 150 \times 10^{-3} \times 2.0 \text{ min} \times \frac{60 \text{ s}}{1 \text{ min}}$$
$$= 150 \times 10^{-3} \times 2.0 \times 60$$
$$Q = 18 \text{ C}$$

Use Equation 8-2. Substitute 108 J for W and 18 C for Q:

$$E = W/Q = 108/18 = 6.0 \text{ V}$$

$\therefore$ The source voltage is 6.0 V.

Resistance

The resistance of an electrical circuit opposes the flow of electric current. Resistance is related to voltage drop and current as in Equation 8-3.

$$R = V/I \quad \text{(ohms)} \tag{8-3}$$

where R = resistance in ohms (Ω)
      V = voltage drop across R in volts (V)
      I = current through R in amperes (A)

Because all parts of an electrical circuit offer some opposition to current flow, the wires connecting the voltage source to the resistive load are selected to have a very small resistance. These connecting wires are referred to as conductors because of their high *conductance* (or low resistance). Copper is usually used as a conductor in electrical circuits because of its high *conductivity*.

When computing the resistance of a circuit, we often neglect the small resistance of the conductors. We may assume the connecting wires to be perfect conductors. This assumption is allowable because the electrical load (see Figure 8-6) presents a much larger opposition to the flow of current than do the conductors.

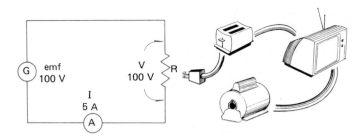

FIGURE 8–6
An electrical circuit with several loads shown. $R = V/I = 100/5$ = 20 Ω. The resistance of the connecting wires has been neglected because it is much smaller than the load resistance.

EXAMPLE 8-4 The television set of Figure 8-7 is connected to a 120-V source and the current is 2.50 A. Compute the resistance of the set.

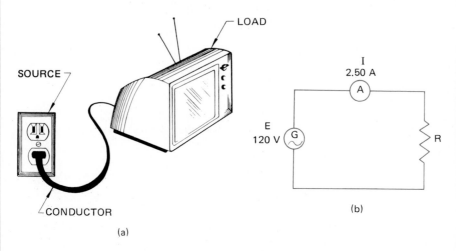

FIGURE 8-7
Circuit for Example 8-4: (a) pictoral diagram; (b) schematic diagram.

Solution Use Equation 8-3. Substitute 120 V for *V* and 2.50 A for *I*:

$$R = V/I = 120/2.50 = 48.0 \ \Omega$$

∴ The resistance of the load (TV set) is 48.0 Ω.

Because electronics is concerned with amplifying, attenuating, shaping, and generating signals for communication, computation, and control, known amounts of resistance are added to the circuit to control the flow of current. This resistance is in the form of a resistive device called a *resistor*. Figure 8-8 shows two types of fixed resistors used in electronic circuits. The parameters of electrical circuits discussed in this section are summarized in Table 8-1.

EXERCISE 8-1

1. Express 100 coulombs per hour (100 C/h) in amperes.
2. Express 8.00 mC/min in amperes.
3. A certain battery is charged with a steady current of 500 mA for 5.00 min. How much charge is accumulated in the battery?

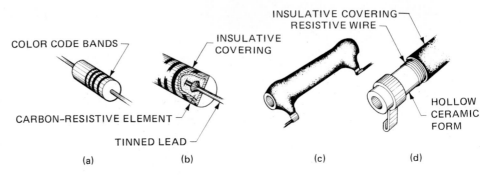

COLOR CODE BANDS

INSULATIVE COVERING

INSULATIVE COVERING
RESISTIVE WIRE

HOLLOW CERAMIC FORM

CARBON-RESISTIVE ELEMENT

TINNED LEAD

(a) (b) (c) (d)

FIGURE 8–8

Two common types of fixed resistors used to control current flow: (a) carbon composition resistor; (b) cutaway of a carbon resistor; (c) wire-wound resistor; (d) cutaway of a wire-wound resistor.

TABLE 8-1

PARAMETERS OF AN ELECTRICAL CIRCUIT

Physical Quantity	*Unit*	*Unit Symbol*	*Schematic Symbol*	*Formula*
Current	ampere	A	I	$I = Q/t$
Potential difference rise	volt	V	E	$E = W/Q$
Potential difference drop	volt	V	V	$V = W/Q$
Resistance	ohm	Ω	R	$R = V/I$

4. How many minutes will it take 1.00 kC to pass through an electrical conductor if the current is 2.50 A?

5. Express 100 joules per 8.33 coulombs (100 J/8.33 C) in volts.

6. Express 6.52 kJ/4.35 kC in volts.

7. A certain battery can move 14.0 C of charge through a circuit by supplying 126 J of energy. What is the voltage of the battery?

8. How much voltage drop will be measured across the terminals of a dc motor if 62.2 C of charge is moved by expending 1.49 kJ of energy?

9. Express 440 V/15 A in ohms.

10. What is the resistance of an electrical circuit that has a measured source voltage of 120 V and a current of 227 mA?

8-2 OHM'S LAW

Over a century ago a German physicist, Georg Simon Ohm, observed that the intensity of the current in an electrical conductor was dependent on the electromotive force producing it. He observed that doubling the source voltage doubled the current and that tripling the source voltage tripled the current. From these observations, a formula was written that describes the relationship between the emf and the current flow. This formula is called *Ohm's law*.

$$E = IR \qquad \text{(volts)} \qquad\qquad (8\text{-}4)$$

Other forms of Ohm's law are:

$$I = E/R \qquad \text{(amperes)} \qquad\qquad (8\text{-}5)$$
$$R = E/I \qquad \text{(ohms)} \qquad\qquad (8\text{-}6)$$

EXAMPLE 8-5 How much voltage will be measured across the terminals of the battery of Figure 8-9 if 150 mA is passing through the 150-Ω resistor?

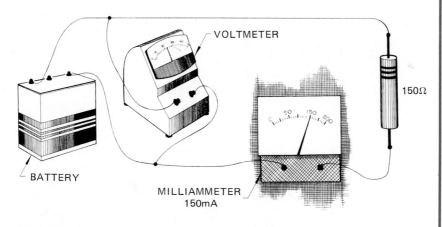

FIGURE 8–9
Pictorial diagram of the connection of the components and meters in the electrical circuit for Example 8– 5.

Solution Draw a schematic diagram and label it with the information given. See Figure 8-10.
Select the appropriate equation: $E = IR$.

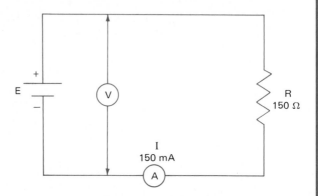

FIGURE 8–10
Schematic diagram of the circuit for Example 8–5.

Substitute 150 mA for I and 150 Ω for R:

$$E = 150 \times 10^{-3} \times 150$$
$$\therefore \quad E = 22.5 \text{ V}$$

EXAMPLE 8-6   In the circuit pictured by Figure 8-9, the source voltage is 22.5 V. Determine the circuit current when the resistance is increased from 150 Ω to 2.70 kΩ.

Solution   Draw a schematic diagram and label it with the information given. See Figure 8-11.

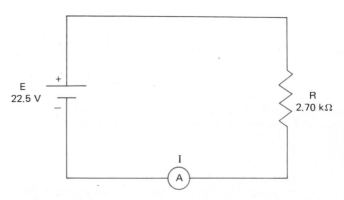

FIGURE 8–11
Schematic diagram of the circuit for Example 8–6.

Select the appropriate equation: $I = E/R$.
Substitute 22.5 V for E and 2.70 kΩ for R:

$$I = \frac{22.5}{2.70 \times 10^3}$$
$$= 0.008\ 33$$
$$\therefore \quad I = 8.33 \text{ mA}$$

EXAMPLE 8-7   In the circuit pictured by Figure 8-9 the emf is increased to 90 V. What must the circuit resistance be in order to maintain the current at 150 mA?

Solution   Draw a schematic diagram and label it with the information given. See Figure 8-12.

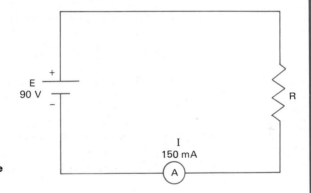

FIGURE 8-12
Schematic diagram for the circuit for Example 8-7.

Select the appropriate equation: $R = E/I$.
Substitute 90.0 V for E and 150 mA for I:

$$R = \frac{90}{150 \times 10^{-3}}$$
$$\therefore \quad R = 600 \text{ Ω}$$

EXERCISE 8-2
In solving each of the following, it is suggested that you (1) draw a schematic diagram and label it with the information given, (2) state the equation being used, and (3) show the solution to the problem.

1. Calculate each voltage drop (V) given the following information:
   (a) 10 A passing through 100 Ω
   (b) 6.06 mA passing through 3.30 kΩ
   (c) 15 μA passing through 1.9 MΩ
   (d) 0.732 A passing through 320 Ω
2. Calculate each current (I) given the following information:
   (a) 100 V across 10 Ω
   (b) 15.0 V across 560 Ω
   (c) 2.2 kV across 4.7 MΩ
   (d) 176 μV across 22.0 Ω
3. A milliammeter connected in a circuit reads 74 mA and a voltmeter connected across the voltage source reads 45 V. What is the resistance of the circuit?
4. A voltmeter is connected across the leads of a 10-kΩ resistor. What should the voltmeter reading be if 9.3 mA of current is flowing through the resistor?
5. The voltage across the heating element of an electric heater is 120 V. If the resistance of the heater is 15 Ω, what should an ammeter indicate when it is placed in the circuit?
6. The terminal of an automobile battery is severely corroded, which causes resistance at the terminals. A voltmeter placed from the terminal post of the battery to the cable (see Figure 8-13) indicates 9.5 V

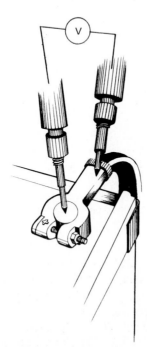

FIGURE 8–13
Measuring the voltage drop across the terminal of a battery.

when 15 A of current is flowing through the cable. Compute the resistance of the connection.

7. Two wires were spliced (connected) together by soldering. However, the connection was poorly made and resulted in a resistive connection (a cold solder joint). Compute the resistance of the connection if 2.3 V is measured across the splice when 0.92 A is flowing.

8. A 20-horsepower motor that drives a pump is connected to the source voltage with wiring that has a total resistance of 432 mΩ. Compute the voltage drop across the wiring when 40 A of current flows.

9. Compute the *hot* resistance of the soldering iron pictured in Figure 8-14 when it is plugged into a 117-V source. The current rating is 300 mA at 117 V.

10. A digital clock requires 84 mA for operation when plugged into a 120-V outlet. Compute the resistance of the clock.

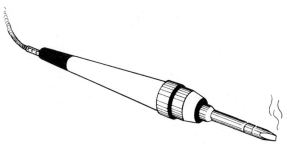

FIGURE 8–14
Soldering iron for Exercise 8–2, problem 9.

8-3 RESISTANCE IN A SERIES CIRCUIT

The circuits we have worked with up until now have been concerned with a single resistive load. We will now consider several resistive loads connected so that they form a *series circuit.*

Series Circuit Defined

A series circuit is formed when each component is successively connected so that one end of a component is connected to an end of the next component, and so forth, until a complete path is formed from the beginning to the end. An example of a series circuit is shown in Figure 8-15.

One way to determine if a circuit is connected in series is to *trace* the flow of current from the voltage source through the loads and back to the

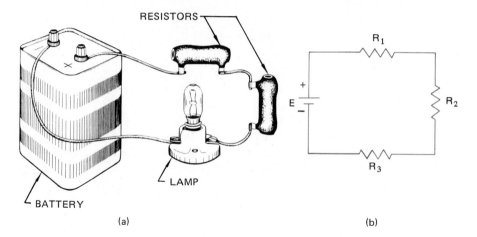

R_1

E +
−

R_2

R_3

LAMP

BATTERY

(a)

(b)

FIGURE 8–15
Series circuit consisting of a battery, two wire-wound resistors, and a lamp: (a) pictorial diagram of the connection; (b) schematic diagram with the resistive loads shown and labeled with a subscript variable.

source. The components are in series if the current flows through each component in turn (one after the other) before returning to the voltage source.

An important characteristic of a series circuit is that the circuit current, I, is the same throughout the entire series circuit at the same time. This is stated mathematically as Equation 8-7.

$$I = I_1 = I_2 = I_3 \quad \text{(amperes)} \tag{8-7}$$

Total Resistance in a Series Circuit

The total resistance of a series circuit, R_T, is found by summing (adding up) each of the resistances in the circuit. Thus,

$$R_T = R_1 + R_2 + R_3 \quad \text{(ohms)} \tag{8-8}$$
$$R_T = \sum R \quad \text{(ohms)} \tag{8-9}$$

R_T is the total resistance, and ΣR is read "the sum of the resistances."

EXAMPLE 8-8   Determine the total resistance, R_T, of the circuit pictured in Figure 8-16. $R_1 = 50\ \Omega$, $R_2 = 20\ \Omega$, and $R_3 = 130\ \Omega$.

Solution   Sum the resistances: $R_T = \Sigma R$.

$$R_T = 50 + 20 + 130$$
$$\therefore \quad R_T = 200\ \Omega$$

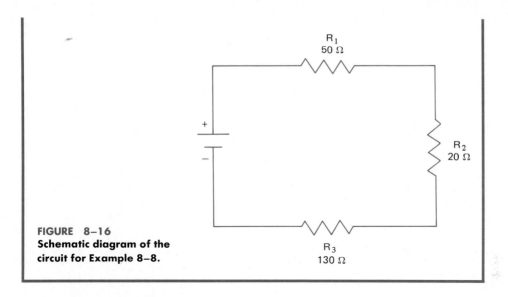

FIGURE 8–16
Schematic diagram of the circuit for Example 8–8.

R_1
50 Ω

R_2
20 Ω

R_3
130 Ω

Series Equivalent Circuit

A series circuit made up of several resistances may be simplified to a circuit with a single resistance having the value of R_T. This single resistance is the *equivalent resistance* of the equivalent circuit. It is labeled R_{eq} for the equivalent resistance. The *equivalent series circuit* has the same source voltage, current, and total resistance as the series circuit from which it was derived.

EXAMPLE 8-9   Draw and label an equivalent series circuit for the circuit of Figure 8-17.

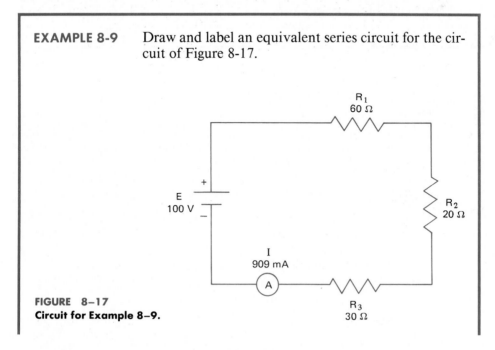

FIGURE 8–17
Circuit for Example 8–9.

R_1
60 Ω

E
100 V

R_2
20 Ω

I
909 mA

A

R_3
30 Ω

Solution   Use Equation 8-9.

$$R_T = 60 + 20 + 30$$
$$= 110 \ \Omega$$

Since R_{eq} is equal to R_T:

$$R_{eq} = 110 \ \Omega$$

Draw and label the schematic of the equivalent circuit. See Figure 8-18.

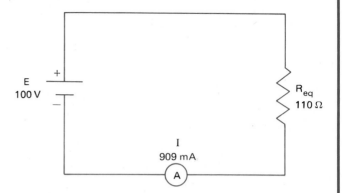

FIGURE 8–18
Equivalent circuit of Figure 8–17; $R_{eq} = R_T$.

EXERCISE 8-3

1. Three resistances, $R_1 = 4.7 \ \text{k}\Omega$, $R_2 = 12 \ \text{k}\Omega$, and $R_3 = 6.8 \ \text{k}\Omega$, are connected in series across an ac source of voltage. Determine the total resistance, R_T, of the circuit.
2. Find the total resistance, R_T, of the following series-connected carbon resistors: 3.0 kΩ, 0.220 MΩ, 15 kΩ, 180 kΩ, 22 kΩ, and 0.560 MΩ.
3. A wire-wound resistor is connected in series with a lamp. What is the combined resistance (R_T) of the two resistances if the lamp resistance is 2.0 kΩ and the resistance of the resistor is 750 Ω?
4. Draw and label a schematic of an equivalent circuit of a series circuit consisting of a generator of 6.3 V and two resistors, one of 20 Ω and the other of 10 Ω. The circuit current is 210 mA.

5. Three carbon-composition resistors ($R_1 = 120$ kΩ, $R_2 = 91$ kΩ, and $R_3 = 62$ kΩ) are connected in series across a 220-V dc source. If $I = 806$ μA:
   (a) Determine R_T.
   (b) Determine current in R_3.
   (c) Draw and label the schematic of the equivalent circuit.
6. To limit the circuit current to 262 mA, a 100-Ω wire-wound resistor is connected in series with a 40-W, 358-Ω soldering iron. If $E = 120$ V:
   (a) Determine R_T.
   (b) Draw and label the schematic of the circuit.
   (c) Draw and label the schematic of the equivalent circuit.

8-4 APPLYING OHM'S LAW

Kirchhoff's Voltage Law

In a series circuit the sum of the voltage drops is equal to the voltage source. Gustav Kirchhoff is credited with the discovery of this important circuit characteristic. Stated mathematically,

$$E = V_1 + V_2 + V_3 \quad \text{(volts)} \tag{8-10}$$
$$E = \sum V \quad \text{(volts)} \tag{8-11}$$

where $\sum V$ is read "the sum of the voltage drops."

EXAMPLE 8-10   Verify Kirchhoff's voltage law for the circuit of Figure 8-19 (next page), where $V_1 = 0.360$ V, $V_2 = 1.05$ V, $V_3 = 2.00$ V, $V_4 = 2.87$ V, $V_5 = 3.72$ V, and $E = 10.0$ V.

Solution   Use Equation 8-10: $E = V_1 + V_2 + V_3 + V_4 + V_5$.

$$10.0 = 0.360 + 1.05 + 2.00 + 2.87 + 3.72$$
$$10.0 = 10.0$$

∴   The sum of the voltage drops equals the voltage source

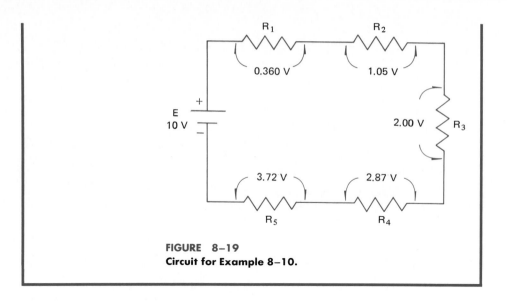

FIGURE 8–19
Circuit for Example 8–10.

Applying Ohm's Law

When Ohm's law is applied to a series circuit, the symbol for resistance in the Ohm's law formula is R_T, which indicates that it is the summation of *all* the resistances in the series circuit.

$$E = IR_T \quad \text{(volts)} \tag{8-12}$$

where E = source voltage in volts (V)
I = circuit current in amperes (A)
R_T = summation of the circuit resistance in ohms (Ω)

EXAMPLE 8-11   Compute the circuit current (I) in a series circuit that has an applied voltage of 22.5 V and resistance of R_1 = 50 Ω, R_2 = 150 Ω, and R_3 = 100 Ω. Draw and label a schematic of the equivalent circuit.

Solution   Draw and label a schematic diagram for the given conditions. See Figure 8-20.
Compute the total resistance, R_T:

$$R_T = \sum R$$
$$= 50 + 150 + 100$$
$$R_T = 300 \ \Omega$$

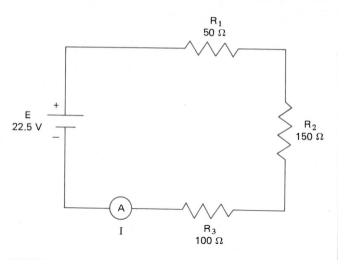

FIGURE 8–20
Circuit for Example 8–11.

Compute the circuit current, I, using Equation 8-12:

$$I = 22.5/300 = 0.075$$
$$\therefore \quad I = 75 \text{ mA}$$

Draw and label the schematic for the equivalent circuit. See Figure 8-21.

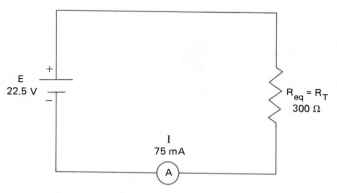

FIGURE 8–21
Equivalent circuit of Figure 8–20.

EXAMPLE 8-12 A 6.30-V, 150-mA lamp is to be operated from a 120-V source as shown in Figure 8-22. Compute the value of the series resistor (R_1) needed to drop the excess voltage and limit the current to 150 mA.

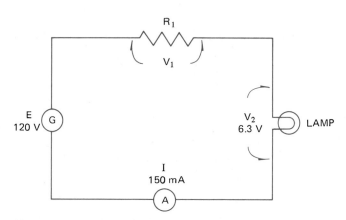

FIGURE 8–22
Circuit for Example 8–12.

Solution To solve for R_1, the values of V_1 and I are needed. Solve for V_1 using Equation 8-10: $E = V_1 + V_2$.

$$V_1 = E - V_2$$

Substitute $E = 120$ V and $V_2 = 6.30$ V:

$$V_1 = 120 - 6.30$$
$$= 114 \text{ V}$$

Solve for R_1. I is given as 150 mA:

$$R_1 = V_1/I$$
$$= 114/(150 \times 10^{-3})$$
$$R_1 = 760 \ \Omega$$

∴ 760 Ω is needed to limit the current to 150 mA in the circuit of Figure 8-22.

Computing the Voltage Drop Across One Series Resistance

The voltage drop across one resistance in a series circuit is computed by Ohm's law as Equation 8-13.

$$V_n = IR_n \quad \text{(volts)} \tag{8-13}$$

where V_n = voltage in volts (V) across selected resistance, R_n
I = circuit current in amperes (A)
R_n = selected resistance in ohms (Ω)
n = number of the selected resistor

EXAMPLE 8-13 Two carbon-composition resistors are connected in series with a generator, as shown in the schematic of Figure 8-23. The current throughout the circuit is 500 μA.
(a) Compute the voltage drop across each resistor.
(b) Compute E, the generator voltage.

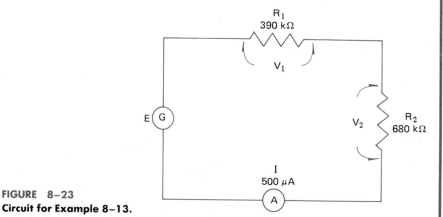

FIGURE 8–23
Circuit for Example 8–13.

Solution (a) Use Equation 8-13: $V_1 = IR_1$. Substitute $I = 500$ μA and $R_1 = 390$ kΩ:

$$V_1 = 500 \times 10^{-6} \times 390 \times 10^3$$
$$V_1 = 195 \text{ V}$$

$V_2 = IR_2$. Substitute $I = 500\ \mu A$ and $R_2 = 680\ k\Omega$:

$$V_2 = 500 \times 10^{-6} \times 680 \times 10^3$$
$$V_2 = 340\ V$$
$$\therefore\quad V_1 = 195\ V, \quad V_2 = 340\ V$$

(b) Use Equation 8-10: $E = V_1 + V_2$.

$$E = 195 + 340$$
$$= 535\ V$$
$$\therefore\quad E = 535\ V$$

EXERCISE 8-4

1. Determine the source voltage of a series circuit having voltage drops of 12, 7.5, and 22.5 V.
2. Determine the source voltage of a series circuit having voltage drops of 238 mV, 686 mV, 52 mV, and 1.024 V.
3. Two resistors are connected in series across a generator; determine the voltage drop across R_2 if the source voltage is 65 V and V_1 is 28 V.
4. Two resistors, $R_1 = 6\ \Omega$ and $R_2 = 24\ \Omega$, are connected in series with a battery. The circuit current is 400 mA. Determine V_1 and E.
5. Three resistors, $R_1 = 200\ \Omega$, $R_2 = 500\ \Omega$, and $R_3 = 800\ \Omega$, are connected in series across a 120-V source. The circuit current is 80 mA. Show that $E = V_1 + V_2 + V_3$.
6. Two resistors, $R_1 = 100\ \Omega$ and $R_2 = 50\ \Omega$, are connected across an ac source. The voltage dropped by R_2 is 28 V. Determine the current through R_1.

TABLE 8-2

CHARACTERISTICS OF A SERIES CIRCUIT

Equation	Comment
$I = I_1 = I_2 = I_3$	The current (I) is the same throughout the series circuit
$R_T = R_1 + R_2 + R_3$	Total resistance (R_T) is the sum of the individual resistances (ΣR)
$E = V_1 + V_2 + V_3$	Source voltage (E) is the sum of the voltage drops (ΣV)
$E = IR_T$	Ohm's law for a series circuit
$V_n = IR_n$	Voltage across the selected resistance, R_n
$R_{eq} = R_T$	Equivalent resistance (R_{eq}) of a series circuit is the total resistance

8-5 SUMMARY OF THE SERIES CIRCUIT

The concepts of the preceding sections of this chapter are summarized in Table 8-2. Refer to this table while studying the following examples and working the additional exercises.

EXAMPLE 8-14 Four resistors, 270, 220, 330, and 180 Ω, are connected in series across a 10-V source. Draw a schematic diagram of the series circuit and determine:
(a) The total resistance (R_T)
(b) The circuit current (I)
(c) The voltage drop across each resistor
(d) That the source voltage equals the sum of the voltage drops
Draw and label a schematic of the equivalent circuit.

Solution Draw and label a schematic diagram. See Figure 8-24.
(a) Determine the total resistance: $R_T = \Sigma R$.

$$R_T = 270 + 220 + 330 + 180$$
$$= 1000 \ \Omega$$
$$\therefore \quad R_T = 1 \ k\Omega$$

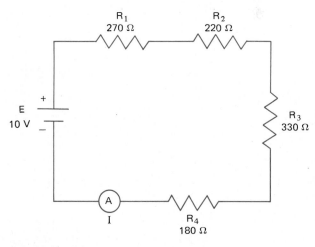

FIGURE 8–24
Schematic diagram for Example 8–14.

(b) Determine the circuit current: $I = E/R_T$.

$$I = 10/1 \times 10^3 = 10 \times 10^{-3} = 10 \text{ mA}$$
$$\therefore \quad I = 10 \text{ mA}$$

(c) Determine the voltage drop across each resistance.

$$V_1 = IR_1$$
$$= 10 \times 10^{-3} \times 270$$
$$\therefore \quad V_1 = 2.70 \text{ V}$$
$$V_2 = IR_2$$
$$= 10 \times 10^{-3} \times 220$$
$$\therefore \quad V_2 = 2.20 \text{ V}$$
$$V_3 = IR_3$$
$$= 10 \times 10^{-3} \times 330$$
$$\therefore \quad V_3 = 3.30 \text{ V}$$
$$V_4 = IR_4$$
$$= 10 \times 10^{-3} \times 180$$
$$\therefore \quad V_4 = 1.80 \text{ V}$$

(d) Determine that the source voltage equals the sum of the voltage drops: $E = \Sigma V$.

$$10 = 2.70 + 2.20 + 3.30 + 1.80$$
$$10 \text{ V} = 10 \text{ V}$$

They are equal.
Draw and label a schematic diagram of the equivalent series circuit. See Figure 8-25.

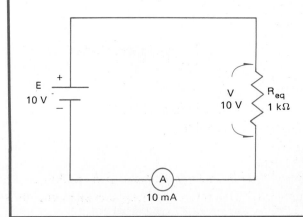

FIGURE 8–25
Schematic diagram of the equivalent circuit for Example 8–14.

1. Three carbon-composition resistors (1.5 kΩ, 3.0 kΩ, and 10 kΩ) are connected in series across a 145-V source. Draw a schematic diagram of the series circuit and determine:
   (a) The total resistance
   (b) The current throughout the circuit
   (c) The current through the 10-kΩ resistor
   (d) The voltage drop across the 3.0-kΩ resistor
2. Repeat problem 1 for a source voltage of 1.2 kV.
3. Four identical indicator lights of the type shown in Figure 8-26 are connected in series across a 112-V source. When lit, each lamp has a resistance of 700 Ω. Determine:
   (a) The total circuit resistance
   (b) The circuit current
   (c) The voltage across one of the lamps

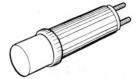

FIGURE 8–26
Subminiature indicator light.

4. A lamp similar to the one pictured in Figure 8-26 is to be operated from a 120-V source. If the lamp is specified as 28 V, 40 mA, what resistance is needed to limit the current to the rated value?
5. Two resistors are connected in series across a 90-V source. If a voltmeter connected across R_1 reads 22 V and the current through R_2 is 20 mA, determine:
   (a) The current through R_1
   (b) The voltage across R_2
   (c) The resistance of R_1
   (d) The total resistance
6. Three resistors are connected in series across a 100-V source. If $V_1 = 32$ V, $V_2 = 18$ V, and $R_3 = 47.0$ kΩ, determine:
   (a) The circuit current
   (b) The value of R_2
   (c) The total resistance across the voltage source
7. In the circuit of Figure 8-27, determine:
   (a) The circuit current (I)
   (b) The source voltage (E)
8. Repeat problem 7 for $R_1 = 220$ Ω, $R_2 = 620$ Ω, and $R_3 = 390$ Ω.
9. Four resistors are connected in series. The voltage across R_1 and R_2 is

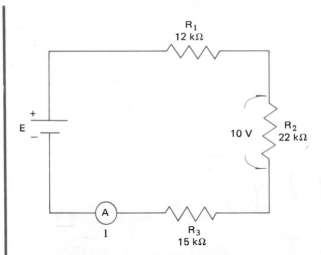

R_1
12 kΩ

E

+

−

10 V

R_2
22 kΩ

A

I

R_3
15 kΩ

FIGURE 8-27
*Circuit for Exercise 8-5,
problem 7.*

the same. The resistance of R_3 is 15 kΩ, and the voltage drop across R_4 is 20 V. If the source voltage is 150 V and the circuit current is 2.0 mA, determine R_1, R_2, and R_4.

10. Three resistors are connected in series across a generator. It is known that $R_1 = 110\ \Omega$. The resistance of R_3 is twice the resistance of R_1, R_2 has a resistance that is three times that of R_1, and the voltage drop across R_2 is 15 V. Compute the generator voltage.

SELECTED TERMS

conductivity   Indication of the ability of a material to pass an electric current.

current   Flow of electric charge along a conductor in a closed circuit.

equivalent resistance   The combining of several resistances in a circuit into a single resistance called the equivalent resistance, R_{eq}.

KVL   Kirchhoff's voltage law; the algebraic sum of the voltages around a closed circuit is equal to zero.

Ohm's law   Relationship between voltage and current formulated by Georg Simon Ohm; $E = IR$.

resistance   Ratio of voltage to current; that is, $R = V/I$. A substance has a resistance of 1 ohm when the application of 1 volt results in a current of 1 ampere.

voltage   Name given to energy added per unit of electric charge as the charge passes through a source.

9

APPLYING MATHEMATICS TO ELECTRICAL CONCEPTS

This chapter will be concerned with applying mathematics to electrical concepts, particularly power, efficiency, and cost of electrical energy.

9-1 POWER

Power was defined in Chapter 7 as the *rate* of producing or using energy. Power is also the *rate* of doing work. We learned that

$$P = W/t \quad \text{(watts)} \tag{9-1}$$

where P = power in watts (W)
$\quad W$ = work in joules (J)
$\quad t$ = time in seconds (s)

Remember that a joule per second (J/s) is a watt (W), and that power is the rate of doing work or using energy. Thus,

$$1 \text{ watt (W)} = 1 \text{ joule per second (J/s)}$$

EXAMPLE 9-1   Five kilojoules of energy was used in moving a mass of 170 kg a vertical distance of 3 m in 1.25 min. At what rate is energy being used?

Solution   Use Equation 9-1: $P = W/t$.

$$P = \frac{5.00k}{1.25} \frac{\text{J}}{\text{min}}$$

Convert joules per minute to joules per second:

$$P = \frac{5.00 \times 10^3}{1.25} \frac{\text{J}}{\text{min}} \times \frac{1 \text{ min}}{60 \text{ s}}$$
$$= 66.7 \text{ J/s}$$
$$\therefore \quad P = 66.7 \text{ W}$$

Power in Electrical Circuits

Equation 9-1 is not the best equation to use to compute power in an electrical circuit. A more useful formula is Equation 9-2.

$$P = \text{EI or } P = VI \quad \text{(watts)} \tag{9-2}$$

where P = power in watts (W)

$\quad\quad E$ = voltage rise in volts (V) or V = voltage drop in volts (V)

$\quad\quad I$ = current in amperes (A)

Since the product of E and I is an expression for power, it must have the units of power (joules per second). Exploring:

$$P = EI$$
$$E = W/Q$$
$$I = Q/t$$

Substituting into Equation 9-2:

$$P = \frac{W}{Q} \times \frac{Q}{t}$$

Simplify:

$$P = \frac{W}{\cancel{Q}} \times \frac{\cancel{Q}}{t} = \frac{W}{t}$$
$$P = W/t \quad \text{(joules/second)}$$
$$P = EI = W/t \quad \text{(watts)}$$

Additional forms of the formula for power may be derived by substituting forms of Ohm's law ($E = IR$ and $I = E/R$) into Equation 9-2. Thus, substitute IR for E in $P = EI$:

$$P = IR \times I$$
$$P = I^2R \quad \text{(watts)} \quad\quad\quad\quad (9\text{-}3)$$

Substitute E/R for I in $P = EI$:

$$P = E \times E/R$$
$$P = E^2/R \quad \text{(watts)} \quad\quad\quad\quad (9\text{-}4)$$

The power formulas are summarized in Table 9-1.

EXAMPLE 9-2 An electric iron is attached to a 120-V ac source. Determine the power rating of the iron if 11 A of current passes through the iron.

Solution Use Equation 9-2, $P = EI$:

$$P = 120 \times 11$$
$$= 1320 \text{ W}$$
$$\therefore \quad P = 1.32 \text{ kW}$$

TABLE 9-1

POWER FORMULAS	
Formula	**Equation Number**
$P = W/t$	(9-1)
$P = EI$ or $P = VI$	(9-2)
$P = I^2R$	(9-3)
$P = E^2/R$ or $P = V^2/R$	(9-4)

Power Considerations for Resistors

Resistors are rated by their resistive properties in ohms and by their rate of converting electric energy to heat energy in watts (commonly called power dissipation). Thus, a carbon-composition resistor may have a rating of 470 Ω, $\frac{1}{2}$ W, while a wire-wound resistor might be rated as 500 Ω, 10 W.

Although resistors may be operated near their power ratings, it generally is a good practice to operate them well below their rated value.

EXAMPLE 9-3   A 1.0-kΩ, $\frac{1}{2}$-W resistor has 25 V dropped across it. Determine:
(a) The power being dissipated by the resistor
(b) Whether the resistor is being operated within its rating

Solution   (a) Select the equation from Table 9-1 that relates power to voltage and resistance:

$$P = V^2/R$$
$$= 25^2/(1 \times 10^3)$$
$$\therefore \quad P = 0.625 \text{ W}$$

(b) Determine whether the resistor is being operated within its rating:

$$625 \text{ mW} > 500 \text{ mW}$$
$$\therefore \quad \text{No. The wattage rating is exceeded by 125 mW.}$$

EXAMPLE 9-4   Select the wattage rating for an 8-kΩ wire-wound resistor that has 39 mA passing through it. This resistor is available as an 8-, 12-, or 20-W unit.

Solution Select the equation from Table 9-1 that relates power to current and resistance:

$$P = I^2R$$
$$= (39 \times 10^{-3})^2 \times 8 \times 10^3$$
$$= 12.2 \text{ W}$$

$\therefore$ The 20-W unit is selected.

Total Power of Several Loads

The power dissipated in an electrical circuit may be computed by adding the power dissipation of each of the loads connected to the source. Thus,

$$P_T = P_1 + P_2 + P_3 \qquad \text{(watts)} \qquad (9\text{-}5)$$

where P_T is the total power dissipation in watts (W)

EXAMPLE 9-5 Two resistances ($R_1 = 200\ \Omega$ and $R_2 = 100\ \Omega$) are connected in series with a 180-V source. Determine:
(a) The power dissipation of each resistance
(b) The power dissipation of the total circuit
(c) That $P_T = P_1 + P_2$

Solution (a) Solving for I in Ohm's law, $I = E/R_T$:

$$I = 180/300 = 600 \text{ mA}$$

Substituting into Equation 9-3:

$$P_1 = I^2R_1$$
$$= (600 \times 10^{-3})^2 \times 200$$
$$\therefore \quad P_1 = 72.0 \text{ W}$$
$$P_2 = I^2R_2$$
$$= (600 \times 10^{-3})^2 \times 100$$
$$\therefore \quad P_2 = 36.0 \text{ W}$$

(b) Compute the total power (P_T) from the total resistance (R_T) and the source voltage:

$$P_T = E^2/R_T$$
$$= (180)^2/300$$
$$\therefore \quad P_T = 108 \text{ W}$$

(c) Compute P_T from Equation 9-5:

$$P_T = P_1 + P_2$$
$$= 72 + 36$$
$$P_T = 108$$

∴   The two methods for calculating P_T give the same answer.

EXERCISE 9-1

1. Determine the rate in watts at which 380 J of energy is converted into heat if the energy is delivered over a 3.00-min period.
2. Determine the time needed to deliver 1000 J of energy to a lamp if the power to the lamp is 40 W.
3. Determine the energy delivered to a 1000-W toaster for one week if it is used 4 min/day.
4. Determine the rate at which a resistor converts electric energy into heat if 3.0 kJ of heat is produced in 5.0 min.
5. Determine the energy delivered to a TV set for one month (30 days) if it is operated for 100 h/month. The set is supplied from a 120-V ac source, and it requires 1.30 A for operation.
6. Determine the power use in joules per second (J/s) of a lamp rated at 7.50 W, 120 V.
7. Determine the power rating of a fluorescent fixture that has a voltage drop of 117 V across it when 1.37 A of current is passing through it.
8. If carbon-composition resistors are available with $\frac{1}{2}$-W, 1-W, and 2-W ratings.
   (a) Determine the wattage dissipation for a 330-Ω resistor passing 56 mA.
   (b) Select one of the three specified ratings for this resistor.
9. Determine the maximum voltage that can be dropped across a 8.0-W, 1.25-kΩ resistor without overheating.
10. Determine the maximum current that can be passed through a $\frac{1}{2}$-W, 560-Ω resistor without exceeding its wattage rating.
11. The service entrance (as pictured in Figure 9-1) to a home is rated at 220 V, 50 A. Determine:
    (a) How much power is available to the home
    (b) If the following loads may be operated at the same time:
        (1) A 3.20-kW electric range
        (2) Refrigerator/freezer, 120 V, 3.30 A

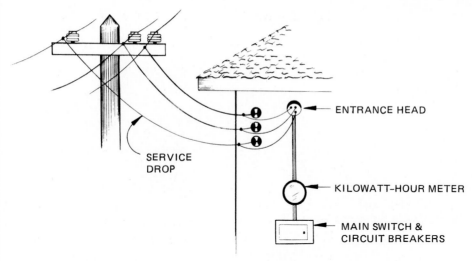

ENTRANCE HEAD

SERVICE
DROP

KILOWATT-HOUR METER

MAIN SWITCH &
CIRCUIT BREAKERS

FIGURE 9–1
Electrical service entrance for Exercise 9–1, problem 11.

   (3) A 1100-W toaster
   (4) A 4850-W clothes dryer
   (5) Electric motor, 220 V, 8.20 A

12. Three resistors, $R_1 = 62\ \Omega$, $R_2 = 91\ \Omega$, and $R_3 = 51\ \Omega$, are connected in series across a 12-V generator. Draw and label a schematic and determine:
 (a) The circuit current
 (b) The power dissipated by each resistor
 (c) The power provided by the generator
 (d) That $P_T = P_1 + P_2 + P_3$

9-2 EFFICIENCY

Evaluating Energy Conversion

Many devices have been invented to convert or transform electrical energy to other forms of energy. In the process of converting energy from one form to another, some energy is wasted, usually as heat. A basic principle of physics is the *law of conservation of energy,* which states that energy may be changed from form to form, but energy cannot be created nor destroyed. Stated mathematically, energy input = energy output + energy wasted. Figure 9-2 illustrates this concept. The motor converts electrical energy (W_{in}) to mechanical

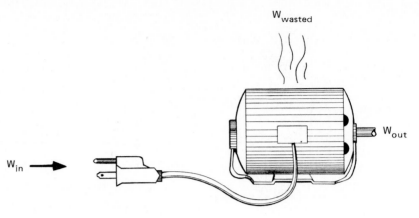

FIGURE 9–2
Conservation of energy.

energy (W_{out}). However, some energy is converted to heat, in this case a nonuseful form of energy (W_{wasted}), which is *lost* to the output. The *efficiency* of an energy-converting device indicates how well the device performs as an energy converter. Equation 9-6 is used to compute the efficiency of the device as an energy converter.

$$\eta = \frac{W_{out}}{W_{in}}$$

(9-6)

where η = efficiency of the device
W_{out} = *useful* output energy
W_{in} = input energy

Efficiency can also be calculated from the power input and the power output.

$$\eta = \frac{P_{out}}{P_{in}}$$

(9-7)

where P_{out} = output power
P_{in} = input power

Percent of Efficiency

Efficiency, η, may be expressed as a percentage (%). Thus,

$$\eta = \frac{P_{out}}{P_{in}} \times 100\%$$

(9-8)

EXAMPLE 9-6 A kettle containing 1 liter of water is brought to a boil in 5.0 min by an electric heating element rated at 1200 W. Determine the percent of efficiency of the heating element if the water has obtained 300 kJ of heat energy from the heating element.

Solution Use Equation 9-8:
Find the value for P_{in}:

$$P_{in} = 1200 \text{ W}$$

Find the value for P_{out}:

$$P_{out} = 300 \text{ kJ}/5 \text{ min}$$
$$= \frac{300 \text{ kJ}}{5 \text{ min}} \times \frac{1 \text{ min}}{60 \text{ s}}$$
$$= 1.0 \text{ kJ/s}$$
$$P_{out} = 1000 \text{ W}$$

Substitute into Equation 9-8:

$$\eta = \frac{1000}{1200} \times 100\%$$
$$\therefore \quad \eta = 83.3\%$$

Mechanical Power

The watt is the SI unit of power for both electrical and mechanical power. However, horsepower (hp) is also a commonly used unit for mechanical power. To convert horsepower to watts, use Equation 9-9:

$$1 \text{ hp} = 746 \text{ W} \qquad (9\text{-}9)$$

EXAMPLE 9-7 Determine the percent of efficiency of a 120-V electric motor that develops 0.50 hp when 3.57 A passes through the motor.

Solution Evaluate P_{in}:

$$P_{in} = EI$$
$$= 120 \times 3.57$$
$$\therefore \quad P_{in} = 428 \text{ W}$$

Evaluate P_{out}:

$$P_{out} = 0.50 \cancel{hp} \times \frac{746 \text{ W}}{1 \cancel{hp}}$$

$$\therefore \quad P_{out} = 373 \text{ W}$$

Substitute into Equation 9-8:

$$\eta = 373/428 \times 100\%$$

$$\therefore \quad \eta = 87.2\%$$

EXAMPLE 9-8   A certain electrical motor delivers 0.25 hp at 1800 revolutions per minute (rpm). The motor is attached to a 120-V source and is 78.5% efficient. What current is used during operation of the motor?

Solution   Use the efficiency equation to find the input power:

$$\eta = \frac{P_{out}}{P_{in}} \times 100\%$$

Solve for P_{in}:

$$P_{in} = \frac{P_{out}}{\eta} \times 100\%$$

Substitute:

$$P_{in} = \frac{0.25 \text{ hp}}{78.5\%} \times 100\%$$

Convert to watts:

$$P_{in} = \frac{0.25 \cancel{hp}}{78.5} \times 100 \times 746 \frac{\text{W}}{\cancel{hp}}$$

$$P_{in} = 238 \text{ W}$$

Find current from $P = EI$:

$$I = P/E$$

$$I = 238/120$$

$$\therefore \quad I = 1.98 \text{ A}$$

Efficiency of a System

When several energy-converting devices are connected together, they form a system. The generating and distribution system of Figure 9-3 is an example of an electrical system. The efficiency of each component in a system contributes to the overall percent of efficiency in the following manner:

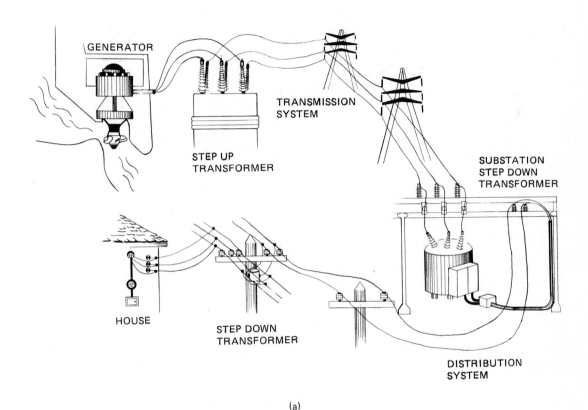

(a)

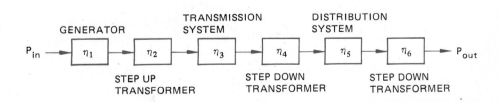

(b)

FIGURE 9–3

Generation and distribution system: (a) hydroelectric generation and distribution system; (b) cascaded system block diagram of the hydroelectric generation system.

$$\eta_{ov} = \eta_1 \times \eta_2 \times \eta_3 \times 100\% \qquad (9\text{-}10)$$

where η_{ov} = overall percent of efficiency of the system
η_n = efficiency of component n (NOTE: There may be any number of components in a system.)

In Figure 9-3(a), the hydro(water) energy is converted into mechanical energy by the turbine, which then drives the generator that converts mechanical energy to electrical energy. The overall percent of efficiency of any system decreases each time an additional conversion device is added to the system.

EXAMPLE 9-9   Determine the overall percent of efficiency of the generating system of Figure 9-3(b) if $\eta_1 = 0.89$, $\eta_2 = 0.98$, $\eta_3 = 0.96$, $\eta_4 = 0.98$, $\eta_5 = 0.97$, and $\eta_6 = 0.98$.

Solution

$$\eta_{ov} = \eta_1 \times \eta_2 \times \eta_3 \times \eta_4 \times \eta_5 \times \eta_6 \times 100\%$$
$$= 0.89 \times 0.98 \times 0.96 \times 0.98 \times 0.97$$
$$\times 0.98 \times 100\%$$
$$\therefore \quad \eta_{ov} = 78\%$$

Example 9-9 points out how dependent η_{ov} is on each efficiency in the system. We see that the overall efficiency is a product of each efficiency in a system, and it will always be smaller than the smallest efficiency in the system.

EXERCISE 9-2

1. Convert 58.0 hp to (a) watts and (b) kilowatts.
2. Convert 0.75 hp to (a) watts and (b) kilowatts.
3. A system for charging batteries is 82% efficient when delivering 14.4 V at 10 A to a battery being charged. Determine the input current taken from a 117-V ac source.
4. Determine the percent of efficiency of a 1.0-hp electric motor if 850 W of power is required to operate it.
5. A power supply for a solid-state amplifier has a 76% overall efficiency. Determine the power supplied to the amplifier by the power supply if 62 W is taken in by the power supply for its operation.
6. How much power must be supplied to an audio system having an overall efficiency of 8.0% if the audio power is equal to 5.0-W output?
7. Determine the power rating in watts of a stove heating element if 70% of the energy produced in 2 min went into heating a kettle of water, and 100 kJ of energy was delivered to the water during the 2-min period.

8. Two energy-converting devices with $\eta_1 = 88\%$ and $\eta_2 = 84\%$ are cascaded. Determine:
   (a) The output energy if 530 J of energy is put into the system
   (b) The overall efficiency of the cascaded system

9. The overall efficiency of three energy-converting devices is 0.582. If $\eta_1 = 0.910$ and $\eta_2 = 0.820$, determine η_3.

10. A 2-hp electric motor, 120-V, 60-Hz, having an efficiency of 0.86 directly drives a 440-V, 400-Hz alternator having an efficiency of 0.83. Draw a cascaded system block diagram and determine:
    (a) The power (in watts) into the motor
    (b) The overall efficiency of the system
    (c) The power out of the system

11. Two differently shaped containers hold equal volumes of water. Each amount of water takes 670 kJ to bring it to a boil. Each container is heated to a boil by identical 1500-W heating elements. Determine:
    (a) The efficiency of each if one takes 9.20 min and the other takes 11.0 min to come to a boil
    (b) The amount of energy wasted (lost) by each

9-3 COST OF ELECTRICAL ENERGY

Practical Unit of Energy

As you have learned, electrical energy is measured in joules (J). The joule, however, is not a practical unit for the sale of electrical energy because it is very small. A more useful unit is the kilowatt-hour (kWh). It was shown in Chapter 7 that 1.0 kWh = 3.6 MJ. Since energy is stated mathematically as $W = Pt$, where W = energy in watt-seconds (Ws) or joules (J), P = power in joules per second (J/s) or watts (W), and t = time in seconds, it would follow that energy in kilowatt-hours would be expressed by the same formula. Thus,

$$W = Pt \quad \text{(kWh)} \tag{9-11}$$

where W = energy in kilowatt-hours (kWh)
P = power in kilowatts (kW)
t = time in hours (h)

EXAMPLE 9-10 Determine the amount of energy (in kilowatt-hours) used to operate a 300-W television for a month (30 days), if it is in use 4 h/day.

Solution   Use Equation 9-11:
Express power in kilowatts:

$$P = 300 \text{ W}$$
$$P = 0.300 \text{ kW}$$

Express time in hours:

$$t = 30 \cancel{d} \times \frac{4 \text{ h}}{\cancel{d}}$$
$$t = 120 \text{ h}$$

Substitute into Equation 9-11:

$$W = Pt$$
$$= 0.300 \text{ kW} \times 120 \text{ h}$$
$$\therefore \quad W = 36 \text{ kWh}$$

Measuring Electrical Energy Use

Electrical energy usage is measured by an instrument called a kilowatt-hour meter. This instrument records the energy use on a series of dials, as seen in Figure 9-4.

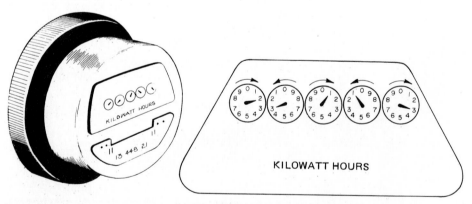

FIGURE 9-4

(a) Kilowatt-hour meter used in residential or commercial buildings to measure energy supplied. (b) The dials record the use. The dials are read from left to right and they rotate in the direction indicated. The reading is 23 113 or 23 113 kWh.

Determining the Cost of Electrical Energy

The cost of electrical energy is given as a rate in cents per kilowatt-hour (¢/kWh) and varies from utility company to utility company. Cost may be computed with the aid of Equation 9-12.

$$\text{Cost} = \text{rate} \times \text{number of kilowatt-hours} \qquad (9\text{-}12)$$

EXAMPLE 9-11 If the rate is 14¢/kWh, then compute the cost of operating the television in Example 9-10.

Solution The Example 9-10 solution indicates that 36 kWh were used in one month of operation.
Use Equation 9-12 and substitute:

$$\text{Cost} = \text{rate} \times \text{number of kilowatt-hours}$$
$$= \frac{14¢}{\cancel{\text{kWh}}} \times 36 \,\cancel{\text{kWh}}$$
$$= 504¢$$
$$\therefore \quad \text{Cost} = \$5.04$$

Table 9-2 lists the wattage ratings of several major household appliances.

EXAMPLE 9-12 Determine the cost of operating the microwave oven listed in Table 9-2 for 30 days if the oven is used for an average of 30 min/day. Assume an electrical rate of 14.4¢/kWh.

Solution Use Equation 9-12.
Rate = 14.4¢/kWh
Determine the number of kilowatt-hours:

$$\text{kWh} = 1.50 \text{ kW} \times \frac{30 \,\cancel{\text{min}}}{\cancel{d}} \times 30 \,\cancel{d} \times \frac{1 \text{ h}}{60 \,\cancel{\text{min}}}$$
$$= 1.50 \text{ kW} \times 30 \times 30 \times \frac{1 \text{ h}}{60}$$
$$= 22.5 \text{ kWh}$$

Determine cost using Equation 9-12:

$$\text{Cost} = 14.4 \times 22.5$$
$$\therefore \quad \text{Cost} = 324¢ = \$3.24$$

TABLE 9-2

WATTAGE RATINGS OF MAJOR HOUSEHOLD APPLIANCES

Appliance	Wattage Rating (W)	Typical Energy Use per Month (kWh)
Clothes washer	500	10
Clothes dryer	5000	90
Dishwasher	1200	40
Iron	1000	5
Microwave oven	1500	15
Refrigerator/freezer		
Frost free	400	130
Manual defrost	300	100
TV (color)		
Portable	200	35
Console	300	50

EXERCISE 9-3

1. Determine the amount of energy (in kilowatt-hours) supplied to an electric motor in 5 h if the rate of supplying energy is 500 J/s.

2. Determine the amount of energy expended in kilowatt-hours by a 100-W lamp if the lamp is operated 7 h/day for 80 days.

3. Determine the cost of operating a stereo for 1 month at 12¢/kWh. Assume that the average monthly use of electrical energy is 9 kWh.

4. Determine the cost of operating a clothes dryer for 60 days at 14.8¢/kWh if 120 kWh is used per month.

5. Determine the monthly cost (30 days) of operating three 100-W lamps for 5.2 hours per day at 15¢/kWh.

6. Determine the cost of operating a 115-V, 10-A toaster for 1 week at 9.5¢/kWh if the toaster is used 5 min each day.

7. When the rate is 10¢/kWh, determine:
   (a) The cost of operating both a 100-W and a 60-W lamp over the 5000-h life of the lamp
   (b) The savings in dollars by using the 60-W lamp instead of the 100-W lamp

8. Determine the difference in operating cost between the frost-free refrigerator/freezer and the manual-defrost refrigerator/freezer listed in Table 9-2 over a 10-year period if each runs 12 h/day. Assume the average rate to be 6¢/kWh.

9. A 10-hp motor operating at 78% efficiency has its input monitored by a kilowatt meter. Determine:
   (a) The reading of the kilowatt meter

(b) The cost at 11.4¢/kWh of the wasted energy (lost to heat) for one week of continuous operation

10. A 120-V, 0.75-hp motor has an efficiency of 86%. Determine:
  (a) The output power in kilowatts
  (b) The circuit current
  (c) The cost of operation for 2 weeks at 10¢/kWh if the motor is operated 6 h/day

---

SELECTED TERMS

conservation of energy Energy may be changed in form but energy can neither be created nor destroyed.

efficiency Indicates how well an energy-converter performs.

joule SI unit for energy and work; one joule of work is done when an applied force of one newton moves an object one meter.

kilowatt-hour Unit of energy used by power companies in the sale of energy to customers. One kilowatt-hour equals 3.6 mega joules.

power Rate of producing or using energy; rate of doing work; J/s.

10

FRACTIONS

In this chapter we are concerned with both arithmetic fractions, such as $\frac{1}{2}$, $\frac{3}{5}$, and $\frac{5}{8}$, and algebraic fractions, such as $3/a$, x/y, and $(x-2)/4$. Fractions are used in many equations that describe electrical and electronic principles and circuits.

10-1 INTRODUCTORY CONCEPTS

The following concepts apply to both arithmetic and algebraic fractions. These concepts are basic to working with fractions.

- A fraction indicates division. Thus, $8 \div 4$, 8/4, and $\frac{8}{4}$ all indicate division. Furthermore, $\frac{8}{4}$ and 8/4 are the same fraction printed in different styles.
- The fraction $9y/3$ is read "$9y$ divided by 3" or "$9y$ over 3."
- The numerator is the quantity above the bar.
- The denominator is the quantity below the bar.
- The numerator and denominator of a fraction may be called the *parts* of the fraction.
- The bar separating the parts of the fraction is called the *vinculum*. The vinculum is a sign of grouping for the numerator and the denominator.
- Fractions may be written in two forms, $\dfrac{xy}{4}$ and $\dfrac{a-2b}{5}$ or $\frac{1}{4}xy$ and $\frac{1}{5}(a-2b)$. It is important to recognize that the two forms are equal. Thus, $\dfrac{xy}{4} = \dfrac{1}{4}xy$ and $\dfrac{a-2b}{5} = \dfrac{1}{5}(a-2b)$.
- A fraction has three signs: the sign of the fraction, the sign of the numerator, and the sign of the denominator. When the sign is positive, it usually is not written—it is implied. Thus, $\dfrac{1}{2}$ means $+\dfrac{+1}{+2}$.
- Division by zero is not permitted. If the denominator evaluates to zero, then the fraction is not defined. Thus, the fraction $\dfrac{7+x}{x-3}$ is *undefined* when $x = 3$, because the denominator evaluates to zero. $\dfrac{7+3}{3-3} = \dfrac{10}{0}$. Try dividing zero into ten with your calculator.
- The numerator and denominator of a fraction may be multiplied or divided by the same number or expression without changing the value of the fraction. Thus,

$$\frac{1}{4} = \frac{1 \times 3}{4 \times 3} = \frac{3}{12} \qquad \therefore \frac{1}{4} \text{ and } \frac{3}{12} \text{ are equivalent fractions}$$

$$\frac{5}{15} = \frac{5 \div 5}{15 \div 5} = \frac{1}{3} \qquad \therefore \frac{5}{15} \text{ and } \frac{1}{3} \text{ are equivalent fractions}$$

- A very important set of equivalent fractions has the value 1. Some of its members are

$$1 = \frac{1}{1} = \frac{2}{2} = \frac{3}{3} = \frac{1059}{1059} = \frac{-64.2}{-64.2} = \frac{14a}{14a} \text{ , and so forth}$$

- Any fraction may be multiplied or divided by 1 without changing its value. Thus,

$$\frac{2a}{9} = \frac{2a}{9} \times 1 = \frac{2a}{9}\left(\frac{5\beta}{5\beta}\right)$$

$$\frac{3}{T-5} = \frac{3}{T-5} \times 1 = \frac{3}{T-5}\left(\frac{29}{29}\right)$$

EXAMPLE 10-1   In the expression $\dfrac{x}{21 - 3x}$, determine the value of x for which the fraction is undefined.

Solution   Set the denominator equal to zero and solve the resulting equation for x:

$$21 - 3x = 0$$
$$-3x = 0 - 21$$
$$3x = 21$$
$$x = 7$$

$\therefore$ When $x = 7$, the denominator is zero, and the fraction is undefined.

EXERCISE 10-1

Determine the value of the variable in each of the following fractions that causes the fraction to be undefined. Do this by setting the denominator equal to zero and solving the resulting equation.

1. $7/x$ 2. $c/5c$ 3. $8/(y + 1)$

4. $\dfrac{t}{t + 2}$ 5. $\dfrac{a + 1}{2a - 1}$ 6. $\dfrac{2}{2x + 4}$

7. $\dfrac{10a}{15 - 3a}$ 8. $\dfrac{m - 6}{m - 6}$ 9. $\dfrac{3x + 4}{10 - 2x}$

10-2 FORMING EQUIVALENT FRACTIONS

A fraction may have its form changed by multiplying both parts of the fraction by the same number. This is done to shape the denominator into a desired form. The value of the fraction that results from multiplying both numerator and denominator by the same number has not been changed—only the form has been changed. Thus, $\frac{1}{3}$, $\frac{2}{6}$, and $\frac{3}{9}$ are *equivalent fractions;* that is, they have the same value.

EXAMPLE 10-2   Change $\frac{2}{5}$ to an equivalent fraction with -15 as the denominator.

    Solution   Multiply both 2 and 5 by -3:

$$\text{M:} \qquad \frac{2}{5}\left(\frac{-3}{-3}\right) = \frac{-6}{-15}$$

$$\therefore \qquad \frac{2}{5} = \frac{-6}{-15}$$

RULE 10-1. MULTIPLICATION PROPERTY OF A FRACTION

Multiplying both parts of a fraction by the same number (except zero) results in an equivalent fraction. Thus,

$$\frac{a}{b} = \frac{a}{b}\left(\frac{c}{c}\right) = \frac{ac}{bc} \qquad c \neq 0$$

EXAMPLE 10-3   Form $\frac{3}{8}$ into an equivalent fraction having a denominator of $24 - 16x$.

    Solution

$$\frac{3}{8} = \frac{?}{24 - 16x}$$

Factor a common monomial of 8 out of $24 - 16x$:

$$8(3 - 2x)$$

Multiply both 3 and 8 by $(3 - 2x)$:

$$\frac{3}{8} = \frac{3}{8}\left(\frac{3 - 2x}{3 - 2x}\right) = \frac{9 - 6x}{24 - 16x}$$

$$\therefore \quad \frac{3}{8} = \frac{9 - 6x}{24 - 16x}$$

We see in the preceding examples that multiplying a fraction by one, $(-3/-3)$ in Example 10-2 and $(3 - 2x)/(3 - 2x)$ in Example 10-3, changed the way the fraction looks, but it has not changed the value of the fraction.

EXERCISE 10-2

Form the given fraction into equivalent fraction(s) having the specified denominator(s).

1. $\dfrac{1}{5} = \dfrac{?}{10} = \dfrac{?}{60}$

2. $\dfrac{1}{7} = \dfrac{?}{49} = \dfrac{?}{35}$

3. $\dfrac{1}{13} = \dfrac{?}{-39} = \dfrac{?}{52}$

4. $\dfrac{1}{2} = \dfrac{?}{32} = \dfrac{?}{-16}$

5. $\dfrac{-3}{7} = \dfrac{?}{77} = \dfrac{?}{-21}$

6. $\dfrac{27}{37} = \dfrac{?}{74} = \dfrac{?}{111}$

7. $\dfrac{m}{n} = \dfrac{?}{5n} = \dfrac{?}{-8n}$

8. $\dfrac{x}{y} = \dfrac{?}{cy} = \dfrac{?}{2yb}$

9. $\dfrac{s}{t} = \dfrac{?}{t^2} = \dfrac{?}{-2t^3}$

10. $\dfrac{-2a}{b} = \dfrac{?}{4ab} = \dfrac{?}{-3b^2}$

11. $\dfrac{2y}{-5x} = \dfrac{?}{30x^2} = \dfrac{?}{15xy^2}$

12. $\dfrac{3}{x - y} = \dfrac{?}{-5x + 5y}$

13. $\dfrac{b}{a - b} = \dfrac{?}{2b^2 - 2ab}$

14. $\dfrac{2x + y}{x^2 - y} = \dfrac{?}{3y^2z - 3x^2yz}$

10-3 SIMPLIFYING FRACTIONS

A fraction is in its simplest form when the numerator and denominator have no factors in *common* except the number 1. The process of simplifying fractions is called *reducing the fraction to its lowest terms*. To reduce a fraction to its lowest terms, divide the numerator and denominator by their *common factors*. If the *same* factor occurs in both parts of the fraction, then it is a common factor. Rule 10-2 shows the steps in reducing fractions to their lowest terms.

> **RULE 10-2. REDUCING FRACTIONS TO LOWEST TERMS**
>
> To reduce a fraction to lowest terms:
>   1. Factor each part of the fraction.
>   2. Group the common factors.
>   3. Remove the common factors from each part of the fraction.

EXAMPLE 10-4   Reduce $\dfrac{4a + 8}{a + 2}$ to lowest terms.

Solution   Use Rule 10-2.

Step 1:   Factor:

$$\frac{4(a + 2)}{a + 2}$$

Step 2:   Group the common factors:

$$\frac{4(a + 2)}{1(a + 2)}$$

Step 3:   Remove the common factors:

$$4/1$$

$$\therefore \quad \frac{4a + 8}{a + 2} = 4$$

EXAMPLE 10-5   Simplify $\dfrac{3x^2y}{6xy^2}$.

Solution   Use Rule 10-2.

Step 1:   Factor:

$$\frac{3 \cdot x \cdot x \cdot y}{2 \cdot 3 \cdot x \cdot y \cdot y}$$

Step 2: Group factors:

$$\frac{1}{2}\left(\frac{3}{3}\right)\left(\frac{x}{x}\right)\frac{x}{1}\left(\frac{y}{y}\right)\frac{1}{y}$$

Step 3: Remove the common factors:

$$\frac{1}{2}\cdot\frac{x}{1}\cdot\frac{1}{y}=\frac{x}{2y}$$

$$\therefore \quad \frac{3x^2y}{6xy^2}=\frac{x}{2y}$$

EXERCISE 10-3

Reduce each fraction to its lowest terms.

1. 16/32

2. 12/18

3. 14/35

4. x/x^2

5. $15b/10$

6. $27c^2/18$

7. $\dfrac{26a^2b}{26ab}$

8. $\dfrac{6xy}{15y^3}$

9. $\dfrac{22d^2c}{121d}$

10. $\dfrac{m^2n^2}{mn}$

11. $\dfrac{x^8}{x^3}$

12. $\dfrac{y^5}{y^{13}}$

13. $\dfrac{39a^4b^5}{26a^2b^2}$

14. $\dfrac{72x^2y}{18x^2y}$

15. $\dfrac{(a+1)(a-1)}{a+1}$

16. $\dfrac{8m+24}{7m+21}$

17. $\dfrac{t^2-2t}{4t-8}$

18. $\dfrac{3a+a^2}{15a+5a^2}$

19. $\dfrac{x}{x^2+x}$

20. $\dfrac{(3x-3)(2x+2)}{6x-6}$

21. $\dfrac{\theta\beta^3+\beta^2\theta^2}{\beta+\theta}$

10-4 MULTIPLYING FRACTIONS

When two or more fractions are multiplied, the product that results is formed by multiplying the numerators and dividing by the product of the denominators. Rule 10-3 will guide you in this process.

RULE 10-3. MULTIPLYING FRACTIONS

The product of two or more fractions results in a new fraction, which is formed by:

1. Multiplying the numerators together, forming the numerator of the new fraction.
2. Multiplying the denominators together, forming the denominator of the new fraction.
3. Multiplying the signs of the fractions together, forming the sign of the new fraction.
4. Simplifying the new fraction by reducing to lowest terms.

EXAMPLE 10-6 Multiply -15 by $\frac{3}{5}$.

Solution Use Rule 10-3.

Steps 1-2: Multiply numerators and denominators together:

$$\frac{-15}{1} \times \frac{3}{5} = \frac{-15 \times 3}{1 \times 5}$$

Remove common factors of 5 and multiply:

$$\frac{-3 \cdot 5 \times 3}{1 \times 5} = \frac{-3 \times 3}{1}\left(\frac{5}{5}\right) = -9$$

$$\therefore \quad -15 \times \tfrac{3}{5} = -9$$

EXAMPLE 10-7 Multiply $\frac{4}{5}$ by $-\frac{2}{3}$.

Solution

$$\frac{4}{5}\left(-\frac{2}{3}\right) = (+ \; -)\frac{4 \times 2}{5 \times 3} = -\frac{8}{15}$$

$$\therefore \quad \frac{4}{5}\left(-\frac{2}{3}\right) = -\frac{8}{15}$$

EXAMPLE 10-8 Multiply $\dfrac{4xy^2}{-6}$ by $\dfrac{18}{8x^2}$.

Solution Perform the first three steps of Rule 10-3 and factor:

$$\frac{4xy^2}{-6} \times \frac{18}{8x^2} = \frac{4 \cdot x \cdot y^2 \cdot 6 \cdot 3}{-6 \cdot 4 \cdot 2 \cdot x \cdot x}$$

Simplify by removing common factors:

$$\left(\frac{1}{-1}\right)\left(\frac{6}{6}\right)\left(\frac{4}{4}\right)\left(\frac{x}{x}\right)\left(\frac{y^2}{1}\right)\left(\frac{3}{2}\right)\left(\frac{1}{x}\right) = \frac{3y^2}{-2x}$$

$$\therefore \quad \frac{4xy^2}{-6} \times \frac{18}{8x^2} = \frac{3y^2}{-2x}$$

EXAMPLE 10-9   Multiply $-\dfrac{5x^2 - 25x}{3x - 15}$ by $\dfrac{12xy^2}{10x^3\,y + 10x^2}$.

Solution   Factor both numerator and denominator. Remember that the numerator and the denominator are grouped by the vinculum. Simplify by removing common factors:

$$-\frac{5x(x-5)}{3(x-5)} \frac{3 \cdot 4(xy^2)}{10x^2(xy+1)}$$

$$= -\left(\frac{5x}{10x^2}\right)\left(\frac{x-5}{x-5}\right)\left(\frac{3}{3}\right)\left(\frac{4}{1}\right)\left(\frac{xy^2}{xy+1}\right)$$

$$= -\frac{20x^2y^2}{10x^2(xy+1)}$$

Remove the common factors of $10x^2$:

$$-\frac{10x^2(2y^2)}{10x^2(xy+1)} = -\frac{2y^2}{xy+1}$$

EXERCISE 10-4

Multiply the following fractions.

1. $\dfrac{2}{3} \times \dfrac{6}{8}$

2. $\dfrac{5}{9} \times \dfrac{3}{5}$

3. $\dfrac{-3}{4} \times \dfrac{8}{15}$

4. $\dfrac{5}{7} \times \dfrac{3}{4} \times \dfrac{14}{15}$

5. $\dfrac{-7}{13} \times \dfrac{10}{35} \times \dfrac{39}{40}$

6. $\dfrac{-5}{8} \times \dfrac{4}{17} \times \dfrac{51}{60}$

7. $\dfrac{-a}{b} \cdot \dfrac{b}{a}$

8. $\dfrac{4m}{9} \cdot \dfrac{1}{2m}$

9. $\dfrac{2x}{y} \cdot \dfrac{-3y}{x}$

10. $\dfrac{n}{2m} \cdot \dfrac{6}{p}$

11. $\dfrac{2x + y}{6z} \cdot \dfrac{12z}{y}$

12. $\dfrac{3a + 2a}{2 + 7} \cdot \dfrac{3a}{5a^2}$

13. $\dfrac{-a}{c + b} \cdot \dfrac{c - b}{a}$

14. $\dfrac{x - y}{x + y} \cdot \dfrac{x + y}{x - y}$

15. $\dfrac{15}{a^4} \cdot \dfrac{4ab}{9}$

16. $\dfrac{m^3}{12} \cdot \dfrac{1}{m^4} \cdot 3m$

17. $\dfrac{2\pi r^2}{1} \cdot \dfrac{1}{8r}$

18. $\dfrac{2x - 10}{3x + 9} \cdot \dfrac{3 + x}{x - 5}$

19. $\dfrac{x^2 - x}{2xy - 2y} \cdot \dfrac{2y}{3x}$

20. $\dfrac{6a - 12}{4a + 4} \cdot \dfrac{(a^2 + a)(a - 1)}{2a^2 - 4a}$

10-5 DIVIDING FRACTIONS

To divide one fraction by another, *invert the divisor and multiply.* That is, multiply the dividend by the reciprocal of the divisor.

EXAMPLE 10-10

Solve $\frac{2}{6} \div \frac{2}{3}$.

Solution Take the reciprocal of the divisor, $\frac{2}{3}$, and multiply the dividend, $\frac{2}{6}$:

$$\frac{2}{6} \div \frac{2}{3} = \frac{2}{6} \times \frac{3}{2}$$

Multiply and remove the common factor:

$$\frac{6}{6 \times 2} = \frac{1}{2}$$

$$\therefore \quad \frac{2}{6} \div \frac{2}{3} = \frac{1}{2}$$

EXAMPLE 10-11

Solution

Divide $\dfrac{2a + 2b}{6}$ by $\dfrac{a + b}{3}$.

Invert and multiply:

$$\frac{2a + 2b}{6} \div \frac{a + b}{3} = \frac{2a + 2b}{6} \cdot \frac{3}{a + b}$$

Factor and simplify:

$$\frac{2(a + b)}{6} \cdot \frac{3}{a + b} = \frac{6(a + b)}{6(a + b)} = 1$$

EXAMPLE 10-12

Solution

Solve $\dfrac{2mn}{(m - n)^2} \div \dfrac{4n}{2m - 2n}$

Invert and multiply:

$$\frac{2mn}{(m - n)^2} \cdot \frac{2m - 2n}{4n}$$

Factor and simplify:

$$\frac{2mn}{(m - n)(m - n)} \cdot \frac{2(m - n)}{4n}$$

$$\frac{4mn(m - n)}{4n(m - n)(m - n)} = \frac{m}{m - n}$$

EXERCISE 10-5

Divide the following fractions.

1. $\dfrac{3}{8} \div \dfrac{1}{4}$

2. $\dfrac{4}{9} \div \dfrac{4}{3}$

3. $\dfrac{3}{10} \div \dfrac{2}{5}$

4. $\dfrac{-3}{16} \div \dfrac{9}{2}$

5. $\dfrac{-3}{4} \div \dfrac{-5}{7}$

6. $\dfrac{x}{y} \div z$

204

7. $\dfrac{2a}{b} \div \dfrac{a}{2b}$ 8. $\dfrac{-6m}{7n} \div \dfrac{3m}{-14n}$

9. $\dfrac{3}{2x} \div \dfrac{3y}{2x}$ 10. $\dfrac{-4ab}{6cb} \div \dfrac{8b}{12ac}$

11. $\dfrac{2x^2}{y^2} \div \dfrac{2xy}{y^2}$ 12. $\dfrac{b^2}{6a^2} \div \dfrac{3b}{2a}$

13. $\dfrac{-(3a-6)}{9} \div \dfrac{1}{3}$ 14. $\dfrac{(x-3)(x+3)}{15} \div \dfrac{x-3}{5}$

15. $\dfrac{a^2-ab}{b} \div \dfrac{a}{b}$ 16. $\dfrac{b}{b-3} \div \dfrac{b^3}{2}$

10-6 COMPLEX FRACTIONS

In the preceding section we learned that division of fractions is carried out by multiplying with the reciprocal of the divisor. The arithmetic operator $\div$ was used to indicate division.

    In this section we will indicate division of fractions by forming a *complex fraction*. A complex fraction is one that has a fraction in the numerator or the denominator or in both the numerator and the denominator.

$$\dfrac{\frac{2}{3}}{\frac{3}{4}}, \quad \dfrac{\frac{a}{2}}{\frac{b}{4}} \quad \text{and} \quad \dfrac{a+b}{\frac{a}{b}}$$

are complex fractions.

**EXAMPLE
10-13**

Write $\dfrac{x-y}{3} \div \dfrac{x-y}{6}$ as a complex fraction.

Solution

$$\dfrac{x-y}{3} \div \dfrac{x-y}{6} = \dfrac{\dfrac{x-y}{3}}{\dfrac{x-y}{6}}$$

Notice that the vinculum of the main fraction is made longer and bolder to distinguish the numerator from the denominator.

To simplify a complex fraction having a simple fraction as the numerator and denominator, divide the denominator into the numerator by:
1. Reciprocating the denominator.
2. Multiplying the numerator by the reciprocal of the denominator.
3. Reducing to lowest terms.

EXAMPLE 10-14

Simplify: $\dfrac{\dfrac{xy}{5}}{\dfrac{2x}{10y}}$.

Solution   Use Rule 10-4.

Step 1:   Reciprocate the denominator:

$$\frac{2x}{10y} \Longrightarrow \frac{10y}{2x}$$

Step 2:   Multiply:

$$\frac{xy}{5} \cdot \frac{10y}{2x}$$

Step 3:   Remove common factors:

$$\frac{xy}{5}\frac{10y}{2x} = y^2$$

EXAMPLE 10-15

Form $\dfrac{mn + m}{n} \div \dfrac{n + 1}{n^2}$ into a complex fraction, and then simplify.

Solution

$$\frac{mn + m}{n} \div \frac{n + 1}{n^2} = \frac{\dfrac{mn + m}{n}}{\dfrac{n + 1}{n^2}}$$

Take the reciprocal of the denominator $\dfrac{n + 1}{n^2}$ and multiply:

$$\frac{mn + m}{n} \cdot \frac{n^2}{n + 1}$$

Factor and remove common factors:

$$\frac{m(n + 1)}{n} \cdot \frac{n \cdot n}{n + 1}$$

$$mn \left(\frac{n}{n}\right)\left(\frac{n + 1}{n + 1}\right) = mn$$

EXERCISE 10-6

Form each of the following expressions into a complex fraction.

1. $\dfrac{8}{3} \div \dfrac{4}{5}$

2. $\dfrac{3}{10} \div \dfrac{2}{5}$

3. $\dfrac{-9}{2} \div \dfrac{3}{16}$

4. $z \div \dfrac{x}{y}$

5. $\dfrac{-6x}{7y} \div \dfrac{3x}{14y}$

6. $\dfrac{4ab}{6cb} \div \dfrac{ab}{cb}$

7. $\dfrac{R^2}{-t} \div \dfrac{2tR}{t^2}$

8. $\dfrac{1}{3} \div \dfrac{3a - b}{9}$

9. $\dfrac{b}{b - 3} \div \dfrac{b^2}{2}$

10. $\dfrac{3x + 3}{2y} \div \dfrac{6 + 6x}{y}$

Simplify the following complex fractions.

11. $\dfrac{\dfrac{3}{3}}{\dfrac{3}{10}}$

12. $\dfrac{\dfrac{-39}{5}}{\dfrac{13}{10}}$

13. $\dfrac{\dfrac{18}{20}}{\dfrac{6}{5}}$

14. $\dfrac{\dfrac{x + 1}{x^2}}{\dfrac{yx + y}{x}}$

15. $\dfrac{\dfrac{3a + 3b}{a - b}}{\dfrac{6b + 6a}{a - b}}$

16. $\dfrac{\dfrac{a + b}{b}}{\dfrac{a - b}{b}}$

$$17. \frac{\dfrac{x-y}{y}}{\dfrac{x+y}{y}} \qquad\qquad 18. \frac{a}{\dfrac{a-1}{a}}$$

$$19. \frac{a+5}{\dfrac{(a+5)(a-5)}{a}} \qquad 20. \frac{\dfrac{2a}{x-3y}}{\dfrac{2ab}{2x-6y}}$$

10-7 ADDING AND SUBTRACTING FRACTIONS

Every fraction has three signs: (1) the sign of the numerator, (2) the sign of the denominator, and (3) the sign of the vinculum (also called the sign of the fraction). When no sign is written, it is understood to be positive. This is an *implied* sign.

It is easier to add fractions when the sign of the vinculum is positive. To change the sign of the vinculum of the fraction, apply this rule:

> **RULE 10-5. CHANGING SIGN**
>
> To change the signs of a fraction, change any two of the three signs: the vinculum and the numerator; or the vinculum and the denominator; or the numerator and the denominator.

We see from Rule 10-5 that the sign of the vinculum and one other sign must be changed. When the sign of the vinculum is changed, either the sign of the numerator or the denominator must also be changed. Thus, $-\dfrac{a}{2b}$ may be written with a positive vinculum if we change either the sign of the numerator, $+\dfrac{-a}{2b}$, or the sign of the denominator, $+\dfrac{a}{-2b}$. So

$$-\frac{a}{2b} = \frac{-a}{2b} = \frac{a}{-2b}$$

**EXAMPLE
10-16**

Change the sign of the vinculum of $-\dfrac{x-2y}{5}$.

Solution Change two of the three signs: the vinculum and the numerator; leave the sign of the denominator unchanged.

$$-\frac{x-2y}{5} = +\frac{-(x-2y)}{5}$$

Distribute the sign change through the numerator:

$$\frac{-x+2y}{5} \quad \text{or} \quad \frac{2y-x}{5}$$

In Example 10-16 we chose to change the sign of the numerator along with the sign of the vinculum. The choice was made through the application of the following guidelines.

GUIDELINES 10-1. CHANGING THE SIGN OF THE VINCULUM

When changing the sign of the vinculum, select:

1. The numerator for the other sign to change, if both the denominator and the numerator are positive.
2. The denominator for the other sign to change, if both the numerator and the denominator are negative.
3. The one that is negative, if either the numerator or denominator is negative.

EXAMPLE 10-17

Change the sign of the vinculum of the following fractions by applying Guidelines 10-1:

$$(a) \ -\frac{4}{9} \qquad (b) \ -\frac{1}{-2} \qquad (c) \ -\frac{-5}{-8}$$

Solution Using Guidelines 10-1:

$$(a) \ -\frac{4}{9} = +\frac{-4}{9}$$

$$(b) \ -\frac{1}{-2} = +\frac{1}{+2}$$

$$(c) \ -\frac{-5}{-8} = +\frac{-5}{+8}$$

Change the sign of the vinculum in each of the following fractions by applying Guidelines 10-1.

1. $-\dfrac{1}{4}$

2. $-\dfrac{-3}{8}$

3. $-\dfrac{-4}{-5}$

4. $-\dfrac{-7}{-9}$

5. $-\dfrac{x-y}{3a}$

6. $-\dfrac{5t}{-2-s}$

7. $-\dfrac{3y+4}{2x+1}$

8. $-\dfrac{-a^2+a-2}{3b}$

9. $-\dfrac{-2\pi R}{-E-1}$

10. $-\dfrac{-5a-3(b-1)}{a+b}$

11. $-\dfrac{5\omega+\beta}{-6-\beta}$

12. $-\dfrac{(x+y)}{-5}$

Common Denominator

Before fractions may be added, they must have the *same* (or a common) denominator. Each of the fractions to be added is first changed to an equivalent fraction having a common denominator.

EXAMPLE 10-18

Change each fraction to an equivalent fraction having a common denominator of $36a^2$.

$$\text{(a)}\ \frac{3}{12a} \qquad \text{(b)}\ \frac{8}{18a^2} \qquad \text{(c)}\ \frac{5a}{6}$$

Solution Form equivalent fractions having $36a^2$ as the denominator.
(a) Multiply 3/12a by 1 in the form of 3a/3a:

$$\frac{3}{12a}\left(\frac{3a}{3a}\right) = \frac{9a}{36a^2}$$

(b) Multiply 3/18a² by 1 in the form of 2/2:

$$\frac{8}{18a^2}\left(\frac{2}{2}\right) = \frac{16}{36a^2}$$

(c) Multiply $5a/6$ by 1 in the form of $6a^2/6a^2$:

$$\frac{5a}{6}\left(\frac{6a^2}{6a^2}\right) = \frac{30a^3}{36a^2}$$

EXERCISE 10-8

Form each of the following sets of fractions into equivalent fractions having the given common denominators.

Common Denominator

1. $\dfrac{1}{6}, \dfrac{2}{3}, \dfrac{5}{12}$ 12

2. $\dfrac{7}{15}, \dfrac{9}{20}, \dfrac{11}{30}$ 60

3. $\dfrac{-3}{4}, \dfrac{1}{3}, \dfrac{5}{-9}$ 36

4. $\dfrac{2}{5}, -\dfrac{2}{3}, \dfrac{-7}{15}$ 15

5. $\dfrac{x}{3}, \dfrac{2b}{7}, \dfrac{5}{6}$ 42

6. $\dfrac{2}{13}, \dfrac{5x}{b}, \dfrac{10}{3}$ $39b$

7. $\dfrac{a}{x}, \dfrac{3}{xy}, ax$ xy

8. $\dfrac{a}{b}, \dfrac{2a + b}{a - b}$ $b(a - b)$

9. $\dfrac{x - b}{a - b}, -\dfrac{x}{b - a}$ $(a - b)$

10. $\dfrac{RE}{Z + 1}, \dfrac{E}{R}$ $R(Z + 1)$

Adding Fractions Having Common Denominators

To add fractions, each must have the same or a common denominator. Once they have a common denominator, they may be combined by Rule 10-6.

To add or subtract fractions having a common denominator:
  1. Make the signs of the vinculums positive.
  2. Place the sum (or difference) of the numerators over the common
     denominator.
  3. Simplify by combining like terms and reducing to lowest terms.

**EXAMPLE
10-19**

Combine $\dfrac{2a}{10a} - \dfrac{7a}{10a}$.

Solution   Use Rule 10-6.
Step 1:   Make the sign of the vinculum positive:

$$\frac{2a}{10a} + \frac{-7a}{10a}$$

Step 2:   Place the sum of the numerators over the common
           denominator:

$$\frac{2a - 7a}{10a}$$

Step 3:   Simplify by combining like terms and reducing to low-
           est terms:

$$\frac{-5a}{10a} = -\frac{1}{2}$$

$$\therefore \quad \frac{2a}{10a} - \frac{7a}{10a} = -\frac{1}{2}$$

**EXAMPLE
10-20**

Combine $\dfrac{-5a}{2a^2 + 4a} - \dfrac{5 - a}{2a^2 + 4a} - \dfrac{3}{2a^2 + 4a}$.

Solution   Use Rule 10-6.

$$\frac{-5a - (5 - a) - 3}{2a^2 + 4a} = \frac{-5a - 5 + a - 3}{2a^2 + 4a}$$

$$= \frac{-4a - 8}{2a^2 + 4a} = \frac{-4(a + 2)}{2a(a + 2)} = \frac{-4}{2a} = \frac{-2}{a}$$

EXERCISE 10-9
Add or subtract each of the following as indicated.

1. $\dfrac{3}{4} + \dfrac{5}{4}$

2. $\dfrac{2}{12} - \dfrac{-8}{12}$

3. $-\dfrac{11}{13} - \dfrac{8}{13} + \dfrac{-7}{13}$

4. $\dfrac{3}{11} - \dfrac{5}{11} + \dfrac{-8}{11}$

5. $\dfrac{3}{a} + \dfrac{7}{a}$

6. $\dfrac{2}{3a} - \dfrac{8}{3a}$

7. $\dfrac{5}{x + 1} - \dfrac{3}{x + 1}$

8. $\dfrac{x}{x - y} - \dfrac{y}{x - y}$

9. $\dfrac{2a}{a + 1} + \dfrac{2}{a + 1}$

10. $\dfrac{3b}{b + 1} + \dfrac{b + 4}{b + 1}$

11. $\dfrac{3x + y}{3xy} - \dfrac{3y - 2x}{3xy} + \dfrac{4x + 2y}{3xy}$

12. $\dfrac{3}{3x^2 - 2x} - \dfrac{2x + 5}{3x^2 - 2x} + \dfrac{5x}{3x^2 - 2x}$

Finding Common Denominators

To add fractions, the numerators must be placed over a common denominator. How do we find a common denominator? Let's explore the process of finding a common denominator through the following examples.

**EXAMPLE
10-21**

Find some possible common denominators for 1/3 and a/5 and select the "simplest."

Solution Since the common denominator *must* be divisible by both of the denominators, one common denominator could be the product of 3 and 5, or 15. Or any multiple of the product of 3 and 5, such as 30, 45, 60, and 75 could be a common denominator.

∴   By inspection, 15 is the *simplest* common denominator.

EXAMPLE 10-22

Find some possible common denominators for $y/2x$ and $7/yz$, and select the *simplest*.

Solution   First select the product of $2x$ and yz, which is $2xyz$. Some other possible denominators are $4xyz$, $2x^2yz$, $8wxyz$, and 2 "whatever" xyz; however, the simplest is $2xyz$.

EXAMPLE 10-23

Find some possible common denominators for $\frac{5}{12}$ and $\frac{11}{18}$ and select the simplest.

Solution   First, select the product of 12 and 18, which is 216, as a possible common denominator. Although 216 is a common denominator, it is bigger than we need. Find the common factors of 12 and 18:

$$2, 3, \text{ and } 6$$

Divide 216 by 6, the largest common factor:

$$\frac{216}{6} = 36$$

Observe that 36 is a multiple of both 12 and 18.

∴   36 is the simplest common denominator.

EXAMPLE 10-24

Find the simplest common denominator for $\dfrac{2a}{9y}$ and $\dfrac{b}{3y^2}$.

Solution Find the common factors of $9y$ and $3y^2$:

$$3, \ y, \ 3y$$

The simplest common denominator is the product of the two denominators $9y$ and $3y^2$, divided by their largest common factor, $3y$:

$$\frac{9y(3y^2)}{3y} = 9y^2$$

$\therefore$ $9y^2$ is the simplest common denominator.

RULE 10-7. FORMING A COMMON DENOMINATOR

To form a common denominator:
1. Find the common factors of each denominator.
2. Form the product of the denominators.
3. Divide this product by the largest common factor. This results in the *simplest* common denominator.

EXAMPLE 10-25

Find the simplest common denominator for $\dfrac{1}{3c - 1}$ and $\dfrac{7}{6ac - 2a}$.

Solution Use Rule 10-7.

Step 1: Find the common factors of each denominator:

Factors of $3c - 1$ are 1 and $3c - 1$.
Factors of $6ac - 2a$ are 2, a, $3c - 1$, $2a$, $6c - 2$, and $3ac - a$.
The largest common factor is $3c - 1$.

Step 2: Form the product of $3c - 1$ and $6ac - 2a$:

$$(3c - 1)(6ac - 2a)$$

Step 3: Divide the product by the largest common factor, $3c - 1$:

$$\frac{(3c - 1)(6ac - 2a)}{3c - 1} = 6ac - 2a$$

$$\therefore \quad 6ac \quad - \quad 2a \quad \text{is} \quad \text{the} \quad \text{simplest} \quad \text{common}$$
denominator.

EXERCISE 10-10

Determine the simplest common denominator for each of the following pairs of fractions:

1. $\dfrac{2}{3}, \dfrac{5}{6}$

2. $\dfrac{5}{52}, \dfrac{7}{13}$

3. $\dfrac{3}{25}, \dfrac{1}{15}$

4. $\dfrac{5}{12}, \dfrac{19}{32}$

5. $\dfrac{B}{2A}, \dfrac{B^2}{A}$

6. $\dfrac{4}{3C}, \dfrac{2}{3D}$

7. $\dfrac{150}{\pi R}, \dfrac{35}{2\pi R}$

8. $\dfrac{7}{3V}, \dfrac{\lambda}{V^2}$

9. $\dfrac{\mu}{4\alpha\beta}, \dfrac{\sigma}{6\beta}$

10. $\dfrac{7J}{12K}, \dfrac{9}{4JK}$

11. $\dfrac{7}{8\alpha(\beta + 1)}, \dfrac{3\beta}{\alpha^2(2\beta + 2)}$

12. $\dfrac{8}{21W^2}, \dfrac{3}{63W}$

13. $\dfrac{AB}{49K - 7}, \dfrac{NO}{14K - 2}$

14. $\dfrac{CB^2}{3A(B - 6)}, \dfrac{13B}{3B - 18}$

15. $\dfrac{3}{st + s}, \dfrac{a}{2t + 2}$

16. $\dfrac{P}{15uV - 5u}, \dfrac{17R}{10u^2}$

Adding Fractions with Unequal Denominators

When adding fractions, remember that each fraction must have a common denominator before it can be added. Rule 10-8 shows how to combine fractions with unequal denominators.

216

EXAMPLE 10-26

Combine $\dfrac{a}{3} - \dfrac{3a}{6}$.

Solution   The common denominator of 3 and 6 is 6. Use Rule 10-8.

Step 1:   Write equivalent fractions:

$$\frac{a}{3} \cdot \frac{2}{2} + \frac{-3a}{6} = \frac{2a}{6} + \frac{-3a}{6}$$

Step 2:

$$\frac{2a - 3a}{6}$$

Step 3:

$$\frac{-a}{6}$$

$$\therefore \quad \frac{a}{3} - \frac{3a}{6} = \frac{-a}{6}$$

EXAMPLE 10-27

Combine $\dfrac{x-1}{3} + \dfrac{2x+1}{2x} - \dfrac{3x+8}{9}$.

Solution   The common denominator is $18x$. Use Rule 10-8.

Step 1:   Modify each fraction:

$$\left(\frac{6x}{6x}\right)\frac{x-1}{3} + \left(\frac{9}{9}\right)\frac{2x+1}{2x} + \left(\frac{2x}{2x}\right)\frac{-3x-8}{9}$$

Step 2: Perform multiplication:

$$\frac{6x(x-1)}{18x} + \frac{9(2x+1)}{18x} + \frac{2x(-3x-8)}{18x}$$

$$\frac{6x^2 - 6x + 18x + 9 - 6x^2 - 16x}{18x}$$

Step 3: Combine like terms:

$$\frac{-4x+9}{18x}$$

$$\therefore \quad \frac{x-1}{3} + \frac{2x+1}{2x} - \frac{3x+8}{9} = \frac{-4x+9}{18x}$$

EXAMPLE 10-28

Combine $\dfrac{3x+2}{x^2-x} + \dfrac{3}{x} - \dfrac{5}{x-1}$.

Solution The common denominator is $x(x-1)$. Modify each fraction.

$$\frac{3x+2}{x(x-1)} + \left(\frac{x-1}{x-1}\right)\frac{3}{x} + \left(\frac{x}{x}\right)\frac{-5}{x-1}$$

$$\frac{3x+2}{x(x-1)} + \frac{3x-3}{x(x-1)} + \frac{-5x}{x(x-1)}$$

$$\frac{3x+2+3x-3+(-5x)}{x(x-1)}$$

Combine like terms and simplify:

$$\frac{x-1}{x(x-1)} = \frac{1}{x}$$

$$\therefore \quad \frac{3x+2}{x^2-x} + \frac{3}{x} - \frac{5}{x-1} = \frac{1}{x}$$

EXERCISE 10-11

Combine and simplify the following fractions.

1. $2 + \dfrac{c}{d}$

2. $\dfrac{x}{y} - 3$

3. $\dfrac{1}{2} + \dfrac{5}{6x}$

4. $8y + \dfrac{1}{3}$

5. $\dfrac{1}{x} + \dfrac{1}{y}$

6. $\dfrac{a}{b} + \dfrac{1}{c}$

7. $n + \dfrac{1}{m}$

8. $\dfrac{4x}{3} + \dfrac{5x}{4} - \dfrac{x}{5}$

9. $\dfrac{3ab}{4} - \dfrac{2ab}{3} + \dfrac{5ab}{8}$

10. $\dfrac{3x}{a} - \dfrac{y}{b}$

11. $\dfrac{m}{x} + \dfrac{2n}{y}$

12. $\dfrac{a}{x} - \dfrac{5}{x^2}$

13. $\dfrac{5t}{4} - \dfrac{3t}{12} - \dfrac{t}{3}$

14. $\dfrac{7}{n^2} + \dfrac{5}{n} + \dfrac{9}{n^3}$

15. $\dfrac{7a - 3b}{3} + \dfrac{2a + 5b}{2}$

16. $\dfrac{5x - 4}{12} - \dfrac{3x - 8}{18}$

17. $\dfrac{3x - 1}{3x} - \dfrac{2x - 4}{2}$

18. $\dfrac{m + n - 5}{m} - \dfrac{2m - n + 1}{2m}$

19. $\dfrac{2a - b}{4b} - \dfrac{a - 3b}{6a}$

20. $\dfrac{x - y}{y} + \dfrac{y^2 - x}{y^2}$

21. $\dfrac{6}{6x - 12} - \dfrac{4}{x - 2}$

22. $\dfrac{5}{6a + 6} - \dfrac{3}{2a + 2}$

23. $\dfrac{2b - 4}{9b^2 - 9b} - \dfrac{5}{6b - 6}$

24. $\dfrac{2a - 3}{16a^2} - \dfrac{2 - a}{8a} + \dfrac{3}{4a}$

25. $\dfrac{3}{m^2 + 5m} - \dfrac{3}{m} - \dfrac{3}{m - 5}$

26. $\dfrac{-3a}{-x + y} + \dfrac{2}{x - y} + \dfrac{1}{x + y}$

10-8 CHANGING A MIXED EXPRESSION TO A FRACTION

A mixed expression is formed by adding (or subtracting) a polynomial and a fraction. Thus, $3x - \dfrac{2}{5x}$ is a mixed expression. A mixed expression may be changed to a fraction by adding both terms, as demonstrated in the following examples.

EXAMPLE 10-29

Write $3x - \dfrac{2}{5x}$ as a fraction.

Solution

$$3x - \frac{2}{5x} = \frac{3x}{1}\left(\frac{5x}{5x}\right) + \frac{-2}{5x}$$

$$= \frac{15x^2 - 2}{5x}$$

EXAMPLE 10-30

Write $2a + b - \dfrac{2}{a + 1}$ as a fraction.

Solution The common denominator is $a + 1$.

$$\frac{2a}{1} \cdot \frac{a + 1}{a + 1} + \frac{b}{1} \cdot \frac{a + 1}{a + 1} - \frac{2}{a + 1}$$

$$\frac{2a^2 + 2a + ab + b - 2}{a + 1}$$

EXERCISE 10-12

Change each of the following mixed expressions into a fraction.

1. $5\frac{3}{4}$

2. $6\frac{2}{5}$

3. $2 + \dfrac{1}{3}$

4. $7 + \dfrac{3}{5}$

5. $3 - \dfrac{2}{m}$

6. $n - \dfrac{5}{n}$

7. $3a + \dfrac{2}{3}$

8. $\dfrac{a}{b} + 5$

9. $4x - \dfrac{7}{2x}$

10. $2\beta - \dfrac{\pi}{2a}$

11. $\omega C + 3 - \dfrac{Z}{L}$

12. $2b + 1 - \dfrac{5}{b}$

13. $x + \dfrac{2}{x + 3}$ 14. $\dfrac{a + 2b}{a - b} + 7$

15. $x + 5 + \dfrac{3x - 1}{x + 2}$ 16. $y + x + c - \dfrac{3}{yx}$

10-9 ADDITIONAL WORK WITH COMPLEX FRACTIONS

In this section we will work with simplifying complex fractions. This section uses many of the concepts already presented in this chapter.

RULE 10-9. SIMPLIFYING COMPLEX FRACTIONS

To change a complex fraction to a simple fraction:
1. Simplify the numerator to a fraction.
2. Simplify the denominator to a fraction.
3. Divide the resulting numerator by the resulting denominator.
4. Reduce to lowest terms.

**EXAMPLE
10-31**

Change the complex fraction to a simple fraction:

$$\dfrac{3 - \dfrac{1}{x + 2}}{5 + \dfrac{1}{x + 2}}$$

Solution Use Rule 10-9 to simplify the complex fraction.
Step 1: Simplify the numerator:

$$\dfrac{3(x + 2) - 1}{x + 2} = \dfrac{3x + 6 - 1}{x + 2} = \dfrac{3x + 5}{x + 2}$$

Step 2: Simplify the denominator:

$$\dfrac{5(x + 2) + 1}{x + 2} = \dfrac{5x + 10 + 1}{x + 2} = \dfrac{5x + 11}{x + 2}$$

Step 3: Divide:

$$\frac{\dfrac{3x + 5}{x + 2}}{\dfrac{5x + 11}{x + 2}}$$

Invert and multiply:

$$\frac{3x + 5}{x + 2} \cdot \frac{x + 2}{5x + 11}$$

Step 4:

$$\frac{3x + 5}{5x + 11}$$

**EXAMPLE
10-32**

Change the complex fraction to a simple fraction:

$$\frac{2 - \dfrac{2m}{n}}{1 - \dfrac{m}{n}}$$

Solution Use Rule 10-9 to simplify. Simplify the numerator and denominator:

$$\frac{\dfrac{2n - 2m}{n}}{\dfrac{n - m}{n}}$$

Invert and multiply:

$$\frac{2(n - m)}{n} \cdot \frac{n}{n - m} = \frac{2}{1}$$

$$\therefore \quad \frac{2 - \dfrac{2m}{n}}{1 - \dfrac{m}{n}} = 2$$

Change each of the complex fractions to a simple fraction.

1. $\dfrac{7 + \dfrac{4}{5}}{\dfrac{1}{5} + 5}$

2. $\dfrac{6 - \dfrac{1}{3}}{2 + \dfrac{5}{6}}$

3. $\dfrac{a - \dfrac{1}{b}}{\dfrac{1}{b} + a}$

4. $\dfrac{ax + \dfrac{x}{b}}{ax - \dfrac{x}{b}}$

5. $\dfrac{3 - \dfrac{2}{a}}{6a - 4}$

6. $\dfrac{a + \dfrac{a}{b}}{\dfrac{1}{b} + \dfrac{1}{b^2}}$

7. $\dfrac{2 + \dfrac{a + b}{a - b}}{1 - \dfrac{a + b}{a - b}}$

8. $\dfrac{x - 2 + \dfrac{3}{x}}{1 + \dfrac{1}{x}}$

9. $3 - \dfrac{1}{1 - \dfrac{1}{x - 1}}$

10. $5 - \dfrac{2a}{3 + \dfrac{a}{2}}$

11. $\dfrac{\dfrac{1}{a + b} - \dfrac{1}{a - b}}{\dfrac{1}{a - b} + \dfrac{1}{a + b}}$

12. $\dfrac{\dfrac{x}{x + y} - \dfrac{y}{x - y}}{\dfrac{x}{x - y} + \dfrac{y}{x + y}}$

13. $\dfrac{\dfrac{2}{b + 3} + \dfrac{1}{b - 2}}{\dfrac{2}{b - 2} - \dfrac{1}{b + 3}}$

14. $\dfrac{\dfrac{3}{y - 4} - \dfrac{5}{y + 3}}{\dfrac{2}{y + 3} + \dfrac{7}{y - 4}}$

15. $\dfrac{2 - \dfrac{2}{x + 4}}{\dfrac{3}{x + 4} - \dfrac{1}{x - 1}}$

16. $\dfrac{\dfrac{4}{y - 3} + 5}{\dfrac{4}{y - 3} - \dfrac{7}{y + 3}}$

SELECTED TERMS

common denominator   Equal denominators of several fractions.

complex fraction   Fraction with a numerator or denominator (or both) that contains a fraction.

equivalent fractions   Fractions that have different forms but have the same value.

mixed expression   Sum (difference) of a fraction and a polynomial.

undefined fraction   If the denominator evaluates to zero in a fraction, then the fraction is undefined.

vinculum   Sign of grouping; a bar drawn over terms to show they are treated as a unit.

11

EQUATIONS CONTAINING FRACTIONS

Many equations used by electronic technicians to solve problems involve fractions. In this chapter you will have an opportunity to expand your understanding of algebra by working with fractions, fractional equations, and literal equations containing fractions and by evaluating formulas.

11-1 SOLVING EQUATIONS CONTAINING FRACTIONS

Equations having numerical coefficients that are fractions are transformed and solved by the methods in earlier chapters. Equations with fractional coefficients are best solved by transforming them into equations without fractions. This is done by multiplying each member of the equation by the common denominator.

EXAMPLE 11-1 Solve $\frac{3}{8}y = \frac{3}{4}$.

Solution Multiply each member by the common denominator, 8; then simplify:

$$(8)\tfrac{3}{8}y = \tfrac{3}{4}(8)$$
$$3y = 6$$
$$y = 2$$

Check Does $\frac{3}{8}y = \frac{3}{4}$ when $y = 2$?

$$\tfrac{3}{8}(2) = \tfrac{3}{4}$$
$$\tfrac{3}{4} = \tfrac{3}{4} \qquad \text{Yes!}$$

$$\therefore \qquad y = 2 \text{ is the solution.}$$

EXAMPLE 11-2 Solve $\dfrac{2x}{5} - 3 = \dfrac{5 - 3x}{10}$.

Solution Multiply each member by the common denominator of 10:

$$10\left(\frac{2x}{5} - 3\right) = 10\left(\frac{5 - 3x}{10}\right)$$

$$10\left(\frac{2x}{5}\right) + 10(-3) = 10\left(\frac{5 - 3x}{10}\right)$$

$$2(2x) - 30 = 5 - 3x$$
$$4x + 3x = 5 + 30$$
$$7x = 35$$
$$x = 5$$

Check   Does $\dfrac{2x}{5} - 3 = \dfrac{5 - 3x}{10}$ when $x = 5$?

$$\frac{2(5)}{5} - 3 = \frac{5 - 3(5)}{10}$$
$$2 - 3 = \frac{5 - 15}{10}$$
$$-1 = -1 \qquad \text{Yes!}$$

$\therefore \qquad x = 5$ is the solution.

EXAMPLE 11-3   Solve $\dfrac{3x + 3}{12} = \dfrac{x - 3}{18} + \dfrac{x}{6}$.

Solution   Multiply each member by the common denominator of 36:

$$36\left(\frac{3x + 3}{12}\right) = 36\left(\frac{x - 3}{18} + \frac{x}{6}\right)$$
$$36\left(\frac{3x + 3}{12}\right) = 36\left(\frac{x - 3}{18}\right) + 36\left(\frac{x}{6}\right)$$
$$3(3x + 3) = 2(x - 3) + 6x$$
$$9x + 9 = 2x - 6 + 6x$$
$$9x - 8x = -15$$
$$x = -15$$

Check   Does $\dfrac{3x + 3}{12} = \dfrac{x - 3}{18} + \dfrac{x}{6}$ when $x = -15$?

$$\frac{3(-15) + 3}{12} = \frac{-15 - 3}{18} + \frac{-15}{6}$$
$$\frac{-45 + 3}{12} = -1 + \frac{-15}{6}$$

$$\frac{-42}{12} = \frac{-6 - 15}{6}$$

$$\frac{-7}{2} = \frac{-21}{6}$$

$$\frac{-7}{2} = \frac{-7}{2} \qquad \text{Yes!}$$

$$\therefore \qquad x = -15 \text{ is the solution.}$$

EXERCISE 11-1

Solve the following equations and check your solutions.

1. $\dfrac{1}{8}x = \dfrac{1}{4}$

2. $\dfrac{-x}{2} = \dfrac{1}{6}$

3. $\dfrac{a}{15} = \dfrac{1}{10}$

4. $\dfrac{5}{4} = \dfrac{4}{5}y$

5. $\dfrac{1}{3}R = -\dfrac{15}{7}$

6. $\dfrac{x}{3} = \dfrac{4}{5}$

7. $\dfrac{m - 7}{4} = 2m$

8. $\dfrac{2b - 5}{3} = \dfrac{45}{3}$

9. $\dfrac{4}{7} = \dfrac{12V}{16}$

10. $\dfrac{a}{3} - \dfrac{a}{2} = 5$

11. $\dfrac{3I + 4}{7} = I + 2$

12. $\dfrac{3}{4} + E = \dfrac{5}{4}E$

13. $5 - \dfrac{2}{3}x = \dfrac{-x}{2}$

14. $\dfrac{V}{2} = \dfrac{4V}{5} - 3$

15. $\dfrac{1}{6} = \dfrac{a - 15}{3} + \dfrac{7 - a}{2}$

16. $\dfrac{m - 3}{3} = \dfrac{m + 4}{5} - 3$

17. $\dfrac{3x}{10} - \dfrac{2}{10} = \dfrac{2x - 7}{5} + \dfrac{1}{5}$

18. $\dfrac{2R}{5} = \dfrac{4R - 1}{4} - \dfrac{3}{16}$

19. $\dfrac{2I - 3}{3} - 3 = -\dfrac{3}{6} - \dfrac{I}{2}$ 20. $\dfrac{8y - 1}{3} = 3 - \dfrac{10y}{6} - \dfrac{7}{6}$

21. $\dfrac{3x - 3}{4} = \dfrac{2x}{3} - \dfrac{x - 5}{2}$ 22. $\dfrac{4a - 3}{2} = \dfrac{4a - 3}{5}$

11-2 SOLVING FRACTIONAL EQUATIONS

A fractional equation is an equation that has a variable in the denominator of one or more terms of the equation. The solution of fractional equations is carried out in the same way as the solution of equations containing fractions by multiplying both members of the equation by the common denominator.

EXAMPLE 11-4 Solve $\dfrac{2}{x} = \dfrac{5}{x} - \dfrac{3}{5}$.

Solution Multiply each member by the common denominator of $5x$:

$$5x\left(\dfrac{2}{x}\right) = 5x\left(\dfrac{5}{x}\right) + 5x\left(-\dfrac{3}{5}\right)$$

$$10 = 25 - 3x$$
$$3x = 15$$
$$x = 5$$

Check Does $\dfrac{2}{x} = \dfrac{5}{x} - \dfrac{3}{5}$ when $x = 5$?

$$\dfrac{2}{5} = \dfrac{5}{5} - \dfrac{3}{5}$$

$$\dfrac{2}{5} = \dfrac{2}{5} \qquad \text{Yes!}$$

$\therefore \quad x = 5$ is the solution.

It is important to check the roots of fractional equations. When we multiply both members of an equation by an expression containing the unknown, we run the risk of introducing an *extraneous root*. An extraneous root will *check* in the derived equation, but **not** in the original equation. Extraneous roots result from multiplying by zero.

EXAMPLE 11-5   Solve $\dfrac{5x}{x-1} = \dfrac{5}{x-1}$.

Solution   Multiply each member by $(x - 1)$, the common denominator:

$$(x - 1)\frac{5x}{x-1} = (x - 1)\frac{5}{x-1}$$
$$5x = 5$$
$$x = 1$$

Check   Does $\dfrac{5x}{x-1} = \dfrac{5}{x-1}$ when $x = 1$?

$$\frac{5(1)}{1-1} = \frac{5}{1-1}$$
$$\frac{5}{0} = \frac{5}{0}$$

No! Both members of the equation are undefined.

Since division by zero is not permitted, this equation has no root.

$\therefore \quad x = 1$ does not satisfy the equation.

EXAMPLE 11-6   Solve $\dfrac{3x}{x+1} = 3 - \dfrac{4}{2x}$.

Solution   Multiply each member by the common denominator, which is $2x(x + 1)$:

$$\frac{2x\cancel{(x+1)}(3x)}{\cancel{x+1}} = 2x(x + 1)(3) - \frac{\cancel{2x}(x + 1)(4)}{\cancel{2x}}$$
$$6x^2 = 6x^2 + 6x - 4x - 4$$
$$6x^2 - 6x^2 = 6x - 4x - 4$$
$$0 = 2x - 4$$
$$2x = 4$$
$$x = 2$$

Check Does $\dfrac{3x}{x+1} = 3 - \dfrac{4}{2x}$ when $x = 2$?

$$\frac{3(2)}{2+1} = 3 - \frac{4}{2(2)}$$

$$\frac{6}{3} = 3 - 1$$

$$2 = 2 \qquad \text{Yes!}$$

$$\therefore \qquad x = 2 \text{ is the solution.}$$

EXERCISE 11-2
Solve the following equations and check your solutions.

1. $\dfrac{3}{x} + \dfrac{5}{x} = 2$

2. $\dfrac{8}{x} = \dfrac{4}{x} - \dfrac{2}{3}$

3. $\dfrac{8}{y} - 1 = \dfrac{7}{y}$

4. $\dfrac{3}{x} - 4 = \dfrac{7}{2}$

5. $\dfrac{8}{x+1} = 4$

6. $\dfrac{x+3}{x-3} = 4$

7. $\dfrac{3y}{4y-5} = 2$

8. $\dfrac{2x+5}{x+4} = 1$

9. $\dfrac{4}{y-3} = \dfrac{2}{y}$

10. $\dfrac{3x-2}{5x-10} = \dfrac{4}{10}$

11. $\dfrac{4x-2}{4x} - \dfrac{1}{2} = \dfrac{3}{x}$

12. $\dfrac{3}{2} - \dfrac{5}{2y} = 2 - \dfrac{3}{y}$

13. $\dfrac{x-2}{5x} = \dfrac{4x-8}{30x}$

14. $\dfrac{1}{y-1} = \dfrac{3}{y-3}$

15. $\dfrac{12}{x-6} = \dfrac{2x}{x-6}$

16. $5 - \dfrac{5x}{x+1} = \dfrac{15}{4x}$

17. $\dfrac{2y}{y+1} = 2 - \dfrac{8}{2y}$

18. $\dfrac{3}{y} = \dfrac{3y}{y-4} - 3$

19. $\dfrac{5}{3+y} = \dfrac{3y}{y+3} - \dfrac{3y-4}{2y+6}$

20. $\dfrac{3}{x} + \dfrac{6}{x^2+5x} = \dfrac{6x}{x^2+5x}$

11-3 LITERAL EQUATIONS CONTAINING FRACTIONS

Equations that have one or more numbers represented by letters are *literal equations*. The formulas used to solve electrical problems are literal equations.

Variables and Constants

In literal equations, such as $\dfrac{x}{a} - 2 = b$, $\dfrac{c}{3} = y$, and $\dfrac{x}{m} - n = a$, the letters a, b, c, m, and n are *standing in* for numbers that have a fixed or constant value. The letters x and y are *standing in* for numbers that may change (or could be made to change) or vary.

In literal equations, it is conventional to represent constants by the letters a, b, c, g, h, k, l, m, and n. It is conventional to represent variables by the letters s, t, u, v, w, x, y, and z.

Constants are divided into two types. Those represented by letters (a, b, c, etc.) are called *arbitrary constants*, whereas numbers and symbols representing unvarying values are called *numerical constants*. Numerical constants include integers (1, 2, 3, etc.), fractions ($\frac{1}{2}$, $\frac{2}{3}$, etc.), and irrational numbers ($\pi = 3.14159\ldots$, $e = 2.71828\ldots$, $\sqrt{2} = 1.414\ldots$, etc.).

EXAMPLE 11-7 Solve the literal equation $\dfrac{x}{a} + 2 = b$ for x.

Solution

$$\frac{x}{a} + 2 = b$$

$$\frac{x}{a} = b - 2$$

$$\therefore \quad x = a(b - 2)$$

EXAMPLE 11-8 Solve $\dfrac{7}{b} - \dfrac{y}{a} = 4$ for y.

Solution

$$\frac{7}{b} - \frac{y}{a} = 4$$

$$7a - yb = 4ab$$

$$-yb = 4ab - 7a$$

Change sign:

$$yb = 7a - 4ab$$

$$\therefore \quad y = \frac{7a - 4ab}{b}$$

EXAMPLE 11-9 Solve $\dfrac{xm}{2} - \dfrac{m}{3} = x$ for x.

Solution

$$\frac{xm}{2} - \frac{m}{3} = x$$

$$\frac{xm}{2} = x + \frac{m}{3}$$

$$\frac{xm}{2} - x = \frac{m}{3}$$

$$6\left(\frac{xm}{2} - x\right) = 6\left(\frac{m}{3}\right)$$

$$3xm - 6x = 2m$$

Factor the left member:

$$x(3m - 6) = 2m$$

$$\therefore \quad x = \frac{2m}{3m - 6}$$

EXERCISE 11-3
Solve each equation for the *conventional* variable s, t, u, v, w, x, y, or z.

1. $\dfrac{a}{x} = b$

2. $\dfrac{3}{\pi} = \dfrac{a}{W}$

3. $\dfrac{1}{z} - \dfrac{1}{a} = -\dfrac{2}{z}$

4. $\dfrac{hu}{m} + \dfrac{ku}{m} = 5$

5. $\dfrac{y}{2m} + \dfrac{y}{n} = 1$

6. $2g = \dfrac{hke + V}{aV}$

7. $c = \dfrac{b - W}{\pi^2 kW}$

8. $z - 1 - m = \dfrac{5z - 8}{m}$

9. $\dfrac{c}{s} = \dfrac{c}{3}(D - c)$

10. $\dfrac{\pi x + b}{\pi x - b} = 3$

11. $\dfrac{m - n}{ay} = b + 1$

12. $\dfrac{2u + 2}{au} - \dfrac{3}{u} = -\dfrac{7 - u}{u}$

13. $\dfrac{5x}{g - h} - 3 = \dfrac{3x}{k}$

14. $\dfrac{1}{4(a - b)} = \dfrac{1}{5(x - a)}$

15. $\dfrac{1}{u} - \dfrac{1}{m} = \dfrac{1}{n} - \dfrac{1}{u}$

16. $\dfrac{1}{w} = \dfrac{1}{a} + \dfrac{1}{b} + \dfrac{1}{c}$

17. $\dfrac{b}{a} = \dfrac{x - b}{x} + \dfrac{2b}{ax}$

18. $\dfrac{4k + 1}{4k} = \dfrac{1}{2z} + \dfrac{2k}{z}$

19. $\dfrac{4g}{3g} - \dfrac{6}{6V} = \dfrac{4g}{V} - \dfrac{1}{3g}$

20. $\dfrac{z}{ab} - \dfrac{1}{ac} = \dfrac{z}{bc}$

11-4 EVALUATING FORMULAS

Evaluating a formula is the process of finding the value of one variable in a formula when all the other variables are known. When evaluating a formula for only one set of values, it is simpler to first substitute the known values into the formula and then solve for the variable representing the desired quantity. This technique is demonstrated in Example 11-10.

EXAMPLE 11-10

Using the formula $F = \frac{9}{5}C + 32$, determine the temperature in Celsius (C) when the temperature in Fahrenheit (F) is 77°F.

Solution First substitute 77 for F and then solve for C:

$$F = \tfrac{9}{5}C + 32$$
$$77 = \tfrac{9}{5}C + 32$$
$$\tfrac{9}{5}C = 45$$
$$C = 45\left(\tfrac{5}{9}\right)$$

$$\therefore \quad C = 25°C$$

In Example 11-10, we first substituted into the formula, and then solved the problem. When we wish to find several values for a particular variable, it is best to first solve the formula for the desired variable and then substitute into the formula. This concept is demonstrated in Example 11-11.

EXAMPLE 11-11

Using $F = \frac{9}{5}C + 32$, determine the temperature in Celsius (C) for temperatures in Fahrenheit (F) of: (a) 105°F (b) 93°F (c) 85°F

Solution   First solve for C:

$$F = \frac{9}{5}C + 32$$
$$\frac{9}{5}C = F - 32$$
$$C = \frac{5(F - 32)}{9} \ {}^{\circ}C$$

Now, substitute each of the given temperatures.
(a) F = 105°F

$$C = \frac{5(105 - 32)}{9} = 40.6{}^{\circ}C$$

(b) F = 93°F

$$C = \frac{5(93 - 32)}{9} = 33.9{}^{\circ}C$$

(c) F = 85°F

$$C = \frac{5(85 - 32)}{9} = 29.4{}^{\circ}C$$

In the study of electronics, formulas are routinely evaluated. Because of this, it is very important that you develop the skill of successfully working with formulas. The following exercise will give you some practice.

EXERCISE 11-4
In the following formulas, solve for the indicated variable.

Formula:	*Solve for:*
1. $C = \frac{5}{9}(F - 32)$	F

Formula:	Solve for:

2. $A = \frac{1}{2}h(b + c)$ $\qquad c$

3. $X_C = \dfrac{1}{2\pi f C}$ $\qquad f$

4. $C = \dfrac{1}{C_1} + \dfrac{1}{C_2}$ $\qquad C_2$

5. $k = \dfrac{wv^2}{2g}$ $\qquad w$

6. $\beta = \dfrac{\alpha}{1 - \alpha}$ $\qquad \alpha$

7. $\theta_{JA} = \dfrac{T_J - T_A}{P_D}$ $\qquad T_A$

8. $A_v' = \dfrac{A_v}{1 + \beta A_v}$ $\qquad A_v$

9. $C = \dfrac{E}{R_1 + R_2}$ $\qquad R_2$

10. $\dfrac{V_1}{V_2} = \dfrac{P_2}{P_1}$ $\qquad P_1$

11. $\theta = \dfrac{108(n - 2)}{n}$ $\qquad n$

12. $s = t\left(\dfrac{V_0 + V_1}{2}\right)$ $\qquad V_1$

13. $\eta = \dfrac{A_2 S_2}{F_1 S_1}$ $\qquad S_2$

14. $Q = \dfrac{t K A(T_0 - T_1)}{L}$ $\qquad T_0$

15. $F = \dfrac{q_1 q_2}{4\pi \epsilon_0 d^2}$ $\qquad d^2$

16. $V = \dfrac{R_2(V_{CC} - R_1 I_{\text{sat}} - 1)}{R_1 + R_2}$ $\qquad R_1$

17. $R_T = \dfrac{R_1 R_2}{R_1 + R_2}$ $\qquad R_2$

Formula	Solve for:

18. $I_1 = \dfrac{IR_2}{R_1 + R_2}$ R_2

19. $S = \dfrac{R_1 + R_2}{R_E + R_\beta(1 - \alpha)}$ α

20. $R_2 = \dfrac{Z_Z(k + 1)}{k - 1} - R_3$ k

Evaluate each of the following formulas. First assign the indicated values to the variables and then solve.

21. Determine the resistance (R) of a coil in ohms from

$$Q_0 = \frac{2\pi f_{ar}L}{R}$$

when f_{ar} = 310 kHz, L = 5 mH, Q_0 = 200, and $\pi \approx 3.14$.

22. Determine the collector current (I_c) in amperes from

$$I_{CBO} = \frac{I_c - \beta I_B}{\beta + 1}$$

when I_{CBO} = 5μA, I_B = 60 μA, and β = 100.

23. Determine the total resistance of a parallel network (R_T) in ohms from

$$\frac{1}{R_T} = \frac{1}{R_1} + \frac{1}{R_2}$$

when R_1 = 2.20 kΩ and R_2 = 1.50 kΩ.

24. Determine the impedance of the load (Z_2) in ohms from

$$\frac{V_1^2}{V_2^2} = \frac{Z_1}{Z_2}$$

when V_1 = 60 V, V_2 = 2 V, and Z_1 = 7.2 kΩ. Refer to Figure 11-1.

25. Determine the frequency (f) in hertz from

$$X_C = \frac{1}{2\pi fC}$$

when X_C = 410 Ω, C = 0.05 μF, and $\pi \approx 3.14$.

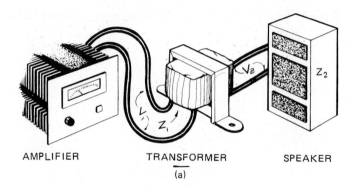

AMPLIFIER   TRANSFORMER   SPEAKER
                          (a)

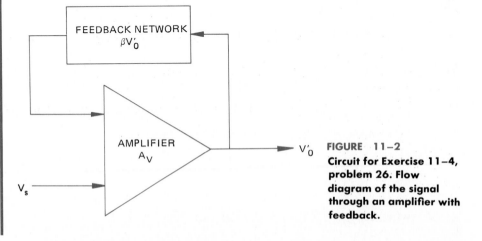

V_1 Z_1 → V_2 Z_2

FIGURE 11–1   (b)
Circuit for Exercise 11–4, problem 24: (a) pictorial diagram showing an output transformer being used to match the load impedance to the amplifier; (b) schematic diagram of the circuit.

26. Determine the output voltage (V_o') in volts from

$$\frac{V_o'}{V_s} = \frac{A_v}{1 - \beta A_v}$$

when $\beta = 0.05$, $A_v = 60$, and $V_s = 1.0$ V. Refer to Figure 11-2.

FEEDBACK NETWORK
$\beta V_0'$

AMPLIFIER
A_V

V_s

V_0'

FIGURE 11–2
Circuit for Exercise 11–4, problem 26. Flow diagram of the signal through an amplifier with feedback.

27. Determine the forward current transfer ratio of a transistor (β) from

$$\alpha = \frac{\beta}{\beta + 1}$$

when $\alpha = 0.985$.

28. Determine the load resistance (R_L) in ohms from

$$A_v = \frac{\mu R_L}{r_p + R_L}$$

when $\mu = 18$, $r_p = 0.2$ kΩ, and $A_v = 12$.

29. Determine the value of the bias resistance (R_1) in ohms from

$$I_B = \frac{V_{cc}}{R_1 + (\beta + 1)R_E}$$

when $I_B = 40$ μA, $V_{cc} = 18$ V, $\beta = 100$, and $R_E = 470$ Ω.

30. Determine the temperature of the case of a power transistor (T_c) in degrees Celsius from

$$P_D = \frac{T_c - T_s}{\theta_{cs}}$$

when $P_D = 75$ W, $T_s = 55$°C, and $\theta_{cs} = 0.35$°C/W.

SELECTED TERMS

extraneous root   A solution of a mathematical equation that is not a solution to the physical problem.

fractional equation   Equation that has a variable in the denominator of one or more terms.

literal equation   Equation with one or more numbers represented by letters.

12

APPLICATION OF FRACTIONS

This chapter relates the application of fractions to electrical concepts, covering ratio, percent, and parts per million; proportion; and electrical conductors.

12-1 RATIO, PERCENT, AND PARTS PER MILLION

We often compare two quantities by dividing one by the other. To compare the cost of two resistors, a ratio is used. If R_1 costs 60¢ and R_2 costs 15¢, we could say that R_1 costs four times as much as R_2. This comparison is made by computing the quotient of 60¢ and 15¢; $60/15 = 4$. This price comparison can be stated as a ratio of "4 to 1." This is written 4:1 using the ratio sign (:).

Ratio

A ratio of one number to another is the *quotient* of the *first* number divided by the *second* number. You may express a ratio in several forms, as shown in Table 12-1.

TABLE 12-1

EXPRESSING RATIOS	
The ratio of one number to another may be expressed as a quotient.	
Quotient Indicated by:	**Ratio Expressed as:**
÷ Divide sign	$5 \div 3$
: Ratio sign	$5:3$
/ Vinculum	$\dfrac{5}{3}$
Decimal fraction	1.667 to 1

The two numbers in a ratio **must have the same units.** We may express the ratio of 3 meters to 1 meter, but we cannot express a ratio between 5 meters and 2 kilograms. If two numbers express the same quantity, but have different units, then change each to the same unit. So, when computing the ratio of 1 m to 800 cm, first change 1 m to 100 cm, and then find the ratio of 100 cm to 800 cm, shown as 100:800.

RULE 12-1. FINDING THE RATIO OF TWO LIKE QUANTITIES

To find the ratio of two like quantities:
1. Express the quantities in the same units.
2. Write the ratio as a fraction.
3. Reduce the fraction to its lowest terms.

EXAMPLE 12-1   Express the ratio of 1.2 m to 80 cm in lowest terms.
     Solution   Use Rule 12-1.
     Step 1:   Change 1.2 m to centimeters:

$$1.2 \text{ m} \Rightarrow 120 \text{ cm}$$

     Step 2:   Write the ratio as a fraction:

$$\frac{120}{80}$$

     Step 3:   Reduce to lowest terms:

$$\frac{120}{80} = \frac{3}{2}$$

$\therefore$ The ratio of 1.2 m to 80 cm = 3/2.

Because a ratio is a quotient between two numbers, it has all the properties of a fraction and may be formed into an equivalent ratio.

EXAMPLE 12-2   Express the ratio 15:25 as an equivalent ratio.
     Solution

$$15:25 = \frac{15}{25}$$

$$\text{D: 5} \quad \frac{15}{25} = \frac{3}{5}$$

$$\text{M: 20} \quad \frac{3}{5} = \frac{60}{100}$$

$\therefore$ 15:25, 3:5, and 60:100 are equivalent ratios.

EXERCISE 12-1

Reduce each of the following ratios to their lowest terms.

1. 3:12	2. 8:2	3. 14 to 21
4. $7x$ to $13x$	5. 4 cm/6 cm	6. 30 s to 1.5 min
7. 8.42 to 2	8. 0.6 to 6	9. 15 mA to 2.0 A
10. $20k$ to $313.15k$	11. $1.15 to 95¢	12. 440 V ÷ 110 V

Percent

Percent is shorthand for a ratio in which the second number in the ratio is 100. Thus, 8% is the ratio of 8:100 and 58% is the ratio of 58:100.

    Errors in electrical measurements are often expressed in percent to indicate the difference between the true and measured values. Equation 12-1 is used to compute the percent error in measurements.

$$\text{Percent error} = \frac{\text{measured value} - \text{true value}}{\text{true value}} \times 100 \qquad (12\text{-}1)$$

Percent error is used to indicate the degree of *accuracy* of a measurement. The accuracy of a measurement refers to the amount of difference between the measured value and the true value.

EXAMPLE 12-3   Determine the percent error between a DVM and a 50.000-V laboratory standard if the DVM reads 48.3 V when connected to the standard power supply.

Solution   Use Equation 12-1 with your calculator.

$$\text{Percent error} = \frac{48.3 - 50.0}{50.0} \times 100$$
$$= -3.4\%$$

∴   The percent error is -3.4%.

Manufacturers of electronic components and equipment use percent to specify component and equipment tolerances. The percent tolerance of carbon-composition resistors (pictured in Figure 12-1) is indicated by the fourth *color band*. The percent tolerance is the maximum permitted amount of deviation from *nominal* or named value. Percent deviation from the nominal value is used to indicate the deviation between the actual value of an electrical component and the named or nominal value. Equation 12-2 is used to compute percent deviation.

$$\text{Percent deviation} = \frac{\text{actual value} - \text{nominal value}}{\text{nominal value}} \times 100 \qquad (12\text{-}2)$$

FIGURE 12-1

Color coding of a carbon composition resistor. The fourth color band is the tolerance band. The fourth band colors are gold, ± 5%; silver, ± 10%; black (or no color), ± 20%.

TOLERANCE BAND

EXAMPLE 12-4   A 180-Ω (nominal value) resistor with a gold fourth band was measured and found to have an actual value of 172.8 Ω. Determine if the resistor is within the specified tolerance.

Solution   Use Equation 12-2 and then compare the specified tolerance to the percent deviation.

$$\text{Percent deviation} = \frac{172.8 - 180}{180} \times 100$$
$$= -4.00\%$$

A gold fourth band indicates a tolerance of ±5%. Does −4% lie between +5% and −5%? Yes.

∴   The resistor is within the specified tolerance.

EXERCISE 12-2

1. Determine the percent error between a VOM (volt ohm milliammeter) that indicates 46.5 V and a meter calibrator that is set to 50.000 V.

2. A 5-mA panel meter under test indicates 4.2 mA when the laboratory standard is 5.000 mA. Determine the percent error.

3. The dc function of a DVM is being calibrated. The instrument manufacturer has specified the meter to be ±1.0% of reading for all dc range settings. Determine whether the instrument is within the specified tolerance when it indicates 7.2 V. It is connected to an 8.000-V standard source.

4. A radio-frequency (RF) oscillator has a 0.03% error between the true value of 2.5000 MHz and the measured value. Determine the measured value.

5. The operating frequency of a commercial broadcasting station was measured and found to be 800.020 kHz. Determine the assigned frequency if the measured frequency was +0.0025% in error.

6. A 2.2-kΩ resistor was measured and found to have an actual value of 2862.0 Ω. Determine the percent deviation from the nominal value.

7. A 5000-Ω wire-wound resistor was measured and found to have an actual value of 4732.0 Ω. Determine the percent deviation from the nominal value.

8. The actual value of a 47-Ω carbon resistor is 51.43 Ω. What color should the tolerance band be?

9. A voltage standard has a nominal value of 10.000 V. Determine the actual value of voltage from the standard if the deviation is −0.008%.

244

10. A 1.82-Ω, ±1% carbon film precision resistor has an actual value of 1.8018 Ω. Determine if the resistor is within the specified tolerance of ±1%.

Parts Per Million

Technicians who work in standards laboratories make measurements that result in very high accuracy. Deviations of 0.0001% are frequently encountered. The notation used to show such *high accuracy* is *parts per million* (ppm). Parts per million is shorthand for a ratio used to express accuracies, errors, deviations from a nominal (name) value, and temperature coefficients in which the second number of the ratio is 1 000 000. The procedure for converting percent to parts per million is outlined in Rule 12-2.

RULE 12-2. CONVERTING FROM PERCENT TO PARTS PER MILLION

To convert from percent to parts per million, drop the percent sign and:
  1. Move the decimal point four places to the right.
  2. Write ppm after the resulting number.

EXAMPLE 12-5   Express 0.0052% in parts per million.
  Solution   Use Rule 12-2.
  Step 1:   Move the decimal four places to the right:

$$0.0052 \Rightarrow 52.$$

$$1\ 2\ 3\ 4$$

  Step 2:   Write ppm after the number:

$$52 \text{ ppm}$$
$$\therefore \quad 0.0052\% = 52 \text{ ppm}$$

When to use parts per million or percent will become clear to you with use. Table 12-2 will aid in determining when to use percent and when to use parts per million in expressing the accuracy of a measurement. Table 12-2 is based on the generally accepted and widely used practice in metrology, the art and science of measurement.

From Table 12-2, we can see that *lower accuracy* is best expressed in percent, such as 1% or 2%. *Higher accuracy* is conveniently expressed in parts per million, such as 1 ppm, rather than 0.0001%.

TABLE 12-2

GUIDELINES FOR APPLYING PERCENT AND PARTS PER MILLION

To Express:	Use:	To Express:	Use:
1.0%	1.0%	0.001%	10 ppm
0.1%	0.1%	0.0001%	1.0 ppm
0.01%	0.01% or 100 ppm	0.000 01%	0.1 ppm

EXAMPLE 12-6   A precision resistor has a nominal value of 1000.000 Ω. If its true value is 1000.013 Ω, state the deviation from the nominal value in parts per million.

Solution   Use Equation 12-2 with your calculator.

$$\text{Percent deviation} = \frac{1000.013 - 1000.000}{1000.000} \times 100$$
$$= 0.0013\%$$

Use Rule 12-2 to change to parts per million:

$$0.0013\% \Rightarrow 13 \text{ ppm}$$

∴   The deviation is 13 ppm.

Frequently, we need to reverse the process and express parts per million as percent. Rule 12-3 explains how this is done.

RULE 12-3. CONVERTING PARTS PER MILLION TO PERCENT

To convert from parts per million to percent, drop the "ppm" and:
  1. Move the decimal point four places to the left.
  2. Write % after the resulting number.

EXAMPLE 12-7   Express 180 ppm as a percent.
Solution   Using Rule 12-3, move the decimal point four places to the left and express as percent:

∴ $180 \text{ ppm} \Rightarrow 0.018\%$

Express each of the following as parts per million (ppm).

  1. 0.0015%   2. 0.0032%   3. 0.0136%
  4. 0.0197%   5. 0.022%   6. 0.0002%

Express each of the following as percent (%).

  7. 1000 ppm   8. 318 ppm   9. 100 ppm
 10. 635 ppm   11. 4050 ppm   12. 1272 ppm

13. A metal film resistor has a temperature coefficient of ± 50 ppm/°C. Restate the temperature coefficient as a percent.

14. A digital counter, using a very stable crystal oscillator, has a time-base accuracy of 0.000 025%. Restate the accuracy in parts per million.

12-2 PROPORTION

An equation made up of ratios is a proportion. The electrical formula $N_1/N_2 = E_1/E_2$ is a proportion. This proportion is read "N_1 is to N_2 as E_1 is to E_2."

EXAMPLE 12-8 Solve the proportion $9/36 = x/16$ for the unknown term x.

Solution

$$\frac{9}{36} = \frac{x}{16}$$

$$x = \frac{9(16)}{36}$$

$$\therefore \quad x = 4$$

EXERCISE 12-4
Solve the following proportions for x.

1. $\dfrac{x}{5} = \dfrac{8}{10}$ 2. $\dfrac{x}{3} = \dfrac{1}{6}$ 3. $\dfrac{9}{x} = \dfrac{12}{4}$

4. $\dfrac{7}{6} = \dfrac{4}{x}$ 5. $\dfrac{1}{x} = \dfrac{7}{13}$ 6. $\dfrac{5}{9} = \dfrac{x}{8}$

Determine which of the following proportions are true. Remember that a proportion is a statement that two ratios (fractions) are equal.

7. $2:3 = 10:15$ 8. $\dfrac{3}{8} = \dfrac{5}{9}$ 9. $13:11 = 5:4$

10. $\dfrac{5}{7} = \dfrac{7}{5}$ 11. $\dfrac{19}{72} = \dfrac{133}{504}$ 12. $3:6 = 18:34$

12-3 ELECTRICAL CONDUCTORS

Electrical conductors are manufactured in various forms depending upon the application. Most conductors are made from copper; however, aluminum, nichrome, and tungsten are other metals that are in common use. See Figure 12-2.

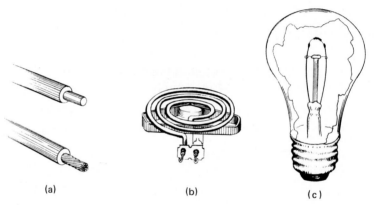

(a)   (b)   (c)

FIGURE 12–2
Metals used in electrical conductors: (a) copper in solid and stranded wire; (b) nichrome in heating elements; (c) tungsten in lamp filaments.

The resistance of a conductor depends upon the length, cross-sectional area, temperature, and kind of material. Table 12-3 demonstrates factors governing the resistance of metal conductors.

Length

Two copper wires of equal cross-sectional area and temperature will have different resistances when their **lengths** are different. The longer wire will have more resistance than the shorter wire. We see that the resistance of a conductor increases as the length increases. Mathematically, this is stated as "the resistance of a conductor is *directly proportional* to the length of the conductor." This is written in a direct proportion as Equation 12-3.

TABLE 12-3

FACTORS GOVERNING THE RESISTANCE OF METAL CONDUCTORS

Factor	Relative Amount of Resistance			
	More Resistance		**Less Resistance**	
Length		Long		Short
Cross-sectional area		Small		Large
Kind of material		Nichrome		Copper
Temperature		Hot		Cold

$$\frac{R_1}{R_2} = \frac{l_1}{l_2} \qquad\qquad (12\text{-}3)$$

where R_1 = resistance of the first conductor
R_2 = resistance of the second conductor
l_1 = length of the first conductor
l_2 = length of the second conductor

EXAMPLE 12-9   The resistance of 10 m of a particular copper wire is 842 mΩ. Determine the resistance of 38 m of the same wire.

Solution   Use Equation 12-3. Substitute 842 mΩ for R_1, 10 m for l_1, and 38 m for l_2:

$$\frac{R_1}{R_2} = \frac{l_1}{l_2}$$

$$\frac{842 \times 10^{-3}}{R_2} = \frac{10}{38}$$

Solve for R_2:

$$R_2 = \frac{38(842 \times 10^{-3})}{10}$$

$$\therefore \quad R_2 = 3.2 \ \Omega$$

EXAMPLE
12-10

An RF coil is made from copper wire that has a resistance of 340 Ω/km. Determine the length (in meters) of the wire used to manufacture the coil if the finished coil has a measured resistance of 1.57 Ω. See Figure 12-3.

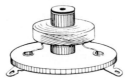

FIGURE 12–3
Radio-frequency coil for Example 12–10.

Solution   Use Equation 12-3. Substitute 340 Ω for R_1, 1 km for l_1, and 1.57 Ω for R_2:

$$\frac{R_1}{R_2} = \frac{l_1}{l_2}$$

$$\frac{340}{1.57} = \frac{1 \times 10^3}{l_2}$$

Solve for l_2:

$$l_2 = \frac{1.57(1 \times 10^3)}{340}$$

$$\therefore \quad l_2 = 4.62 \text{ m}$$

EXERCISE 12-5

In each problem it is assumed that all the conductors are round, solid, copper wire of equal cross-sectional area and temperature:

1. The resistance of 6 m of copper wire is 0.52 Ω. Determine the resistance of 17.4 m of the same wire.
2. The radio-frequency coil of Figure 12-3 has a measured resistance of 7.38 Ω. Determine the length of wire in the coil if the wire has a resistance of 213 Ω/km.
3. A 100-m coil of copper wire has a resistance of 2.10 Ω. If 20 m of wire is cut from the coil, what is the resistance of the remaining 80 m of wire?
4. Determine the resistance of the 20 m of wire removed from the coil in problem 3.

5. An unknown amount of wire has been removed from a 1.0-km spool of wire. The remaining wire has a measured resistance of 39.4 Ω. Determine the length of the remaining wire on the spool if a 1.0-km length of wire has a measured resistance of 84.2 Ω.

6. Complete the following table by determining the missing information:

Length	20 cm	100 cm	5.0 m	20 m	100 m
Resistance		0.0910 Ω			

Cross-sectional Area

Two copper wires of equal length and temperature will have different resistances when their *cross-sectional areas* are different. The wire with larger cross-sectional area will have a smaller resistance than the wire of smaller cross-sectional area. We see that the resistance of a conductor increases as the wire becomes smaller in cross-sectional area. Mathematically stated, the "resistance is *inversely* proportional to the cross-sectional area." This is written as an *indirect proportion* in Equation 12-4.

$$\frac{R_1}{R_2} = \frac{A_2}{A_1} \qquad (12\text{-}4)$$

where R_1 = resistance of the first conductor
R_2 = resistance of the second conductor
A_1 = cross-sectional area of the first conductor
A_2 = cross-sectional area of the second conductor

Equation 12-4 may be restated in terms of the conductors' diameters instead of their cross-sectional area, as in Equation 12-5.

$$\frac{R_1}{R_2} = \frac{d_2^2}{d_1^2} \qquad (12\text{-}5)$$

where R_1 = resistance of the first conductor
R_2 = resistance of the second conductor
d_1 = diameter of the first conductor
d_2 = diameter of the second conductor

EXAMPLE 12-11

The resistances of two conductors of equal length were measured and found to be 4.14 and 16.6 Ω. Determine the diameter in millimeters of the conductor having a resistance of 4.14 Ω if the other conductor is 1.15 mm in diameter.

Solution Use Equation 12-5. Substitute 4.14 Ω for R_1, 16.6 Ω for R_2, and 1.5 mm for d_2.

$$\frac{R_1}{R_2} = \frac{d_2^2}{d_1^2}$$

$$\frac{4.14}{16.6} = \frac{(1.15)^2}{d_1^2}$$

Solve for d_1:

$$d_1^2 = \frac{16.6(1.15)^2}{4.14}$$

$$d_1 = \sqrt{\frac{16.6(1.15)^2}{4.14}}$$

$$\therefore \quad d_1 = 2.30 \text{ mm}$$

The American Wire Gauge Table

The American wire gauge is used to standardize wire sizes. Standard copper wire is designated by one of the 44 standard gauge numbers of the American wire gauge (AWG). The AWG is a sequence of 44 gauge numbers extending from no. 40, the smallest (0.7988 mm), through no. 1 (7.348 mm), to the largest no. 4/0, also written no. 0000 (11.68 mm).

Although the SI metric system is used when calculating wire measurements, the wire itself is sold by the foot with the diameter specified in mils (one-thousandth of an inch). For purposes of calculation, mils may be converted to millimeters by the conversion factor

$$1 \text{ mil} = 0.0254 \text{ mm}$$

For convenience, many recent editions of engineering handbooks include tables for solid, annealed copper wire with the AWG listed in both metric and English units. Table 12-4 is a partial listing of the American wire gauge. The complete table is given in Appendix A.

**EXAMPLE
12-12**

A certain length of no. 24 gauge copper wire having a diameter of 0.511 mm and a resistance of 12 Ω is

TABLE 12-4

PARTIAL LIST OF AMERICAN WIRE GAUGE FOR SOLID ANNEALED COPPER
CONDUCTORS AT 20°C

| Gauge No. | SI Metric Units | | English Units | |
	Dia. (mm)	Ω/km	Dia. (mils)*	Ω/1000 ft
10	2.588	3.277	101.9	0.9989
11	2.305	4.132	90.74	1.260
12	2.053	5.211	80.81	1.588
13	1.828	6.571	71.96	2.003
14	1.628	8.285	64.08	2.525
15	1.450	10.45	57.07	3.184
16	1.291	13.17	50.82	4.016
17	1.150	16.61	45.26	5.064
18	1.024	20.95	40.30	6.385
19	0.912	26.42	35.89	8.051
20	0.812	33.31	31.96	10.15
21	0.723	42.00	28.45	12.80
22	0.644	52.96	25.35	16.14
23	0.573	66.79	22.57	20.36
24	0.511	84.21	20.10	25.67
25	0.455	106.2	17.90	32.37
26	0.405	133.9	15.94	40.81
27	0.361	168.9	14.20	51.47
28	0.321	212.9	12.64	64.90
29	0.286	268.5	11.26	81.83
30	0.255	338.6	10.03	103.2

*1 mil = 0.001 inch.

replaced with a no. 20 gauge copper wire. Determine
the resistance of the 20-gauge wire.

Solution Use Table 12-4 to determine the diameter of the 20-
gauge wire:

$$20 \text{ gauge} \Rightarrow 0.812 \text{ mm diameter}$$

Use Equation 12-5 and substitute 12 Ω for R_1, 0.511
mm for d_1, and 0.812 mm for d_2:

$$\frac{R_1}{R_2} = \frac{d_2^2}{d_1^2}$$

$$\frac{12}{R_2} = \frac{(0.812)^2}{(0.511)^2}$$

Solve for R_2:

$$R_2 = \frac{12(0.511)^2}{(0.812)^2}$$

$$\therefore \quad R_2 = 4.75 \ \Omega$$

**EXAMPLE
12-13**

Determine the diameter in millimeters of a conductor needed to replace a no. 30 gauge wire so that the resistance is one-fourth that of the 30-gauge wire. The resistance of the 30-gauge wire is 8 Ω.

Solution Use Table 12-4 to determine the diameter of the 30-gauge wire:

$$30 \text{ gauge} \Rightarrow 0.255 \text{ mm diameter}$$

Use Equation 12-5 and substitute 8 Ω for R_1, 0.255 mm for d_1, and 2 Ω for R_2:

$$\frac{8}{2} = \frac{d_2^2}{(0.225)^2}$$

Solve for d_2:

$$d_2^2 = \frac{8(0.255)^2}{2}$$

$$d_2 = \sqrt{\frac{8(0.255)^2}{2}}$$

$$\therefore \quad d_2 = 0.51 \text{ mm; this is 24-gauge wire.}$$

EXERCISE 12-6

In each problem it is assumed that all the conductors are round, solid, annealed copper wire of equal length and at 20°C.

1. A certain wire has a diameter of 0.079 mm and a resistance of 10 Ω. Determine the resistance when the diameter is increased to 0.254 mm.

2. A no. 32 wire has a diameter of 202 μm and a resistance of 52.7 Ω. Determine the resistance when the diameter is 0.102 mm.

3. One hundred meters of no. 10 wire, which has a diameter of 2.59 mm, has a resistance of 0.3277 Ω. Determine the resistance of 100 m of no. 6 wire, which has a diameter of 4.12 mm.
4. It is desired to reduce the resistance of 0.333 km of no. 18 wire to 3 Ω by replacing it with a new wire of a larger diameter.
   (a) Determine the diameter of the new wire.
   (b) Select a wire gauge from Table 12-4 that will meet the requirements of the new wire.
5. Repeat problem 4 for 140 m of no. 22 wire.
6. One hundred meters of no. 10 wire, which has a diameter of 2.59 mm, has a resistance of 0.3277 Ω. Determine the resistance of 100 m of no. 18 wire, which has a diameter of 1.024 mm.

Kind of Material

Conductors made of different metals are compared to one another by their *resistivity* (sometimes called specific resistance). Resistivity is measured in ohm-meters (Ωm) and is based on a standard conductor, which is 1 m in length and 1 m² in cross section at a temperature of 20°C. The Greek letter ρ (rho) is the symbol used to indicate resistivity, as in Equation 12-6.

$$\rho = \frac{RA}{l} \quad \text{(ohm-meters)} \tag{12-6}$$

where ρ = resistivity of the material in ohm-meters (Ωm)
R = resistance in ohms (Ω)
A = cross-sectional area in square meters (m²)
l = length in meters (m)

Table 12-5 lists the resistivity of several metals; silver, the best conductor, is listed first.

TABLE 12-5

RESISTIVITY OF SOME METALS AT 20°C			
Metal	*Resistivity (Ωm)*	*Metal*	*Resistivity (Ωm)*
Silver	0.0164×10^{-6}	Tungsten	0.0552×10^{-6}
Copper	0.0172×10^{-6}	Nichrome	1.00×10^{-6}
Aluminum	0.0283×10^{-6}		

EXAMPLE 12-14

Determine the resistance at 20°C of a nichrome wire of 0.404-mm diameter and 2.0 m long.

Solution   Express the diameter in meters:

$$0.404 \text{ mm} = 0.404 \times 10^{-3} \text{ m}$$

Compute the area of the conductor in square meters:

$$A = \frac{\pi}{4} d^2$$
$$= \frac{3.14}{4} (0.404 \times 10^{-3})^2$$
$$A = 0.128 \times 10^{-6} \text{ m}^2$$

Use Equation 12-6 and substitute 1.00×10^{-6} Ωm for ρ, 0.128×10^{-6} m^2 for A, and 2.0 m for l:

$$\rho = \frac{RA}{l}$$
$$1.00 \times 10^{-6} = \frac{R(0.128 \times 10^{-6})}{2.0}$$

Solve for R:

$$R = \frac{1.00 \times 10^{-6} \times 2.0}{0.128 \times 10^{-6}}$$
$$\therefore \quad R = 15.6 \ \Omega$$

**EXAMPLE
12-15**

Determine the resistance of a solid-copper rectangular conductor that is 10.0 cm long, 2.0 cm wide, and 0.50 cm thick at 20°C.

Solution   Compute the cross section in square meters:

$$A = \text{width} \times \text{thickness}$$
$$= 2.0 \times 10^{-2} \text{ m} \times 0.50 \times 10^{-2} \text{ m}$$
$$A = 1 \times 10^{-4} \text{ m}^2$$

Determine ρ for copper from Table 12-5:

$$\rho = 0.0172 \times 10^{-6} \text{ Ωm}$$

Express length in meters:

$$10.0 \text{ cm} = 10.0 \times 10^{-2} \text{ m}$$

Substitute 0.0172×10^{-6} Ωm for ρ, 10.0×10^{-2} m for l, and 1×10^{-4} m² for A in Equation 12-6:

$$\rho = \frac{RA}{l}$$

$$0.0172 \times 10^{-6} = \frac{R(1 \times 10^{-4})}{10 \times 10^{-2}}$$

Solve for R:

$$R = \frac{(0.0172 \times 10^{-6})(10 \times 10^{-2})}{1 \times 10^{-4}}$$
$$= 1.72 \times 10^{-5}\ \Omega$$
$$\therefore \quad R = 17.2\ \mu\Omega$$

EXERCISE 12-7

1. Determine the resistance of 15 m of aluminum wire that has a diameter of 2.588 mm at 20°C.
2. Determine the resistance of 0.2 m of nichrome wire that has a diameter of 0.255 mm at 20°C.
3. What is the length of a 24-gauge tungsten wire that has a resistance of 0.85 Ω at 20°C?
4. How long is an 18-gauge copper wire that has a resistance of 13.0 Ω at 20°C?
5. Determine the resistivity (ρ) of an unknown metal conductor that is 10.0 m long and has a diameter of 0.643 mm. The measured resistance is 1.61 Ω at 20°C.
6. Determine the resistance of the shunt pictured in Figure 12-4. It is made of constantan, which has a $\rho = 0.49 \times 10^{-6}$ Ωm at 20°C. Use the central cross section in Figure 12-4(b) to compute the area.

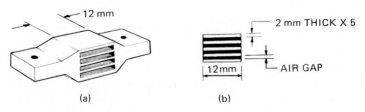

FIGURE 12–4
Constantan shunt for Exercise 12–7, problem 6.

Temperature

So far, we have investigated the effects that *length, cross-sectional area,* and *kind of material* have on electrical conductors. Now let us consider temperature.

Metals increase in resistance as temperature rises. A copper conductor at 20°C will have a substantial increase in resistance at 80°C. The amount of increase in resistance for each degree Celsius rise in temperature varies from metal to metal. The effect of temperature on the resistance of a copper conductor is computed by Equation 12-7.

$$\frac{R_2}{R_1} = \frac{234.5 + T_2}{234.5 + T_1} \qquad\qquad (12\text{-}7)$$

where R_1 = resistance of the copper conductor at T_1
R_2 = resistance of the copper conductor at T_2
$-234.5°C$ = extrapolated temperature, where copper has a resistance of $0\ \Omega$

EXAMPLE 12-16

Determine the resistance of a copper conductor at 80°C if it has a resistance of 10 Ω at 20°C.

Solution Solve Equation 12-7 for R_2:

$$R_2 = R_1 \frac{234.5 + T_2}{234.5 + T_1}$$

Substitute 10 Ω for R_1, 20°C for T_1, and 80°C for T_2:

$$R_2 = 10 \times \frac{234.5 + 80}{234.5 + 20}$$

$$\therefore \quad R_2 = 12.4\ \Omega$$

EXAMPLE 12-17

Number 28 gauge copper wire has a resistance of 213 Ω/km at 20°C. Determine the resistance at $-10°C$.

Solution Use Equation 12-7 and solve for R_2:

$$R_2 = R_1 \frac{234.5 + T_2}{234.5 + T_1}$$

Substitute 213 Ω for R_1, 20°C for T_1, and −10°C for T_2:

$$R_2 = 213 \times \frac{234.5 + (-10)}{234.5 + 20} = \frac{213 \times 224.5}{254.5}$$

$$\therefore \quad R_2 = 188 \ \Omega$$

EXERCISE 12-8

1. The resistance of 50 m of copper wire at 30°C is 1.62 Ω. Determine its resistance at 57°C.
2. The resistance of an RF coil wound with copper wire is 12.8 Ω at 42°C. Determine the resistance at −6.0°C.
3. The copper winding of a motor was measured at a temperature of 22°C and found to be 0.48 Ω. While in use, the resistance of the winding was found to be 0.55 Ω. Determine the temperature of the winding.
4. The resistance of the copper wire within an inductor was measured and found to be 2.78 Ω at a temperature of 28°C. After several hours of operation, the resistance increased to 3.34 Ω. Determine the temperature of the wire.

SELECTED TERMS

AWG American Wire Gauge; a table of standard wire diameters indicated by gauge numbers.

direct proportion Two ratios are in direct proportion when one ratio in the proportion increases and the other ratio increases or one ratio decreases and the other ratio also decreases.

indirect proportion Two ratios are in indirect proportion when one ratio in the proportion increases and the other ratio decreases.

parts per million Per one million; divided by 1 000 000.

percent Per one hundred; divided by 100.

proportion Equation made up of two ratios.

resistivity Resistance between opposite parallel faces of a 1-meter cube of a material. Resistivity depends only on the properties of the material; whereas, resistance depends on the properties of the material as well as length and cross-sectional area.

13

APPLYING FRACTIONS TO ELECTRICAL CIRCUITS

Many of the formulas used with electrical circuits are fractional in form. We will look at voltage division in series circuits and several concepts associated with parallel circuits. These topics involve both fractions and electronics.

13-1 VOLTAGE DIVISION IN THE SERIES CIRCUIT

In Chapter 8 we learned about the characteristics of the series circuit. One important characteristic is that the circuit current, I, is the same throughout the entire series circuit. In the series circuit of Figure 13-1, the circuit current may be determined by Ohm's law as

$$I = \frac{E}{R_T} \text{ or } I = \frac{V_1}{R_1} \text{ or } I = \frac{V_2}{R_2} \quad \text{(amperes)}$$

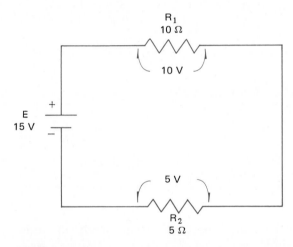

FIGURE 13–1
The source voltage, E, divides between resistance R_1 and R_2. The larger resistance has the greater voltage drop.

Since the current is constant in a series circuit, an equation may be written using the fractions of E/R_T and V_1/R_1. This equation states that the applied voltage, E, is to the total resistance, R_T, as the voltage drop across resistance one, V_1, is to the resistance one, R_1. Thus,

$$\frac{E}{R_T} = \frac{V_1}{R_1} \tag{13-1}$$

Equation 13-1 may be stated more generally as Equation 13-2.

$$\frac{E}{\Sigma R} = \frac{V_n}{R_n} \qquad\qquad (13\text{-}2)$$

where E = applied voltage
ΣR = summation of all the series resistances $(R_T = \Sigma R)$
V_n = voltage drop across R_n
R_n = one resistance in the series circuit
n = subscript of the selected resistor

EXAMPLE 13-1   Determine the voltage drop, V_2, across resistor R_2 in Figure 13-1 when $R_2 = 2.2$ kΩ, $R_T = 2.7$ kΩ, and $E = 40$ V.

Solution   Use Equation 13-2 and substitute 2.2 kΩ for R_2, 2.7 kΩ for ΣR, and 40 V for E:

$$\frac{E}{\Sigma R} = \frac{V_2}{R_2}$$

$$\frac{40}{2.7 \times 10^3} = \frac{V_2}{2.2 \times 10^3}$$

Solve for V_2:

$$V_2 = \frac{40 \times 2.2 \times 10^3}{2.7 \times 10^3}$$

$$\therefore \quad V_2 = 33 \text{ V}$$

EXERCISE 13-1
Use Equation 13-2 and Figure 13-1 to determine the indicated quantity in each of the following.

1. Find V_1 when $R_1 = 1.4$ kΩ, $\Sigma R = 22.5$ kΩ, and $E = 14.5$ V.
2. Find V_2 when $R_2 = 860$ Ω, $\Sigma R = 2.2$ kΩ, and $E = 16$ V.
3. Find V_2 when $R_2 = 1.5$ MΩ, $\Sigma R = 3.6$ MΩ, and $E = 20$ V.
4. Find E when $\Sigma R = 860$ kΩ, $R_2 = 560$ kΩ, and $V_2 = 12.5$ V.
5. Find E when $\Sigma R = 775$ kΩ, $R_1 = 7.00$ kΩ, and $V_1 = 0.600$ V.
6. Find E when $\Sigma R = 560$ Ω, $R_1 = 270$ Ω, and $V_1 = 375$ mV.
7. Find ΣR when $E = 46$ V, $R_1 = 375$ kΩ, and $V_1 = 12.4$ V.
8. Find ΣR when $E = 28$ V, $R_2 = 47$ kΩ, and $V_2 = 16.5$ V.
9. Find R_1 when $E = 15$ V, $\Sigma R = 3.23$ kΩ, and $V_1 = 4.04$ V.
10. Find R_2 when $E = 28$ V, $\Sigma R = 2.95$ kΩ, and $V_2 = 11.4$ V.

Voltage Divider

Since the applied voltage is divided between the resistances in a series circuit, a series circuit may be called a *voltage divider*. The voltage drop across any resistance in a series circuit may be determined by Equation 13-3, which results from solving for V_n in Equation 13-2.

$$V_n = \frac{ER_n}{\Sigma R} \quad \text{(volts)} \qquad (13\text{-}3)$$

where V_n = voltage drop across R_n
$\quad E$ = applied voltage
$\quad R_n$ = one resistance in the voltage divider
$\quad \Sigma R$ = summation of all the resistances in the divider

---

EXAMPLE 13-2 Compute the voltage drop across each of the resistances of Figure 13-2. Use Equation 13-3 and show that $E = V_1 + V_2 + V_3$.

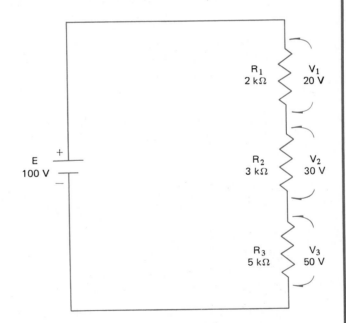

FIGURE 13-2
Series voltage divider. Each voltage drop is directly proportional to the resistance it is across. Larger resistances have larger voltage drops than smaller resistances.

Solution   Note that E is 100 V and ΣR is 10 kΩ in the solution for V_1, V_2, and V_3. Solve for V_1; substitute 2 kΩ for R_1:

$$V_1 = \frac{ER_1}{\Sigma R}$$
$$= \frac{100 \times 2 \times 10^3}{10 \times 10^3}$$
$$\therefore \quad V_1 = 20 \text{ V}$$

Solve for V_2; substitute 3 kΩ for R_2:

$$V_2 = \frac{ER_2}{\Sigma R}$$
$$= \frac{100 \times 3 \times 10^3}{10 \times 10^3}$$
$$\therefore \quad V_2 = 30 \text{ V}$$

Solve for V_3; substitute 5 kΩ for R_3:

$$V_3 = \frac{ER_3}{\Sigma R}$$
$$= \frac{100 \times 5 \times 10^3}{10 \times 10^3}$$
$$\therefore \quad V_3 = 50 \text{ V}$$

Show that the summation of the voltage drops equals the voltage rise:

$$E = V_1 + V_2 + V_3$$
$$100 = 20 + 30 + 50$$
$$100 \text{ V} = 100 \text{ V}$$

$\therefore$ The voltage rise equals the sum of the voltage drops.

EXERCISE 13-2

Use Equation 13-3 to determine the voltage drop across the indicated resistor of Figure 13-2, given the following information.

1. Find V_2 when $E = 700$ V, $R_1 = 2.2$ kΩ, $R_2 = 3.0$ kΩ, $R_3 = 1.8$ kΩ.
2. Find V_3 when $E = 36$ V, $R_1 = 390$ Ω, $R_2 = 110$ Ω, $R_3 = 500$ Ω.

3. Find V_2 when $E = 50$ V, $R_1 = 82$ kΩ, $R_2 = 300$ kΩ, $R_3 = 120$ kΩ.
4. Find V_1 when $E = 9$ V, $R_1 = 10$ kΩ, $R_2 = 27$ kΩ, $R_3 = 15$ kΩ.
5. Find V_3 when $E = 80$ V, $R_1 = 4.7$ kΩ, $R_2 = 5.6$ kΩ, $R_3 = 6.2$ kΩ.
6. Find V_1 when $E = 220$ V, $R_1 = 750$ kΩ, $R_2 = 1.2$ MΩ, $R_3 = 2.0$ MΩ.
7. Find V_2 when $E = 50$ V, $R_1 = 12$ kΩ, $R_2 = 22$ kΩ, $R_3 = 9.1$ kΩ.
8. Find V_1 when $E = 480$ V, $R_1 = 39$ Ω, $R_2 = 56$ Ω, $R_3 = 20$ Ω.
9. Find V_3 when $E = 1.2$ kV, $R_1 = 1.8$ MΩ, $R_2 = 2.7$ MΩ, $R_3 = 820$ kΩ.
10. Find V_3 when $E = 6$ V, $R_1 = 30$ kΩ, $R_2 = 56$ kΩ, $R_3 = 75$ kΩ.

13-2 CONDUCTANCE OF THE PARALLEL CIRCUIT

We have previously learned that resistance, measured in ohms (Ω), indicates the amount of *opposition* offered by a circuit or component to current flow. *Conductance,* measured in siemens (S), indicates the ease with which current *passes* through a circuit or component. Conductance is inversely proportional to resistance. This is stated as Equation 13-4.

$$G = 1/R \quad \text{(siemens)} \tag{13-4}$$

where G = conductance in siemens (S)
R = resistance in ohms (Ω)

EXAMPLE 13-3   Determine the conductance of a 500-Ω resistance.
   Solution   Use Equation 13-4 and your calculator.

$$G = 1/R$$
$$\boxed{1/x} \quad G = 1/500$$
$$\therefore \quad G = 2.00 \text{ mS}$$

EXERCISE 13-3

Calculator Exercise
Use the reciprocal key on your calculator to compute the conductance of the following resistances.

1. 56 Ω   2. 1.8 kΩ   3. 910 Ω
4. 390 kΩ   5. 6.2 kΩ   6. 3.3 MΩ

266

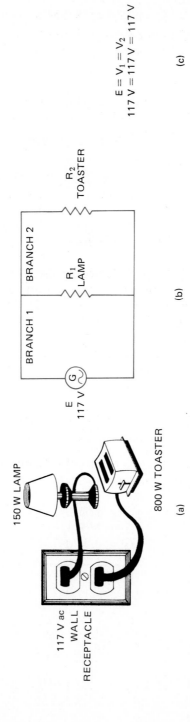

FIGURE 13-3

Parallel circuit consisting of a lamp and a toaster connected to a 117-V wall receptacle: (a) pictorial diagram of the two-branch parallel circuit; (b) schematic diagram of the two-branch parallel circuit; (c) the voltages (V_1 and V_2) across each branch are the same in a parallel circuit and they are equal to the applied voltage, *E*.

$$7. \ 2.7 \ k\Omega \qquad 8. \ 22 \ \Omega \qquad 9. \ 8.2 \ M\Omega$$
$$10. \ 750 \ \Omega \qquad 11. \ 560 \ m\Omega \qquad 12. \ 3.7 \ m\Omega$$

Parallel Circuit

When two or more resistances are connected in parallel, as shown in Figure 13-3, the voltage drop across each branch is the same, and it is equal to the source voltage. This is expressed as Equation 13-5.

$$E = V_1 = V_2 = V_3 \qquad \text{(volts)} \qquad\qquad (13\text{-}5)$$

In the parallel circuit, the total conductance of all the branches is determined by summing the conductance of each branch. Thus,

$$G_T = G_1 + G_2 + G_3 \qquad \text{(siemens)} \qquad\qquad (13\text{-}6)$$

where G_T = total conductance of the entire circuit in siemens (S)
       G_1 = conductance of the first branch
       G_2 = conductance of the second branch
       G_3 = conductance of the third branch

Equation 13-6 may be stated more generally as Equation 13-7:

$$G_T = \Sigma \ G \qquad \text{(siemens)} \qquad\qquad (13\text{-}7)$$

where $\Sigma \ G$ = summation of the branch conductances in siemens (S).

EXAMPLE 13-4 Compute the total conductance of the parallel circuit pictured by Figure 13-4.

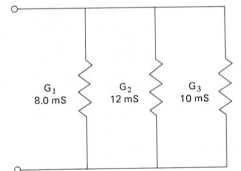

FIGURE 13–4
Circuit for Example 13–4.

Solution   Use Equation 13-7.

$$G_T = \Sigma \, G$$
$$\boxed{+} \quad G_T = 8.0 \times 10^{-3} + 12 \times 10^{-3} + 10 \times 10^{-3}$$
$$\therefore \quad G_T = 30 \text{ mS}$$

EXAMPLE 13-5   Compute the total conductance of the parallel circuit pictured by Figure 13-5.

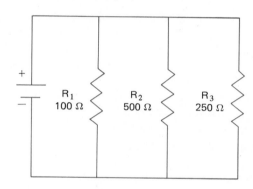

FIGURE   13–5
Circuit for Example 13–5.

Solution   Compute the conductances of each branch by Equation 13-4:

$$\boxed{1/x} \quad G_1 = 1/R_1 = 1/100 = 0.010 \text{ S}$$
$$\boxed{1/x} \quad G_2 = 1/R_2 = 1/500 = 0.002 \text{ S}$$
$$\boxed{1/x} \quad G_3 = 1/R_3 = 1/250 = 0.004 \text{ S}$$

Use Equation 13-7 to compute the total circuit conductance:

$$G_T = \Sigma \, G$$
$$\boxed{+} \quad G_T = 0.010 + 0.002 + 0.004$$
$$= 0.016 \text{ S}$$
$$\therefore \quad G_T = 16 \text{ mS}$$

Observation   These computations can be performed in the chain calculation mode on your calculator by recognizing that $G_T = \Sigma \, 1/R$.

EXERCISE 13-4
Determine the total conductances of each parallel circuit given the following information.

1. $G_1 = 3.0$ mS $\quad G_2 = 15$ mS
2. $G_1 = 0.30$ S $\quad G_2 = 0.80$ S
3. $G_1 = 150$ mS $\quad G_2 = 325$ mS
4. $G_1 = 1.0$ mS $\quad G_2 = 17$ mS $\quad G_3 = 9.0$ mS
5. $G_1 = 25$ μS $\quad G_2 = 15$ μS $\quad G_3 = 2.0$ μS
6. $R_1 = 5$ Ω $\quad R_2 = 10$ Ω $\quad R_3 = 20$ Ω
7. $R_1 = 150$ Ω $\quad R_2 = 300$ Ω $\quad R_3 = 100$ Ω
8. $R_1 = 2$ kΩ $\quad R_2 = 5$ kΩ $\quad R_3 = 3$ kΩ
9. $R_1 = 750$ Ω $\quad R_2 = 680$ Ω $\quad R_3 = 420$ Ω
10. $R_1 = 15$ Ω $\quad R_2 = 4.7$ Ω $\quad R_3 = 33$ Ω
11. $R_1 = 1.5$ kΩ $\quad R_2 = 3.9$ kΩ $\quad R_3 = 820$ Ω
12. $R_1 = 9.1$ kΩ $\quad R_2 = 3.0$ kΩ $\quad R_3 = 5.6$ kΩ
13. $R_1 = 180$ kΩ $\quad R_2 = 470$ kΩ $\quad R_3 = 270$ kΩ
14. $R_1 = 7.2$ kΩ $\quad R_2 = 1.1$ kΩ $\quad R_3 = 2.7$ kΩ
15. $R_1 = 2.2$ MΩ $\quad R_2 = 6.2$ MΩ $\quad R_3 = 12$ MΩ
16. $G_1 = 22$ μS $\quad R_2 = 12$ kΩ
17. $R_1 = 820$ Ω $\quad G_2 = 1.8$ mS
18. $R_1 = 10$ kΩ $\quad R_2 = 4.7$ kΩ $\quad G_3 = 220$ μS
19. $G_1 = 9.0$ μS $\quad R_2 = 91$ kΩ $\quad G_3 = 12$ μS
20. $G_1 = 18$ mS $\quad R_2 = 27$ Ω $\quad G_3 = 68$ mS

13-3 EQUIVALENT RESISTANCE OF THE PARALLEL CIRCUIT

A parallel circuit of two or more branches may be simplified to an *equivalent series circuit,* as pictured in Figure 13-6, by computing the total resistance (R_T) of the parallel circuit. The resulting equivalent circuit has the same source voltage, total current, and total resistance as the parallel circuit from which it was derived.

Computing the Total Resistance

The total resistance (R_T) of a parallel circuit is determined by first adding the conductance of each branch to get the total conductance, and then reciprocating the total conductance to get the total resistance. This is stated as Equation 13-8.

$$R_T = \frac{1}{\Sigma G} \quad \text{(ohms)} \qquad (13\text{-}8)$$

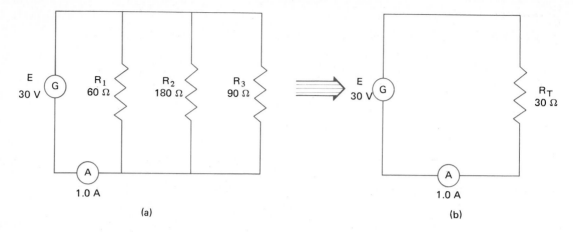

FIGURE 13-6
Equivalent series circuit of a parallel circuit: (a) parallel circuit with three branches
(R_1, R_2, R_3); (b) equivalent series circuit with the loads (R_1, R_2, R_3) reduced to R_T.

where R_T = total equivalent resistance of the parallel resistances in ohms
$\Sigma\, G$ = summation of the branch conductances in siemens

Equation 13-8 is used when the conductances are known. When the resistances are known, Equation 13-9 is more convenient.

$$R_T = \frac{1}{\dfrac{1}{R_1} + \dfrac{1}{R_2} + \dfrac{1}{R_3}} \quad \text{(ohms)} \qquad (13\text{-}9)$$

where R_T = total equivalent resistance of the parallel resistance
R_1, R_2, R_3 = resistances of the branches of the parallel circuit

Equation 13-9 is a very important equation as it is the one most often used with the calculator to compute the total resistance of a parallel circuit. The following guidelines are presented to aid in evaluating Equation 13-9.

GUIDELINES 13-1. COMPUTING TOTAL RESISTANCE

To compute R_T using Equation 13-9 and a calculator:
1. Take the reciprocal of R_1.
2. Add in the reciprocal of R_2 and the reciprocal of R_3.
3. Reciprocate the sum of the reciprocals.

EXAMPLE 13-6 Determine the total resistance (R_T) of the circuit of Figure 13-7.

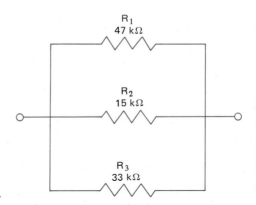

R₁
47 kΩ

R₂
15 kΩ

R₃
33 kΩ

FIGURE 13–7
Circuit for Example 13–6.

Solution Use Equation 13-9 and the previous guidelines:

$$R_T = \cfrac{1}{\cfrac{1}{R_1} + \cfrac{1}{R_2} + \cfrac{1}{R_3}}$$

Substitute 47 kΩ for R_1, 15 kΩ for R_2, and 33 kΩ for R_3:

$$R_T = \cfrac{1}{\cfrac{1}{47 \times 10^3} + \cfrac{1}{15 \times 10^3} + \cfrac{1}{33 \times 10^3}}$$

Use your calculator to evaluate the denominator:

$$\boxed{1/x} \quad R_T = \frac{1}{118 \times 10^{-6}} = 8.46 \times 10^3$$

$$\therefore \quad R_T = 8.5 \text{ k}\Omega$$

EXAMPLE 13-7 Determine the total resistance (R_T) of the circuit of Figure 13-8.

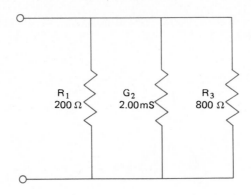

FIGURE 13–8
Circuit for Example 13–7.

Solution   Reciprocate R_1, add in G_2, and then add in the reciprocal of R_3 (G_2 is the reciprocal of R_2):

$$R_T = \cfrac{1}{\cfrac{1}{200} + 2.00 \times 10^{-3} + \cfrac{1}{800}}$$

$\boxed{1/x}$ $R_T = \cfrac{1}{8.25 \times 10^{-3}}$

$\therefore$ $R_T = 121\ \Omega$

EXERCISE 13-5

Compute the total resistance of each parallel circuit given the following information:

1. $R_1 = 100\ \Omega$ $R_2 = 300\ \Omega$
2. $R_1 = 750\ \Omega$ $R_2 = 420\ \Omega$
3. $R_1 = 33\ \Omega$ $R_2 = 15\ \Omega$
4. $R_1 = 9.1\ k\Omega$ $R_2 = 5.6\ k\Omega$ $R_3 = 3.9\ k\Omega$
5. $R_1 = 7.5\ k\Omega$ $R_2 = 2.2\ k\Omega$ $R_3 = 6.2\ k\Omega$
6. $R_1 = 4.7\ M\Omega$ $R_2 = 2.0\ M\Omega$ $R_3 = 3.0\ M\Omega$
7. $G_1 = 3.0\ \mu S$ $R_2 = 120\ k\Omega$ $R_3 = 91\ k\Omega$
8. $R_1 = 4.3\ k\Omega$ $G_2 = 800\ \mu S$ $R_3 = 2.7\ k\Omega$
9. $R_1 = 50\ \Omega$ $R_2 = 18\ \Omega$ $G_3 = 33\ mS$
10. $G_1 = 37\ \mu S$ $R_2 = 82\ k\Omega$ $G_3 = 17.9\ \mu S$

13-4 CURRENT DIVISION IN THE PARALLEL CIRCUIT

The total circuit current (I_T) in a parallel circuit divides between the various branches of the circuit, as shown in Figure 13-9. The value of each branch current depends directly upon the conductance of that branch. The branch with the largest conductance will have the largest current, while the branch with the smallest conductance will have the smallest current passing through it.

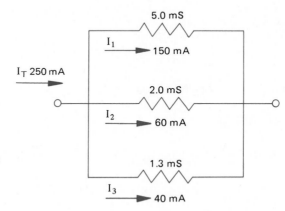

FIGURE 13–9
The division of the total current in a parallel circuit depends directly upon the conductance of each branch.

The total circuit current (I_T) in a parallel circuit is equal to the summation of the branch currents. This can be written two ways, as expressed in Equations 13-10 and 13-11.

$$I_T = I_1 + I_2 + I_3 \quad \text{(amperes)} \tag{13-10}$$

where I_1, I_2, I_3 = branch currents

$$I_T = \Sigma I \quad \text{(amperes)} \tag{13-11}$$

where I_T = total current in the parallel circuit
ΣI = summation of the branch currents

Current Divider

Since the total current (I_T) is divided between the branches of the parallel circuit, a parallel circuit may be called a *current divider*. The current passing through any branch of a parallel circuit may be determined by Equation 13-12.

$$I_n = \frac{I_T G_n}{\Sigma\ G} \qquad \text{(amperes)} \qquad\qquad (13\text{-}12)$$

where I_n = current through branch n in amperes (A)

I_T = total circuit current in amperes (A)

G_n = conductance of branch n in siemens (S)

$\Sigma\ G$ = summation of all the branch conductances in siemens (S)

EXAMPLE 13-8   Determine the current in each branch of the circuit of Figure 13-10.

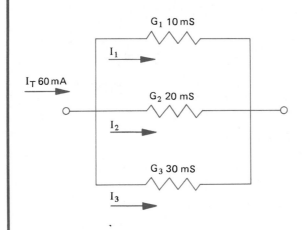

FIGURE 13–10
Circuit for Example 13–8.

Solution   Compute the total conductance of the circuit pictured in Figure 13-10:

$$\Sigma\ G = G_1 + G_2 + G_3$$
$$= 10 \times 10^{-3} + 20 \times 10^{-3} + 30 \times 10^{-3}$$
$$= 60\ \text{mS}$$

Note that $I_T = 60$ mA and $\Sigma\ G$ is 60 mS in the solution for I_1, I_2, and I_3. Solve for I_1 when $G_1 = 10$ mS:

$$I_1 = \frac{I_T G_1}{\Sigma\ G}$$
$$= \frac{60 \times 10^{-3} \times 10 \times 10^{-3}}{60 \times 10^{-3}}$$
$$\therefore \quad I_1 = 10\ \text{mA}$$

Solve for I_2 when $G_2 = 20$ mS:

$$I_2 = \frac{I_T G_2}{\Sigma\, G}$$

$$= \frac{60 \times 10^{-3} \times 20 \times 10^{-3}}{60 \times 10^{-3}}$$

$\therefore\quad I_2 = 20$ mA

Solve for I_3 when $G_3 = 30$ mS:

$$I_3 = \frac{I_T G_3}{\Sigma\, G}$$

$$= \frac{60 \times 10^{-3} \times 30 \times 10^{-3}}{60 \times 10^{-3}}$$

$\therefore\quad I_3 = 30$ mA

Check Does $I_T = I_1 + I_2 + I_3$ when $I_1 = 10$ mA, $I_2 = 20$ mA, $I_3 = 30$ mA, and $I_T = 60$ mA?

60 mA = 10 mA + 20 mA + 30 mA
60 mA = 60 mA   Yes!

EXERCISE 13-6
Use Equation 13-12 to determine the current through the indicated branch of the parallel circuits given the following information. It is suggested that a schematic be drawn and labeled for each problem.

1. Find I_3 when $I_T = 10$ A, $G_1 = 10$ mS, $G_2 = 25$ mS, $G_3 = 15$ mS.
2. Find I_1 when $I_T = 3.0$ mA, $G_1 = 80$ μS, $G_2 = 105$ μS, $G_3 = 150$ μS.
3. Find I_2 when $I_T = 12$ mA, $G_1 = 21.3$ μS, $G_2 = 66.7$ μS, $G_3 = 37.0$ μS.
4. Find I_1 when $I_T = 600$ mA, $G_1 = 17.9$ mS, $G_2 = 12.2$ mS, $G_3 = 30.3$ mS.
5. Find I_3 when $I_T = 38$ mA, $G_1 = 667$ μS, $G_2 = 370$ μS, $G_3 = 147$ μS.
6. Find I_2 when $I_T = 20$ μA, $G_1 = 0.833$ μS, $G_2 = 1.47$ μS, $G_3 = 2.13$ μS.
7. Find I_3 when $I_T = 2.5$ A, $G_1 = 37$ mS, $R_2 = 10\ \Omega$, $G_3 = 66.7$ mS.
8. Find I_1 when $I_T = 150$ mA, $R_1 = 910\ \Omega$, $G_2 = 455$ μS, $G_3 = 0.625$ mS.

9. Find I_3 when $I_T = 50$ mA, $R_1 = 2.0$ kΩ, $R_2 = 3.3$ kΩ, $G_3 = 556$ μS.
10. Find I_2 when $I_T = 300$ μA, $R_1 = 910$ kΩ, $R_2 = 680$ kΩ, $R_3 = 1.0$ MΩ.

13-5 SOLVING PARALLEL CIRCUIT PROBLEMS

The solution of parallel circuit problems involves the application of all the concepts covered so far in this chapter, as well as Ohm's law. The current in each branch of a parallel circuit may be computed by Ohm's law when the applied voltage and the branch resistance or conductance are known. Since $I = E/R$, we may write an equation specifically for computing the branch current in a parallel circuit.

$$I_n = \frac{E}{R_n} \quad \text{(amperes)} \quad (13\text{-}13)$$

where I_n = current in branch n of the parallel circuit
E = applied voltage
R_n = resistance of the branch n
n = branch index

EXAMPLE 13-9   Compute (a) the current in each branch of the circuit of Figure 13-11 and (b) the total current (I_T).

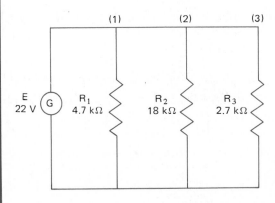

FIGURE   13–11
Circuit for Example 13–9.

Solution   (a) Use Equation 13-13:

$$I_1 = 22/(4.7 \times 10^3) = 4.7 \text{ mA}$$
$$I_2 = 22/(18 \times 10^3) = 1.2 \text{ mA}$$
$$I_3 = 22/(2.7 \times 10^3) = 8.2 \text{ mA}$$

(b) Use Equation 13-11: $I_T = \Sigma I$.

$$I_T = 4.7 + 1.2 + 8.2$$
$$\therefore \quad I_T = 14.1 \text{ mA}$$

Equation 13-13 is used to compute the branch currents when voltage and resistance are known. However, in analyzing solid-state circuits, conductance is often known instead of resistance, so Equation 13-13 is transformed into Equation 13-14, which is used to compute branch current when conductance is known.

$$I_n = EG_n \quad \text{(amperes)} \tag{13-14}$$

where I_n = current in branch n of the parallel circuit
E = applied voltage
G_n = conductance of branch n
n = branch index

EXAMPLE 13-10

Determine the current in branch 2 of the parallel circuit of Figure 13-12.

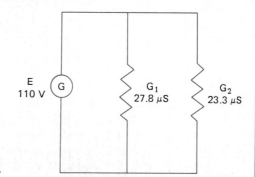

FIGURE 13-12
Circuit for Example 13-10.

Solution   Use Equation 13-14 and substitute 110 V for E and 23.3 μS for G_2:

$$I_2 = EG_2$$
$$= 110 \times 23.3 \times 10^{-6}$$
$$\therefore \quad I_2 = 2.56 \text{ mA}$$

Table 13-1 is a summary of the formulas used to solve parallel circuit problems.

TABLE 13-1

FORMULAS FOR PARALLEL CIRCUITS

Equation	Comment
$G = \dfrac{1}{R}$	Conductance G is the reciprocal of resistance R
$G_T = \Sigma\, G$	Total conductance G_T is the sum of the branch conductances
$E = V_1 = V_2$	Voltage drops across branches V_1 and V_2 are the same and are equal to the source voltage E
$R_T = \dfrac{1}{\Sigma\, G}$	Equivalent resistance R_T is formed by taking the reciprocal of the total conductance
$R_T = \dfrac{1}{\dfrac{1}{R_1} + \dfrac{1}{R_2} + \dfrac{1}{R_3}}$	Equivalent resistance R_T is the reciprocal of the sum of the reciprocals of the branch resistances
$I_T = \Sigma\, I$	Total current I_T is the summation of the individual branch currents
$I_n = \dfrac{I_T G_n}{\Sigma\, G}$	Current in one branch I_n is computed with the current-divider equation
$I_n = \dfrac{E}{R_n}$	Ohm's law is used to find the current in a branch when resistance and voltage are known
$I_n = E G_n$	Current in a branch is found by multiplying voltage and conductance

EXAMPLE 13-11

Two resistances, R_1 and R_2, are connected in parallel across a 5.0-V source. The total circuit current is 0.50 A. The current through R_1 is 200 mA.

(a) Draw a schematic of the circuit.
(b) Determine the value of V_1 and V_2.
(c) Determine the value of R_1 and R_2.
(d) Compute the equivalent resistance, R_T.

Solution (a) Draw and label a schematic. See Figure 13-13.
(b) V_1 and V_2 are equal to the applied voltage.

$$E = 5.0 \text{ V}$$
$$\therefore \quad V_1 = V_2 = 5.0 \text{ V}$$

(c) Determine the value of R_1:

$$R_1 = V_1/I_1$$
$$= 5.0/(200 \times 10^{-3})$$
$$\therefore \quad R_1 = 25 \ \Omega$$

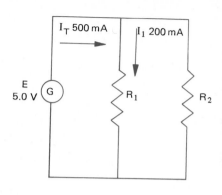

FIGURE 13-13
Schematic diagram for the circuit for Example 13-11.

Determine the value of R_2:

$$I_2 = I_T - I_1$$
$$= 500 \text{ mA} - 200 \text{ mA}$$
$$I_2 = 300 \text{ mA}$$
$$R_2 = V_2/I_2$$
$$= 5.0/(300 \times 10^{-3})$$
$$\therefore \quad R_2 = 16.7 \ \Omega$$

(d) Compute the equivalent resistance, R_T.

$$R_T = \cfrac{1}{\cfrac{1}{R_1} + \cfrac{1}{R_2}}$$

$$= \cfrac{1}{\cfrac{1}{25} + \cfrac{1}{16.7}}$$

$$\therefore \quad R_T = 10.0 \ \Omega$$

EXERCISE 13-7
Solve the following problems.

1. Determine the current through R_2 of Figure 13-14.
2. Compute the total conductance and the equivalent resistance of the circuit of Figure 13-14.
3. If the source voltage E of Figure 13-14 becomes 12.0 V, determine the current through R_1.
4. Determine the current through resistor R_1 in Figure 13-15 when $I_2 = 80 \text{ mA}$.

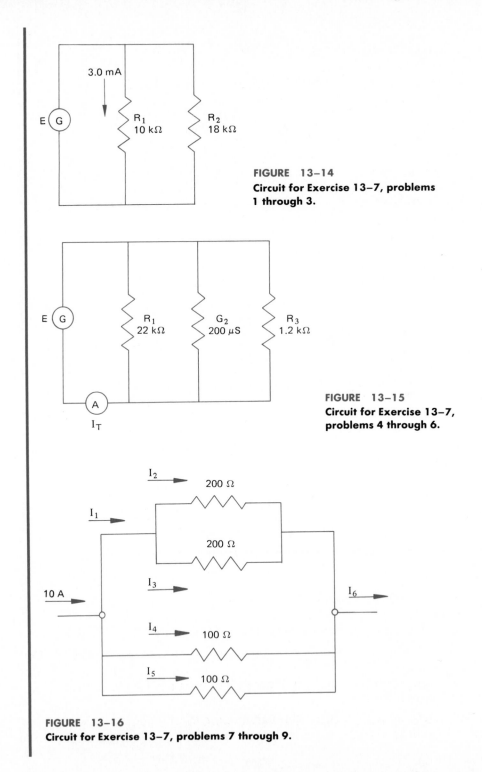

FIGURE 13–14
Circuit for Exercise 13–7, problems 1 through 3.

FIGURE 13–15
Circuit for Exercise 13–7, problems 4 through 6.

FIGURE 13–16
Circuit for Exercise 13–7, problems 7 through 9.

5. Compute the total circuit current, I_T, for Figure 13-15 when the source voltage, E, is 100 V.
6. Calculate the equivalent resistance of the circuit of Figure 13-15.
7. Find I_1 and I_2 in Figure 13-16.
8. Determine I_5 and I_6 in Figure 13-16.
9. Determine I_3 and I_4 in Figure 13-16.
10. Three resistors are connected in parallel. The total circuit current is 300 mA, and R_1 has a conductance of 1.25 mS. The second resistor has a voltage drop of 28.0 V across it, and the third resistor has a current of 112 mA passing through it. Determine the resistance of each of the three resistors.

13-6 USING NETWORK THEOREMS TO FORM EQUIVALENT CIRCUITS

The analysis of an electric circuit is simplified when the *network theorems* are used to form an equivalent circuit. In this section you will be introduced to Thévenin's theorem and Norton's theorem.

In electronics we use mathematical formulation to *model* electrical circuits. When the mathematical model is a good approximation of the actual circuit, then both the original circuit and its model will yield valid results and the two representations are *equivalent*.

Any linear two-terminal circuit may be represented by an electrically equivalent circuit consisting of:

1. A voltage source and a series resistance (Thévenin's)
2. A current source and a parallel resistance (Norton's)

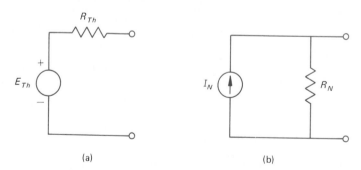

(a)

(b)

FIGURE   13–17
Equivalent circuits: (a) Thévenin's equivalent circuit; (b) Norton's equivalent circuit.

Figure 13-17 pictures each of these equivalent circuits. Figure 13-17(a) shows a **Thévenin's equivalent circuit** consisting of a *Thévenin's equivalent voltage source, E_{Th},* and a series *Thévenin's equivalent resistance, R_{Th}.* Figure 13-17(b) pictures a **Norton's equivalent circuit** consisting of a *Norton's equivalent current source, I_N,* and a parallel *Norton's equivalent resistance, R_N.*

Since both Thévenin's and Norton's theorems deal with equivalent models of circuits connected between two terminals, the circuit of Figure 13-18 may be represented by either one of these two *network theorems.* Each of the theorems will be used to model the circuit of Figure 13-18.

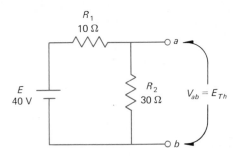

FIGURE 13–18
Circuit for Example 13–12.

EXAMPLE 13-12

Form the Thévenin's and Norton's equivalent circuits for the circuit of Figure 13-18.

Solution   View the circuit of Figure 13-18 from terminals *a-b.* Thévenin's equivalent voltage (E_{Th}) is V_{ab} of Figure 13-18. Compute E_{Th} using the voltage-divider equation, Equation 13-3.

$$E_{Th} = V_{ab} = \frac{ER_2}{R_1 + R_2} = \frac{(40)(30)}{40} = 30 \text{ V}$$

Observation V_{ab} is the voltage across R_2.

Norton's equivalent current (I_N) is the current that passes through a short circuit placed across terminals *a-b,* as pictured in Figure 13-19(a).

$$I_N = E/R_1 = 40/10 = 4\text{A}$$

Observation R_2 is bypassed by the short circuit.

Compute Thévenin's equivalent resistance by setting the voltage source to zero (short circuit) as

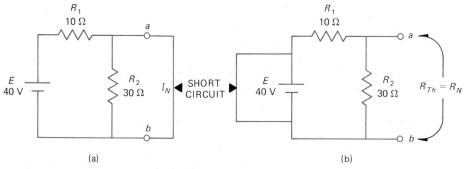

FIGURE 13–19

(a) Norton's equivalent current, I_N, is the current passing in the short circuit between terminals a–b.
(b) The equivalent resistance ($R_{Th} = R_N$) is calculated by first setting the voltage source to zero (short circuit) and then looking across terminals a–b.

	shown in Figure 13-19(b) and calculate the resistance between terminals a-b.
Observation	R_1 is now in parallel with R_2. Use Equation 13-9 to compute R_{Th}.

$$R_{Th} = \frac{1}{\dfrac{1}{R_1} + \dfrac{1}{R_2}} = \frac{1}{\dfrac{1}{10} + \dfrac{1}{30}} = 7.5\ \Omega$$

Compute Norton's equivalent resistance by setting the voltage source to zero (short circuit) as pictured in Figure 13-19(b) and calculate the resistance between terminals a-b.

Observation The Norton's equivalent resistance is computed in the same manner as the Thévenin's equivalent resistance. Thus:

$$R_{Th} \equiv R_N \qquad\qquad (13\text{-}15)$$

$$R_N = R_{Th} = \frac{1}{\dfrac{1}{R_1} + \dfrac{1}{R_2}} = \frac{1}{\dfrac{1}{10} + \dfrac{1}{30}} = 7.5\ \Omega$$

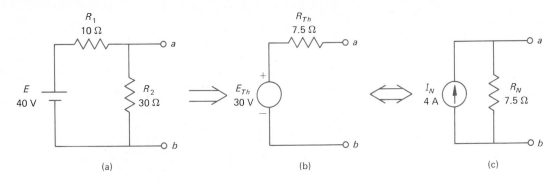

FIGURE 13-20
(a) Circuit for Example 13-12; (b) Thévenin's equivalent circuit of circuit (a); (c) Norton's equivalent circuit of circuit (a).

∴   The Thévenin's equivalent circuit is formed as shown in Figure 13-20(b) and the Norton's equivalent circuit is formed as shown in Figure 13-20(c).

Observation   The original circuit is pictured as Figure 13-20(a).

Before moving on to another example, a word of caution is in order. In forming the Norton's equivalent circuit, the terminals were short circuited to determine the short-circuit current, I_N. You must realize that this is a mathematical technique (pencil-and-paper process) that must not be attempted in the laboratory with real laboratory equipment. This is also true of the technique used to set the voltage source to zero.

EXAMPLE 13-13

Form the Thévenin's and Norton's equivalent circuits for the circuit pictured in Figure 13-21(a).

Solution   Determine the equivalent resistance by setting the current source to zero (open circuit) and calculating the resistance between terminals a-b.

$$R_{ab} = R_{Th} = R_N = \cfrac{1}{\cfrac{1}{R_1} + \cfrac{1}{R_2}} = \cfrac{1}{\cfrac{1}{10} + \cfrac{1}{15}} = 6\ \Omega$$

Norton's equivalent current is the current that passes through a short circuit across terminals a-b. Because

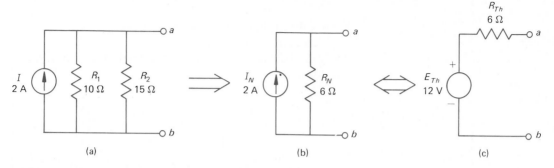

FIGURE 13–21
(a) Circuit for Example 13–13; (b) Norton's equivalent circuit of circuit (a); (c) Thévenin's equivalent circuit of circuit (a).

the short bypasses R_1 and R_2, the Norton's current is equal to the source current of 2 A.

$$I_N = 2 \text{ A}$$

Observation Because the two network theorems are equivalent, the Thévenin's equivalent voltage, E_{Th}, may be determined from the Norton's equivalent current, I_N, and, conversely, the Norton's equivalent current, I_N, may be determined from the Thévenin's equivalent voltage, E_{Th}. Stated mathematically,

$$I_N = E_{Th}/R_{Th} \qquad \text{where } R_{Th} = R_N \quad (13\text{-}16)$$
$$E_{Th} = I_N R_N \qquad \text{where } R_N = R_{Th} \quad (13\text{-}17)$$

Compute Thévenin's equivalent voltage using Equation 13-17 and Figure 13-21(b).

$$E_{Th} = I_N R_N = (2)(6) = 12 \text{ V}$$

∴ The Norton's equivalent is formed as shown in Figure 13-21(b) and the Thévenin's equivalent is formed as shown in Figure 13-21(c).

Observation The two network theorems are equally valid since both yield the same results. The choice of which theorem to use depends on how you choose to model the circuit.

EXAMPLE
13-14

Demonstrate that the two circuits pictured in Figure 13-22 are equivalent; that is, the parameters of one circuit may be derived from the parameters of the other circuit.

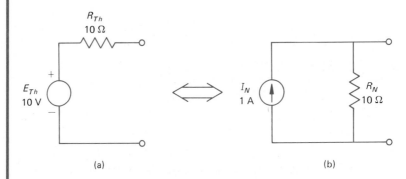

(a)   (b)

FIGURE 13–22
Conversion from Thévenin's (a) to
Norton's (b) equivalent circuit for Example
13–14.

Solution   Derive the Norton's equivalent circuit from the Thévenin's equivalent circuit in Figure 13-22(a). Use Equations 13-15 and 13-16.

$$\therefore \quad R_\text{N} = R_\text{Th} = 10 \ \Omega$$

and

$$I_\text{N} = E_\text{Th}/R_\text{Th} = 10/10 = 1 \ \text{A}$$

Derive the Thévenin's equivalent circuit from the Norton's equivalent circuit of Figure 13-22(b). Use Equations 13-15 and 13-17.

$$\therefore \quad R_\text{Th} = R_\text{N} = 10 \ \Omega$$

and

$$E_\text{Th} = I_\text{N}R_\text{N} = (1)(10) = 10 \ \text{V}$$

Summary

Using network theorems to form equivalent circuits is a straightforward procedure. First, calculate the equivalent resistance ($R_N = R_{Th}$) by looking into the two terminals of the selected circuit. Set all sources to zero; that is, short-circuit voltage sources and open-circuit current sources. Second, calculate either the open-circuit voltage (E_{Th}) or the short-circuit current (I_N) at the selected terminals.

> EXERCISE 13-8
> Solve the following problems using Equations 13-15, 13-16, and 13-17.
>
> 1. Determine the parameters of Norton's equivalent circuit of Figure 13-17(b) when $E_{Th} = 25$ V and $R_{Th} = 150$ Ω in Figure 13-17(a).
> 2. Repeat problem 1 when $E_{Th} = 16$ V and $R_{Th} = 200$ Ω.
> 3. Determine the parameters of the Thévenin's equivalent circuit of Figure 13-17(a) when $I_N = 0.3$ A and $R_N = 47$ Ω in Figure 13-17(b).
> 4. Repeat problem 3 when $I_N = 28$ mA and $R_N = 1.2$ kΩ.
> 5. Determine the Thévenin's equivalent circuit of a power supply with a specified open-circuit voltage of 20 V (E_{Th}) and a short-circuit current (I_N) of 2 A.
> 6. Repeat problem 5 for an open-circuit voltage of 12 V and a short-circuit current of 5 A.
> 7. Determine the Norton's equivalent circuit of Figure 13-18 when $E = 15$ V, $R_1 = 68$ Ω, and $R_2 = 56$ Ω.
> 8. Determine the Norton's equivalent circuit of Figure 13-18 when $E = 6.3$ V, $R_1 = 180$ Ω, and $R_2 = 220$ Ω.
> 9. Determine the Thévenin's equivalent circuit of Figure 13-21(a) when $I = 45$ mA, $R_2 = 2.7$ kΩ, and $R_2 = 3.0$ kΩ.
> 10. Determine the Thévenin's equivalent circuit of Figure 13-21(a) when $I = 85$ mA, $R_1 = 3.9$ kΩ, and $R_2 = 4.7$ kΩ.

SELECTED TERMS

conductance   Measure of the ease with which a conductor carries an electric current; the reciprocal of resistance.

current divider   Two or more resistances (loads) connected in parallel across a source will cause the source current to divide.

voltage divider   Two or more resistances connected in series with a voltage source form a voltage divider.

14

RELATIONS AND FUNCTIONS

A function is a mathematical statement about the relationship between numerical quantities. Such mathematical statements are used to bring the power of mathematics to bear on physical problems in electronics. The topics in this chapter have been selected to introduce functions and functional notation.

14-1 MEANING OF A FUNCTION

By studying the cause and effect relationship between quantities, laws are formulated. Ohm's law was formed by such observations. In the last chapters, many equations were used to solve problems. In applying these equations you may have observed that the quantities represented by the variables are interrelated with one another. A change in the value of one quantity caused a change in another quantity.

When one condition in a relationship depends on one or more other conditions in the relationship, it is said that the first condition is the *function* of the other conditions. In the equation $y = 2x$, y is a function of x. We can say that y is a function of x if for each value of x there is a corresponding value of y.

EXAMPLE 14-1   Determine if the following statement conforms to the definition of a function. The voltage drop across a resistor is a *function* of the current through the resistance.

Solution   Use Ohm's law: $V = IR$. Let R remain constant. If various values of I are substituted into the formula, a value of V results in each case, even when zero is substituted. So voltage V is a *function* of current I.

14-2 VARIABLES AND CONSTANTS

Dependent and Independent Variables

A dependence of one variable on another is indicated by the definition of a function. In the equation $y = 5x + 2$, the value of y *depends* upon the selection of the value for x. For example, when $x = 2$, $y = 5(2) + 2 = 12$.

In the equation $y = 5x + 2$, x is the *independent variable*. We may let x assume any value that we wish independent of the equation. In an equation the independent variable appears in the right member.

In the equation $y = 5x + 2$, y is the *dependent variable* because the value of y depends on the value chosen for x. In an equation the dependent variable is the left member.

EXAMPLE 14-2 In the following equations, identify the independent and the dependent variables.

$$\text{(a)} \; I_1 = \frac{I_T G_1}{\Sigma \, G}$$

(b) $R_T = R_1 + R_2 + R_3$

(c) $E = V_1 + V_2$

Solution (a) I_1 is the dependent variable.
I_T, G_1, and $\Sigma \, G$ are independent variables.
(b) R_T is the dependent variable.
R_1, R_2, and R_3 are the independent variables.
(c) E is the dependent variable.
V_1 and V_2 are the independent variables.

Constants

There are three types of constants used in mathematics. The first type of constant is a regular number such as 23 and 57.9. The second type of constant is represented by a special symbol such as π (3.14159) and e (2.71828). The third type of constant is a variable with an arbitrary value that remains fixed during a problem or discussion. The use of an *arbitrary constant* is illustrated in the following.

When current is determined from the applied voltage and the circuit resistance, Ohm's law is written as $I = E/R$. If R is a fixed resistor of 100 Ω and E is varied, then the current is the dependent variable, the voltage is the independent variable, and the resistance is an arbitrary constant.

EXAMPLE 14-3 Identify the symbols representing the constants and variables in the following equations.

(a) $u = 3v + b$, where $b = 16.4$

(b) $A = \pi r^2$

(c) $y = 8x^2 + 6x + e$

Solution (a) b and 3 are the constants.
u and v are the variables.

(b) π is the constant.

r and A are the variables.

(c) 8, 6, and e are the constants.

x and y are the variables.

EXERCISE 14-1
In each of the following equations, state which symbols represent constants, dependent variables, and independent variables.

1. The force (F) required to keep motion in a circular path is a function of mass (m), speed (v), and radius (r). The equation for centripetal force is $F = kmv^2/r$, where $k = 1$.
2. The distance traveled by a freely falling body(s) is a function of time (t) and the acceleration due to gravity (g). The equation for displacement of a freely falling body is $s = \frac{1}{2}gt^2$, where g is approximately 9.81 m/s^2 at the surface of the earth.
3. The attractive force (F) between two particles is a function of their masses (m_1 and m_2) and the distance between them (d). The equation for the attractive force is $F = Gm_1m_2/d^2$, where G is the gravitational constant.
4. The brake horsepower (P) developed by a motor is measured by a device called a *pony brake.* The brake horsepower is a function of the rotational speed (ω) and the difference in force between two balances (ΔF). The equation for brake horsepower is $P = 2\pi r\omega(\Delta F)/33\ 000$, where r is constant.
5. The speed of sound (v) through a gaseous medium is a function of the absolute temperature (T). The equation for the speed of sound is $v = RT/M$, where R is the gas constant.
6. The electrical resistance (R) of a conductor is a function of the length of the conductor (l) and the cross-sectional area (A). The equation for the resistance is $R = \rho l/A$, where ρ is the resistivity of the conductor—a constant.

14-3 FUNCTIONAL NOTATION

In the equation $y = 5x + 2$, y is a *function of x.* The statement that y is a function of x is written in functional notation as $y = f(x)$, where $f(x)$ is read "function of x," "f of x," or "f function of x." Thus, $y = 5x + 2$ may be written $f(x) = 5x + 2$. In this case, remember that $y = f(x) = 5x + 2$ means that y, $f(x)$, and $5x + 2$ are all the same function of x.

EXAMPLE 14-4   Use functional notation to state that:

(a) e is a function of i.
(b) u is a function of v.
(c) c is a function of r.

Solution   (a) $e = f(i)$
(b) $u = f(v)$
(c) $c = f(r)$

In the preceding examples, it is important that you understand that the notation $f(i)$, $f(v)$, $f(r)$ indicates functional relationship and not that f is multiplied by some variable. The letter f is usually used in functional notation. However, other letters may also be used, as shown in the following example.

EXAMPLE 14-5   Use functional notation to write the following expressions:

(a) y equals f of x.
(b) w is the g function of y.
(c) v is the h function of u.
(d) s equals Q of t.

Solution   (a) $y = f(x)$
(b) $w = g(y)$
(c) $v = h(u)$
(d) $s = Q(t)$

The functional notation $y = f(x)$ does not show the specific function of x that y represents; $y = f(x)$ could indicate that y is any function of x. For example,

$$y = f(x) = 5x + 2$$
$$y = f(x) = x^2$$
$$y = f(x) = \sin(x)$$
$$y = f(x) = 3x^3 - 8$$

To indicate a given function of x, such as $5x + 2$, we should simply use $f(x) = 5x + 2$. Functional notation also provides a way to show the specific value of the independent variable to be used. Example 14-6 shows several instances of finding the value of a function for a particular value of the independent variable. This process is called *evaluating a function*.

EXAMPLE 14-6   Given that $f(x) = 3x + 2$, evaluate:

(a) $f(5)$ (b) $f(-3)$ (c) $f(0)$ (d) $f(a)$ (e) $f(m)$

Solution   (a) To find $f(5)$, substitute 5 for x:

$$f(x) = 3x + 2$$
$$f(5) = 3(5) + 2 = 17$$

(b) To find $f(-3)$, substitute -3 for x:

$$f(x) = 3x + 2$$
$$f(-3) = 3(-3) + 2 = -7$$

(c) To find $f(0)$, substitute 0 for x:

$$f(x) = 3x + 2$$
$$f(0) = 3(0) + 2 = 2$$

(d) To find $f(a)$, substitute the constant a for x:

$$f(x) = 3x + 2$$
$$f(a) = 3a + 2$$

(e) To find $f(m)$, substitute the constant m for x:

$$f(x) = 3x + 2$$
$$f(m) = 3m + 2$$

EXAMPLE 14-7   Evaluate $f(x) = 2x^2 - 5x$ for $x = 5$.
Solution

$$f(x) = 2x^2 - 5x$$
$$f(5) = 2(5)^2 - 5(5)$$
$$\therefore \quad f(5) = 25$$

EXERCISE 14-2
Given that $f(e) = 5e - 3$, find:

1. $f(2)$	2. $f(-5)$	3. $f(4)$	4. $f(-1)$
5. $f(-2)$	6. $f(0)$	7. $f(-b)$	8. $f(3a)$

Use functional notation to express each of the following four statements.

9. x is the g function of y.
10. u is the f function of x.
11. z is f of v.
12. y equals g of w.
13. $f(x) = 4x^2 - x + 2$; find $f(2)$, $f(-3)$, $f(0.5)$.
14. $g(r) = \pi r^2$; find $g(3)$, $g(5)$, $g(0.316)$.
15. $Q(e) = \dfrac{1}{e - 2} + \dfrac{1}{e - 3}$; find $Q(1)$, $Q(0.5)$, $Q(5)$.

14-4 FUNCTIONAL VARIATION

When applying a formula to determine the value for a circuit component or a circuit condition, it is often necessary to make several calculations before a satisfactory result is reached. In this process it is very helpful to have an understanding of how the formula affects the selected value in producing the solution.

Suppose that you have constructed a series circuit with a resistive load. You have energized the circuit and measured the source voltage and found it to be too low for your application. You are about to increase the voltage, but you now wonder if the load will be able to dissipate an increase in the power. Your question, then, is, "how does power vary with voltage?" By investigating the functional relationship between power and voltage ($P = E^2/R$), you will have an understanding of how voltage affects power. From the relationship of power and voltage, you may see that an increase by a factor of 10 in voltage produces a *hundredfold* increase in power.

This same basic question can arise in many settings. It will be useful to understand how changing the independent variable will affect the dependent variable without repeating the whole calculation each time.

EXAMPLE 14-8   In the formula $E = IR$, let R be constant, and determine how E changes when I is:

(a) Doubled   (b) Halved   (c) Multiplied by 10
(d) Divided by 10

Solution   (a) $E(I) = IR$
$E(2I) = 2IR = 2E(I)$
E is doubled when I is doubled.

(b) E is halved when I is halved.
(c) E is multiplied by 10 when I is multiplied by 10.
(d) E is divided by 10 when I is divided by 10.

In Example 14-8 we see that the change in I due to multiplying or dividing produces the same change in E. In this example, E *varies directly* as I varies. This behavior is called *direct variation*.

EXAMPLE 14-9

In the formula $X_C = \dfrac{1}{\omega C}$ assume that ω is a constant.
Determine how X_C changes when C is:

(a) Multiplied by 5
(b) Divided by 5
(c) Multiplied by 10
(d) Divided by 10

Solution

(a) $X_C(C) = \dfrac{1}{\omega C}$

$$X_C(5C) = \dfrac{1}{\omega 5C} = \dfrac{1}{5}\left(\dfrac{1}{\omega C}\right) = \dfrac{1}{5}X_C(C)$$

X_C is divided by 5 when C is multiplied by 5.
(b) X_C is multiplied by 5 when C is divided by 5.
(c) X_C is divided by 10 when C is multiplied by 10.
(d) X_C is multiplied by 10 when C is divided by 10.

In Example 14-9 we see that the change in C due to multiplying or dividing produces an inverse change in X_C. In this example, X_C *varies inversely* as C varies; this is called an *inverse variation*.

**EXAMPLE
14-10**

In the formula $P = I^2R$, let R be a constant. Determine how P changes when I is:

(a) Doubled
(b) Halved
(c) Multiplied by 10
(d) Divided by 10

Solution (a) *I* is used as a factor twice; thus RI^2 is multiplied by 2 twice when *I* is multiplied by 2 once. So *P* is multiplied by 4 when *I* is doubled.

(b) RI^2 is multiplied by $\frac{1}{2}$ twice when *I* is multiplied by $\frac{1}{2}$ once. So *P* is multiplied by $\frac{1}{4}$ when *I* is halved.

(c) RI^2 is multiplied by 10 twice when *I* is multiplied by 10 once. So *P* is multiplied by 100 when *I* is multiplied by 10.

(d) RI^2 is multiplied by $\frac{1}{10}$ twice, when *I* is multiplied by $\frac{1}{10}$ once. So *P* is multiplied by $\frac{1}{100}$ when *I* is divided by 10.

In Example 14-10 we see that, when *I* is multiplied by a number, *P* is multiplied by the square of the number. Thus, *P* varies directly with the square of *I*.

EXAMPLE 14-11

In the formula $F = \dfrac{Wv^2}{gr}$, *W* and *g* are constant. Determine the effect on *F* when:

(a) *r* is halved and *v* is doubled.
(b) *r* is doubled and *v* is tripled.

Solution (a) Since *W* and *g* are constant, assume them to be worth 1.

$$F(v,r) = \frac{v^2}{r}$$

Halve *r* and double *v*:

$$F(2v, r/2) = \frac{(2v)^2}{1/2\ r}$$
$$= \frac{(2v)^2\,(2)}{r}$$
$$= \frac{(4v^2)(2)}{r}$$
$$= \frac{8v^2}{r}$$

$\therefore$ F is multiplied by 8 when r is halved and v is doubled.

(b) Double r and triple v:

$$F(3v, 2r) = \frac{(3v)^2}{2r}$$
$$= \frac{9v^2}{2r}$$

$\therefore$ F is multiplied by $\frac{9}{2}$ when r is doubled and v is tripled.

EXERCISE 14-3

1. In the conductance formula $G = 1/R$, how does G change when R is:
    (a) Halved?   (b) Tripled?   (c) Divided by 5?
    (d) How does G vary with R?

2. The inductance of a coil (L) is determined by the formula $L = \frac{\mu N^2 A}{l}$.
    (a) If μ, N^2, and l are constants, how does L vary with A?
    (b) If μ, N^2, and A are constants, how does L vary with l?
    (c) How does L change when N is tripled and μ, A, and l remain the same?

3. In the Ohm's law formula $I = E/R$:
    (a) How is I changed if E is one quarter as large and R is unchanged?
    (b) How is I changed if R is halved and E is constant?
    (c) How is I changed if both E and R are doubled?

4. The average power dissipated by a resistance is determined by the formula $P = E^2/R$. How is P affected when:
    (a) E is doubled and R remains constant?
    (b) R is one quarter as large and E is the same?
    (c) Both E and R are doubled?

5. The bandwidth (BW) of a resonant circuit is determined by the formula $BW = \frac{f_0 R}{X_L}$. If f_0 is constant, then how does BW change when:
    (a) X_L is doubled and R remains the same?
    (b) R is doubled and X_L remains the same?
    (c) R is halved and X_L is doubled?

6. The resistance of a conductor is determined by the formula $R = \rho \dfrac{l}{A}$.

If ρ is constant, then how is R changed when:
- (a) l is divided by 4 and A remains the same?
- (b) Both l and A are divided by 6?
- (c) l is multiplied by 3 and A is divided by 2?

7. The mutual inductance (M) between two coils is determined by the formula $M = k\sqrt{L_1 L_2}$.
- (a) How is M changed if both L_1 and L_2 are doubled and k is constant?
- (b) How is M changed if L_1 is one quarter as large, L_2 is tripled, and k is constant?

8. The closing force developed by the electromagnet of a relay is determined by the formula $F = AB^2/2\mu$. If μ is a constant, then how is F affected when:
- (a) B is halved and A is doubled?
- (b) Both B and A are doubled?
- (c) B is one and a half times as large and A is four times as large?

14-5 SIMPLIFYING FORMULAS

Many formulas used in electronic technology may be simplifed by making some assumptions about the value of one or more of the independent variables. These assumptions can only be made with an understanding of the conditions of the circuit under consideration. The formula for finding the equivalent resistance of two resistors in series is such an equation.

EXAMPLE 14-12

Investigate when the formula for equivalent resistance of a series circuit, $R_T = R_1 + R_2$, may be simplified to $R_T \approx R_1$.

Solution Select values of 1.0 Ω for R_1 and R_2. Substitute and solve

$$R_T = 1 + 1 = 2\,\Omega$$

Notice that the ratio of 2 Ω and 1 Ω is not very close to 1. Increase the value of R_1 an *order of magnitude* from 1.0 Ω to 10 Ω, and let R_2 remain at 1.0 Ω. Substitute and solve:

$$R_T = 10 + 1 = 11\,\Omega$$

Notice that the ratio of 11 Ω to 10 Ω is much, much closer to 1 than the ratio of 2 Ω to 1 Ω.

∴ When the ratio of R_1 to R_2 is equal to or greater than (≥) 10 to 1, we may simplify the equation $R_T = R_1 + R_2$ to $R_T \approx R_1$.

EXAMPLE 14-13

Investigate when the formula for the equivalent resistance of a parallel circuit, $R_T = \dfrac{R_1 R_2}{R_1 + R_2}$ may be simplified to $R_T \approx R_1$.

Solution Select values of 1.0 Ω for R_1 and R_2. Substitute and solve:

$$R_T = \frac{R_1 R_2}{R_1 + R_2}$$

$$= \frac{1.0 \times 1.0}{1.0 + 1.0} = \frac{1}{2} = 0.5 \ \Omega$$

Increase the value of R_2 an *order of magnitude* from 1.0 Ω to 10 Ω, and let R_1 remain at 1.0 Ω. Substitute and solve:

$$R_T = \frac{1.0 \times 10}{1.0 + 10} = \frac{10}{11} = 0.91 \ \Omega$$

Once again increase the value of R_2 an *order of magnitude* from 10 Ω to 100 Ω, and let R_1 remain at 1.0 Ω. Substitute and solve:

$$R_T = \frac{1.0 \times 100}{1.0 + 100} = \frac{100}{101} = 0.99 \ \Omega$$

Finally, increase the value of R_2 an *order of magnitude* from 100 Ω to 1000 Ω, and let R_1 remain at 1.0 Ω. Substitute and solve:

$$R_T = \frac{1.0 \times 1000}{1.0 + 1000} = \frac{1000}{1001} = 0.999 \ \Omega$$

∴ $R_T \approx R_1$ when $R_2 : R_1 \geq 10 : 1$

The approximation is even better when $R_2:R_1 > 10:1$. Table 14-1 is a summary of the investigation carried out in this example.

TABLE 14-1

SUMMARY OF EXAMPLE 14-13			
Ratio of R_1 to R_2	$R_T = \dfrac{R_1R_2}{R_1 + R_2}$	R_1	*% error* $= \dfrac{R_1 - R_T}{R_T} \times 100\%$
1:1	0.50	1.0	100%
1:10	0.91	1.0	9.9%
1:100	0.99	1.0	1.0%
1:1000	0.999	1.0	0.10%

Ratios of 10:1 or greater are symbolized by $\gg$, which is read "much greater than," as in $R_2 \gg R_1$. Ratios of 1:10 or smaller are symbolized by $\ll$, which is read "much less than," as in $R_1 \ll R_2$.

EXAMPLE 14-14

Use the symbol $\ll$ or $\gg$ to indicate when

$$R_T = \frac{R_1R_2}{R_1 + R_2} \text{ may be stated as } R_T \approx R_1.$$

Solution

$$R_T = \frac{R_1R_2}{R_1 + R_2} \Rightarrow R_T \approx R_1$$

when $R_1 \ll R_2$ or $R_2 \gg R_1$

EXAMPLE 14-15

Simplify the formula $C_T = \dfrac{C_1C_2}{C_1 + C_2}$ for capacitors in parallel when $C_2 \gg C_1$

Solution Looking at the denominator, $C_1 + C_2 \Rightarrow C_2$ when $C_2 \gg C_1$. Substitute C_2 for the denominator:

$$C_T \approx \frac{C_1C_2}{C_2}$$

Simplify:

$$\therefore \quad C_T \approx C_1 \text{ when } C_2 \gg C_1$$

EXAMPLE 14-16

If the low-frequency voltage gain of a junction field-effect transistor (JFET) is given by the formula

$$A_v = \frac{Y_{fs}r_{os}R_L}{r_{os} + R_L}$$

then investigate the condition of r_{os} that will allow the formula to be simplified to $A_v \approx Y_{fs}R_L$. R_L is "on the order of" 10 kΩ.

Solution   Select r_{os} to be an order of magnitude greater than R_L ($r_{os} \gg R_L$). $R_L = 10$ kΩ and $r_{os} = 100$ kΩ. Suppose that $Y_{fs} = 5.0$ mS. Substitute and solve:

$$A_v = \frac{5.0 \times 10^{-3} \times 100 \times 10^3 \times 10 \times 10^3}{100 \times 10^3 + 10 \times 10^3}$$

$$= 45.5$$

Substitute and solve in the simplified formula:

$$A_v \approx Y_{fs}R_L$$
$$\approx 5.0 \times 10^{-3} \times 10 \times 10^3 = 50$$

Is $50 \approx 45.5$? Yes, it is within 10%.

$$\therefore \quad A_v = \frac{Y_{fs}r_{os}R_L}{r_{os} + R_L} \Rightarrow A_v \approx Y_{fs}R_L \text{ when } r_{os} \gg R_L$$

EXERCISE 14-4

Determine the condition for each formula to yield the simplified formula.

1. $A_v = -\dfrac{g_m r_p R_0}{r_p + R_0} \Rightarrow A_v \approx -g_m R_0$

2. $A_v = \dfrac{g_m R_k}{\dfrac{R_k}{r_p} + g_m R_k + 1} \Rightarrow A_v \approx \dfrac{g_m R_k}{g_m R_k + 1}$

3. $I_1 = \dfrac{IG_1}{G_1 + G_2} \Rightarrow I_1 \approx I$

4. $V_2 = \dfrac{ER_2}{R_1 + R_2} \Rightarrow V_2 \approx E/2$

5. $R_2 = \dfrac{(K + 1)R_{th}}{K} \Rightarrow R_2 \approx R_{th}$

6. $Z_1 = \dfrac{-j}{\omega C_{gp}}\left(\dfrac{k}{k - 1}\right) \Rightarrow Z_1 \approx \dfrac{-j}{\omega C_{gp}}$

7. $I_L = \dfrac{-AE_i}{A'R_f - (r_0 + R_f + R_L)} \Rightarrow I_L \approx -\dfrac{AE_i}{A'R_f}$

8. $I_0 = -\dfrac{Ae_i}{A'r_f} \Rightarrow I_0 \approx -\dfrac{e_i}{r_f}$

9. $y = 1 - \dfrac{4k^2}{4k^2 - 1}\,e^{-x} \Rightarrow y \approx 1 - e^{-x}$

10. $V' = \dfrac{VR_{dc}}{R_{dc} + R} \Rightarrow V' \approx \dfrac{VR_{dc}}{R}$

11. $g = \dfrac{I + I'}{\eta V} \Rightarrow g \approx \dfrac{I}{\eta V}$

12. $\Delta_V = (V' - V)\left(\dfrac{R + R'}{R}\right)\left(\dfrac{r_0}{r - r_0}\right) \Rightarrow \Delta_V \approx (V' - V)\dfrac{r_0}{r}$

SELECTED TERMS

constant   A numeric quantity which has a fixed value for a given problem.

dependent variable   A variable which is the output of a function.

function   A mathematical relationship between a set of input values and a resulting output value.

independent variable   A variable which is the input to a function.

variable   A numeric quantity represented by a letter which can take on more than one value in a problem; also a numeric quantity with an unknown value.

GRAPHS
AND GRAPHING
TECHNIQUES

A graph is a visual representation of the relationship between two or more quantities. By the information pictured in the graph, we are able to interpret circuit conditions, values, and trends. Graphs, like pictures, "are worth a thousand words." The topics presented here have been selected to help you gain an understanding of how to construct graphs and how to interpret them.

15-1 RECTANGULAR COORDINATES

On a piece of graph paper, construct both a horizontal and a vertical number line. Draw the lines so that their intersection is at the zero point of each, as shown in Figure 15-1.

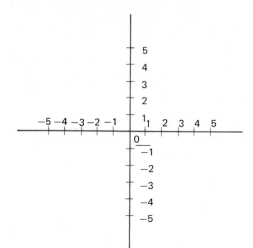

FIGURE 15-1
The horizontal and vertical axes of a graph are formed by two number lines.

The horizontal and vertical axes of a graph are formed by two number lines. The horizontal line is called the *x-axis;* the vertical line is called the *y-axis.* The *x*-axis (horizontal) and the *y*-axis (vertical) divide the graph paper into four parts called *quadrants.* The quadrants are numbered, with Roman numerals, in a counterclockwise (ccw) direction, as shown in Figure 15-2. The intersection of the *x*-axis with the *y*-axis is called the *origin.*

Each point on the graph paper is indicated by a pair of numbers called *coordinates.* The coordinates of the point on the graph paper tell how far the point is from the origin. The first number in the pair of coordinates tells how far the point is along the *x*-axis. This number is called the *x-coordinate* or *abscissa.* The second number, called the *y-coordinate* or *ordinate,* tells how far the point is along the *y*-axis. Together the two coordinates locate the point on the graph paper.

When writing the coordinates of a point, the *x*-coordinate is listed first, and then the *y*-coordinate, as in (*x, y*). Because of this agreed-upon order, the

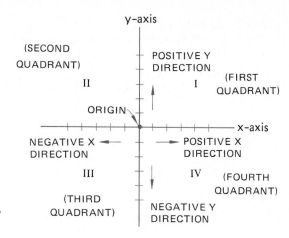

FIGURE 15–2
The graph paper is divided into quadrants.

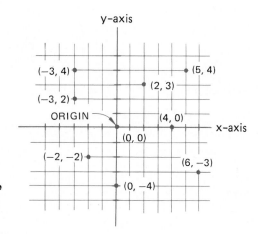

FIGURE 15–3
The points of the plane are numbered by *rectangular coordinates* with the *x*-coordinate given first followed by the *y*-coordinate (*x, y*). The notation (*x, y*) is called an *ordered pair*.

coordinates of a point are called an *ordered pair* of numbers. Figure 15-3 shows several points with their coordinates as ordered pairs of numbers noted.

When constructing a graph, the independent variable is *usually* assigned to the *x*-coordinate, and the dependent variable is *usually* assigned to the *y*-coordinate.

The coordinates of the points in a plane may be negative or positive depending upon the quadrant in which they are located. As seen in Figures 15-1 and 15-2, the *x*-coordinate is positive to the right of the *y*-axis and negative to the left of the *y*-axis. The *y*-coordinate is positive above the *x*-axis and negative below the *x*-axis. Table 15-1 summarizes the signs of the coordinates in each quadrant.

The terms used in the rectangular coordinate system are summarized in Table 15-2. This system of coordinates is sometimes referred to as the Cartesian coordinate system, in honor of its developer René Descartes (1596–1650).

TABLE 15-1

SUMMARY OF SIGNS IN EACH QUADRANT		
Quadrant	x-Coordinate (Abscissa)	y-Coordinate (Ordinate)
I	+	+
II	−	+
III	−	−
IV	+	−

TABLE 15-2

SUMMARY OF TERMS IN RECTANGULAR COORDINATE SYSTEM		
Term	Definition	Figure
Coordinates	Made up of two numbers that describe the location of each point in the plane	15-3
Ordered pair	Coordinates written as (x, y)	15-3
x-Coordinate	Number in the ordered pair that tells how far to move along the x-axis	15-3
y-Coordinate	Number in the ordered pair that tells how far to move along the y-axis	15-3
x-Axis	Horizontal axis	15-2
y-Axis	Vertical axis	15-2
Abscissa	x-coordinate	
Ordinate	y-coordinate	
Origin	Point where the x- and y-axes cross	15-2
Quadrants	Four quarters of the plane created by the crossing of the x- and y-axes	15-2

EXAMPLE 15-1   Using Figure 15-4:

(a) Determine the coordinates of points *A*, *B*, and *C*.

(b) State in which quadrant each point is located.

Solution   *Point A:* coordinates $(-7, 6)$ located in the second quadrant.

Point B: coordinates $(3, -7)$ located in the fourth quadrant.

Point C: coordinates $(-3, -2)$ located in the third quadrant.

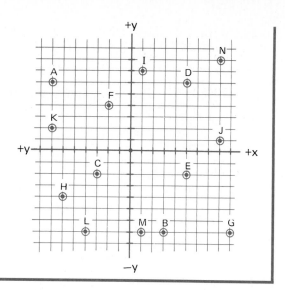

FIGURE 15-4
Letterd points for Example 15-1 and Exercise 15-1.

EXERCISE 15-1
Using Figure 15-4, answer the following.

1. In which quadrants are each of the points A through H located?
2. In which quadrants are each of the points I through N located?
3. Determine the coordinates of points A through H.
4. Determine the coordinates of points I through N.
5. What is the y-coordinate of any point on the x-axis?
6. What is the x-coordinate of any point on the y-axis?
7. What is another name for abscissa?
8. What is another name for ordinate?
9. In which quadrant is a point located if its x-coordinate is negative and its y-coordinate is positive?
10. In which quadrant is a point located if its abscissa is positive and its ordinate is negative?
11. Draw a pair of axes and plot the points $(2, -1)$, $(-3, 4)$, and $(0, -4)$. Label each with its ordered pair.
12. Draw a pair of axes and plot the points $(-5, -2)$, $(6, 0)$, and $(3, -4)$. Label each with its ordered pair.

15-2 GRAPHS OF EQUATIONS

In constructing graphs of equations and functions, we must first make a table of ordered pairs. To do this, selected values of the independent variable (x) are substituted into the equation, and then values of the dependent variable

PROGRAMMABLE

(y) are computed. The ordered pairs are next plotted, and then a smooth curve is passed through the points.

EXAMPLE 15-2   Plot the graph of the equation $y = \frac{1}{2}x^2$.

Solution   Select values of x and solve for y in the equation given. Some calculations are:

$$x = 0 \qquad y = \frac{1}{2}(0)^2 = 0$$
$$x = 1 \qquad y = \frac{1}{2}(1)^2 = \frac{1}{2}$$
$$x = -1 \quad \cdot y = \frac{1}{2}(-1)^2 = \frac{1}{2}$$

Make a table of ordered pairs:

x	-4	-2	-1	0	1	2	4
y	8	2	$\frac{1}{2}$	0	$\frac{1}{2}$	2	8

Plotting these points results in the *parabola* of Figure 15-5.

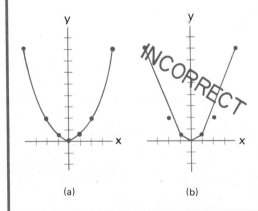

(a)   (b)

FIGURE 15-5
(a) Graph of the equation $y = \frac{1}{2}x^2$ for Example 15-2 correctly drawn with a smooth curve; (b) the graph incorrectly drawn without a smooth curve.

Since a graph is a visual representation of all the solutions of an equation, the coordinates of any point on the graph will satisfy the conditions of the equation. The coordinates of any point *not* on the graph will *not* satisfy the conditions of the equation.

EXAMPLE 15-3 Select several coordinates from the graph of $y = x + 2$ as shown in Figure 15-6, and check them by substituting into the equation.

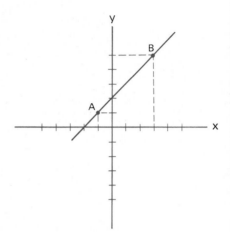

FIGURE 15–6
Graph of the equation $y = x + 2$
for Example 15–3.

Solution From the graph of Figure 15-6, select points $A(-1, 1)$ and $B(3, 5)$. Substitute and check:

Point A	*Point B*
$y = x + 2$	$y = x + 2$
$1 = -1 + 2$	$5 = 3 + 2$
$1 = 1$	$5 = 5$

∴ Both points satisfy the equation.

The graph of a straight line may be located with only two points as in Figure 15-6. However, when plotting graphs that are not straight lines, more points are needed. It is good to remember that the plot of the graph of an equation is only an approximation of the shape of the "real" graph of the equation. We do not need to spend a great deal of time in computing countless coordinates for the graph. In general, fewer points are needed for the straight portions of the graph than for the curved.

EXAMPLE 15-4 Plot the graph of the equation $y = x^3 - 2x^2 - 5x + 6$.

Solution From the equation, compute a preliminary table of ordered pairs:

x	−2	−1	0	1	2	3
y	0	8	6	0	−4	0

Plotting these points results in the graph of Figure 15-7(a). Using the graph of Figure 15-7(a), select additional values of x and compute y:

x	−2.5	−1.5	−0.5	0.5	1.5	2.5	3.5
y	−9.6	5.6	7.9	3.1	−2.6	−3.4	9.9

Plotting these points results in the graph of Figure 15-7(b).

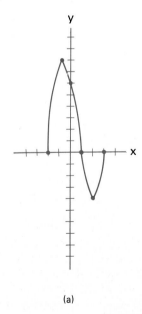

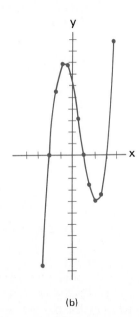

(a)   (b)

FIGURE 15-7

(a) Preliminary graph of the equation $y = x^3 - 2x^2 - 5x + 6$ for Example 15-4; (b) a more complete graph with additional points.

How can functions be graphed? A function may be graphed by first forming an equation. Let $y = f(x)$. Then graph the equation, as demonstrated in the following example.

310

EXAMPLE 15-5   Plot the graph of the function $-x^3$.

Solution

$$y = f(x) = -x^3$$

Select values of x and compute y, and then set up a table of ordered pairs:

x	-2.0	-1.8	-1.5	-1.2	-1.0	-0.5	0.0
y	8.0	5.8	3.4	1.7	1.0	0.1	0.0

x	0.0	0.5	1.0	1.2	1.5	1.8	2.0
y	0.0	-0.1	-1.0	-1.7	-3.4	-5.8	-8.0

Plotting these points results in the graph of Figure 15-8.

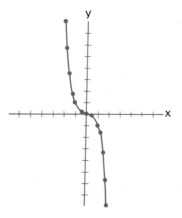

FIGURE 15–8
Graph of $y = -x^3$ for Example 15–5.

Besides showing the shape of an equation, graphs may be used to solve equations. In many cases the graphical solution is actually easier than the mathematical solution. Graphs, then, provide us with an important alternative method of solving equations.

The roots (solutions) of an equation are the coordinates that result when the graph intersects the x-axis. The roots of an equation may be found by *reading* the graph.

EXAMPLE 15-6   Graph the function $x^2 - 4x - 5$ and determine the roots of the equation $y = f(x) = x^2 - 4x - 5$.

Solution Select values of x and compute y, and then set up a table of ordered pairs:

x	−3	−2	−1	0	1	2	3	5	6	7
y	16	7	0	−5	−8	−9	−8	0	7	16

Plotting these points results in the graph of Figure 15-9. The roots of the equation are the intersection of the x-axis and the curve. Reading from the curve, the intersections are $(-1, 0)$ and $(5, 0)$. The roots are $x = -1$ and $x = 5$.

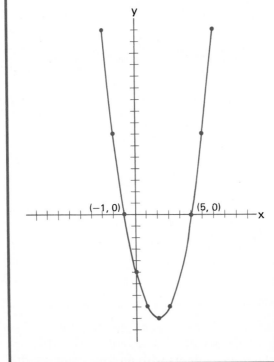

FIGURE 15–9
Graph of $x = x^2 - 4x - 5$ for
Example 15–6. The roots of the
equation $y = x^2 - 4x - 5$ are the
intersections of the curve and the
x-axis; $x = -1$ and $x = 5$ when
$y = 0$.

PROGRAMMABLE

EXERCISE 15-2
Sketch the graph representing each of the following functions. Determine the roots of the functions [when $y = f(x) = 0$].

1. $x + 2$ 2. $2x - 3$
3. $-3x + 1$ 4. $-5x - 6$
5. $-3x^2$ 6. x^3

7. $x^3 - 3x^2 - x + 3$ 8. $-x^2 - 10x - 16$

Sketch the graph representing each of the following equations.

9. $y = x$
10. $y = -5x - 6$
11. $4x - 2y = 3$
12. $6 = 3y - 5x$
13. $y = x^2 + 1$
14. $y = x + 3$
15. $x^2 + y^2 = 9$
16. $x^2 + y^2 = 25$
17. $x^2 + 6y = 0$
18. $2y - 2x^2 = 4x - 6$

15-3 GRAPHS OF LINEAR EQUATIONS

The graph of a linear equation is a straight line. A linear equation has the general form

$$y = mx + b \qquad (15\text{-}1)$$

where m = coefficient of x
b = a constant
x = independent variable
y = dependent variable

Slope

In Figure 15-10 the straight line has two points, P_1 and P_2, with coordinates (x_1, y_1) and (x_2, y_2). The *slope* of a line is found by taking the difference between the y-coordinates and dividing by the difference between the x-coor-

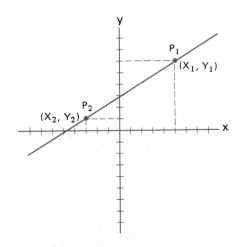

FIGURE 15–10
The straight line has a slope, *m*,
equal to $(y_1 - y_2)/(x_1 - x_2)$.

dinates, as stated in Equation 15-2. The slope is equal to m, the coefficient of x in the general linear equation $y = mx + b$.

$$\text{Slope} = m = \frac{y_1 - y_2}{x_1 - x_2} \qquad\qquad (15\text{-}2)$$

EXAMPLE 15-7 Determine the slope of the lines of Figure 15-11(a) and (b).

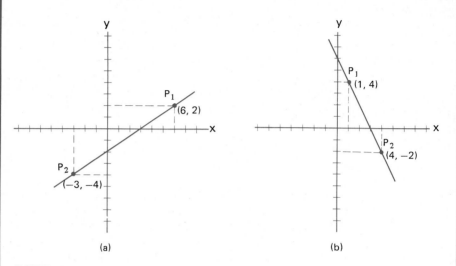

(a)   (b)

FIGURE 15–11
Graphs for Example 15–7: (a) positive slope; (b) negative slope.

Solution Substitute the coordinates of points P_1 and P_2 of Figure 15-11(a) into Equation 15-2:

$$m = \frac{y_1 - y_2}{x_1 - x_2}$$

$$= \frac{2 - (-4)}{6 - (-3)}$$

$$\therefore \quad m = \frac{2}{3}$$

Substitute the coordinates of points P_1 and P_2 of Figure 15-11(b) into Equation 15-2:

$$m = \frac{y_1 - y_2}{x_1 - x_2}$$

$$= \frac{4 - (-2)}{1 - 4}$$

$$\therefore \quad m = -2$$

From Example 15-7 we see that the line of Figure 15-11(a) has a *positive* slope and that the line of Figure 15-11(b) has a *negative* slope.

RULE 15-1. SLOPE OF A STRAIGHT LINE

The slope of a straight line is:

1. Positive—if a point on the line *rises* as the point moves from left to right along the line.
2. Negative—if a point on the line *falls* as the point moves from left to right along the line.

Intercept

The point where the line intersects the *x*-axis is the *x*-intercept. The *x*-coordinate of the *x*-intercept is the root of the equation.

In Equation 15-1, $y = mx + b$, b is the *y-intercept*. Thus, b is the *y*-coordinate of the point where the line and the *y*-axis intersect. The point where the line intersects the *y*-axis is $(0, b)$.

Graphing Linear Equations

The graph of a straight line can be made with just two points. One of these points can be the *y*-intercept $(x = 0, y = b)$. The other point is determined by selecting an appropriate value for the *x*-coordinate and then computing the *y*-coordinate. This procedure is summarized as Rule 15-2.

RULE 15-2. GRAPHING LINEAR EQUATIONS

To graph a straight line:

1. Put the equation in the form of $y = mx + b$.
2. Let $x = 0$ and solve for the *y*-intercept $(y = b)$.
3. Select a value for *x* that is away from the origin and solve for *y*.
4. Plot the two points on the graph paper.
5. Draw a straight line through the two points.
6. As a check, select a point from the line and substitute the coordinates into the equation.

EXAMPLE 15-8   Graph $y = 2x + 3$.

Solution   Apply Rule 15-2. Let $x = 0$ and solve for y:

$$y = 2x + 3$$
$$= 2(0) + 3$$
$$= 3$$

∴   (0, 3) is a point on the line.

Let $x = 3$ and solve for y:
$$y = 2(3) + 3$$
$$= 9$$

∴   (3, 9) is a point on the line.
Construct the graph of $y = 2x + 3$ as shown in Figure 15-12.

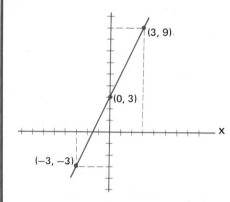

FIGURE   15–12
Graph of $y = 2x + 3$ for Example 15–8.

Check   Does the ordered pair $(-3, -3)$ satisfy $y = 2x + 3$?

$$-3 = 2(-3) + 3$$
$$= -6 + 3$$
$$-3 = -3 \quad \text{Yes!}$$

EXAMPLE 15-9   Graph $-x/2 + y = -5$.

Solution   Apply Rule 15-2.

Step 1:   Put $-x/2 + y = -5$ into the form $y = mx + b$:

$$y = x/2 - 5$$

316

Step 2: Let $x = 0$ and solve for y:

$$y = -5$$

$(0, -5)$ is a point on the line.

Step 3: Let $x = 6$ and solve for y:

$$y = \tfrac{6}{2} - 5$$
$$= 3 - 5$$
$$= -2$$

$(6, -2)$ is a point on the line.

Steps 4–5: Construct the graph of $y = x/2 - 5$ as shown in Figure 15-13.

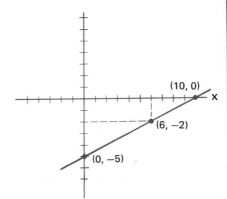

FIGURE 15–13
Graph of $y = (x/2) - 5$ for
Example 15–9.

Check Does the ordered pair $(10, 0)$ satisfy $y = x/2 - 5$?

$$0 = \tfrac{10}{2} - 5$$
$$= 5 - 5$$
$$0 = 0 \qquad \text{Yes!}$$

EXAMPLE
15-10

Graph $E_{th} - I_b(0.2) - 0.7 - 2I_b(0.2) = 0$. Consider $E_{th} = f(I_b)$.

Solution Put the equation in the form $y = mx + b$:

$$E_{th} = (0.2)I_b + (0.4)I_b + 0.7$$
$$E_{th} = (0.6)I_b + 0.7$$

Let $I_b = 0$ and solve for E_{th}:

$$E_{th} = 0.7$$

∴ (0, 0.7) is a point on the line.
Let $I_b = 10$ and solve for E_{th}:

$$E_{th} = (0.6)(10) + 0.7$$
$$= 6 + 0.7$$
$$E_{th} = 6.7$$

∴ (10, 6.7) is a point on the line.
Construct the graph of $E_{th} = (0.6)I_b + 0.7$ as shown in Figure 15-14.

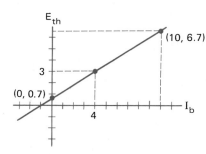

FIGURE 15–14
Graph of E_{th} = (0.6)I_b + 0.7 for Example 15–10.

Check   Does the ordered pair (4, 3) satisfy the equation?

$$3 = (0.6)(4) + 0.7$$
$$= 2.4 + 0.7$$
$$= 3.1$$

Observation   Notice that the ordered pair did not check exactly. However, it is as close as the graph will allow, and it is quite satisfactory.

EXERCISE 15-3
In each of the following, state (*a*) whether the slope is positive or negative, (b) the slope, and (c) the coordinates of the *y*-intercept.

1. $y = -x + 5$ 2. $y = x + 3$
3. $y = -2x - 4$ 4. $y = x/3 + 1$
5. $y = 4x$ 6. $f(x) = -\frac{1}{4}x + 2$

7. $f(x) = -2x/3 + 12$ 8. $f(x) = \frac{5}{2}x - 2$

9. $f(x) = x$ 10. $f(x) = x - 3$

Graph each of the following equations.

11. $V = 10I + 3$ 12. $E = 6R$

13. $A = \frac{1}{2}h$ 14. $E_0 = -2I + 3$

15. $I = E/2$ 16. $R = E/5 - 4$

17. $R_{th} = -3(E_{th}/6 - 2)$ 18. $I_T = 2V + 4$

19. $R_{int} = 5R_s - 2$ 20. $Q_T = 10E + 12$

15-4 DERIVING A LINEAR EQUATION FROM A GRAPH

The equation of a straight line may be derived from its graph by determining its y-intercept and the slope. The general equation $y = mx + b$ is made into a specific equation for the graph by substituting values for m and b.

RULE 15-3. DERIVING AN EQUATION FROM ITS GRAPH

To determine an equation of the graph of a straight line:

1. Locate the coordinates of the y-intercept.
2. Determine the slope of the line.
3. Substitute values of m and b into $y = mx + b$.

EXAMPLE 15-11

Derive the equation of the line pictured in Figure 15-15.

Solution Apply Rule 15-3.

Step 1: Determine the y-intercept:

$$y\text{-intercept} = (0, -3)$$
$$b = -3$$

Step 2: Determine the slope by selecting points P_1 and P_2 as in Figure 15-15. Substitute into Equation 15-2:

$$m = \frac{y_1 - y_2}{x_1 - x_2}$$

$$= \frac{3 - (-6)}{-8 - 4} = \frac{9}{-12} = -\frac{3}{4}$$

$$\therefore \quad m = -\tfrac{3}{4}$$

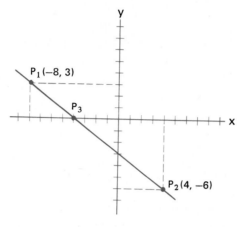

FIGURE 15-15
Graph for Example 15-11.

Step 3:   Substitute values of m and b into $y = mx + b$:

$$y = -\tfrac{3}{4}x - 3$$

Check   Do the coordinates $(-4, 0)$ of P_3 satisfy the derived equation?

$$y = -\tfrac{3}{4}x - 3$$
$$0 = -\tfrac{3}{4}(-4) - 3$$
$$= 3 - 3$$
$$0 = 0 \quad \text{Yes.}$$

$$\therefore \quad y = -\tfrac{3}{4}x - 3 \text{ is the equation of the line.}$$

**EXAMPLE
15-12**

Derive the equation of the line shown in Figure 15-16 that has $(4, 2)$ and $(-2, -3)$ as coordinates of points on the line.

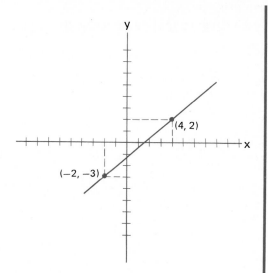

FIGURE 15–16
Graph for Example 15–12.

Solution   Determine the slope using Equation 15-2:

$$m = \frac{2 - (-3)}{4 - (-2)} = \frac{5}{6}$$

Determine the y-intercept, b, by substituting the coordinates of one point (4, 2) of the curve of Figure 15-16 along with the slope into the general equation $y = mx + b$.
Thus,

$$2 = \tfrac{5}{6}(4) + b$$
$$12 = 20 + 6b$$
$$6b = -8$$
$$b = \frac{-8}{6} = \frac{-4}{3}$$

Substitute the values for m and b into $y = mx + b$:

$$y = \tfrac{5}{6}x - \tfrac{4}{3}$$

∴   The equation of the line is $y = \tfrac{5}{6}x - \tfrac{4}{3}$.

Use Figure 15-17 and derive the equations of the following.

1. The line connecting points A and B
2. The line connecting points C and D
3. The line connecting points E and F

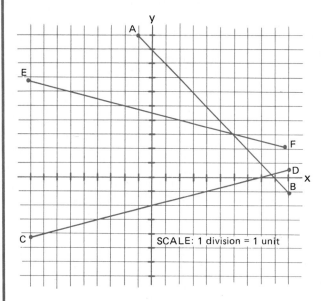

FIGURE 15–17
Graph for Exercise 15–4, problems 1–3.

Derive the equation of the straight line determined by the following points.

4. $(0, 13)$ and $(8, -3)$	5. $(-10, -6)$ and $(10, 2)$
6. $(0, 6)$ and $(12, 0)$	7. $(-2, 9)$ and $(8, -1)$
8. $(2, 4)$ and $(-4, -2)$	9. $(-5, 3)$ and $(6, -3)$
10. $(9, -4)$ and $(0, 0)$	11. $(-4, 5)$ and $(4, 5)$

15-5 GRAPHING EMPIRICAL DATA

Data that are gathered by making measurements of circuit conditions are called *empirical data.* As you know, measurements made with test equipment are *not perfect;* instead, a measurement has a range of possible values that depends upon the accuracy statement of the test equipment. Table 15-3 is a table of ordered pairs of measurements for the circuit of Figure 15-18. Notice that each measurement has been given an uncertainty.

TABLE 15-3

EMPIRICAL DATA TAKEN FROM THE CIRCUIT OF FIGURE 15-18			
V (volts)	I (mA)	V (volts)	I (mA)
0.0	0.0	6.0 ± 0.2	75.9 ± 5
1.0 ± 0.2	12.0 ± 5	7.0 ± 0.2	82.0 ± 5
2.0 ± 0.2	25.0 ± 5	8.0 ± 0.2	98.0 ± 5
3.0 ± 0.2	40.0 ± 5	9.0 ± 0.2	105.0 ± 5
4.0 ± 0.2	46.0 ± 5	10.0 ± 0.2	118.0 ± 5
5.0 ± 0.2	60.0 ± 5		

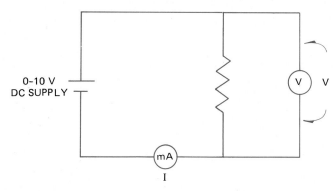

FIGURE 15–18
Resistive load under test.

The graph of the data in Table 15-3 is shown in Figure 15-19, with shaded rectangles indicating regions of uncertainty about each point. How do we draw a *smooth curve* through these data points? From the circuit in Figure 15-18 and our understanding of Ohm's law, we expect the *curve* to be a straight line passing through the origin. But which straight line? One that has about as many points above as below and as near all the points as possible. Getting the line as near as possible to all the points is called *averaging*.

When constructing the *smooth* curve, it is important that the curve be *averaged* through the plotted points. This may result in many of the plotted points not lying on the curve, as seen in Figure 15-19. However, the curve does pass through the region of uncertainties of each point. It is because of the uncertainty of the actual points that the smooth curve is averaged through the plotted points.

Notice that the graph in Figure 15-19 shows only the first quadrant. This is because the ordered pairs were all in the first quadrant. In this case, it was not necessary to show any of the other quadrants. In some graphs where the range is very small but the values are large, even the origin is not included! Figure 15-20 is an example of such a case.

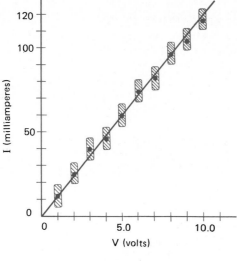

FIGURE 15–19
Graph of the data of Table 15–3 with the uncertainties shown as the shaded area.

f (kHz)	V (volts)
65.1	50.0
65.4	50.2
65.5	50.4
65.7	50.6
65.9	50.7

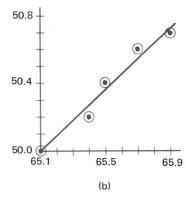

(a)   (b)

FIGURE 15–20
(a) Table of ordered pairs that have a small range and large value; (b) graph of the points on a modified axis. Notice that the origin is not shown.

PROCEDURE 15-1. GRAPHING EMPIRICAL DATA

To graph measured data (empirical data):

1. Make a table of ordered pairs.
2. Determine the range of each variable. From this, decide on the scale for each axis.
3. Construct and label each axis with the scale and units of the variables.
4. Carefully plot the points on the graph. Draw a small circle around each point.

5. Draw a *smooth* curve *averaging* the plotted points. Use as simple a curve as possible; usually a straight line will work.

EXAMPLE
15-13

Graph the following empirical data:

R	0.5	0.75	1.0	2.0	3.0	4.0
I	10	7	5.2	3	1.8	1.5

R	5.0	6.0	7.0	8.0	9.0	10
I	1.2	1.0	0.9	0.7	0.5	0.6

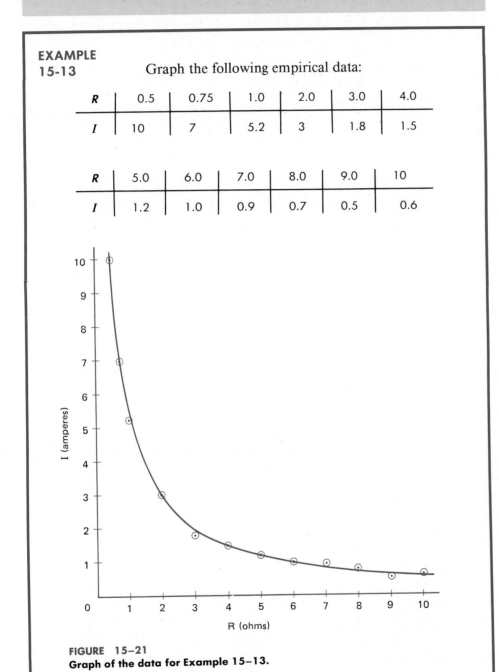

FIGURE 15–21
Graph of the data for Example 15–13.

Solution Use Procedure 15-1.
Step 1: Let R be the abscissa and I the ordinate.
Step 2: Select the range of 0 to 10 for each coordinate.
Step 3: Construct and label each axis with the scale and units of the variables. See Figure 15-21.
Step 4: Plot the points. See Figure 15-21.
Step 5: Draw a smooth curve averaging the plotted points. See Figure 15-21.

EXERCISE 15-5

Graph the following ordered pairs of data. Use Procedure 15-1.

1.

E (volts)	0	1	2	3	4	5	6	7	8
I (milliamps)	10	8	7	7	6	4	3	2	0

2.

E (volts)	0	1	2	3	4	5	6	7	8	9	10
I (amps)	0	1	1	2	2	2	3	4	4	4	5

3.

ω (rev/min)	2000	2200	2400	2600	2800	3000
P (watts)	75 000	155 000	185 000	190 000	160 000	75 000

4.

V_F (volts)	0	0.20	0.35	0.50	0.55	0.60	0.64	0.65	0.70	0.75
I_F (milliamps)	0	5	10	20	30	40	50	60	70	100

5.

R (ohms)	0.5	0.75	1.0	2.0	5.0	10.0
I (amps)	3.0	1.9	1.4	0.75	0.30	0.15

6.

V (volts)	0	0.4	1.5	2.0	2.3	2.8	3.3	3.5	3.9	4.2	4.4	4.5	4.6
I (amps)	0	0.2	0.8	1.0	1.2	1.4	1.8	1.8	2.4	2.6	3.0	3.6	3.9

SELECTED TERMS

axis   A number line used as a basis for a coordinate system.

empirical data   Numerical results from measurements, as opposed to results from calculations.

graph   A visual representation of the relationship between two variables; also the set of coordinates of the points which make up that visual representation.

intersect   Two lines or curves *intersect* where they have a point in common.

roots   The abscissas where a graph intersects the x-axis.

slope   The amount of change produced in the ordinate by a unit change of the abscissa.

16

SOLVING SYSTEMS OF LINEAR EQUATIONS

This chapter deals with two or more linear equations called a *system* of linear equations. Specifically, we will be interested in the coordinates that satisfy each of the equations in the system *simultaneously.* Two equations that have one common solution are called *simultaneous equations.* Several methods are used to solve systems of linear equations. We will study some of these in this chapter.

16-1 GRAPHIC METHOD

We are introducing this method to you so that you may *see* the solution to a system of linear equations. When two lines are graphed on the same coordinate system, there are three possible conditions, as shown in Figure 16-1. The three possible conditions are:

- The two lines may have the same slope and not cross because they are parallel to each other [Figure 16-1(a)]. This results in *no* points in common and *no* solution to the system of linear equations.
- The two lines may coincide with each other [Figure 16-1(b)]. This results in all points on both lines being in common and all points on both lines being solutions to the system of linear equations.
- The two lines intersect at one point [Figure 16-1(c)]. This results in one point in common and one solution to the system of linear equations.

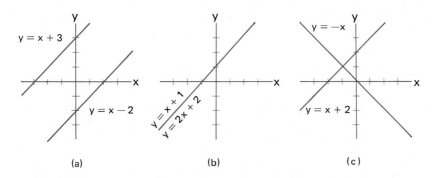

FIGURE 16–1
Graphs of two linear equations may result in one of three conditions:
(a) parallel; (b) coincident; (c) intersect.

EXAMPLE 16-1   Graphically solve the following system of equations for the intersection $y = 2x + 3$ and $y = -(\frac{3}{2})x - 4$.

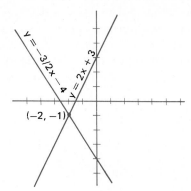

FIGURE 16-2
Graph of the system of equations for Example 16-1.

Solution Graph each equation as shown in Figure 16-2. Read the coordinate (x, y) of the intersection:

$$(-2, -1) \qquad x = -2 \text{ and } y = -1$$

Check Substitute -2 for x in both equations and evaluate y:

$$y = 2(-2) + 3$$
$$= -4 + 3$$
$$y = -1$$

This is the value read from the graph. Now check the second equation.

$$y = -\tfrac{3}{2}(-2) - 4$$
$$= +3 - 4$$
$$y = -1$$

This is also the value read from the graph.
$$\therefore \quad x = -2, y = -1 \text{ is the solution of the system of equations.}$$

┌ EXERCISE 16-1
Graph each system of linear equations and determine the solutions of the system.

1. $\begin{cases} y = x \\ y = -x + 2 \end{cases}$ 2. $\begin{cases} 2x + y = 9 \\ x + 2y = 0 \end{cases}$

3. $\begin{cases} y = 2x \\ y = x - 3 \end{cases}$
4. $\begin{cases} x + 3y = 3 \\ x + y = 5 \end{cases}$

5. $\begin{cases} x + y = 4 \\ x - y = 2 \end{cases}$
6. $\begin{cases} y = x - 3 \\ -2x - 2y = 6 \end{cases}$

7. $\begin{cases} x + y = 5 \\ x - y = -1 \end{cases}$
8. $\begin{cases} x = y + 1 \\ y = 3x + 2 \end{cases}$

9. $\begin{cases} x + y = 0 \\ x - 2y = 6 \end{cases}$
10. $\begin{cases} x + 3y = 11 \\ x - 3y = 5 \end{cases}$

16-2 ADDITION OR SUBTRACTION METHOD

This method is an algebraic method that involves eliminating one variable in the system by adding or subtracting equations, as shown in the following three examples.

EXAMPLE 16-2   Solve $\begin{cases} x + 3y = 11 \\ x - 3y = 5 \end{cases}$

Solution   Add the two equations to eliminate the terms containing y:

$$\begin{array}{r} x + 3y = 11 \\ \underline{x - 3y = 5} \\ 2x + 0 = 16 \end{array}$$

Solve for x:

$$2x = 16$$
$$x = 8$$

Solve for y by substituting the value of x into:

$$x + 3y = 11$$
$$8 + 3y = 11$$
$$y = 1$$

Check Substitute $x = 8$ into the second equation:

$$x - 3y = 5$$
$$8 - 3y = 5$$
$$-3y = -3$$
$$y = 1$$

This is the value computed in the solution.

∴ $x = 8$ and $y = 1$ is the solution of the system of equations.

EXAMPLE 16-3

Solve $\begin{cases} \dfrac{5x}{6} + \dfrac{y}{4} = 7 \\[2mm] \dfrac{2x}{3} - \dfrac{y}{8} = 3 \end{cases}$

Solution Clear the equations of fractions before adding or subtracting:

$$\text{M: 12} \qquad \frac{5x}{6} + \frac{y}{4} = 7$$
$$10x + 3y = 84$$

$$\text{M: 24} \qquad \frac{2x}{3} - \frac{y}{8} = 3$$
$$16x - 3y = 72$$

Add the two new equations to eliminate the terms containing y and solve for x:

$$10x + 3y = 84$$
$$\underline{16x - 3y = 72}$$
$$26x + 0 = 156$$
$$x = 6$$

Solve for y by substituting the value 6 for x into:

$$10x + 3y = 84$$
$$60 + 3y = 84$$
$$y = 8$$

Check Substitute $x = 6$ and solve for y in each original equation:

$$\frac{5x}{6} + \frac{y}{4} = 7$$

$$\frac{5(6)}{6} + \frac{y}{4} = 7$$

$$\frac{y}{4} = 2$$

$$y = 8$$

This is the value computed in the solution.

$$\frac{2x}{3} - \frac{y}{8} = 3$$

$$\frac{2(6)}{3} - \frac{y}{8} = 3$$

$$4 - y/8 = 3$$

$$-y/8 = -1$$

$$y = 8$$

This is the value computed in the solution.

∴ $x = 6$ and $y = 8$ is the solution of the system of equations.

EXAMPLE 16-4   Solve $\begin{cases} 6x + 5y = 2 \\ 4x - 2y = 12 \end{cases}$

Solution   Eliminate x by multiplying the first equation by 4 and the second equation by 6:

$$
\begin{array}{ll}
\text{M: 4} & 6x + 5y = 2 \\
& 24x + 20y = 8 \\
\text{M: 6} & 4x - 2y = 12 \\
& 24x - 12y = 72
\end{array}
$$

Subtract the two new equations to eliminate x:

$$24x + 20y = 8$$
$$\underline{-(24x - 12y = 72)}$$
$$0 + 32y = -64$$
$$y = -2$$

Substitute the value of y into the first equation:

$$6x + 5y = 2$$
$$6x + 5(-2) = 2$$
$$6x = 12$$
$$x = 2$$

Check   Substitute $y = -2$ into the second equation:

$$4x - 2y = 12$$
$$4x + 4 = 12$$
$$4x = 8$$
$$x = 2$$

This is the value computed in the solution.
$\therefore$ $x = 2$ and $y = -2$ is the solution.

The techniques for solving a system of linear equations explored in the three previous examples are summarized in Guidelines 16-1.

GUIDELINES 16-1. SOLVING A SYSTEM OF LINEAR EQUATIONS

For solving a system of linear equations using addition or subtraction:
1. The goal is to eliminate one of the variables.
2. Clear both equations of any fractions before proceeding.
3. Choose which variable to eliminate.
4. If necessary, form equivalent equations in the following manner. Multiply the first equation by the coefficient of the variable being eliminated in the second equation. Multiply the second equation by the coefficient of the variable being eliminated in the first equation. (Do not multiply by zero in either case!)

EXERCISE 16-2
Solve each system of equations by addition or subtraction. Use Guidelines 16-1.

1. $\begin{cases} x + y = 4 \\ x - y = 6 \end{cases}$ 2. $\begin{cases} 2x - 3y = -12 \\ x - 3y = 3 \end{cases}$

3. $\begin{cases} x + y = 9 \\ -x + y = 5 \end{cases}$ 4. $\begin{cases} x/3 - y = 10 \\ x/5 + 2y/5 = 2 \end{cases}$

5. $\begin{cases} 2x + y = 17 \\ 2x - y = -5 \end{cases}$ 6. $\begin{cases} 2x/5 + y = -2 \\ 2x/5 - y/2 = 1 \end{cases}$

7. $\begin{cases} x + 3y = 28 \\ x - 3y = -20 \end{cases}$ 8. $\begin{cases} x + y = 3 \\ 3x - 5y = 17 \end{cases}$

9. $\begin{cases} x - 5y = -5 \\ x + 3y = 3 \end{cases}$ 10. $\begin{cases} x/6 + y/4 = 3/2 \\ 2x/3 - y/2 = 0 \end{cases}$

16-3 SUBSTITUTION METHOD

As with the addition-subtraction method, this technique of solving the system of equations involves eliminating one of the variables to obtain one equation in one unknown. The approach in substitution is to solve either of the equations for one variable in terms of the other. This expression is then *substituted* into the other equation, resulting in a third equation in one unknown. This principle is shown in the following examples.

EXAMPLE 16-5 Solve $\begin{cases} x + 2y = 1 \\ 2x + 5y = 5 \end{cases}$

Solution   Use substitution. Solve $x + 2y = 1$ for x:

$$x = 1 - 2y$$

Solve for y by substituting $(1 - 2y)$ for x into:

$$2x + 5y = 5$$
$$2(1 - 2y) + 5y = 5$$
$$2 - 4y + 5y = 5$$
$$y = 3$$

Solve for x by substituting $y = 3$ into $x = 1 - 2y$:

$$x = 1 - 2(3)$$
$$x = -5$$

Check   Substitute $x = -5$ into the second equation and solve for y:

$$2x + 5y = 5$$
$$2(-5) + 5y = 5$$
$$5y = 15$$
$$y = 3$$

This checks the solution.

∴ $x = -5$ and $y = 3$ is the solution.

EXAMPLE 16-6

Solve $\begin{cases} \dfrac{x}{3} - \dfrac{y}{2} = \dfrac{7}{6} \\ \dfrac{x}{3} - \dfrac{y}{6} = \dfrac{1}{6} \end{cases}$

Solution   Clear the equations of fractions:

$$\text{M: 6} \qquad \frac{x}{3} - \frac{y}{2} = \frac{7}{6}$$
$$2x - 3y = 7$$

$$\text{M: 6} \qquad \frac{x}{3} - \frac{y}{6} = \frac{1}{6}$$
$$2x - y = 1$$

Solve $2x - 3y = 7$ for x:

$$x = \frac{7 + 3y}{2}$$

Substitute for x in the second equation, $2x - y = 1$:

$$2\left(\frac{7 + 3y}{2}\right) - y = 1$$

Solve for y:

$$7 + 3y - y = 1$$
$$2y = -6$$
$$y = -3$$

Solve for x by substituting $y = -3$ into

$$x = \frac{7 + 3y}{2} :$$

$$x = \frac{7 + 3(-3)}{2} = \frac{-2}{2}$$
$$x = -1$$

Check   Substitute $y = -3$ into each equation and solve for x. (We should get -1.)

$$x/3 - y/2 = 7/6$$
$$x/3 + 3/2 = 7/6$$
$$x/3 = -2/6$$
$$x = (-2/6)(3)$$
$$x = -1$$

$$x/3 - y/6 = 1/6$$
$$x/3 + 3/6 = 1/6$$
$$x/3 = -2/6$$
$$x = (-2/6)(3)$$
$$x = -1$$

This completes the check.

$\therefore \quad x = -1$ and $y = -3$ is the solution.

EXERCISE 16-3

Solve each system of equations by substitution.

1. $\begin{cases} x - y = 0 \\ x + y = 2 \end{cases}$ 2. $\begin{cases} 3x - 4y = 13 \\ 2x + 3y = 3 \end{cases}$

3. $\begin{cases} -5y = 17 - 3x \\ x + y = 3 \end{cases}$ 4. $\begin{cases} 5x + 2y = 36 \\ 8x - 3y = -54 \end{cases}$

5. $\begin{cases} 4x - 3y - 2 = 2x - 7y \\ x + 5y - 2 = y + 4 \end{cases}$ 6. $\begin{cases} x + y = 0 \\ x - 2y = 6 \end{cases}$

7. $\begin{cases} x + 3y = 11 \\ x - 3y = 5 \end{cases}$ 8. $\begin{cases} 2x + 3(x + y) = 15 \\ 2x - 3y = -1 \end{cases}$

9. $\begin{cases} x - 5y = -5 \\ x + 3y = 3 \end{cases}$ 10. $\begin{cases} 8(y + 1) = 2x \\ 3(x - 3y) = 15 \end{cases}$

11. $\begin{cases} x/3 - y/2 = -8/3 \\ x/7 - y/3 = -39/21 \end{cases}$ 12. $\begin{cases} x/2 + y/4 = 1 \\ 3x/4 + y \;\; = 1/4 \end{cases}$

13. $\begin{cases} \dfrac{x}{3} + \dfrac{y}{6} = \dfrac{1}{2} \\ \dfrac{x}{2} + \dfrac{3y}{10} = 1 \end{cases}$ 14. $\begin{cases} \dfrac{x + y}{2} - \dfrac{x - y}{2} = 2 \\ \dfrac{x - y}{4} + \dfrac{x + y}{2} = 5 \end{cases}$

16-4 DERIVING ELECTRICAL FORMULAS

The techniques of simultaneous equations may be used to derive a formula from two or more known equations. The following examples demonstrate this concept.

EXAMPLE 16-7 Derive a formula for work (W) in terms of current flow (I), time (t), and voltage (E) from the equations $E = W/Q$ and $Q = It$

Solution Eliminate Q in $E = W/Q$ by substituting It for Q:

$$E = W/It$$

Solve for W:

$$\therefore \quad W = EIt$$

EXAMPLE 16-8 Derive a formula for power (P) in watts (W) from $E = IR$ and $P = IE$ in terms of voltage (E) and resistance (R).

Solution Eliminate I in $P = IE$ by solving $E = IR$ for I and substituting:

$$I = E/R$$

Substitute E/R for I in $P = IE$:

$$P = \left(\frac{E}{R}\right) E$$

$$\therefore \qquad P = E^2/R$$

EXAMPLE 16-9   Derive a formula for voltage division across a selected resistor in a series circuit (V_n) in terms of the source voltage (E), the selected resistor (R_n), and the total resistance (R_T) from the equations $V_n = IR_n$ and $E = IR_T$.

Solution   Eliminate I in $V_n = IR_n$ by solving $E = IR_T$ for I:

$$I = E/R_T$$

Substituting E/R_T for I in $V_n = IR_n$:

$$V_n = \left(\frac{E}{R_T}\right) R_n$$

$$\therefore \qquad V_n = \frac{ER_n}{R_T}$$

EXERCISE 16-4

1. Solve for E in terms of E_{av} from the equations $E = 0.707E_{mx}$ and $E_{av} = 0.637E_{mx}$.
2. Solve for P in terms of E and I from the equations $2P = E_{mx}I_{mx}$, $E = E_{mx}/\sqrt{2}$, and $I = I_{mx}/\sqrt{2}$.
3. Solve for L in terms of ω and X_L from the equations $L/T = 0.1592X_L$, $f = 1/T$, and $\omega = 6.283f$.
4. Solve for P in terms of E, I, R, and Z from the equations $P_A = P/P_F$, $P_A = EI$, and $P_F = R/Z$.
5. Solve for W in terms of L and I from the equations $W = EI/\omega$, $I = E/X_L$, and $X_L = \omega L$.
6. Solve for f^2 in terms of L, C, and the numerical constant of 2π from the equations $X_L = X_C$, $X_L = \omega L$, $X_C = 1/(\omega C)$, and $\omega = 2\pi f$.

7. Solve for Q_0 in terms of R and X_L from the equations $P_q = PQ_0$, $P = I^2R$, and $P_q = I^2X_L$.

8. Solve for BW in terms of f_0 and Q_0 from the equations $2\pi L = R/\text{BW}$, $Q_0 = \omega_0 L/R$, and $\omega_0 = 2\pi f_0$.

9. Solve for R_L in terms of X_L, Q_0, and Q_0' from the equations
$$R' = \frac{R_L R}{R_L + R}, R = \omega L Q_0, R' = \omega L Q_0' \text{ and } X_L = \omega L.$$

10. Solve for I_{in} in terms of N and I_{out} from the following equations: $P_{in} = E_{in}I_{in}$, $P_{out} = E_{out}I_{out}$, $N = E_{out}/E_{in}$, and $P_{in} = P_{out}$.

16-5 DETERMINANTS OF THE SECOND ORDER

Meaning of a Determinant

Systems of linear equations may be solved by applying the rules and techniques associated with determinants. Programmable calculators have the capacity to solve determinants. You should consult your owner's guide to see if your calculator can solve determinants. Because a special symbol is used for the determinant, you may find the idea of what a determinant is and how it is used a new and unusual experience.

A second-order determinant is a tool that can be used to solve a system of two linear equations. A second-order determinant operates on four numbers to produce a new number. A second-order determinant looks like $\begin{vmatrix} a_1 & b_1 \\ a_2 & b_2 \end{vmatrix}$, where the a's and b's are the four numbers on which the determinant operates. They are called the *elements* of the determinant. The numerical value of a determinant is given by Equation 16-1.

$$\begin{vmatrix} a_1 & b_1 \\ a_2 & b_2 \end{vmatrix} = a_1b_2 - a_2b_1 \tag{16-1}$$

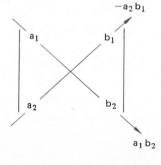

FIGURE 16-3
A 2 by 2 square array is called a determinant of the second order. It is made up of four elements. The diagonal of elements $a_1 b_2$ is called the principal diagonal.

340

From the definition of a determinant, we see that a number results by first taking the product of the two elements in the *principal* diagonal (from top left to bottom right) as shown in Figure 16-3, and then subtracting the product of the two elements in the other diagonal.

EXAMPLE 16-10

Evaluate $\begin{vmatrix} 3 & 2 \\ -6 & 4 \end{vmatrix}$.

Solution Since

$$\begin{vmatrix} a_1 & b_1 \\ a_2 & b_2 \end{vmatrix} = a_1 b_2 - a_2 b_1$$

$$\begin{vmatrix} 3 & 2 \\ -6 & 4 \end{vmatrix} = 3(4) - (-6)2$$

$$= 12 + 12 = 24$$

$$\therefore \qquad \begin{vmatrix} 3 & 2 \\ -6 & 4 \end{vmatrix} = 24$$

This process of evaluating a 2 by 2 determinant is very important and very mechanical (see Figure 16-4). You should practice the process until you master it.

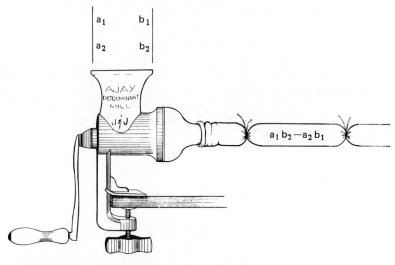

FIGURE 16–4
Turning the crank to expand a determinant.

Evaluate $\begin{vmatrix} -5 & 7 \\ 3 & -2 \end{vmatrix}$.

Solution Use Equation 16-1 to evaluate:

$$\begin{vmatrix} a_1 & b_1 \\ a_2 & b_2 \end{vmatrix} = a_1b_2 - a_2b_1$$

$$\begin{vmatrix} -5 & 7 \\ 3 & -2 \end{vmatrix} = -5(-2) - (3)7$$

$$= 10 - 21 = -11$$

$$\therefore \quad \begin{vmatrix} -5 & 7 \\ 3 & -2 \end{vmatrix} = -11$$

Solving Systems of Linear Equations

Systems of two linear equations are solved by applying special formulas to the generalized system of linear equations.

$$\begin{cases} a_1x + b_1y = k_1 \\ a_2x + b_2y = k_2 \end{cases} \tag{16-2}$$

where a_1 and a_2 = numerical coefficients of the unknown x
b_1 and b_2 = numerical coefficients of the unknown y
k_1 and k_2 = numerical constants
x and y = coordinates of the common solution to the system of linear equations

The solution of the system of equations (16-2) by determinants is carried out with the following formulas:

$$\Delta = \begin{vmatrix} a_1 & b_1 \\ a_2 & b_2 \end{vmatrix} = a_1b_2 - a_2b_1 \tag{16-3}$$

$$x = \frac{\begin{vmatrix} k_1 & b_1 \\ k_2 & b_2 \end{vmatrix}}{\Delta} = \frac{k_1b_2 - k_2b_1}{\Delta} \tag{16-4}$$

$$y = \frac{\begin{vmatrix} a_1 & k_1 \\ a_2 & k_2 \end{vmatrix}}{\Delta} = \frac{a_1k_2 - a_2k_1}{\Delta} \tag{16-5}$$

EXAMPLE 16-12

Solve $\begin{cases} 4x - 5 = y \\ 12 - 2y = 3x \end{cases}$

Solution Write the equation in the form of the system of equations (16-2):

$$4x - y = 5$$
$$3x + 2y = 12$$

Evaluate the determinant, Δ, using Equation 16-3:

$$\Delta = \begin{vmatrix} a_1 & b_1 \\ a_2 & b_2 \end{vmatrix} = \begin{vmatrix} 4 & -1 \\ 3 & 2 \end{vmatrix}$$

$$= 4(2) - (3)(-1) = 8 + 3$$

$$\Delta = 11$$

Solve for x using Equation 16-4:

$$x = \frac{\begin{vmatrix} k_1 & b_1 \\ k_2 & b_2 \end{vmatrix}}{\Delta} = \frac{k_1 b_2 - k_2 b_1}{\Delta}$$

$$= \frac{\begin{vmatrix} 5 & -1 \\ 12 & 2 \end{vmatrix}}{11} = \frac{5(2) - (12)(-1)}{11}$$

$$= \frac{10 + 12}{11} = \frac{22}{11}$$

$$x = 2$$

Solve for y using Equation 16-5:

$$y = \frac{\begin{vmatrix} a_1 & k_1 \\ a_2 & k_2 \end{vmatrix}}{\Delta} = \frac{a_1 k_2 - a_2 k_1}{\Delta}$$

$$= \frac{\begin{vmatrix} 4 & 5 \\ 3 & 12 \end{vmatrix}}{11} = \frac{4(12) - (3)(5)}{11}$$

$$= \frac{48 - 15}{11} = \frac{33}{11}$$

$$y = 3$$

Check Substitute $x = 2$ into each equation and solve for y (we should get 3):

$$4x - y = 5$$
$$4(2) - y = 5$$
$$8 - y = 5$$
$$y = 3$$

$$3x + 2y = 12$$
$$3(2) + 2y = 12$$
$$6 + 2y = 12$$
$$y = 3$$

$\therefore$ $x = 2$ and $y = 3$ is the solution of the system of equations.

EXAMPLE 16-13

Solve $\begin{cases} 3x - y = 10 \\ 4x = 9y - 2 \end{cases}$.

Solution Write the equations in the generalized form of the system of equations (16-2):

$$3x - y = 10$$
$$4x - 9y = -2$$

Evaluate the determinant, Δ, using Equation 16-3:

$$\Delta = \begin{vmatrix} 3 & -1 \\ 4 & -9 \end{vmatrix} = 3(-9) - (4)(-1)$$

$$= -27 + 4$$

$$\Delta = -23$$

Solve for x using Equation 16-4:

$$x = \frac{\begin{vmatrix} 10 & -1 \\ -2 & -9 \end{vmatrix}}{-23} = \frac{10(-9) - (-2)(-1)}{-23}$$

$$= \frac{-90 - 2}{23} = \frac{-92}{-23}$$

$$x = 4$$

Solve for y using Equation 16-5:

$$y = \frac{\begin{vmatrix} 3 & 10 \\ 4 & -2 \end{vmatrix}}{-23} = \frac{(3)(-2) - (4)(10)}{-23}$$

$$= \frac{-6 - 40}{-23} = \frac{-46}{-23}$$

$$y = 2$$

$\therefore$ $x = 4$ and $y = 2$ is the solution of the system of equations.

EXERCISE 16-5

Evaluate each determinant.

1. $\begin{vmatrix} 1 & 2 \\ -1 & 1 \end{vmatrix}$ 2. $\begin{vmatrix} 5 & 0 \\ 3 & 2 \end{vmatrix}$

3. $\begin{vmatrix} -2 & -2 \\ -2 & -2 \end{vmatrix}$ 4. $\begin{vmatrix} 3 & 6 \\ 2 & -4 \end{vmatrix}$

Solve each system of equations by determinants.

PROGRAMMABLE

5. $\begin{cases} x + y = 4 \\ x - y = 2 \end{cases}$ 6. $\begin{cases} 4x + y = 14 \\ -4x + y = -2 \end{cases}$

7. $\begin{cases} y = -x + 8 \\ x = y + 2 \end{cases}$ 8. $\begin{cases} 3x + 2y = 1 \\ 2x = 18 + 3y \end{cases}$

9. $\begin{cases} 3x + y = 11 \\ y = 3x + 5 \end{cases}$ 10. $\begin{cases} -3x + 4y = 1 \\ x - y = 0 \end{cases}$

11. $\begin{cases} x + 3y = 1 \\ x - 5 = 3y \end{cases}$ 12. $\begin{cases} x/2 + y/3 = 4 \\ x - y = 3 \end{cases}$

13. $\begin{cases} x + y = 0 \\ x - 2y = 6 \end{cases}$ 14. $\begin{cases} 2x = y + 4 \\ 3x + y = 5 \end{cases}$

15. $\begin{cases} -x - 5y = 0 \\ -10y - 3x = -10 \end{cases}$

16. $\begin{cases} 3y = 4 + 2x \\ 6x = y + 1 \end{cases}$

17. $\begin{cases} 7x = 56 + 4y \\ 5y - 45 = 3x \end{cases}$

18. $\begin{cases} 2(x + 3y) = 7 \\ 9 = 4x - 3y \end{cases}$

19. $\begin{cases} 9x - 34 = -2y \\ -14 = 6x + 5y \end{cases}$

20. $\begin{cases} 3x/2 - 1 = y/4 \\ x/3 + y/2 = 2 \end{cases}$

16-6 DETERMINANTS OF THE THIRD ORDER

Determinants of the third order have three rows and three columns with nine elements, and have the following form:

$$\begin{vmatrix} a_1 & b_1 & c_1 \\ a_2 & b_2 & c_2 \\ a_3 & b_3 & c_3 \end{vmatrix} = a_1b_2c_3 + b_1c_2a_3 + c_1a_2b_3 - a_3b_2c_1 - b_3c_2a_1 - c_3a_2b_1 \quad (16\text{-}6)$$

PROGRAMMABLE

The expansion of the determinant may be found by copying the first two columns to the right of the determinant, and then forming the downward diagonal products with plus signs and the upward diagonal products with minus signs. Figure 16-5 shows this technique.

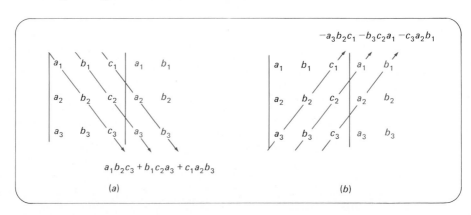

FIGURE   16–5
Forming the expansion of the determinant: (a) downward diagonal products are positive; (b) upward diagonal products are negative.

EXAMPLE 16-14

Evaluate the following determinant:

$$\begin{vmatrix} 1 & 3 & 5 \\ 2 & 1 & 1 \\ 1 & 2 & 4 \end{vmatrix}$$

Solution Repeat the first two columns:

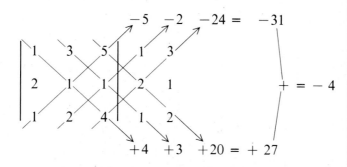

$\therefore$ The expansion of the determinant is $(4 + 3 + 20) + (-5 - 2 - 24) = 27 - 31 = -4$.

EXAMPLE 16-15

Evaluate the following determinant:

$$\begin{vmatrix} 2 & -1 & 4 \\ -1 & -3 & 2 \\ 2 & 0 & -1 \end{vmatrix}$$

Solution Repeat the first two columns:

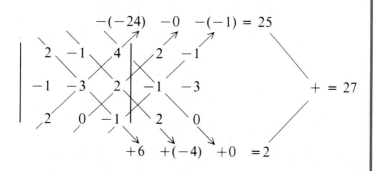

$\therefore$ The expansion of the determinant is $(6 - 4 + 0) + (24 - 0 + 1) = 2 + 25 = 27$.

Systems of three linear equations are solved by applying special formulas to the following generalized system of linear equations

$$\begin{cases} a_1x + b_1y + c_1z = k_1 \\ a_2x + b_2y + c_2z = k_2 \\ a_3x + b_3y + c_3z = k_3 \end{cases} \qquad (16\text{-}7)$$

The solution of the system of equations is carrried out with the following formulas:

$$\Delta = \begin{vmatrix} a_1 & b_1 & c_1 \\ a_2 & b_2 & c_2 \\ a_3 & b_3 & c_3 \end{vmatrix} \qquad (16\text{-}8)$$

$$x = \frac{\begin{vmatrix} k_1 & b_1 & c_1 \\ k_2 & b_2 & c_2 \\ k_3 & b_3 & c_3 \end{vmatrix}}{\Delta} \qquad (16\text{-}9)$$

$$y = \frac{\begin{vmatrix} a_1 & k_1 & c_1 \\ a_2 & k_2 & c_2 \\ a_3 & k_3 & c_3 \end{vmatrix}}{\Delta} \qquad (16\text{-}10)$$

$$z = \frac{\begin{vmatrix} a_1 & b_1 & k_1 \\ a_2 & b_2 & k_2 \\ a_3 & b_3 & k_3 \end{vmatrix}}{\Delta} \qquad (16\text{-}11)$$

EXAMPLE 16-16

Solve $\begin{cases} x - y + z = 2 \\ 2x - y + 3z = 6 \\ x + y + z = 6 \end{cases}$

Solution   Evaluate the determinant, Δ:

$$\Delta = \begin{vmatrix} 1 & -1 & 1 \\ 2 & -1 & 3 \\ 1 & 1 & 1 \end{vmatrix} \begin{matrix} 1 & -1 \\ 2 & -1 \\ 1 & 1 \end{matrix}$$

$$= -1 - 3 + 2 - (-1) - 3 - (-2)$$

$$\Delta = -2$$

Solve for x using Equation 16-9:

$$x = \frac{\begin{vmatrix} 2 & -1 & 1 \\ 6 & -1 & 3 \\ 6 & 1 & 1 \end{vmatrix} \begin{matrix} 2 & -1 \\ 6 & -1 \\ 6 & 1 \end{matrix}}{-2}$$

$$= \frac{-2 - 18 + 6 - (-6) - 6 - (-6)}{-2} = \frac{-8}{-2}$$

$$x = 4$$

Solve for y using Equation 16-10:

$$y = \frac{\begin{vmatrix} 1 & 2 & 1 \\ 2 & 6 & 3 \\ 1 & 6 & 1 \end{vmatrix} \begin{matrix} 1 & 2 \\ 2 & 6 \\ 1 & 6 \end{matrix}}{-2}$$

$$= \frac{6 + 6 + 12 - 6 - 18 - 4}{-2} = \frac{-4}{-2}$$

$$y = 2$$

Solve for z using Equation 16-11:

$$z = \frac{\begin{vmatrix} 1 & -1 & 2 \\ 2 & -1 & 6 \\ 1 & 1 & 6 \end{vmatrix} \begin{matrix} 1 & -1 \\ 2 & -1 \\ 1 & 1 \end{matrix}}{-2}$$

$$= \frac{-6 - 6 + 4 - (-2) - 6 - (-12)}{-2}$$

$$z = 0$$

Check   Substitute $x = 4$, $y = 2$, and $z = 0$ into each equation. The left member and the right member should *balance*.

$$x - y + z = 2$$
$$4 - 2 + 0 = 2$$
$$2 = 2$$

$$2x - y + 3z = 6$$
$$2(4) - 2 + 3(0) = 6$$
$$8 - 2 = 6$$
$$6 = 6$$

$$x + y + z = 6$$
$$4 + 2 + 0 = 6$$
$$6 = 6$$

$\therefore$ $x = 4$, $y = 2$, and $z = 0$ is the solution of the system of equations.

EXAMPLE 16-17   Solve $\begin{cases} y = 4 - x \\ x + 2 = -z \\ z + y = 8 \end{cases}$

Solution   Write the equations in the generalized forms as in Equation 16-7. Add zeros where a term is missing.

$$x + y + 0 = 4$$
$$x + 0 + z = -2$$
$$0 + y + z = 8$$

Evaluate the determinant:

$$\Delta = \begin{vmatrix} 1 & 1 & 0 \\ 1 & 0 & 1 \\ 0 & 1 & 1 \end{vmatrix} \begin{matrix} 1 & 1 \\ 1 & 0 \\ 0 & 1 \end{matrix}$$

$$= 0 + 0 + 0 + 0 - 1 - 1$$

$$\Delta = -2$$

Solve for x using Equation 16-9:

$$x = \frac{\begin{vmatrix} 4 & 1 & 0 \\ -2 & 0 & 1 \\ 8 & 1 & 1 \end{vmatrix} \begin{matrix} 4 & 1 \\ -2 & 0 \\ 8 & 1 \end{matrix}}{-2}$$

$$= \frac{0 + 8 + 0 - 0 - 4 - (-2)}{-2}$$

$$x = -3$$

Solve for y using Equation 16-10:

$$y = \frac{\begin{vmatrix} 1 & 4 & 0 \\ 1 & -2 & 1 \\ 0 & 8 & 1 \end{vmatrix} \begin{matrix} 1 & 4 \\ 1 & -2 \\ 0 & 8 \end{matrix}}{-2}$$

$$= \frac{-2 + 0 + 0 - 0 - 8 - 4}{-2}$$

$$y = 7$$

Solve for z using Equation 16-11:

$$z = \frac{\begin{vmatrix} 1 & 1 & 4 \\ 1 & 0 & -2 \\ 0 & 1 & 8 \end{vmatrix} \begin{matrix} 1 & 1 \\ 1 & 0 \\ 0 & 1 \end{matrix}}{-2}$$

$$= \frac{0 + 0 + 4 - 0 - (-2) - 8}{-2}$$

$$z = 1$$

Check Substitute $x = -3$, $y = 7$, and $z = 1$ into each equation. The left member should equal the right member.

$$y = 4 - x$$
$$7 = 4 - (-3)$$
$$7 = 7$$

$$x + 2 = -z$$
$$-3 + 2 = -1$$
$$-1 = -1$$

$$z + y = 8$$
$$1 + 7 = 8$$
$$8 = 8$$

∴ $x = -3$, $y = 7$, and $z = 1$ is the solution of the system of equations.

PROGRAMMABLE

EXERCISE 16-6
Evaluate each determinant.

1. $\begin{vmatrix} 1 & 2 & 1 \\ 2 & -1 & 3 \\ 2 & 0 & 2 \end{vmatrix}$

2. $\begin{vmatrix} 0 & -1 & 1 \\ -3 & 0 & 2 \\ -2 & 1 & 0 \end{vmatrix}$

3. $\begin{vmatrix} 1 & 1 & 1 \\ 1 & 1 & 1 \\ 1 & 1 & 1 \end{vmatrix}$

4. $\begin{vmatrix} 2 & 3 & 4 \\ 0 & 5 & -1 \\ -1 & -2 & -3 \end{vmatrix}$

Solve each system of equations by determinants.

5. $\begin{cases} 3x + 3y - 2z = 2 \\ 2x - 3y + z = -2 \\ x - 6y + 3z = -2 \end{cases}$

6. $\begin{cases} x - 11 + y = -z \\ -11 + z + 3y = -x \\ 3x + z + y = 13 \end{cases}$

7. $\begin{cases} x + y + z = 6 \\ x + y - z = 0 \\ x - y - z = 2 \end{cases}$

8. $\begin{cases} -x + 3z = 7 \\ 4z - 9 + 3x = y \\ 2x + 3y = 15 \end{cases}$

9. $\begin{cases} x + 3y + 4z = 14 \\ z + x - 7 = 2y \\ -2 + 2z + y = -2x \end{cases}$

10. $\begin{cases} 14 = x - z \\ -y + 10 = -x - z \\ y = 21 - z \end{cases}$

11. $\begin{cases} -45I_A + 30I_B + 10I_C = 10 \\ 30I_A - 50I_B + 20I_C = -20 \\ 10I_A + 20I_B - 50I_C = 0 \end{cases}$

12. $\begin{cases} 13I_1 - 8I_2 - 2I_3 = 100 \\ -2I_1 - 4I_2 + 12I_3 = 0 \\ -8I_1 + 22I_2 - 4I_3 = 0 \end{cases}$

352

SELECTED TERMS

determinant   An operator that combines a square array of numbers into a single number used in solving systems of linear equations.

system of linear equations   Set of linear equations, each involving the same variables.

17

APPLYING GRAPHS TO ELECTRONIC CONCEPTS

Because graphs play an important role in describing the behavior of electronic devices, it is important that you have experience in reading and interpreting graphs of technical information. This chapter will explore the use of graphs in determining device parameters and circuit conditions.

17-1 GRAPHIC ESTIMATION OF STATIC PARAMETERS

Before working through the following examples, you must first have an understanding of what is meant by a *static parameter*. Static means stationary or not moving. So a static parameter is a stationary parameter. In a graph, the static parameter is the dependent or the independent variable. The value of a static parameter can be read directly from the graph.

EXAMPLE 17-1 A 2-W, carbon-composition resistor is operated in an ambient atmosphere of 100°C. What is its *safe power dissipation* at this temperature? Use the temperature derating curve of Figure 17-1.

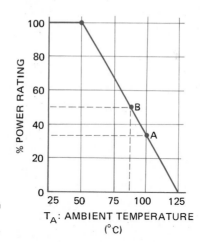

FIGURE 17–1

Percent power-temperature derating curve for determining the safe power dissipation for a carbon resistor operating above 50°C ambient.

Solution Enter the graph at a T_A of 100°C. Project vertically upward, strike the derating curve at point A, and then project horizontally across the vertical axis. Read the vertical axis as ≈35% of the power rating.
Compute the safe power dissipation of the 2-W resistor:

$$\therefore \quad P_D = 2 \times 0.35 = 0.7 \text{ W}$$

The power derating in the preceding example is a static parameter. The derating curve of Figure 17-1 may be used to compute circuit parameters for a given size and value of resistor, as demonstrated by the following example.

EXAMPLE 17-2   Determine the amount of current that can safely flow through a 2-W, 100-Ω resistor operating at an ambient temperature of 88°C.

Solution   Read the percent power rating from Figure 17-1 (point *B*):

$$\text{percent power rating} = 50\%$$

Compute the safe power dissipation of the 2-W resistor:

$$P_D = 2 \times 0.5 = 1 \text{ W}$$

Compute the current through the 100-Ω resistor:

$$P = I^2R$$
$$I = \sqrt{P/R}$$
$$= \sqrt{1/100}$$
$$\therefore \quad I = 100 \text{ mA}$$

The graph of Figure 17-2 relates voltage drop to the resistance of a 75-W incandescent lamp. Without the curve of Figure 17-2 to assist us, we would

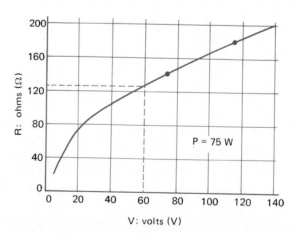

FIGURE   17–2

Voltage-resistance characteristic of a 75-W incandescent lamp.

be unable to determine the resistance of the filament for a given voltage. This is due to the nonlinear resistance characteristic of the lamp filament.

EXAMPLE 17-3 (a) Determine the approximate resistance of a 75-W incandescent lamp when 60 V is dropped across it.
(b) Compute the current passing through the lamp.

Solution (a) Use the voltage-resistance characteristic of Figure 17-2 to determine the resistance:

$$\therefore \quad R \approx 125 \ \Omega$$

(b) Compute the current through the lamp:

$$I = E/R$$
$$= 60/125$$
$$\therefore \quad I = 480 \text{ mA}$$

Manufacturers of electronic devices may publish a series of curves (called a *family* of curves) on the same axis in order to show the effect of a parameter on the operation of the device. Figure 17-3 is such a graph showing how temperature affects the voltage drop across the device.

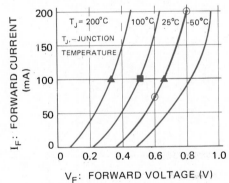

FIGURE 17-3
Family of forward diode characteristics showing the effect of heat on the parameters of the diode.

EXAMPLE 17-4 From the graphs of Figure 17-3, determine:

(a) The voltage drop (V_F) for a current (I_F) of 100 mA at temperatures of 25°C and 200°C

(b) The internal temperature of the device (T_J) when a V_F of 0.5 V is measured at 100 mA I_F

Solution   (a) From Figure 17-3 for $T_J = 25°C$:

$$V_F = 0.65 \text{ V} \quad \text{for an } I_F \text{ of 100 mA}$$

From Figure 17-3 for $T_J = 200°C$:

$$V_F = 0.35 \text{ V} \quad \text{for an } I_F \text{ of 100 mA}$$

(b) From Figure 17-3:

$$T_J = 100°C \quad \text{for } V_F = 0.5 \text{ V and } I_F = 100 \text{ mA}$$

The family of graphs of Figure 17-4 indicates the effect that lead length has on the amount of power that may be dissipated by the semiconductor device. This curve is helpful in determining how long a lead may be to dissipate a specific power when the lead temperature is known.

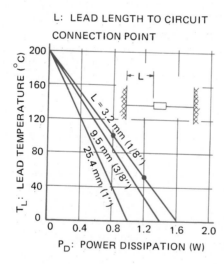

FIGURE 17-4

Temperature-power derating curve for a solid-state device showing the effect of lead length on the parameters of the device.

EXAMPLE 17-5   Determine the longest lead length that allows the device to dissipate 1 W when the lead temperature is 60°C.

Solution   From Figure 17-4:

$$\therefore \qquad L = 9.5 \text{ mm} \quad \text{for } P_D = 1.0 \text{ W and } T_L = 60°C$$

EXERCISE 17-1

1. Using Figure 17-1, determine the *safe power* dissipation of a $\frac{1}{2}$-W carbon resistor operating at an ambient temperature of 75°C.
2. Repeat problem 1 for a 2-W carbon resistor operating at an ambient temperature of 113°C.
3. The 75-W incandescent lamp of Figure 17-2 is operated from a 40-V source; determine the current flowing in the lamp.
4. Repeat problem 3 for a source voltage of 120 V.
5. The diode of Figure 17-3 has a forward voltage drop (V_F) of 0.7 V when a forward current of 50 mA is flowing. Determine the junction temperature (T_J).
6. It is desired to dissipate 1.2 W with the device of Figure 17-4 at a maximum lead temperature of 50°C. From the three curves pictured in Figure 17-4, select the most appropriate lead length.
7. Repeat problem 6 for a lead temperature of 120°C and a power dissipation of 0.5 W.
8. The semiconductor of Figure 17-4 is connected to the circuit with a lead length of 10 mm. When operating, the lead temperature was measured at 100°C. Determine the maximum current through the device if the device has a resistance of 775 Ω.
9. The battery of Figure 17-5 is used to power an AM receiver having an equivalent resistance of 1000 Ω. Using the information of the curve of Figure 17-5 and Ohm's law, determine the current through the receiver after 18 hours of operation.

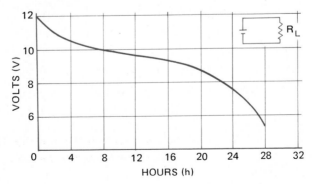

FIGURE   17–5
Service-life curve of a 12-V battery for a load resistance (R_L) of 1000 Ω.

10. The receiver of problem 9 is operated for the same amount of time each day for 7 days. The receiver current was measured at the end of the seventh day and was found to be 6.5 mA. Determine how long the receiver was operated each day. Use Figure 17-5.

17-2 GRAPHIC ESTIMATION OF DYNAMIC PARAMETERS

In this section we will learn to determine various dynamic device parameters by taking the difference between the coordinates of two points of a curve. We will use special notation to indicate that a difference of two coordinates has been taken. The difference between two ordinates ($y_2 - y_1$) is noted as Δy, while the difference between two abscissas ($x_2 - x_1$) is noted at Δx. To summarize:

$$\Delta x = x_2 - x_1$$
$$\Delta y = y_2 - y_1$$

Approximation of Semiconductor Parameters

Semiconductor parameters are often noted as a small difference or *interval* in one parameter over an interval in another parameter, while a third parameter is held constant. For example, the small-signal, common-emitter, forward-current transfer ratio of a transistor, h_{fe} (also called β), is given by Equation 17-1.

$$h_{fe} = \left. \frac{\Delta i_c}{\Delta i_b} \right|_{v_{ce}} \tag{17-1}$$

In Equation 17-1, the vertical bar with v_{ce} written to the right of it indicates that v_{ce} is held constant while the intervals Δi_c and Δi_b are selected from the curve.

EXAMPLE 17-6   Determine the forward-current transfer ratio (h_{fe}) of the transistor having an i_c-i_b curve as shown in Figure 17-6 when $v_{ce} = 5$ V.

Solution   Select an interval from the 5-V curve of Figure 17-6. Draw a horizontal line through the lower point A. See Figure 17-7.

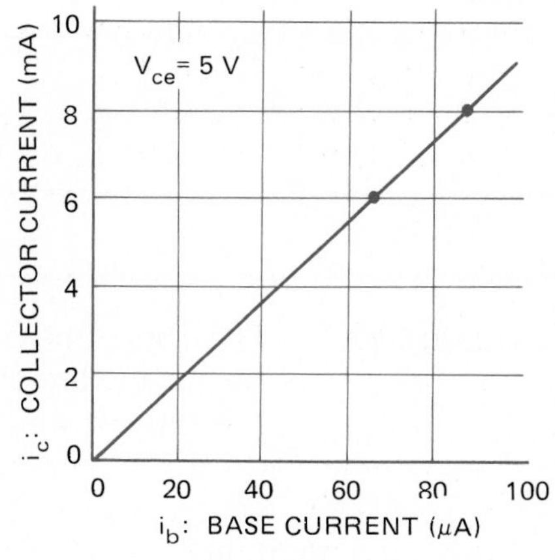

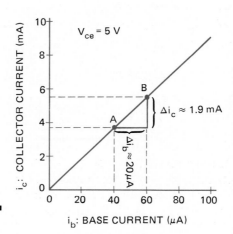

FIGURE 17–6
Transistor $i_c - i_b$ transfer characteristic curve for Example 17-6.

FIGURE 17–7
Graphic estimation of h_{fe} from a transistor $i_c - i_b$ transfer curve.

Draw a vertical line through the upper point *B* and down through the horizontal line. See Figure 17-7. Determine Δi_c from Figure 17-7:

$$\Delta i_c \approx 1.9 \text{ mA}$$

Determine Δi_b from Figure 17-7:

$$\Delta i_b \approx 20 \ \mu\text{A}$$

Compute h_{fe} for $v_{ce} = 5$ V:

$$h_{fe} = \frac{\Delta i_c}{\Delta i_b}$$

$$h_{fe} = \frac{1.9 \times 10^{-3}}{20 \times 10^{-6}}$$

$$\therefore \quad h_{fe} = 95$$

EXAMPLE 17-7   Determine the output admittance (h_{oe}) of the transistor having the characteristic curve of Figure 17-8 for $i_b = 60\ \mu A$, where

$$h_{oe} = \left. \frac{\Delta i_c}{\Delta v_{ce}} \right|_{i_b} \qquad \text{(siemens)} \qquad (17\text{-}2)$$

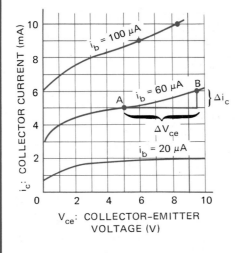

FIGURE 17–8
Graphic estimation of h_{oe} from a transistor collector characteristic curve.

Solution   To hold $i_b = 60\ \mu A$ constant, select an interval from the 60-μA curve of Figure 17-8.

Construct a horizontal line through point A and a vertical line through point B. See Figure 17-8.

Determine the vertical interval Δi_c by counting the squares between point B and the horizontal line. Each square equals 1.0 mA.

$$\Delta i_c \approx 1\ \text{mA}$$

Determine the horizontal interval Δv_{ce} by counting the squares between point A and the vertical line. Each square equals 1.0 V.

$$\Delta v_{ce} \approx 4.5 \text{ V}$$

Compute h_{oe} for $i_b = 60 \ \mu\text{A}$:

$$h_{oe} = \frac{\Delta i_c}{\Delta v_{ce}}$$

$$= \frac{1 \times 10^{-3}}{4.5}$$

$$\therefore \quad h_{oe} = 222 \ \mu\text{S}$$

EXAMPLE 17-8 Estimate the *transadmittance* (y_{fs}) of the field-effect transistor (FET) having the transfer characteristic curve of Figure 17-9 for $v_{ds} = 15$ V, where

$$y_{fs} = \frac{\Delta i_d}{\Delta v_{gs}} \bigg|_{v_{ds}} \qquad \text{(siemens)} \qquad (17\text{-}3)$$

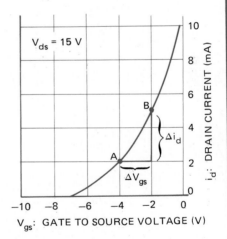

FIGURE 17–9
Graphic estimation of y_{fs} from a FET transfer characteristic curve.

Solution Select an interval from the 15-V curve of Figure 17-9. Determine Δi_d, the vertical interval:

$$\Delta i_d \approx 3 \text{ mA}$$

Determine Δv_{gs}, the horizontal interval:

$$\Delta v_{gs} \approx 2 \text{ V}$$

Compute y_{fs} for $v_{ds} = 15$ V:

$$y_{fs} = \frac{\Delta i_d}{\Delta v_{gs}}$$

$$= \frac{3 \times 10^{-3}}{2}$$

$$\therefore \quad y_{fs} = 1.5 \text{ mS}$$

EXAMPLE 17-9 Determine the dynamic resistance (r_{ac}) of the diode having the forward characteristic curve of Figure 17-10, where

$$r_{ac} = \frac{\Delta v_f}{\Delta i_f} \quad \text{(ohms)} \qquad (17\text{-}4)$$

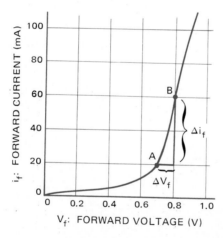

FIGURE 17-10
Graphic estimation of r_{ac} from a diode forward characteristic curve.

Solution Select an interval from the curve of Figure 17-10. Determine Δv_f, the horizontal interval:

$$\Delta v_f \approx 0.1 \text{ V}$$

Determine Δi_f, the vertical interval:

$$\Delta i_f \approx 40 \text{ mA}$$

Compute r_{ac}:

$$r_{ac} = \frac{\Delta v_f}{\Delta i_f}$$

$$= \frac{0.1}{40 \times 10^{-3}}$$

$$\therefore \quad r_{ac} = 2.5 \ \Omega$$

EXERCISE 17-2

1. Approximate the dynamic current (i) from the interval of the curve of Figure 17-2 delineated by the two noted points, where

$$i = \frac{\Delta V}{\Delta R} \quad \text{(amperes)}$$

2. Approximate the dynamic resistance (r_{ac}) from the region of the curve of Figure 17-3 delineated by the two circled points, where

$$r_{ac} = \frac{\Delta V_F}{\Delta I_F}\bigg|_{25°C} \quad \text{(ohms)}$$

3. Approximate the thermal resistance (θ_{lead}) of the lead wire from the region of the curve of Figure 17-4 delineated by the two noted points, where

$$\theta_{\text{lead}} = \frac{\Delta T_L}{\Delta P_D}\bigg|_{3.2 \ \text{mm}} \quad °C/W$$

4. Approximate the h_{fe} or β of a transistor from the region of the curve of Figure 17-6 delineated by the two noted points, where

$$h_{fe} = \frac{\Delta i_c}{\Delta i_b}\bigg|_{5 \ V}$$

5. Approximate the h_{oe} of a transistor from the region of the curve of Figure 17-8 delineated by the two noted points, where

$$h_{oe} = \frac{\Delta i_c}{\Delta v_{ce}}\bigg|_{100 \ \mu A} \quad \text{(siemens)}$$

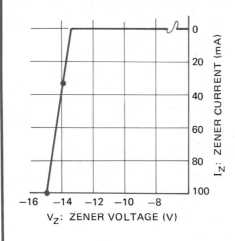

FIGURE 17–11
Zener diode reverse characteristic
curve for Exercise 17–2, problem 6.

6. Approximate the Z_{zt} of the zener diode from the region of the curve of Figure 17-11 delineated by the two noted points, where

$$Z_{zt} = \frac{\Delta V_z}{\Delta i_z} \quad \text{(ohms)}$$

17-3 GRAPHIC ANALYSIS OF NONLINEAR CIRCUITS

Many of the devices used in electronics have nonlinear characteristics. These devices are often used with devices having linear characteristics to form a nonlinear series circuit. Figure 17-12, which consists of a resistor in series with a diode, is an example of a nonlinear series circuit. The rules that govern the

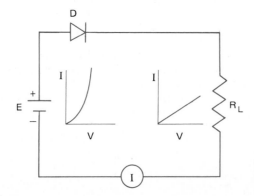

FIGURE 17–12
Nonlinear series circuit with the
linear characteristic of the resistor,
R_L, and the nonlinear characteristic
of the diode, D, shown.

nonlinear series circuit are the same as those that govern the linear series circuit. They are as follows:

- The total circuit current (I) is the same throughout the circuit.
- The sum of the voltage drops (ΣV) equals the voltage source (E).

The solution of the nonlinear circuit is better done by graphic methods rather than by the methods of algebra. The graphic method is usually faster, easier, and clearer than the methods of algebra.

The graphic solution of a nonlinear series circuit is done by constructing both the curve of the linear device and the curve of the nonlinear device on the same set of coordinates. The point of intersection of the two curves has the circuit current as the ordinate, and the voltage drop across the nonlinear device as the abscissa.

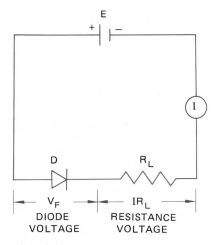

FIGURE 17–13
Simple nonlinear series circuit.

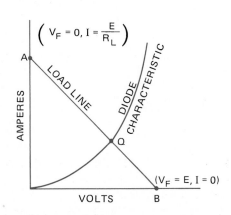

FIGURE 17–14
Graphical solution of the nonlinear circuit of Figure 17–13.

Figures 17-13 and 17-14 are used to develop the mathematical and graphical method of solving a nonlinear series circuit. The nonlinear series circuit of Figure 17-13 has a voltage equation of

$$E = V_F + V_{RL} \qquad (17\text{-}5)$$

Substituting IR_L for V_{RL}:

$$E = V_F + IR_L \qquad (17\text{-}6)$$

When $V_F = 0$ V, Equation 17-6 becomes $E = 0 + IR_L$ and $I = E/R_L$. The conditions ($V_F = 0$ and $I = E/R_L$) form an ordered pair, which is plotted as point A of Figure 17-14.

When $I = 0$ A, Equation 17-6 becomes $E = V_F + 0$ and $E = V_F$. The conditions ($V_F = E$ and $I = 0$) form an ordered pair, which is plotted as point B of Figure 17-14.

The straight line connecting points A and B intersects the diode characteristic curve at Q. The coordinates of point Q are the operating conditions of the nonlinear circuit of Figure 17-13 and are detailed in Figure 17-15.

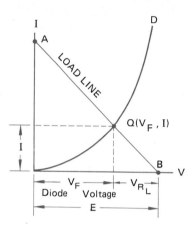

FIGURE 17–15
Graphical solution of the nonlinear circuit of Figure 17–13. The abscissa of the Q point is the voltage across the nonlinear device; the ordinate is the circuit current.

The *load line* of Figure 17-15 has the general equation of $y = mx + b$. In this equation, y is the circuit current (I), x is the voltage across the nonlinear device (V_F), and the slope (m) and the y-intercept (b) are determined by solving Equation 17-6 for I.

$$I = \frac{E - V_F}{R_L} \qquad (17\text{-}7)$$

Place each term in the right member over the common denominator (R_L):

$$I = \frac{E}{R_L} - \frac{V_F}{R_L}$$

and rearrange the terms to form Equation 17-8:

$$I = \frac{-1}{R_L} V_F + \frac{E}{R_L} \qquad (17\text{-}8)$$

Equation 17-8 is the specific equation of the load line where the slope $m = -1/R_L$ and the y-intercept $b = E/R_L$.

The procedure for the graphic solution of a nonlinear series circuit is summarized as Rule 17-1.

RULE 17-1. GRAPHICAL SOLUTION

To solve a simple nonlinear series circuit:

1. Construct the *load line* on the same coordinates as the curve of the nonlinear device by:
   (a) Establishing two points of the load line:
   Point 1: V of the nonlinear device is zero.

$$V = 0, \qquad I = E/R_L$$

   Point 2: I of the circuit is zero.

$$E = V, \qquad I = 0$$

   where V = voltage drop of the nonlinear device
   E = applied voltage of the circuit
   I = circuit current
   R_L = resistance of the linear device

   (b) Constructing the load line as a straight line through the two points.
2. Read the circuit operating condition as the abscissa and the ordinate of the point of intersection of the linear and nonlinear curves.

EXAMPLE 17-10

In the circuit of Figure 17-16, determine the following using the pictured diode characteristics.
(a) The current in the circuit (I)
(b) The voltage across the diode
(c) The voltage across the resistance

Solution Use Rule 17-1.
Step 1(a): To construct the load line, use:

$Point\ 1: V_F = 0$ V, $I = E/R = 3/50 = 60$ mA
$Point\ 2: I = 0$ mA, $E = V_F = 3$ V

Step 1(b): Plot the two points:

$P_1(0$ V, 60 mA) and $P_2(3$ V, 0 mA)

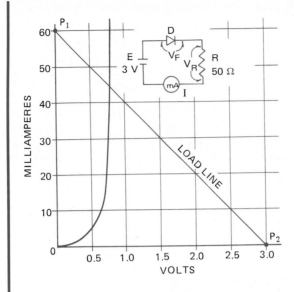

FIGURE 17–16
Circuit and characteristic curve
for Example 17–10.

Draw the load line through these two points. See Figure 17-16.

Step 2: Read the operating condition from the intersection of the two curves:

(a) $I \approx 44.5$ mA
(b) $V_F \approx 0.78$ V
(c) $V_R = E - V_F = 3 - 0.78 = 2.22$ V

EXERCISE 17-3
Using the information given in Figure 17-17:

1. Determine the circuit current (I) and the voltage across R when $R = 75\ \Omega$ and $E = 3$ V.
2. Determine the circuit current (I) and the voltage across R when $R = 40\ \Omega$ and $E = 4$ V.
3. Determine the circuit current (I) and the voltage across R when $R = 50\ \Omega$ and $E = 6$ V.

Using the information given in Figure 17-18:

4. Determine the circuit current (I), the voltage across the lamp (nonlinear device), and the voltage across the resistance when $R = 100\ \Omega$ and $E = 100$ V.

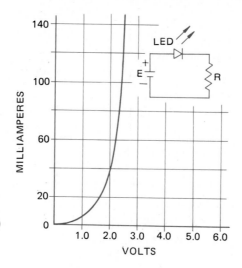

FIGURE 17–17
Circuit and light-emitting diode (LED) forward characteristic curve for Exercise 17–3, problems 1–3.

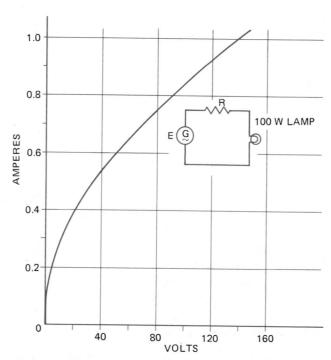

FIGURE 17–18
Circuit and lamp characteristic curve for Exercise 17–3, problems 4–6.

5. Determine the circuit current (I), the voltage across the lamp, and the voltage across the resistance when $R = 80\ \Omega$ and $E = 60$ V.
6. Determine the circuit current (I), the voltage across the lamp, and the voltage across the resistance when $R = 250\ \Omega$ and $E = 150$ V.

SELECTED TERMS

dynamic parameter   A parameter that changes with time.

load line   A line drawn on the characteristic curve of a non-linear device to locate the operating point of the circuit.

static parameter   A parameter that does not change with time.

APPLYING SYSTEMS OF LINEAR EQUATIONS TO ELECTRONIC CONCEPTS

This chapter will deal with the solution of *resistive networks* through the use of a system of linear equations based on Kirchhoff's voltage law.

18-1 APPLYING KIRCHHOFF'S VOLTAGE LAW

Kirchhoff's voltage law (KVL) states that in any complete circuit the algebraic sum of the voltage drops must equal the algebraic sum of the voltage rises. Stated as an equation, Kirchhoff's voltage law becomes:

$$\Sigma V = \Sigma E \qquad (18\text{-}1)$$

or

$$V_1 + V_2 + V_3 = E_1 + E_2 + E_3 \qquad (18\text{-}2)$$

In a complete circuit, as shown in Figure 18-1, we may assume a current (I) passes through the resistive components in the circuit, causing voltage drops. Thus, $V = IR$. Replacing V_1, V_2, and V_3 of Equation 18-2 with their identities, we have a new KVL equation.

$$IR_1 + IR_2 + IR_3 = E_1 + E_2 + E_3 \qquad (18\text{-}3)$$

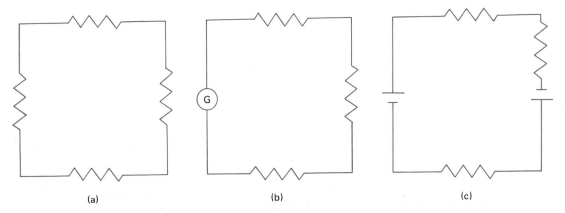

(a) (b) (c)

FIGURE 18–1

A complete electrical circuit is called a *closed loop*. We see that: (a) a closed loop may not have a voltage source; (b) a closed loop may have one voltage source; (c) a closed loop may have several voltage sources.

Conventional Direction for Current

In the study of electricity it is *conventional* to think of the current as flowing from positive ($+$) to negative ($-$). This is opposite to *electron flow*, shown in Figure 18-2. For purposes of mathematical discussion, we will use conven-

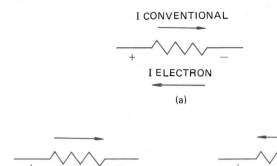

(a)

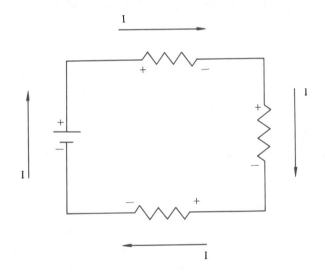

CONVENTIONAL CURRENT FLOW
FROM + TO –
(b)

ELECTRON CURRENT FLOW
FROM – TO +
(c)

FIGURE   18–2
The polarity of the voltage drop across a resistance depends upon the choice of current flow—conventional (b) or electron (c). The conventional direction for current flow is used in this chapter.

FIGURE   18–3
When conventional current passes through a resistance, the point of entering is positive and the point of exit is negative.

tional current flow. Following the conventional current direction through Figure 18-3, you will notice that there is a rise in voltage across the source (that is, − to +) and a drop in voltage across the resistance (that is + to −).

Procedure for Applying KVL Equations

The following series of examples will serve to introduce you to the process of using KVL equations.

EXAMPLE 18-1 Write a KVL equation for the circuit pictured in Figure 18-4 and solve for the loop current (I).

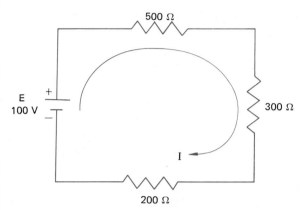

FIGURE 18–4
Circuit for Example 18–1.

Solution Label each resistance and polarize the drops across each resistance using conventional current as shown in Figure 18-5.

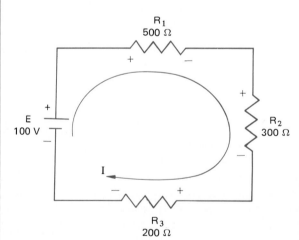

FIGURE 18–5
The voltage drops (across the resistances) are polarized using the conventional direction for current flow.

Write the KVL equation; use Equation 18-3; start with R_1:

$$IR_1 + IR_2 + IR_3 = E_1 + E_2 + E_3$$

Since there is only one source, set E_2 and E_3 to zero:

$$I(500) + I(300) + I(200) = 100$$

Solve for the loop current (I):

$$1000I = 100$$

$$I = \frac{100}{1000} = 100 \text{ mA}$$

$$\therefore \quad I = 100 \text{ mA}$$

Example 18-2 demonstrates what happens when the assumed loop current is selected opposite to that of the actual loop current. The source is assigned a minus sign and the loop current results in a minus quantity.

EXAMPLE 18-2   Repeat Example 18-1, only this time assume the loop current (I) to be in the opposite direction.

Solution   Label each resistance and polarize the voltage drops across each resistance as shown in Figure 18-6.

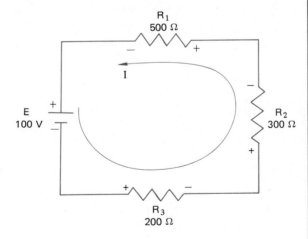

FIGURE 18–6
The voltage drops (across the resistances) are polarized using the conventional direction of current flow.

Write the KVL equation; use Equation 18-3; start with R_3:

$$IR_3 + IR_2 + IR_1 = E_1 + E_2 + E_3$$
$$I(200) + I(300) + I(500) = -100$$

Solve for the assumed loop current (I):

$$1000I = -100$$
$$\therefore \quad I = -100 \text{ mA}$$

Observation   The minus loop current means that the actual direction is opposite to that of the assumed direction.

As demonstrated in the preceding examples, the selection of the direction of the loop current makes no difference, as the direction will be *flagged* by the minus sign. Remember that a negative loop current indicates that the assumed direction is opposite to the actual direction.

Kirchhoff's voltage law may be applied to a simple closed loop through the use of the concepts in the following guidelines.

GUIDELINES 18-1. KVL LOOP PROCEDURE

To write a KVL equation for a simple loop, remember that:

1. The direction of the assumed loop current is always positive. Thus, the conventional current enters the resistance from the positive side and leaves from the negative.
2. The polarity of a voltage source is not changed by the direction of the assumed loop current.
3. Assumed loop current flow through a voltage source from − to + is considered positive and is assigned a + sign.
4. Assumed loop current flow through a voltage source from + to − is considered negative and is assigned a − sign.

EXAMPLE 18-3   Apply KVL and Guidelines 18-1 to the circuit of Figure 18-7. Solve for *I*.

Solution   Assume a loop current direction as shown in Figure 18-8. Label and polarize the resistances and label the voltage sources.
Write the KVL equation applying Guidelines 18-1 and Equation 18-3:

$$IR_1 + IR_2 + IR_3 + IR_4 + IR_5 = E_1 + E_2 + E_3$$

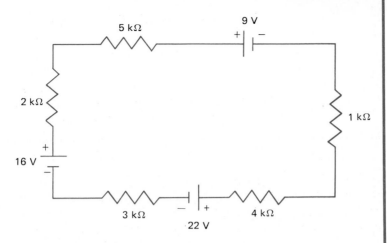

FIGURE 18-7
Circuit for Example 18-3.

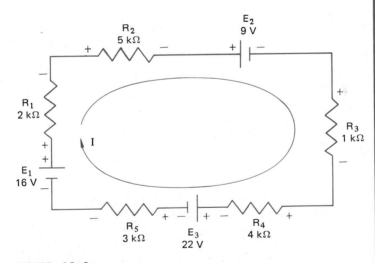

FIGURE 18-8
The resistances of Figure 18-7 are polarized using the conventional direction of current flow.

Substitute; pay particular attention to the sign of the voltage sources (Guidelines 18-1, steps 3 and 4); start with R_1:

$$2000I + 5000I + 1000I + 4000I + 3000I$$
$$= 16 - 9 - 22$$

$$15 \times 10^3 I = -15$$
$$I = -15/(15 \times 10^3)$$
$$I = -1 \text{ mA}$$

∴ The current is − 1 mA, and the actual direction of the current flow is opposite to the assumed direction.

EXAMPLE 18-4 Apply KVL to the circuit of Figure 18-9; solve for V_3.

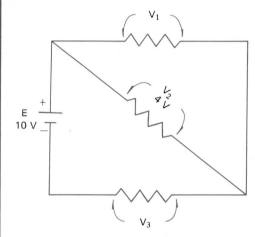

FIGURE 18–9
Circuit for Example 18–4.

Solution Since there are two loops that include V_3, we may solve for V_3 with either one. Assume loop current directions as shown in Figure 18-10, and polarize the resistance using the conventional direction of current. Write the KVL equation for loop 1:

$$\Sigma V = \Sigma E$$
$$V_2 + V_3 = E$$

Substitute and solve for V_3:

$$4 + V_3 = 10$$
$$V_3 = 6 \text{ V}$$

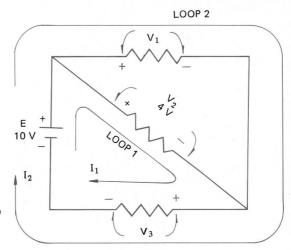

FIGURE 18-10
The resistances of Figure 18-9 are polarized using the conventional direction of current flow.

Write the KVL equation for loop 2:

$$\Sigma V = \Sigma E$$
$$V_1 + V_3 = E$$

Substitute and solve for V_3. $V_1 = V_2 = 4$ V because the resistances are connected in parallel:

$$4 + V_3 = 10$$
$$V_3 = 6 \text{ V}$$

$\therefore$ V_3 may be found with either loop equation.

EXERCISE 18-1

1. Write a KVL equation for the circuit of Figure 18-11 and solve for the loop current.
2. Write a KVL equation for the circuit of Figure 18-11 when R_1 is 200 Ω. Solve for the loop current.
3. Write a KVL equation for the circuit of Figure 18-11 when E_2 is reversed in polarity. Solve for the loop current.
4. Write a KVL equation for the circuit of Figure 18-11 when R_2 is 600 Ω and E_1 is increased to 15 V. Solve for the voltage drop, V_3.
5. Write a KVL equation for the circuit of Figure 18-12 and solve for the loop current.

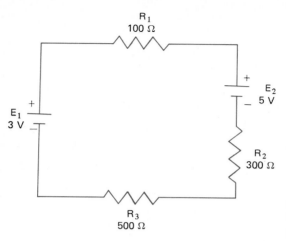

FIGURE 18–11
Circuit for Exercise 18–1, problems 1–4.

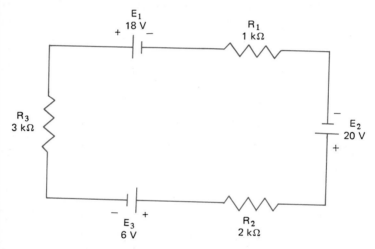

FIGURE 18–12
Circuit for Exercise 18–1, problems 5–8.

6. Write a KVL equation for the circuit of Figure 18-12 when R_1 is 1.8 kΩ. Solve for the voltage drop, V_2.

7. Write a KVL equation for the circuit of Figure 18-12 when E_3 is reversed in polarity. Solve for the loop current.

8. Write a KVL equation for the circuit of Figure 18-12 when R_1 is 4.7 kΩ and E_2 is increased to 36 V. Solve for the voltage drop, V_1.

9. Write a KVL equation and solve for V_3 in Figure 18-13.

10. Write a KVL equation and solve for V_1 in Figure 18-14.

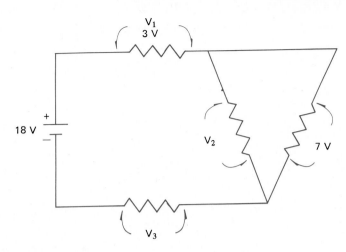

FIGURE 18–13
Circuit for Exercise 18–1, problem 9.

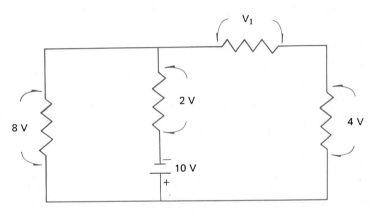

FIGURE 18–14
Circuit for Exercise 18–1, problem 10.

18-2 MESH ANALYSIS

Through the use of mesh analysis, we may solve a *network* (circuit) that requires more than just the rules of series or parallel circuits. *Mesh analysis* is one of several methods used to analyze a network.

A *mesh* is a simple closed loop having no branches. Even a very complicated network can be represented by two or more meshes. Figure 18-15 is an example of a network with two meshes. In solving the network, an equa-

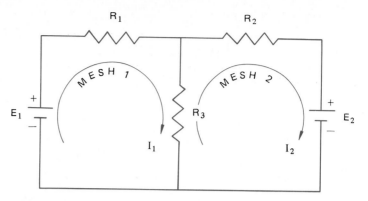

FIGURE 18–15
Network with two meshes. The mesh currents I_1 and I_2 are assumed to flow in a clockwise (cw) direction.

tion (similar to a KVL equation) is written for each mesh. The resulting system of equations is solved for each of the mesh currents. Once the mesh currents are determined, then the actual branch currents and voltage drops may be determined.

Mutual Resistance

An important idea in mesh analysis is that of *mutual resistance.* The mutual resistance is the resistance *shared* by two mesh currents. In Figure 18-15, R_3 is the mutual resistance.

The mesh currents passing through the mutual resistance may be *aiding* or *opposing,* as shown in Figure 18-16. When we write the mesh equation, we must pay close attention to the mesh currents passing through the mutual resistance. **Since each mesh current of the two meshes passes through the mutual resistance, both must be included in each of the mesh equations.** Examples 18-5 and 18-6 will help to clarify this important concept.

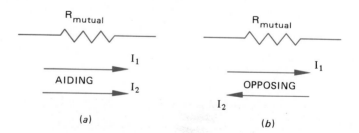

FIGURE 18–16
Mesh currents in the mutual resistance: (a) aiding currents are added; (b) opposing currents are subtracted.

EXAMPLE 18-5 Write the mesh equations for the network of Figure 18-17 and solve for the mesh currents, I_1 and I_2. Assume a clockwise (cw) direction for each current.

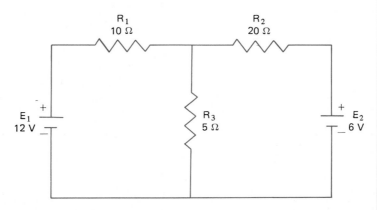

FIGURE 18–17
Network for Examples 18–5 and 18–6.

Solution Draw in mesh currents I_1 and I_2 in a cw direction as shown in Figure 18-18.

Polarize the resistances as was done in the previous KVL section. This is shown in Figure 18-18.

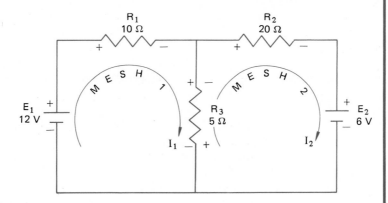

FIGURE 18–18
Network for Figure 18–17 with two meshes. Notice that the mutual resistance R₃ has two polarizations. One polarization is determined by the direction of mesh current I_1; the other polarization is determined by the direction of mesh current I_2.

Write the mesh equation for each mesh. Use KVL Equation 18-3 and Guidelines 18-1.
Mesh 1, Figure 18-19:

$$10I_1 + \underbrace{5I_1 - 5I_2}_{\text{current through } R_3} = 12$$

Notice that the mutual resistance ($R_3 = 5\,\Omega$ of Figure 18-18) has opposing currents passing through it. For mesh 1, mesh current I_2 is negative, as emphasized in Figure 18-19.

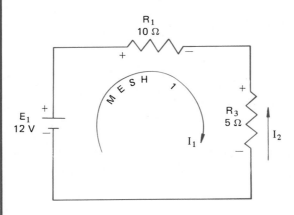

FIGURE 18–19
Mesh 1 of the network for Figure 18–18.

Mesh 2, Figure 18-20:

$$\underbrace{\qquad\qquad}_{\text{current through } R_3} + 20I_2 = -6$$

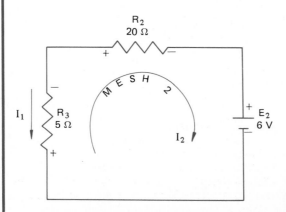

FIGURE 18–20
Mesh 2 of the network for Figure 18–18.

Notice that the mutual resistance ($R_3 = 5 \, \Omega$ of Figure 18-18) has opposing currents passing through it. For mesh 2, mesh current I_1 is negative, as emphasized in Figure 18-20.

$$\text{Mesh 1: } 15I_1 - 5I_2 = 12$$
$$\text{Mesh 2: } -5I_1 + 25I_2 = -6$$

Solve for I_1 and I_2 using determinants:

$$\Delta = \begin{vmatrix} 15 & -5 \\ -5 & 25 \end{vmatrix} = 375 - 25 = 350$$

$$I_1 = \frac{\begin{vmatrix} 12 & -5 \\ -6 & 25 \end{vmatrix}}{\Delta} = \frac{300 - 30}{350} = \frac{270}{350} = 771 \text{ mA}$$

$$I_2 = \frac{\begin{vmatrix} 15 & 12 \\ -5 & -6 \end{vmatrix}}{\Delta} = \frac{-90 + 60}{350} = \frac{-30}{350} = -85.7 \text{ mA}$$

∴ $I_1 = 771$ mA in the direction assumed.
$I_2 = 85.7$ mA in the direction opposite to that assumed.

In Example 18-5 the directions of the mesh currents were selected so that the currents passing through the mutual resistance were opposing. In Example 18-6 the directions of the mesh currents are selected so that the currents passing through the mutual resistance are aiding.

EXAMPLE 18-6   Write the mesh equations for the network of Figure 18-21 and solve for the mesh currents, I_1 and I_2.

Solution

$$\text{Mesh 1: } 10I_1 + \underbrace{5I_1 + 5I_2}_{\text{current through } R_3} = 12$$

$$\text{Mesh 2: } 20I_2 + \underbrace{5I_2 + 5I_1}_{\text{current through } R_3} = 6$$

Since the mutual resistance (R_3 of Figure 18-21) has aiding currents passing through it, I_2 in mesh 1 and I_1

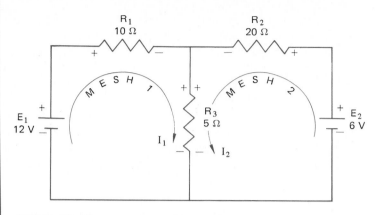

FIGURE 18–21
Network for Example 18–6. Notice that the mutual resistance R₃ has *aiding currents* passing through it.

in mesh 2 are both positive. Simplify each mesh equation:

$$\text{Mesh 1: } 15I_1 + 5I_2 = 12$$
$$\text{Mesh 2: } 5I_1 + 25I_2 = 6$$

Solve for I_1 and I_2 using determinants:

$$\Delta = \begin{vmatrix} 15 & 5 \\ 5 & 25 \end{vmatrix} = 375 - 25 = 350$$

$$I_1 = \frac{\begin{vmatrix} 12 & 5 \\ 6 & 25 \end{vmatrix}}{\Delta} = \frac{300 - 30}{350} = \frac{270}{350} = 771 \text{ mA}$$

$$I_2 = \frac{\begin{vmatrix} 15 & 12 \\ 5 & 6 \end{vmatrix}}{\Delta} = \frac{90 - 60}{350} = \frac{30}{350} = 85.7 \text{ mA}$$

∴ $I_1 = 771$ mA in the direction assumed.
$I_2 = 85.7$ mA in the direction assumed.

Once the mesh currents are computed, the actual branch currents may be determined. Figure 18-22 shows the actual branch currents for the network of Examples 18-5 and 18-6. Notice that the mutual resistance has an actual branch current (I_C) equal to the algebraic sum of the mesh currents (I_1 and I_2). Also notice that the actual branch current I_A is equal to the mesh current I_1, and that branch current I_B is equal to the mesh current I_2.

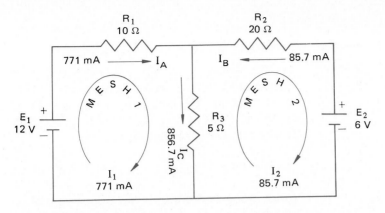

FIGURE 18-22
Network for Examples 18–5 and 18–6 with the actual branch currents
I_A, I_B, and I_C shown and the assumed direction of the mesh current I_1
and I_2 selected to correspond to the actual direction. $I_A = I_1$; $I_B = I_2$;
$I_C = I_1 + I_2$.

EXAMPLE 18-7 From the following sets of mesh currents, determine:

(a) The actual branch current passing through the
mutual resistance, R_3 of Figure 18-23
(b) The actual direction (up or down) when:

Case 1: $I_1 = 15$ A $I_2 = 5$ A
Case 2: $I_1 = -15$ A $I_2 = 5$ A
Case 3: $I_1 = 15$ A $I_2 = -5$ A
Case 4: $I_1 = 5$ A $I_2 = 15$ A

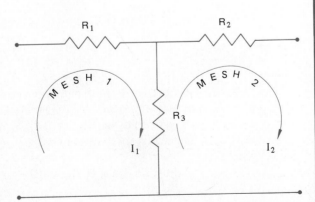

FIGURE 18-23
**Network for Example
18-7 and Exercise 18-2.**

Solution

$$\text{Case 1: } \begin{cases} I_1 = 15 \text{ A} \\ I_2 = 5 \text{ A} \end{cases}$$

Because both mesh currents I_1 and I_2 are positive, the actual direction of the currents through the mutual resistance, R_3, is as pictured in Figure 18-23. The currents are *opposing* one another.

$\therefore$ Case 1: $I_{R_3} = 15 - 5 = 10$ A downward.

Notice when two currents are opposing one another that the direction of the difference is determined by the direction of the larger current.

$$\text{Case 2: } \begin{cases} I_1 = -15 \text{ A} \\ I_2 = 5 \text{ A} \end{cases}$$

Because I_1 is negative, its direction is opposite to that pictured in Figure 18-23. The direction of I_2 is as pictured in Figure 18-23. Thus, both I_1 and I_2 are actually flowing upward through R_3 as *aiding* currents.

$\therefore$ Case 2: $I_{R_3} = -(-15) + 5 = 20$ A upward.

$$\text{Case 3: } \begin{cases} I_1 = 15 \text{ A} \\ I_2 = -5 \text{ A} \end{cases}$$

Because I_2 is negative, its direction is opposite to that pictured in Figure 18-23. The direction of I_1 is as pictured in Figure 18-23. Thus, both I_1 and I_2 are actually flowing downward through R_3 as *aiding* currents.

$\therefore$ Case 3: $I_{R_3} = 15 - (-5) = 20$ A downward.

$$\text{Case 4: } \begin{cases} I_1 = 5 \text{ A} \\ I_2 = 15 \text{ A} \end{cases}$$

As in case 1, both mesh currents are positive. So the actual direction through the mutual resistance, R_3, is as pictured in Figure 18-23, and the currents are *opposing* one another.

$\therefore$ Case 4: $I_{R_3} = 15 - 5 = 10$ A upward.

NOTE: Once again, when two currents oppose one another, the direction of the difference is determined by the direction of the larger current.

EXERCISE 18-2
From the following sets of mesh currents, determine (a) the actual branch current passing through the mutual resistance, R_3 of Figure 18-23, and (b) the actual direction—upward or downward.

1. $\begin{cases} I_1 = 3\ \text{A} \\ I_2 = 2\ \text{A} \end{cases}$ 2. $\begin{cases} I_1 = -10\ \text{A} \\ I_2 = 6\ \text{A} \end{cases}$

3. $\begin{cases} I_1 = 1\ \text{A} \\ I_2 = 3\ \text{A} \end{cases}$ 4. $\begin{cases} I_1 = -6\ \text{A} \\ I_2 = 8\ \text{A} \end{cases}$

5. $\begin{cases} I_1 = -7\text{A} \\ I_2 = 2\ \text{A} \end{cases}$ 6. $\begin{cases} I_1 = -12\ \text{A} \\ I_2 = -5\ \text{A} \end{cases}$

7. $\begin{cases} I_1 = -3\ \text{A} \\ I_2 = -9\ \text{A} \end{cases}$ 8. $\begin{cases} I_1 = 7\ \text{A} \\ I_2 = -4\ \text{A} \end{cases}$

9. $\begin{cases} I_1 = 1\ \text{A} \\ I_2 = -1\ \text{A} \end{cases}$ 10. $\begin{cases} I_1 = 5\ \text{A} \\ I_2 = 5\ \text{A} \end{cases}$

18-3 SOLVING NETWORKS BY MESH ANALYSIS

The concepts of mesh currents, KVL, and mutual resistance are brought together as mesh analysis to solve networks.

Guidelines 18-2 are a summary of the techniques used to solve networks with mesh analysis.

GUIDELINES 18-2. MESH ANALYSIS

To solve a network by mesh analysis:

1. Draw a mesh current within each mesh in a cw direction.
2. Write as many mesh equations as there are meshes, using KVL equations and the concepts of mutual resistance.
3. Solve for the mesh currents with determinants.

Guidelines 18-2 will be applied in the following examples. It is suggested that you work along with pencil, paper, and calculator.

EXAMPLE 18-8 Write the mesh equations for the network of Figure 18-24 and determine the mesh currents.

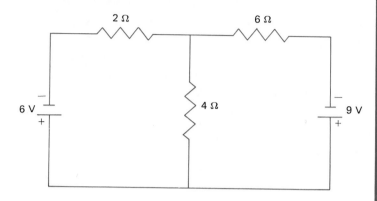

FIGURE 18–24
Network for Example 18–8.

Solution Use Guidelines 18-2.

Step 1: Draw two cw mesh currents within the network as shown in Figure 18-25. Polarize the resistances.

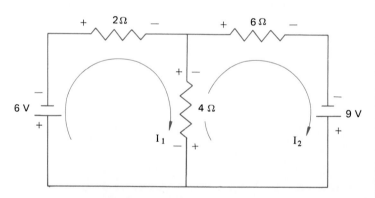

FIGURE 18–25
Network for Example 18–8 with mesh currents (cw) shown and resistance polarized.

Step 2: Write two mesh equations:

$$\text{Mesh 1: } 2I_1 + 4I_1 - 4I_2 = -6$$
$$6I_1 - 4I_2 = -6$$

$$\text{Mesh 2: } 4I_2 - 4I_1 + 6I_2 = 9$$
$$-4I_1 + 10I_2 = 9$$

Step 3: Solve for I_1 and I_2 with determinants:

$$\text{Mesh 1: } 6I_1 - 4I_2 = -6$$
$$\text{Mesh 2: } -4I_1 + 10I_2 = 9$$

$$\Delta = \begin{vmatrix} 6 & -4 \\ -4 & 10 \end{vmatrix} = 60 - 16 = 44$$

$$I_1 = \frac{\begin{vmatrix} -6 & -4 \\ 9 & 10 \end{vmatrix}}{\Delta} = \frac{-60 + 36}{44} = \frac{-24}{44} = -545 \text{ mA}$$

$$I_2 = \frac{\begin{vmatrix} 6 & -6 \\ -4 & 9 \end{vmatrix}}{\Delta} = \frac{54 - 24}{44} = \frac{30}{44} = 682 \text{ mA}$$

$\therefore$ $I_1 = 545$ mA in the direction opposite to that assumed.

        $I_2 = 682$ mA in the direction assumed.

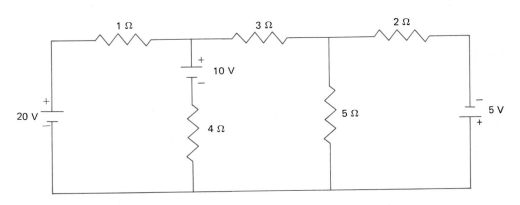

FIGURE 18–26
Network for Example 18–9.

EXAMPLE 18-9   Write the mesh equations for the network of Figure 18-26 and determine the mesh currents.

Solution   Use Guidelines 18-2.

Step 1:   Draw three mesh currents within the network as shown in Figure 18-27. Polarize the resistances.

PROGRAMMABLE

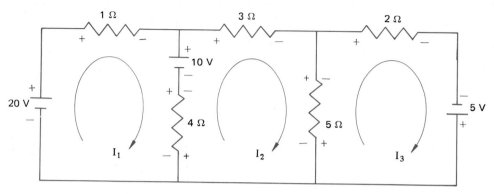

FIGURE 18–27
**Network for Example 18–9 with mesh currents (cw)
shown and resistances polarized.**

Step 2: Write three mesh equations:

$$\text{Mesh 1: } 1I_1 + 4I_1 - 4I_2 = 20 - 10$$
$$5I_1 - 4I_2 = 10$$

$$\text{Mesh 2: } 4I_2 - 4I_1 + 3I_2 + 5I_2 - 5I_3 = 10$$
$$-4I_1 + 12I_2 - 5I_3 = 10$$

$$\text{Mesh 3: } 5I_3 - 5I_2 + 2I_3 = 5$$
$$-5I_2 + 7I_3 = 5$$

Step 3: Solve for I_1, I_2, and I_3 using determinants:

$$\text{Mesh 1: } 5I_1 - 4I_2 + 0I_3 = 10$$
$$\text{Mesh 2: } -4I_1 + 12I_2 - 5I_3 = 10$$
$$\text{Mesh 3: } 0I_1 - 5I_2 + 7I_3 = 5$$

$$\Delta = \begin{vmatrix} 5 & -4 & 0 \\ -4 & 12 & -5 \\ 0 & -5 & 7 \end{vmatrix} = 183$$

$$I_1 = \dfrac{\begin{vmatrix} 10 & -4 & 0 \\ 10 & 12 & -5 \\ 5 & -5 & 7 \end{vmatrix}}{\Delta} = \dfrac{970}{183} = 5.30 \text{ A}$$

$$I_2 = \dfrac{\begin{vmatrix} 5 & 10 & 0 \\ -4 & 10 & -5 \\ 0 & 5 & 7 \end{vmatrix}}{\Delta} = \dfrac{755}{183} = 4.13 \text{ A}$$

$$I_3 = \dfrac{\begin{vmatrix} 5 & -4 & 10 \\ -4 & 12 & 10 \\ 0 & -5 & 5 \end{vmatrix}}{\Delta} = \dfrac{670}{183} = 3.66 \text{ A}$$

$$\therefore \quad I_1 = 5.30 \text{ A}, \ I_2 = 4.13 \text{ A}, \ I_3 = 3.66 \text{ A}$$

EXAMPLE 18-10

Write the mesh equations for the network of Figure 18-28 and determine the mesh currents I_1, I_2, and I_3.

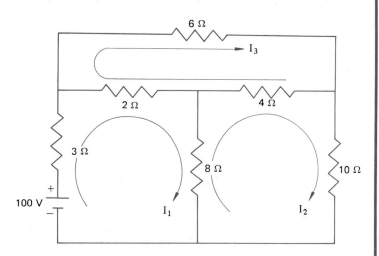

FIGURE 18–28
Network for Exmaple 18–10.

Solution   Write three mesh equations:

$$\text{Mesh 1: } 3I_1 + 2I_1 - 2I_3 + 8I_1 - 8I_2 = 100$$
$$13I_1 - 8I_2 - 2I_3 = 100$$

Mesh 2: $8I_2 - 8I_1 + 4I_2 - 4I_3 + 10I_2 = 0$
$$-8I_1 + 22I_2 - 4I_3 = 0$$
Mesh 3: $4I_3 - 4I_2 + 2I_3 - 2I_1 + 6I_3 = 0$
$$-2I_1 - 4I_2 + 12I_3 = 0$$

Solve for I_1, I_2, and I_3 with determinants:

Mesh 1: $13I_1 - 8I_2 - 2I_3 = 100$
Mesh 2: $-8I_1 + 22I_2 - 4I_3 = 0$
Mesh 3: $-2I_1 - 4I_2 + 12I_3 = 0$

$$\Delta = \begin{vmatrix} 13 & -8 & -2 \\ -8 & 22 & -4 \\ -2 & -4 & 12 \end{vmatrix} = 2240$$

$$I_1 = \frac{\begin{vmatrix} 100 & -8 & -2 \\ 0 & 22 & -4 \\ 0 & -4 & 12 \end{vmatrix}}{\Delta} = \frac{24\,800}{2240} = 11.07 \text{ A}$$

$$I_2 = \frac{\begin{vmatrix} 13 & 100 & -2 \\ -8 & 0 & -4 \\ -2 & 0 & 12 \end{vmatrix}}{\Delta} = \frac{10\,400}{2240} = 4.64 \text{ A}$$

$$I_3 = \frac{\begin{vmatrix} 13 & -8 & 100 \\ -8 & 22 & 0 \\ -2 & -4 & 0 \end{vmatrix}}{\Delta} = \frac{7600}{2240} = 3.39\text{A}$$

EXAMPLE 18-11

Use the mesh currents of Example 18-10 and Figure 18-28 to determine the actual branch currents through the 2-, 4-, and 8-Ω resistances.

Solution Draw the portion of Figure 18-28 in question with the mesh currents shown. See Figure 18-29.

The branch current through the 2-Ω resistance:

$$I_{2\Omega} = I_1 + I_3$$
$$I_{2\Omega} = 11.07 - 3.39 = 7.68 \text{ A to the right}$$

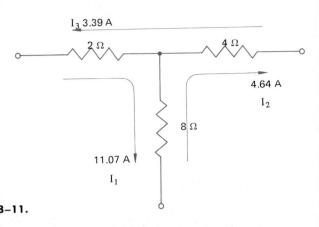

FIGURE 18–29
Network for Example 18–11.

The branch current through the 4-Ω resistance:

$$I_{4\Omega} = I_3 + I_2$$
$$I_{4\Omega} = -3.39 + 4.64 = 1.25 \text{ A to the right}$$

The branch current through the 8-Ω resistance:

$$I_{8\Omega} = I_1 + I_2$$
$$I_{8\Omega} = 11.07 - 4.64 = 6.43 \text{ A downward}$$

The branch currents through the 2-, 4-, and 8-Ω resistances are shown in Figure 18-30.

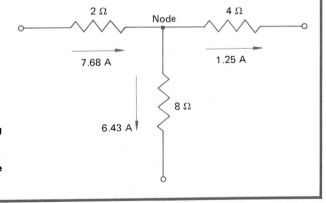

FIGURE 18–30
Branch currents for Example 18–11. Notice that the current entering the node (7.68 A) is equal to the current leaving the node (7.68 = 1.25 + 6.43).

EXERCISE 18-3
Write the mesh equations of Figure 18-31 and solve for I_1 and I_2 when:

1. $R_1 = 2 \ \Omega$ $R_2 = 4 \ \Omega$ $R_3 = 3 \ \Omega$
   $E_1 = 10 \text{ V}$ $E_2 = 20 \text{ V}$

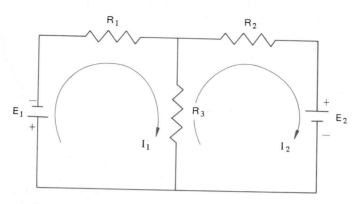

FIGURE 18–31
Network for Exercise 18–3, problems 1–5.

2. $R_1 = 4\ \Omega$ $R_2 = 2\ \Omega$ $R_3 = 1\ \Omega$
    $E_1 = 20\ \text{V}$ $E_2 = 30\ \text{V}$
3. $R_1 = 5\ \Omega$ $R_2 = 10\ \Omega$ $R_3 = 5\ \Omega$
    $E_1 = 9\ \text{V}$ $E_2 = 30\ \text{V}$
4. $R_1 = 500\ \Omega$ $R_2 = 1.0\text{k}\ \Omega$ $R_3 = 300\ \Omega$
    $E_1 = 30\ \text{V}$ $E_2 = 20\ \text{V}$
5. $R_1 = 200\ \Omega$ $R_2 = 700\ \Omega$ $R_3 = 600\ \Omega$
    $E_1 = 15\ \text{V}$ $E_2 = 20\ \text{V}$

Write the mesh equations of Figure 18-32 and solve for I_1, I_2, and I_3 when:

6. $R_1 = 2\ \Omega$ $R_2 = 4\ \Omega$ $R_3 = 8\ \Omega$ $R_4 = 1\ \Omega$ $R_5 = 5\ \Omega$
    $E_1 = 10\ \text{V}$ $E_2 = 10\ \text{V}$ $E_3 = 10\ \text{V}$
7. $R_1 = 10\ \Omega$ $R_2 = 2\ \Omega$ $R_3 = 4\ \Omega$ $R_4 = 1\ \Omega$ $R_5 = 3\ \Omega$
    $E_1 = 6\ \text{V}$ $E_2 = 9\ \text{V}$ $E_3 = 3\ \text{V}$
8. $R_1 = 20\ \Omega$ $R_2 = 60\ \Omega$ $R_3 = 10\ \Omega$ $R_4 = 30\ \Omega$ $R_5 = 50\ \Omega$
    $E_1 = 15\ \text{V}$ $E_2 = 27\ \text{V}$ $E_3 = 12\ \text{V}$

Write the mesh equations of Figure 18-33 and solve for I_1, I_2, and I_3 when:

9. $R_1 = 2\ \Omega$ $R_2 = 7\ \Omega$ $R_3 = 5\ \Omega$ $R_4 = 3\ \Omega$ $R_5 = 6\ \Omega$ $E_1 = 10\ \text{V}$
10. $R_1 = 9\ \Omega$ $R_2 = 1\ \Omega$ $R_3 = 2\ \Omega$ $R_4 = 8\ \Omega$ $R_5 = 5\ \Omega$ $E_1 = 6\ \text{V}$
11. Using the values computed in problem 9, determine the branch current through R_2 of Figure 18-33.

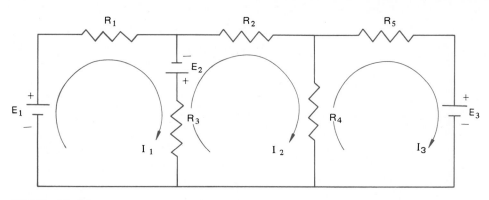

FIGURE 18–32
Network for Exercise 18–3, problems 6–8.

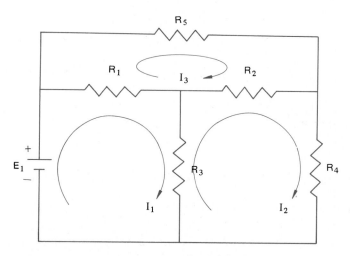

FIGURE 18–33
Network for Exercise 18–3, problems 9–12.

12. Using the values computed in problem 10, determine the branch current through R_1 of Figure 18-33.

13. The circuit of Figure 18-34 is a simplified diagram of an automotive electrical circuit. Using the information given in Figure 18-34, determine if the battery is being charged or discharged, and find the current through R_L.

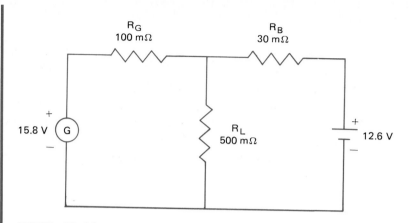

FIGURE 18–34
Network for Exercise 18–3, problem 13.

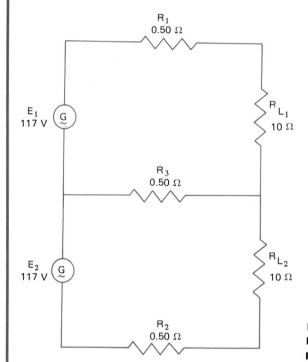

FIGURE 18–35
Network for Exercise 18–3, problem 14.

14. The circuit of Figure 18-35 represents a three-wire distribution system in which R_1, R_2, and R_3 represent the wire resistances and R_{L_1} and R_{L_2} represent the loads. Find the current through R_3 using the information given in Figure 18-35.

15. Change R_{L_1} to 25Ω and leave R_{L_2} at 10Ω. Repeat problem 14.

SELECTED TERMS

conventional current   A way of thinking of the direction of current flow in an electric circuit. Current is thought of as flowing from positive to negative.

mesh   A simple closed loop having no branches.

mesh currents   Currents with an assumed direction used to solve a network using mesh analysis. Once the mesh currents are computed, then the actual branch currents in the network may be computed.

mutual resistance   The resistance shared by two mesh currents.

19

SPECIAL PRODUCTS, FACTORING, AND EQUATIONS

This chapter will give you the opportunity to increase your ability in using the techniques of algebra to solve equations. In your arithmetic courses you learned ways to make your work easier. In this chapter you will learn ways to make your work with equations easier.

19-1 MENTALLY MULTIPLYING TWO BINOMIALS

Because multiplying is so common in arithmetic, you learned the "times tables." In algebra, multiplying two binomials is a common operation and, like the times tables, it may be done at sight. To learn to write the product of two binomials at sight, study the patterns in Figure 19-1.

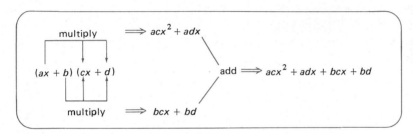

FIGURE 19-1
Pattern for multiplying two binomials.

EXAMPLE 19-1   Find the product of $(5x - 1)(2x - 2)$.
Solution

$$(5x - 1)\quad(2x - 2)$$

$$\Rightarrow 10x^2 - 10x$$
$$\Rightarrow -2x + 2$$
$$+ \Rightarrow 10x^2 - 12x + 2$$

$$\therefore\quad (5x - 1)(2x - 2) = 10x^2 - 12x + 2$$

We may see from the preceding example that the process of multiplying two binomials is done by (1) distributing the first binomial (one term at a time) over the second binomial, (2) adding the two products, and (3) combining like terms.

The result of multiplying two binomials having the same variable is a trinomial product. The trinomial product is made up of three terms, each having a special name as shown in Table 19-1.

TABLE 19-1

TERMS OF THE TRINOMIAL PRODUCT $ax^2 + bx + c$		
Term	**Degree**	**Name**
First ax^2	Two	Quadratic term
Second bx	One	Linear term
Third c	Zero	Constant term

The procedure for mentally multiplying two binomials is summarized as Rule 19-1.

RULE 19-1. MULTIPLYING TWO BINOMIALS

To multiply two binomials mentally, use the distributive law and:

1. Multiply the second binomial by the first term of the first binomial.
2. Multiply the second binomial by the second term of the first binomial.
3. Add the two resulting products and combine the like terms.
4. Form the trinomial product.

EXAMPLE 19-2 Find the product of $(3x - 4)(x + 1)$.

Solution Use Rule 19-1.

Step 1: Distribute (multiply) $3x$ over $(x + 1)$:

$$3x^2 + 3x$$

Step 2: Distribute (multiply) -4 over $(x + 1)$:

$$-4x - 4$$

Step 3: Combine the like terms:

$$3x - 4x = -x$$

Step 4: Form the trinomial product:

$$3x^2 - x - 4$$

$$\therefore \quad (3x - 4)(x + 1) = 3x^2 - x - 4$$

EXAMPLE 19-3 Find the product of $(3y - 5)(-y + 3)$.

Solution Use Rule 19-1.

Step 1: Distribute (multiply) $3y$ over $(-y + 3)$:

$$-3y^2 + 9y$$

Step 2: Distribute (multiply) -5 over $(-y + 3)$:

$$5y - 15$$

Step 3: Combine like terms:

$$9y + 5y = 14y$$

Step 4: Form the trinomial product:

$$-3y^2 + 14y - 15$$

$$\therefore \quad (3y - 5)(-y + 3) = -3y^2 + 14y - 15$$

The following concepts were demonstrated in the preceding example:

- The quadratic term $(-3y^2)$ was formed by multiplying the first term in each binomial: $(3y)(-y) = -3y^2$.
- The constant term (-15) was formed by multiplying the second term in each binomial: $(-5)(3) = -15$.
- The linear term $14y$ was formed by adding the *inner* and *outer* products:

$$(3y - 5)\ (-y + 3) \quad \begin{array}{l} \Rightarrow 5y \\ \Rightarrow 9y \end{array} + \Rightarrow 14y$$

EXERCISE 19-1

Determine the missing terms in the following trinomial products.

1. $(x + 2)(x + 3) = x^2 + 5x + (?)$
2. $(2x + 1)(x - 2) = (?) - 3x - 2$
3. $(5y - 1)(3y - 1) = (?) - 8y + (?)$
4. $(4y - 3)(2y + 2) = 8y^2 + (?) - 6$

Mentally multiply the following pairs of binomials and express the product as a trinomial.

5. $(x + 1)(x + 1)$ 6. $(x + 1)(x + 2)$
7. $(x + 2)(x + 3)$ 8. $(x - 4)(x - 1)$
9. $(x - 5)(x - 3)$ 10. $(x + 5)(x + 3)$
11. $(x + 6)(x + 4)$ 12. $(x - 4)(x - 3)$
13. $(x - 2)(x - 10)$ 14. $(x + 8)(x - 4)$
15. $(x - 5)(x + 2)$ 16. $(2y + 1)(5y - 4)$
17. $(3y - 7)(2y + 3)$ 18. $(4y - 1)(2y - 8)$
19. $(10y - 1)(9y - 2)$ 20. $(2y - 3)(6y - 2)$
21. $(-3x + 4)(2x - 3)$ 22. $(-2x - 5)(x - 2)$
23. $(6x - 4)(-3x + 4)$ 24. $(3x - 7)(-5x - 1)$
25. $(-2x - 4)(-2x + 4)$ 26. $(-5x - 3)(-4x - 6)$
27. $(x + \frac{1}{4})(x + \frac{1}{2})$ 28. $(x - \frac{1}{3})(x + \frac{2}{3})$
29. $(y - \frac{2}{5})(y + \frac{4}{5})$ 30. $(\frac{3}{8}x + 1)(\frac{1}{4}x - 1)$

19-2 PRODUCT OF THE SUM AND DIFFERENCE OF TWO NUMBERS

A special product is formed when the sum of two numbers is multiplied by their difference. The special product formed by the sum and difference of two numbers is called the *difference of two squares,* which is formed by applying Rule 19-2.

RULE 19-2. MULTIPLYING THE SUM AND DIFFERENCE OF TWO NUMBERS

The product of the sum of two numbers and their difference equals the difference of their squares.

$$(a + b)(a - b) = a^2 - b^2$$

EXAMPLE 19-4   Solve $(x + 2)(x - 2)$.
   Solution   Square x:

$$x^2$$

Square 2:

$$2^2 = 4$$

Form the difference of the two squares:

$$x^2 - 4$$

$$\therefore \quad (x + 2)(x - 2) = x^2 - 4$$

EXERCISE 19-2

Mentally multiply each of the following and express the product as a polynomial.

1. $(x + 1)(x - 1)$ 2. $(x + 3)(x - 3)$
3. $(x + 5)(x - 5)$ 4. $(x + 6)(x - 6)$
5. $(2x + 1)(2x - 1)$ 6. $(3x + 2)(3x - 2)$
7. $(5x + 5)(5x - 5)$ 8. $(4x + 3)(4x - 3)$
9. $(-4x + 7)(-4x - 7)$ 10. $(-2x + 8)(-2x - 8)$
11. $(y + a)(y - a)$ 12. $(y + b)(y - b)$
13. $(2y + c)(2y - c)$ 14. $(10y + k)(10y - k)$
15. $(3y + 2t)(3y - 2t)$ 16. $(-5y + 3a)(-5y - 3a)$
17. $(2x + \frac{1}{2})(2x - \frac{1}{2})$ 18. $(\frac{1}{4}x + 3)(\frac{1}{4}x - 3)$
19. $(\frac{2}{5}y + \frac{1}{8})(\frac{2}{5}y - \frac{1}{8})$ 20. $(6x + \frac{1}{3})(6x - \frac{1}{3})$

19-3 SQUARE OF A BINOMIAL

The algebraic expression $(a + b)^2$ means to use $(a + b)$ twice as a factor in a product. The expression $(a + b)^2$ is directing us to square the binomial. This may be done at sight by applying the following rule.

RULE 19-3. SQUARING A BINOMIAL

To form the square of a binomial:

1. Square the first term.
2. Add twice the algebraic product of the two terms.
3. Add the square of the second term.

4. Express the product as a trinomial square with the following pattern:

$$(a + b)^2 = a^2 + 2ab + b^2$$
$$(a - b)^2 = a^2 - 2ab + b^2$$

EXAMPLE 19-5 Solve $(2x + 3)^2$.

Solution Use Rule 19-3.

Step 1: Square the first term, $2x$:

$$4x^2$$

Step 2: Multiply 2 times $2x$ times 3:

$$2(2x)3 = 12x$$

Step 3: Square the second term, 3:

$$9$$

Step 4: Express the product as a trinomial square:

$$4x^2 + 12x + 9$$

$$\therefore \quad (2x + 3)^2 = 4x^2 + 12x + 9$$

The procedure for squaring a binomial is summarized in Figure 19-2. The result is a *perfect trinomial square*.

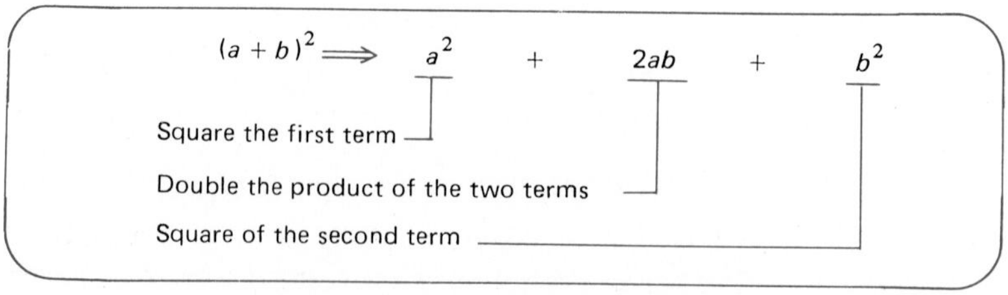

FIGURE 19–2
The square of the binomial $(a + b)$ results in the trinomial square $(a^2 + 2ab + b^2)$.

EXAMPLE 19-6   Solve $(3x - 2)^2$.

   Solution   Use Rule 19-3.

   Step 1:   Square $3x$:

$$9x^2$$

   Step 2:   Multiply 2 times $3x$ times -2:

$$2(3x)(-2) = -12x$$

   Step 3:   Square -2:

$$4$$

   Step 4:   Express the product as a trinomial square:

$$9x^2 - 12x + 4$$

$$\therefore \quad (3x - 2)^2 = 9x^2 - 12x + 4$$

EXERCISE 19-3

Mentally square each of the following binomials. Express the answer as a trinomial square.

1. $(x + 3)^2$ 2. $(x + 4)^2$ 3. $(x - 2)^2$
4. $(x - 3)^2$ 5. $(x + 5)^2$ 6. $(x - 7)^2$
7. $(2x + 2)^2$ 8. $(3x - 1)^2$ 9. $(4x - 2)^2$
10. $(5x + 4)^2$ 11. $(-2x - 1)^2$ 12. $(-x + 6)^2$
13. $(-4x + 3)^2$ 14. $(-3x - 5)^2$ 15. $(x + \frac{1}{5})^2$
16. $(-x - \frac{1}{3})^2$ 17. $(x + \frac{1}{2})^2$ 18. $(\frac{1}{3}x - 6)^2$
19. $(\frac{1}{4}x + \frac{1}{4})^2$ 20. $(\frac{3}{4}x - \frac{4}{3})^2$ 21. $(\frac{4}{5}x + \frac{1}{8})^2$

19-4 FACTORING THE DIFFERENCE OF TWO SQUARES

You know that the product of the sum and difference of two numbers results in the difference of two squares. When the difference of two squares is factored, it factors into the sum and difference of two numbers. Thus, $(a^2 - b^2) = (a + b)(a - b)$. Factoring the difference of two squares is stated as Rule 19-4.

EXAMPLE 19-7   Factor $4x^2 - 25$.
   Solution   Use Rule 19-4.
   Step 1:   Take the square root of $4x^2$ and 25:

$$2x \text{ and } 5$$

   Step 2:   Write the factors as the sum and difference of the two numbers:

$$(2x + 5)(2x - 5)$$

$$\therefore \quad (4x^2 - 25) = (2x + 5)(2x - 5)$$

EXERCISE 19-4

Factor each of the following.

1. $x^2 - 4$	2. $y^2 - 9$	3. $4x^2 - 64$
4. $x^2 - a^2$	5. $y^2 - c^2$	6. $36x^2 - b^2$
7. $1 - 16c^2$	8. $49x^2 - 4a^2$	9. $4c^2 - 9y^2$
10. $25x^2 - 36$	11. $a^2b^2 - x^2$	12. $m^2n^2 - 4y^2$
13. $\frac{1}{16} - x^2$	14. $9y^2 - \frac{1}{4}$	15. $\frac{1}{9} - 25z^2$

19-5 FACTORING A PERFECT TRINOMIAL SQUARE

We have seen that a perfect trinomial square results when a binomial is squared. Thus,

$$(a + b)^2 = a^2 + 2ab + b^2 \tag{19-1}$$

$$(c - d)^2 = c^2 - 2cd + d^2 \tag{19-2}$$

If a trinomial follows the pattern of either Equation 19-1 or 19-2, it is a perfect trinomial square and may be factored as the square of a binomial. The procedure for factoring a perfect trinomial square is summarized as Rule 19-5.

> **RULE 19-5. FACTORING A PERFECT TRINOMIAL SQUARE**
>
> To factor a perfect trinomial square:
>
> 1. Take the square root of each of the monomial squares.
> 2. Form the factors by combining the square roots with the sign of the remaining term.
> 3. Express the factors as the square of the binomial.

EXAMPLE 19-8 Factor $4x^2 - 12x + 9$.
 Solution Use Rule 19-5.
 Step 1: Take the square root of $4x^2$ and 9:

$$2x \text{ and } 3$$

 Step 2: Form the factors by using the sign of $-12x$:

$$2x - 3$$

 Step 3: Express the factors as the square of the binomial:

$$(2x - 3)^2$$

$$\therefore \quad 4x^2 - 12x + 9 = (2x - 3)^2$$

EXAMPLE 19-9 Factor $9y^2 + 30ay + 25a^2$.
 Solution Use Rule 19-5.
 Step 1: Take the square root of $9y^2$ and $25a^2$:

$$3y \text{ and } 5a$$

 Step 2: Form the factors by using the sign of $+30ay$:

$$3y + 5a$$

Step 3: Express the factors as the square of the binomial:

$$(3y + 5a)^2$$

$$\therefore \quad 9y^2 + 30ay + 25a^2 = (3y + 5a)^2$$

EXERCISE 19-5
Factor each of the following.

1. $x^2 - 6x + 9$ 　　　　　 2. $y^2 - 2y + 1$
3. $x^2 - 4x + 4$ 　　　　　 4. $x^2 + 12x + 36$
5. $y^2 + 2y + 1$ 　　　　　 6. $16x^2 + 16x + 4$
7. $9y^2 - 30y + 25$ 　　　 8. $9x^2 - 24ax + 16a^2$
9. $4x^2 - 20cx + 25c^2$ 　 10. $4y^2 + 28yb + 49b^2$
11. $25x^2 - 30xy + 9y^2$ 　 12. $36y^2 + 24xy + 4x^2$

19-6 FACTORING BY GROUPING

In a polynomial expression, several terms may contain common factors. By grouping the terms of the polynomial, you may recognize the form as a type that is factorable. Study the following examples.

**EXAMPLE
19-10**

Factor $2y - ax + ay - 2x$.

Solution Group the terms with common factors:

$$(2y + ay) + (-2x - ax)$$

Factor out the common monomial factor in each:

$$y(2 + a) - x(2 + a)$$

Factor out the common binomial factor $(2 + a)$:

$$(2 + a)(y - x)$$

$$\therefore \quad 2y - ax + ay - 2x = (2 + a)(y - x)$$

EXAMPLE
19-11

Factor $3x + by - 3y - bx$.

Solution   Group the terms with common factors:

$$(3x - bx) + (-3y + by)$$

Factor out the common monomial factor in each:

$$x(3 - b) - y(3 - b)$$

Factor out the common binomial factor $(3 - b)$:

$$(3 - b)(x - y)$$

$$\therefore \qquad 3x + by - 3y - bx = (3 - b)(x - y)$$

EXERCISE 19-6
Completely factor each of the following polynomials.

1. $3(a + 2) - x(a + 2)$ 2. $a^2(x - y) + 3(x - y)$
3. $m(x - 5) - 7(x - 5)$ 4. $ax + ay + bx + by$
5. $cx + cy - ax - ay$ 6. $x^3 - x - x^2 + 1$
7. $z^2 + 1 + z + z^3$ 8. $4x^3 - 8x^2 - 4x + 8$
9. $4b^3 + 5b^2 - 36b - 45$ 10. $c^2y - y - 3c^2 + 3$
11. $bx^2 - b - x^2 + 1$ 12. $y^4 - 8y + y^3z - 8z$

19-7 COMBINING SEVERAL TYPES OF FACTORING

When factoring a polynomial, first remove the common monomial factor and then see if the remaining expression may be factored further. In factoring a polynomial, the factoring is continued until all the steps of Rule 19-6 are completed.

RULE 19-6. FACTORING A POLYNOMIAL

To factor a polynomial:

1. Where possible, remove the common monomial factor from each term.
2. Factor a binomial into the sum and difference of two numbers if it is the difference of two squares.

---

3. Factor a perfect trinomial square into the square of a binomial.
4. Group the terms with common factors, then factor.

EXAMPLE 19-12

Factor $8y^3 - 2y$.

Solution Use Rule 19-6.

Step 1: Factor out $2y$:

$$2y(4y^2 - 1)$$

Step 2: Factor $4y^2 - 1$ into the sum and difference of two numbers:

$$2y(2y - 1)(2y + 1)$$

$$\therefore \quad 8y^3 - 2y = 2y(2y - 1)(2y + 1)$$

EXAMPLE 19-13

Factor $3x^3y - 6x^2y + 3xy$.

Solution Use Rule 19-6.

Factor out $3xy$:

$$3xy(x^2 - 2x + 1)$$

Factor the perfect trinomial square into the square of a binomial:

$$3xy(x - 1)^2$$

$$\therefore \quad 3x^3y - 6x^2y + 3xy = 3xy(x - 1)^2$$

EXAMPLE 19-14

Factor $y^3 - 3y^2 - 9y + 27$.

Solution Group the terms:

$$(y^3 - 3y^2) + (-9y + 27)$$

Factor each group:

$$y^2(y - 3) - 9(y - 3)$$

Factor out the binomial $y - 3$:

$$(y - 3)(y^2 - 9)$$

Factor $y^2 - 9$ into the sum and difference:

$$(y - 3)(y + 3)(y - 3)$$

$$\therefore \quad y^3 - 3y^2 - 9y + 27 = (y + 3)(y - 3)^2$$

EXERCISE 19-7

Factor each of the polynomials.

1. $3a + 3$
2. $x^2 - 16x$
3. $y^4 - c^4$
4. $8x^2 - 18$
5. $3a^2 - 3$
6. $9y^2 - 9$
7. $8y^3 - 32y$
8. $3y^2 - 6y + 3$
9. $5y^2 + 20y + 20$
10. $27x^2 - 18x + 3$
11. $8a^2 + 8ab + 2b^2$
12. $2x^2 - 2(a - b)^2$
13. $5y^2 - 5(m + n)^2$
14. $\frac{1}{4}b^2 - \frac{1}{3}bc + \frac{1}{9}c^2$
15. $(x + 3)^2 - 7x - 21$
16. $5ac - 5ad + 5bc - 5bd$
17. $x^3 + x + x^2 + 1$
18. $m^2n^2 - 4m^2 - a^2n^2 + 4a^2$
19. $4y^2 - 4ay + a^2 - c^2$
20. $x^2 + 2xy + y^2 - a^2$

19-8 LITERAL EQUATIONS

When solving literal equations or formulas, it is sometimes necessary to factor one or more expressions within the equation in order to isolate the desired variable as the left member. The following examples illustrate the application of factoring to the solution of literal equations.

EXAMPLE 19-15

Solve $ax + bx = 3a + 3b$ for x.

Solution Factor both the left and right members:

$$x(a + b) = 3(a + b)$$

D: $(a + b)$ $\quad x = 3$

$$\therefore \quad ax + bx = 3a + 3b \Rightarrow x = 3$$

EXAMPLE 19-16

Solution

Solve $bx + b^2 = a^2 - ax$ for x.

Isolate the terms containing x into the left member:

$$\text{A: } ax - b^2 \qquad bx + ax = a^2 - b^2$$

Factor the left and right members:

$$x(a + b) = (a - b)(a + b)$$

Divide by the coefficient of x:

$$\text{D: } (a + b) \qquad x = a - b$$

$$\therefore \qquad bx + b^2 = a^2 - ax \Rightarrow x = a - b$$

EXAMPLE 19-17

Solution

Solve $\dfrac{1}{R_T} = \dfrac{1}{R_1} + \dfrac{1}{R_2}$ for R_T.

Multiply each member by the common denominator:

$$\text{M: } R_T R_1 R_2 \qquad R_1 R_2 = R_T R_2 + R_T R_1$$

Factor the right member:

$$R_1 R_2 = R_T (R_1 + R_2)$$

Divide by the coefficient of R_T:

$$\text{D: } (R_1 + R_2) \qquad \frac{R_1 R_2}{R_1 + R_2} = R_T$$

Interchange the right and left members:

$$R_T = \frac{R_1 R_2}{R_1 + R_2}$$

$$\therefore \qquad \frac{1}{R_T} = \frac{1}{R_1} + \frac{1}{R_2} \Rightarrow R_T = \frac{R_1 R_2}{R_1 + R_2}$$

EXERCISE 19-8
Solve for x in each of the following equations.

1. $x - ax = 4$
3. $5x - b = cx$
5. $bx - 2 = b - 2x$
7. $ax - bx = 2a - 2b$
9. $bx - 3x = b^2 - 6b + 9$
11. $bx + 16 = 4x + b^2$

2. $bx + 2x = 7$
4. $a - 3x = bx$
6. $cx + 5 = 5x - c$
8. $ax - ma = md - dx$
10. $3x - 9 = ax - a^2$
12. $m^2 - mx + 10m = 5x - 25$

Solve for the indicated variable in each of the following.

Formula	Solve for:	Formula	Solve for:
13. $R = 3E + 2EI$	E	14. $CV = T - C$	C
15. $\beta = \dfrac{\alpha}{1 - \alpha}$	α	16. $L_1a + L_1b = a - b$	a
17. $i_1 = \dfrac{\beta i R}{R + Z}$	R	18. $Z_T = \dfrac{Z_1 Z_2}{Z_1 + Z_2}$	Z_2
19. $I = \dfrac{nV}{Z + nR}$	n	20. $C = \dfrac{V_1 - V_2}{\omega V_1}$	V_1
21. $S = \dfrac{R_E + R_1}{R_E + R_1(1 - \alpha)}$	R_1	22. $S = \dfrac{R_E(R_1 + R_2) + R_1 R_2}{R_E(R_1 + R_2) + R_1 R_2(1 - \alpha)}$	R_2
23. Problem 22	R_E	24. Problem 21	R_E
25. $R_1 = \dfrac{V_{CC} - I_B R_E(\beta + 1)}{I_B}$	I_B		

SELECTED TERMS

difference of two squares   A special product resulting from the multiplication of the sum and difference of two numbers.

factoring   The process of finding the factors of a polynomial or other expression.

perfect trinomial square   A special product resulting from the multiplication of a binomial by itself.

trinomial product   The product of two binomials.

20

SOLVING QUADRATIC EQUATIONS

Recall that equations containing a variable to the first power are first-degree or linear equations. In this chapter, you will study second-degree or quadratic equations. The topics included in this chapter have been selected to introduce you to the solution and application of quadratic equations.

20-1 INTRODUCTION

A quadratic equation may be solved by applying one of the following four methods: (1) square root, (2) factoring, (3) quadratic formula, or (4) graphing.

Square Root: *Incomplete* or *pure quadratic* equations are best solved with the square-root method. An incomplete quadratic contains only the second power of the variable. Thus, $x^2 = 25$, $4x^2 = 36$, and $x^2 - 9 = 0$ are incomplete or pure quadratic equations.

Factoring: *Complete* quadratic equations may be solved by this method when the roots are rational. A complete quadratic contains terms with both the first and second powers of the variable. Thus, $3x^2 + 5x = 2$ and $6x^2 + 4x - 2 = 0$ are complete quadratic equations.

Quadratic Formula: The quadratic formula may be used to solve any quadratic equation. This method is easily done with a calculator.

Graphing: Any quadratic equation can be solved by graphing when the roots of the equation are real. This method provides a visual understanding of why a quadratic may have *two* real roots, *one* real root, or *no* real roots.

Each of these four methods is explored in the following four sections.

20-2 INCOMPLETE QUADRATIC EQUATIONS

Quadratic equations that do not contain a linear term are *incomplete* or *pure quadratic equations*. Quadratics of this type have the general form $ax^2 + b = 0$.

Incomplete quadratics are solved by applying this rule.

RULE 20-1. SOLVING INCOMPLETE QUADRATIC EQUATIONS

To solve an incomplete quadratic equation:

1. Isolate the squared variable in the left member.
2. Take the square root of each member of the equation.
3. For the first root, assign a plus sign. For the second root, assign a minus sign.
4. Check each root.

EXAMPLE 20-1   Solve $3x^2 - 108 = 0$.

Solution   Solve the equation for x^2:

$$3x^2 - 108 = 0$$
$$x^2 = \frac{108}{3}$$
$$x^2 = 36$$

Take the square root of both members:

$$x = \pm \sqrt{36}$$
$$x = \pm 6$$

Check   Does $3x^2 - 108 = 0$ when $x = 6$?

$$3(36) - 108 = 0$$
$$108 - 108 = 0 \quad \text{Yes!}$$

Does $3x^2 - 108 = 0$ when $x = -6$?

$$3(36) - 108 = 0$$
$$108 - 108 = 0 \quad \text{Yes!}$$

$\therefore \quad x = 6$ and $x = -6$ are the roots of $3x^2 - 108 = 0$.

EXAMPLE 20-2   Solve $4x^2 - 7 = 3x^2 + 9$.

Solution   Rearrange terms:

$$4x^2 - 3x^2 = 9 + 7$$
$$x^2 = 16$$
$$x = \pm \sqrt{16} = \pm 4$$

Check   Does $4x^2 - 7 = 3x^2 + 9$ when $x = 4$?

$$4(16) - 7 = 3(16) + 9$$
$$64 - 7 = 48 + 9$$
$$57 = 57 \quad \text{Yes!}$$

Does $4x^2 - 7 = 3x^2 + 9$ when $x = -4$?

$$4(16) - 7 = 3(16) + 9$$
$$64 - 7 = 48 + 9$$
$$57 = 57 \quad \text{Yes!}$$

$\therefore$ $x = 4$ and $x = -4$ are the roots of $4x^2 - 7 = 3x^2 + 9$.

EXERCISE 20-1

Solve the following incomplete quadratic equations. Express radical roots as decimal numbers.

1. $x^2 = 25$
2. $x^2 = 49$
3. $x^2 = 8$
4. $x^2 = 13$
5. $y^2 - 49 = 0$
6. $y^2 - 144 = 0$
7. $x^2 - 18 = 0$
8. $x^2 - 32 = 0$
9. $3x^2 = 300$
10. $4y^2 = 100$
11. $16y^2 = 64$
12. $5x^2 = 100$
13. $72y^2 - 2 = 0$
14. $5x^2 - \frac{1}{20} = 0$
15. $4x^2 + 9 = 45$
16. $3y^2 + 9 = 57$
17. $3x^2 - 15.6 = 0$
18. $3x^2 = 48 - 5x^2$
19. $\frac{x^2}{3} + x = \frac{2x + 12}{2}$
20. $\frac{2x + 1}{3} = \frac{3}{2x - 1}$

20-3 COMPLETE QUADRATIC EQUATIONS

Quadratic equations having both linear and quadratic terms of the variable are *complete quadratic equations*. Quadratics of this type have the general form $ax^2 + bx + c = 0$. Complete quadratic equations having rational roots may be solved by factoring.

Factoring

To solve complete quadratic equations by factoring, you must first learn how to factor *general* trinomials. The result of factoring trinomials of the form $ax^2 + bx + c$ is the product of two binomials.

Because there is no general rule for factoring this type of trinomial, you will learn this method by *trial and error*. Carefully study the following examples.

EXAMPLE 20-3 Factor $x^2 + 3x + 2$.

Solution The factors of this trinomial are two binomials whose product is $x^2 + 3x + 2$:

$$(?)(?) = x^2 + 3x + 2$$

Because both 3 and 2 are positive, the signs of the binomial factors are both $+$:

$$(? + ?)(? + ?) = x^2 + 3x + 2$$

Factor x^2 into x and x; x is the first term of each of the binomials:

$$(x + ?)(x + ?) = x^2 + 3x + 2$$

Factor 2 into 1 and 2; these are the last terms of the binomials:

$$(x + 1)(x + 2) = x^2 + 3x + 2$$

Does the left member of the equation equal the right?

$$x^2 + 3x + 2 = x^2 + 3x + 2 \quad \text{Yes!}$$

$$\therefore \quad x^2 + 3x + 2 = (x + 1)(x + 2)$$

EXAMPLE 20-4 Factor $5x^2 - 17x + 14$.

Solution Because 14 is positive and 17 is negative, the signs of the binomial factors are both negative:

$$(? - ?)(? - ?) = 5x^2 - 17x + 14$$

Factor $5x^2$ into x and $5x$:

$$(x - ?)(5x - ?) = 5x^2 - 17x + 14$$

Factor 14 into 7 and 2 and **test the product:**

$$(x - 7)(5x - 2) \overset{?}{=} 5x^2 - 17x + 14$$
$$5x^2 - 37x + 14 \neq 5x^2 - 17x + 14$$

This yields the wrong product; interchange the 7 and the 2:

$$(x - 2)(5x - 7) = 5x^2 - 17x + 14$$

Test the product:

$$5x^2 - 17x + 14 = 5x^2 - 17x + 14 \quad \text{Correct.}$$

$$\therefore \quad 5x^2 - 17x + 14 = (x - 2)(5x - 7)$$

EXAMPLE 20-5 Factor $6x^2 - 13x - 5$.

Solution Because 5 is negative, the signs of the binomial factors are not the same:

$$(? + ?)(? - ?)$$

Set up a series of binomial products made up of the factors of $6x^2$ and 5:

$$(6x + 1)(x - 5) = 6x^2 - 29x - 5$$
$$(x + 1)(6x - 5) = 6x^2 + x - 5$$
$$(3x + 1)(2x - 5) = 6x^2 - 13x - 5$$
$$(2x + 1)(3x - 5) = 6x^2 - 7x - 5$$
$$(6x - 1)(x + 5) = 6x^2 + 29x - 5$$
$$(x - 1)(6x + 5) = 6x^2 - x - 5$$
$$(3x - 1)(2x + 5) = 6x^2 + 13x - 5$$
$$(2x - 1)(3x + 5) = 6x^2 + 7x - 5$$

By inspection, the third product results in the correct trinomial.

$$\therefore \quad 6x^2 - 13x - 5 = (3x + 1)(2x - 5)$$

Table 20-1 summarizes the sign pattern of the binomial factors implied by the sign pattern of the trinomial to be factored.

TABLE 20-1

SIGN PATTERNS OF BINOMIAL FACTORS

Trinomial	Sign Pattern		Binomial	Factor Sign Pattern
$ax^2 + bx + c$	plus plus	$\Rightarrow$	plus plus	$(+)(+)$
$ax^2 - bx + c$	minus plus	$\Rightarrow$	minus minus	$(-)(-)$
$ax^2 - bx - c$	minus minus	$\Rightarrow$	minus plus	$(-)(+)$
$ax^2 + bx - c$	plus minus	$\Rightarrow$	minus plus	$(-)(+)$

EXERCISE 20-2

Factor the following trinomials.

1. $x^2 + 7x + 10$
2. $y^2 + 6y + 8$
3. $a^2 + 8a + 7$
4. $b^2 + 3b + 2$
5. $x^2 + 2x + 1$
6. $y^2 + 6y + 5$
7. $a^2 + 5a + 4$
8. $x^2 - 6x + 5$
9. $y^2 - 7y + 12$
10. $c^2 - 11c + 18$
11. $x^2 - 10x + 24$
12. $y^2 - 2y - 3$
13. $a^2 - 7a - 18$
14. $b^2 - 5b - 14$
15. $x^2 + 3x - 40$
16. $y^2 + 3y - 4$
17. $2x^2 + 5x + 3$
18. $5y^2 - 7y + 2$
19. $3x^2 + 7x - 6$
20. $3x^2 - 14x - 5$
21. $4b^2 + 4b - 15$
22. $5y^2 - 2y - 7$
23. $5a^2 - 17a + 14$
24. $12x^2 + 11x - 15$
25. $6x^2 + 25x + 14$
26. $9y^2 + 3y - 2$
27. $18a^2 - 19a - 12$
28. $24a^2 + 5a - 36$
29. $4b^2 - 11b + 6$
30. $10x^2 + 11x - 18$

Solving Quadratic Equations

Now that you know how to factor trinomials, you can solve complete quadratic equations. The general procedure for solving complete quadratic equations is summarized as Rule 20-2. Rule 20-2 depends on the following observation: If the product of two factors is zero, then one or the other factor is zero. In symbols: if $A \cdot B = 0$, then $A = 0$ or $B = 0$.

RULE 20-2. SOLVING COMPLETE QUADRATIC EQUATIONS

To solve a complete quadratic equation with rational roots:

1. Express the equation in the form of $ax^2 + bx + c = 0$.
2. Factor the left member into two binomial factors.
3. Set each factor equal to zero.

4. Solve for each root.
5. Check the roots.

EXAMPLE 20-6 Solve $6x^2 - 13x = 5$ by factoring.
    Solution Use Rule 20-2.
    Step 1: Express the equation in the correct form:

$$6x^2 - 13x - 5 = 0$$

    Step 2: Factor the left member into two binomial factors:

$$(3x + 1)(2x - 5) = 0$$

    Steps 3–4: Set each factor to zero and solve for x:

$$
\begin{array}{ll}
3x + 1 = 0 & 2x - 5 = 0 \\
3x = -1 & 2x = 5 \\
x = -\frac{1}{3} & x = \frac{5}{2}
\end{array}
$$

    Step 5: Check.
    Check Does $6x^2 - 13x = 5$ when $x = -\frac{1}{3}$?

$$6(\tfrac{1}{9}) - 13(-\tfrac{1}{3}) = 5$$
$$\tfrac{2}{3} + \tfrac{13}{3} = 5$$
$$\tfrac{15}{3} = 5$$
$$5 = 5 \quad \text{Yes!}$$

Does $6x^2 - 13x = 5$ when $x = \frac{5}{2}$?

$$6(\tfrac{25}{4}) - 13(\tfrac{5}{2}) = 5$$
$$\tfrac{75}{2} - \tfrac{65}{2} = 5$$
$$\tfrac{10}{2} = 5$$
$$5 = 5 \quad \text{Yes!}$$

$\therefore$ The roots of $6x^2 - 13x = 5$ are $x = -\frac{1}{3}$ and $x = \frac{5}{2}$.

EXAMPLE 20-7   Solve $x(5x - 17) = -14$.

    Solution   Use Rule 20-2.

    Step 1:   Express the equation in the correct form:

$$5x^2 - 17x + 14 = 0$$

    Step 2:   Factor the left member into two binomial factors:

$$(x - 2)(5x - 7) = 0$$

    Steps 3–4:   Set each factor to zero and solve for x:

$$x - 2 = 0 \quad 5x - 7 = 0$$
$$x = 2 \qquad x = \tfrac{7}{5}$$

    Step 5:   Check.

    Check   Does $x(5x - 17) = -14$ when $x = 2$?

$$2[5(2) - 17] = -14$$
$$20 - 34 = -14$$
$$-14 = -14 \quad \text{Yes!}$$

Does $x(5x - 17) = -14$ when $x = \tfrac{7}{5}$?

$$\tfrac{7}{5}[5(\tfrac{7}{5}) - 17] = -14$$
$$\tfrac{49}{5} - \tfrac{119}{5} = -14$$
$$-\tfrac{70}{5} = -14$$
$$-14 = -14 \quad \text{Yes!}$$

    $\therefore$ The roots of $x(5x - 17) = -14$ are $x = 2$ and $x = \tfrac{7}{5}$.

EXERCISE 20-3

Use factoring to solve each of the following for its roots.

1. $y^2 + 6y + 8 = 0$ 2. $x^2 + 3x + 2 = 0$

3. $x^2 + 6x + 5 = 0$ 4. $y^2 - 5y + 4 = 0$

5. $y^2 - 6y = -5$ 6. $x^2 - 11x = -18$

7. $x(x + 2) + 1 = 0$ 8. $x(x - 10) = -24$

9. $y^2 - 2y = 3$
10. $x(2x + 5) = -3$
11. $x^2 - 5x = 14$
12. $y(y + 3) = 4$
13. $y(5y - 7) = -2$
14. $4x(x + 1) = 15$
15. $3x^2 - 14x - 5 = 0$
16. $5y^2 - 2y - 7 = 0$
17. $12x^2 + 11x - 15 = 0$
18. $3y(3y + 1) = 2$
19. $x(24x + 5) = 36$
20. $x(10x + 11) = 18$

20-4 THE QUADRATIC FORMULA

The general form of a quadratic equation is

$$ax^2 + bx + c = 0 \qquad (20\text{-}1)$$

This general equation for a quadratic may be solved for x. The result of this solution is called the *quadratic formula*. Developing the formula:

$$ax^2 + bx + c = 0$$
$$\text{S: } c \qquad ax^2 + bx = -c$$
$$\text{D: } a \qquad x^2 + \frac{b}{a}x = \frac{-c}{a}$$

Form the left member into a perfect trinomial square by adding $(b/2a)^2$ to each member. (This technique is called *completing the square*.)

$$x^2 + \frac{b}{a}x + \left(\frac{b}{2a}\right)^2 = \left(\frac{b}{2a}\right)^2 - \frac{c}{a}$$

Factor the left member into a binomial square:

$$\left(x + \frac{b}{2a}\right)^2 = \left(\frac{b}{2a}\right)^2 - \frac{c}{a}$$

Take the square root of each member:

$$x + \frac{b}{2a} = \pm\sqrt{\left(\frac{b}{2a}\right)^2 - \frac{c}{a}}$$

Solve for x:

$$x = -\frac{b}{2a} \pm \sqrt{\left(\frac{b}{2a}\right)^2 - \frac{c}{a}} \quad \text{or} \quad x = \frac{-b}{2a} \pm \sqrt{\left(\frac{-b}{2a}\right)^2 - \frac{c}{a}} \qquad (20\text{-}2)$$

Since $b^2 = (-b)^2$, this is the form best used with the calculator. Simplify the radicand:

$$x = -\frac{b}{2a} \pm \sqrt{\frac{b^2}{4a^2} - \frac{4ac}{4a^2}}$$

$$= -\frac{b}{2a} \pm \sqrt{\frac{b^2 - 4ac}{4a^2}}$$

$$= -\frac{b}{2a} \pm \frac{\sqrt{b^2 - 4ac}}{2a}$$

$$x = \frac{-b \pm \sqrt{b^2 - 4ac}}{2a} \tag{20-3}$$

This is the traditional form of the quadratic formula. The quadratic equatio and the quadratic formulas are summarized in Table 20-2.

TABLE 20-2

PROGRAMMABLE

THE QUADRATIC EQUATION AND FORMULA		
Quadratic equation	$ax^2 + bx + c = 0$	(20-1)
Quadratic formula (calculator)	$x = \dfrac{-b}{2a} \pm \sqrt{\left(\dfrac{-b}{2a}\right)^2 - \dfrac{c}{a}}$	(20-2)
Quadratic formula (traditional)	$x = \dfrac{-b \pm \sqrt{b^2 - 4ac}}{2a}$	(20-3)

where a = numerical coefficient of x^2
b = numerical coefficient of x
c = numerical constant
x = unknown quantity

EXAMPLE 20-8 Solve $x^2 + 3x + 2 = 0$ by the quadratic formula.
Solution Using Equation 20-3:

$$x = \frac{-b \pm \sqrt{b^2 - 4ac}}{2a}$$

Substitute $a = 1$, $b = 3$, $c = 2$:

$$x = \frac{-3 \pm \sqrt{9 - 4(1)(2)}}{2(1)}$$

$$= \frac{-3 \pm \sqrt{9 - 8}}{2}$$

$$= \frac{-3 \pm 1}{2}$$

$$x = \frac{-3 + 1}{2} \text{ or } x = \frac{-3 - 1}{2}$$

$$x = -1 \text{ or } -2$$

Check Does $x^2 + 3x + 2 = 0$ when $x = -1$?

$$1 - 3 + 2 = 0$$
$$0 = 0 \quad \text{Yes!}$$

Does $x^2 + 3x + 2 = 0$ when $x = -2$?

$$4 - 6 + 2 = 0$$
$$0 = 0 \quad \text{Yes!}$$

$\therefore$ The roots of $x^2 + 3x + 2 = 0$ are $x = -1$ and $x = -2$.

EXAMPLE 20-9 Solve $3x^2 - x = 2$ by the quadratic formula.
Solution Put in correct form:

$$3x^2 - x - 2 = 0$$

Substitute $a = 3$, $b = -1$, and $c = -2$ into the calculator form of the quadratic formula:

$$x = \frac{-b}{2a} \pm \sqrt{\left(\frac{-b}{2a}\right)^2 - \frac{c}{a}}$$

$$= \frac{1}{6} \pm \sqrt{\left(\frac{1}{6}\right)^2 - \frac{-2}{3}}$$

$$= 0.1667 \pm 0.8333$$

$$x = 1.0 \text{ or } -\tfrac{2}{3}$$

Check Does $3x^2 - x = 2$ when $x = 1.0$?

$$3 - 1 = 2$$
$$2 = 2 \quad \text{Yes!}$$

Does $3x^2 - x = 2$ when $x = -\tfrac{2}{3}$?

$$3(-\tfrac{2}{3})^2 + \tfrac{2}{3} = 2$$
$$\tfrac{4}{3} + \tfrac{2}{3} = 2$$
$$\tfrac{6}{3} = 2$$
$$2 = 2 \quad \text{Yes!}$$

∴ The roots of $3x^2 - x = 2$ are $x = 1$ and $x = -\tfrac{2}{3}$.

EXAMPLE 20-10

Solve $31.5x^2 - 52x = 23.6$ by the quadratic formula.

Solution Put into the correct form:

$$31.5x^2 - 52x - 23.6 = 0$$

Substitute $a = 31.5$, $b = -52$, and $c = -23.6$ into the calculator form of the quadratic formula:

$$x = \frac{-b}{2a} \pm \sqrt{\left(\frac{-b}{2a}\right)^2 - \frac{c}{a}}$$

$$= \frac{52}{2(31.5)} \pm \sqrt{\left(\frac{52}{2(31.5)}\right)^2 + \frac{23.6}{31.5}}$$

$$x = 2.02 \quad \text{or} \quad -0.37$$

Check   Does $31.5x^2 - 52x = 23.6$ when $x = 2.02$?

$$31.5(2.02)^2 - 52(2.02) = 23.6$$
$$23.6 = 23.6 \quad \text{Yes!}$$

Does $31.5x^2 - 52x = 23.6$ when $x = -0.37$?

$$31.5(-0.37)^2 - 52(-0.37) = 23.6$$
$$23.6 = 23.6 \quad \text{Yes!}$$

$\therefore$ The roots of $31.5x^2 - 52x = 23.6$ are $x = 2.02$ and $x = -0.37$.

EXERCISE 20-4

Use the quadratic formula to solve each of the following for its roots:

PROGRAMMABLE

1. $2x^2 + 5x = 3$
2. $4(y^2 + y) = 15$
3. $3x^2 + 7x = 6$
4. $6y^2 + 25y = -14$
5. $18y^2 - 19y - 12 = 0$
6. $4x^2 - 11x + 6 = 0$
7. $12x^2 + 11x = 15$
8. $24x^2 + 5x - 36 = 0$
9. $5.2y^2 - 7.3y = -2.2$
10. $11.3x^2 + 12.1x = 19$

20-5 GRAPHING QUADRATIC FUNCTIONS

Quadratic equations may be graphed and the roots found (assuming that they are real) by constructing a table of values and plotting the resulting points. Besides solving the quadratic equation, graphs provide a means of understanding the behavior of the roots of the quadratic. In this section, graphing will be used to explore the shape of the curve, the vertex of the curve, and to determine the nature of the roots of a quadratic equation.

Determining the Shape of the Curve

The general quadratic equation $ax^2 + bx + c = 0$ is a function of x. For purposes of graphing, let $y = f(x)$. Thus,

$$y = ax^2 + bx + c \qquad (20\text{-}4)$$

PROGRAMMABLE

The effect of a, the coefficient of x^2, will be investigated in the following example.

**EXAMPLE
20-11**

Determine the effect of a on the shape of the curve of $y = ax^2$ ($b = 0$, $c = 0$).

Solution Plot $y = ax^2$ for the following values of a:

$a = 1$

x	−4	−3	−2	−1	0	1	2	3	4
y	16	9	4	1	0	1	4	9	16

$a = 2$

x	−3	−2	−1	0	1	2	3
y	18	8	2	0	2	8	18

$a = -1$

x	−4	−3	−2	−1	0	1	2	3	4
y	−16	−9	−4	−1	0	−1	−4	−9	−16

$a = -2$

x	−3	−2	−1	0	1	2	3
y	−18	−8	−2	0	−2	−8	−18

From the curves of Figure 20-1, the following can be concluded:

1. A positive a ($a > 0$) results in an *opening upward* parabola.
2. A negative a ($a < 0$) results in an *opening downward* parabola.
3. As a gets larger, the curve gets *narrower* and *steeper*.

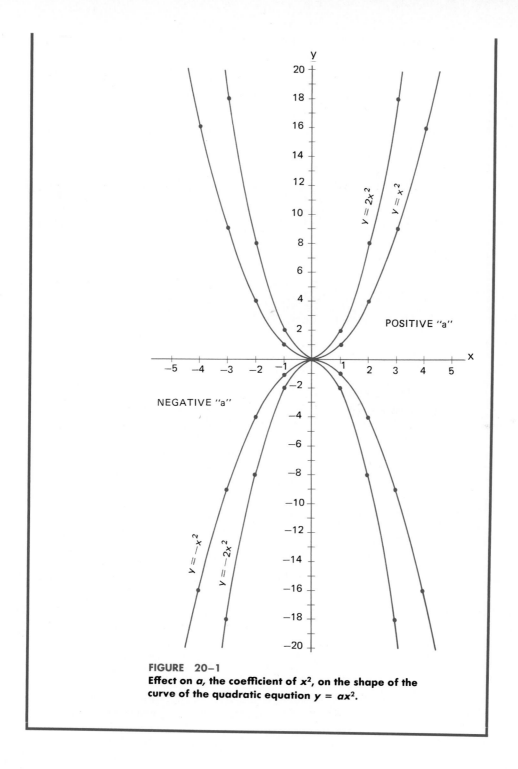

FIGURE 20–1
Effect on *a,* the coefficient of x^2, on the shape of the
curve of the quadratic equation $y = ax^2$.

The effects of c on the position of the curve are investigated in the following example.

EXAMPLE 20-12

Determine the effects of c on the position of the curves of $y = x^2 - 4x + c$ for the following values of c:

$$c = 0$$

x	−2	−1	0	1	2	3	4	5	6
y	12	5	0	−3	−4	−3	0	5	12

$$c = 4$$

x	−2	−1	0	1	2	3	4	5	6
y	16	9	4	1	0	1	4	9	16

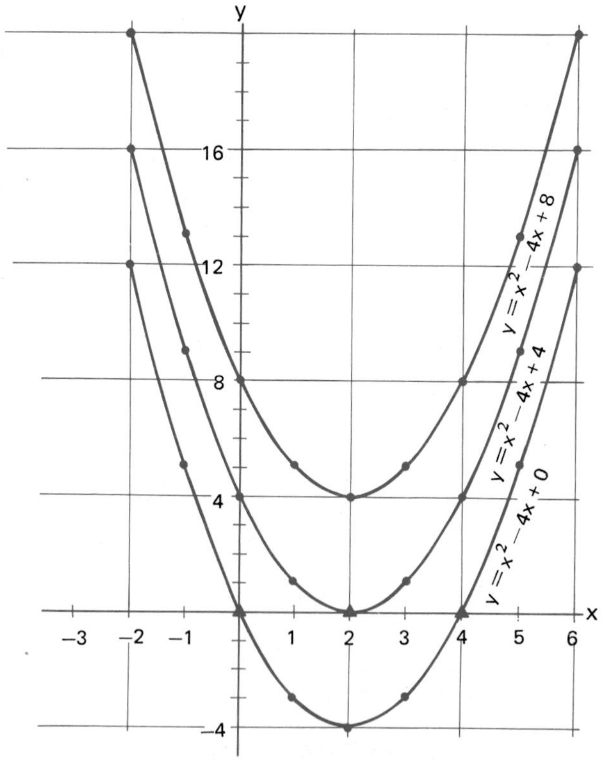

FIGURE 20-2
Effect of c on the position of the curve in the quadratic equation $y = x^2 - 4x + c$.

$$c = 8$$

x	-2	-1	0	1	2	3	4	5	6
y	20	13	8	5	4	5	8	13	20

Solution   From the *family* of curves of Figure 20-2, we see that c determines the vertical position of the curve. As c gets smaller (decreases in size), the curve moves down.

Determining the Vertex of the Curve

The x-coordinate of the vertex (the turning point) of the curve is found by letting $x = -b/2a$. The y-coordinate is then found by substituting the value of x into the general graphic quadratic equation, $y = ax^2 + bx + c$. The ordered pair (x_0, y_0) of the vertex is

$$x_0 = -\frac{b}{2a} \qquad (20\text{-}5)$$

$$y_0 = ax_0^2 + bx_0 + c \qquad (20\text{-}6)$$

EXAMPLE 20-13   Compute the coordinates of the vertex of the curve $y = 2x^2 - 8x - 5$.

Solution   Find the coordinates using Equations 20-5 and 20-6:

$$x_0 = -\frac{b}{2a}$$

$$= -\frac{(-8)}{2(2)}$$

$$x_0 = 2$$

$$y_0 = 2x_0^2 - 8x_0 - 5$$

$$= 2(2)^2 - 8(2) - 5$$

$$y_0 = -13$$

$\therefore$ The vertex coordinates are $(2, -13)$.

EXAMPLE
20-14

Compute the coordinates of the vertex of each of the curves in Figure 20-2.

Solution 1. $y = x^2 - 4x + 0$

$$x_0 = -b/2a \qquad y_0 = (2)^2 - 4(2) + 0$$
$$x_0 = 4/2 = 2 \quad y_0 = -4$$

∴ The vertex coordinates are $(2, -4)$.

2. $y = x^2 - 4x + 4$

$$x_0 = \frac{-b}{2a} \qquad y_0 = (2)^2 - 4(2) + 4$$
$$x_0 = 4/2 = 2 \quad y_0 = 0$$

∴ The vertex coordinates are $(2, 0)$.

3. $y = x^2 - 4x + 8$

$$x_0 = \frac{-b}{2a} \qquad y_0 = (2)^2 - 4(2) + 8$$
$$x_0 = 4/2 = 2 \quad y_0 = 4$$

∴ The vertex coordinates are $(2, 4)$.

Determining the Characteristics of the Roots

There are three possible cases for the number and kinds of roots of the quadratic equation. Referring to the curves of Figure 20-2, notice each of the three cases.

Case 1: $y = x^2 - 4x + 0$. When the curve passes through the x-axis, the two x-intercepts are the two real roots of the quadratic equation. In this curve, the roots are $x = 0$ and $x = 4$.

Case 2: $y = x^2 - 4x + 4$. When the vertex of the curve just touches the x-axis, the two real roots occur together (called a *double root*). In this curve both roots are $x = 2$.

Case 3: $y = x^2 - 4x + 8$. When the entire curve is above (or below) the x-axis as in this case, there are no real roots. The roots are *complex*.

By "reading" the radicand in the quadratic formula ($b^2 - 4ac$), we can tell which of the three cases will result. Because the quantity represented by

$b^2 - 4ac$ tells which one of the three cases holds, it is called the *discriminant* of the quadratic equation. The kind and number of roots are determined by the size of the discriminant in the following manner.

$$b^2 - 4ac > 0 \quad \text{two different real roots}$$
$$b^2 - 4ac = 0 \quad \text{one double real root}$$
$$b^2 - 4ac < 0 \quad \text{no real roots}$$

**EXAMPLE
20-15**

Compute the discriminant of each of the quadratic equations of Figure 20-2.

Solution   1. $y = x^2 - 4x + 0$

$$b^2 - 4ac = 16 - 4(1)(0) = 16$$

$\therefore$ Because the discriminant is greater than zero, there are two different real roots.

2. $y = x^2 - 4x + 4$

$$b^2 - 4ac = 16 - 4(1)(4) = 0$$

$\therefore$ Because the discriminant is equal to zero, there is one double real root.

3. $y = x^2 - 4x + 8$

$$b^2 - 4ac = 16 - 4(1)(8) = -16$$

$\therefore$ Because the discriminant is less than zero, there are no real roots.

Table 20-3 is a summary of the possible number and kinds of roots of a quadratic equation.

TABLE 20-3

THREE CASES OF THE DISCRIMINANT $b^2 - 4ac$

Case	Value of Discriminant	Number of Real Roots	Position of Curve
1	Positive	2	Passes through x-axis
2	Zero	1	Vertex touches x-axis
3	Negative	0	No x-intercept; entire curve above or below the x-axis

**EXAMPLE
20-16**

Solution

Solve $x^2 - 10x + 30$ by the quadratic formula.
Compute the discriminant of the quadratic formula first:

$$b^2 - 4ac = (-10)^2 - 4(1)(30)$$
$$= 100 - 120$$
$$b^2 - 4ac = -20$$

$\therefore$ There are no real roots, because the value of the discriminant is less than zero $(-20 < 0)$. The solution is terminated at this point.

**EXAMPLE
20-17**

Solution

Graph $y = -2x^2 + 4x + 6$.
Because a is negative, the parabola will open downward. Compute the discriminant to determine the nature of the roots:

$$b^2 - 4ac = 16 - 4(-2)(6)$$
$$= 16 + 48$$
$$b^2 - 4ac = 64 > 0$$

$\therefore$ Two different real roots.
Compute the roots:

$$x = \frac{-b \pm \sqrt{b^2 - 4ac}}{2a}$$

$$= \frac{-4 \pm \sqrt{64}}{-4}$$

$$x = \frac{-4 \pm 8}{-4}$$

$\therefore$ $x = 3$ $x = -1$

Compute the coordinates of the vertex:

$$x_0 = -b/2a \quad y_0 = -2x_0^2 + 4x_0 + 6$$
$$= -4/-4 \quad = -2 + 4 + 6$$
$$x_0 = 1 \quad\quad y_0 = 8$$

∴   The vertex coordinates are (1, 8).
Construct a table of values that includes the coordinates of the roots and the vertex:

x	−2	−1	0	1	2	3	4
y	−10	0	6	8	6	0	−10

root   vertex   root

Plot the ordered pairs and construct the curve as shown in Figure 20-3.

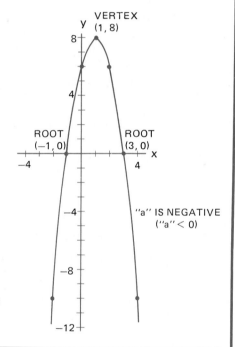

FIGURE 20–3
Graph of the quadratic equation
y = −2x² + 4x + 6, used in
Example 20–17.

EXERCISE 20-5
In each of the following quadratic equations, determine (a) the shape of the curve (opening up or down), (b) the characteristic of the roots, (c) the coordinates of the vertex, and (d) the coordinates of the real roots.

PROGRAMMABLE

Graph each of the functions of x by setting y = f(x).

1. $2x^2$ 2. $x^2 + 1$

3. $2x^2 + 2$ 4. $x^2 + 2x + 1$

5. $-x^2 + 4x$ 6. $-x^2 - 2x + 3$

7. $3x^2 + 5x + 3$ 8. $x^2 - 3x - 4$

9. $-x^2 - 2x - 2$ 10. $-3x^2 + 4x + 1$

11. $\frac{1}{2}x^2 - 6x + 5$ 12. $-\frac{1}{3}x^2 - 3x + 4$

20-6 APPLYING QUADRATIC EQUATIONS TO ELECTRONICS

When the conditions of a problem lead to a quadratic equation, two answers will be obtained for the variable. In many problems, one of the answers can be discarded because it is physically impossible or inappropriate. For example, negative number of turns in a coil, negative dimension of an object, and positive values that seem inappropriate are all discarded.

EXAMPLE 20-18

The power dissipated in a 7-Ω load is 21 W. Determine the current passing through the load.

Solution $P = I^2R$ is an incomplete quadratic equation.
Solve for I:

$$I = \pm \sqrt{P/R}$$

Substitute 7 Ω for R and 21 W for P:

$$I = \pm \sqrt{21/7}$$
$$= \pm 1.73$$

Select the positive root:

$$\therefore \quad I = 1.73 \text{ A}$$

EXAMPLE 20-19

Determine the voltage (V_2) of the circuit shown in Figure 20-4.

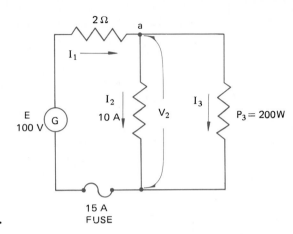

FIGURE 20–4
Circuit for Example 20–19.

15 A
FUSE

Solution   Write a voltage loop equation around the path containing I_1, I_2, and E:

$$(1) \quad E = 2I_1 + V_2$$

Write the current equation for node a:

$$(2) \quad I_1 = I_2 + I_3$$

Substitute equation *2* into equation *1*:

$$(3) \quad E = 2(I_2 + I_3) + V_2$$

Express I_3 in terms of P_3 (200 W) and V_2:

$$P_3 = V_2 I_3$$
$$I_3 = P_3/V_2 = 200/V_2$$

Substitute $200/V_2$ for I_3, 10 for I_2, and 100 for E in equation 3:

$$100 = 2(10 + 200/V_2) + V_2$$

Expand and clear fractions:

$$100 = 20 + 400/V_2 + V_2$$
$$100V_2 = 20V_2 + 400 + V_2^2$$

Place into the general quadratic form $(ax^2 + bx + c = 0)$:

$$V_2^2 - 80V_2 + 400 = 0$$

Solve for V:

$$V_2 = \frac{80 \pm \sqrt{(-80)^2 - 4(1)(400)}}{2}$$

$$V_2 = 74.6 \text{ V} \quad \text{or} \quad V_2 = 5.36 \text{ V}$$

Because the fuse will blow, 5.36 V is inappropriate.

$$\therefore \qquad V_2 = 74.6 \text{ V}$$

**EXAMPLE
20-20**

Graphically determine the value of the load resistance that will cause maximum power to be developed in the load of Figure 20-5.

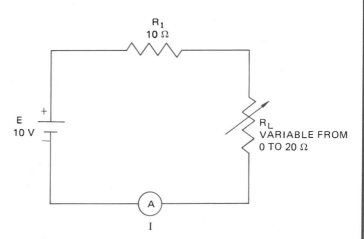

FIGURE 20–5
Circuit for Example 20–20.

Solution Derive an equation that relates load power (P_L) to the resistance of the load (R_L).

$$(1)\ P_L = I^2 R_L$$

also

$$(2) \; I = \frac{E}{R_1 + R_L}$$

Substitute $E/(R_1 + R_L)$ for I in equation 1:

$$P_L = \left(\frac{E}{R_1 + R_L}\right)^2 R_L$$

Substitute $E = 10$ V and $R_1 = 10 \; \Omega$:

$$P_L = \left(\frac{10}{10 + R_L}\right)^2 R_L$$

Load power (P_L) is now a function of load resistance (R_L):

$$P_L = f(R_L) = \left(\frac{10}{10 + R_L}\right)^2 R_L$$

Construct a table of values:

R_L	4	6	8	9	10	11	12	16	20
$f(R_L)$	2.0	2.3	2.47	2.49	2.5	2.49	2.48	2.4	2.2

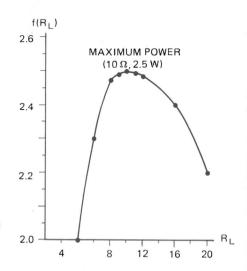

FIGURE 20–6
Power transfer curve for the load resistance of Figure 20–5 and Example 20–20.

Plot the ordered pairs and construct the curve as shown in Figure 20-6.

∴ The maximum power of 2.5 W is developed for $R_L = 10\ \Omega$.

1. Determine the value of two resistors (R_1 and R_2) whose product is 440 if $R_2 = R_1 + 2\ \Omega$.
2. Determine the value of a resistor whose value squared is 30 more than its value.
3. The value of the voltage V (measured across a circuit load) when added to its square results in 12. What is the measured value of the voltage?
4. Two currents in a parallel circuit have a difference of 14 A and a product of 51. What are the numbers?
5. Determine the voltage across a 12-Ω load that is dissipating 500 mW.
6. Determine the current through a 47-Ω resistor that is dissipating 0.625 W.
7. Determine the value of two resistances (R_1 and R_2) connected in parallel. The total equivalent resistance is 200 Ω and $R_2 = R_1 + 300\ \Omega$.
8. Determine the value of R_1 and R_3 in the circuit of Figure 20-7.

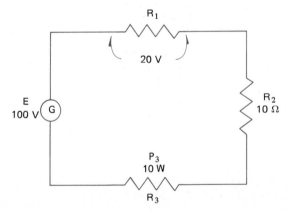

FIGURE 20-7
Circuit for Exercise 20-6, problem 8.

9. Three resistors are connected to form a series-parallel circuit as shown in Figure 20-8. The total equivalent resistance is 60 Ω, $R_1 = R_2 + 10$ Ω, and $R_3 = 2R_1$. What are the values of R_1, R_2, and R_3?

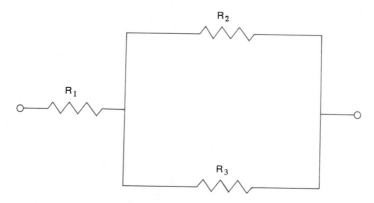

FIGURE 20–8
Circuit for Exercise 20–6, problem 9.

10. Determine the current (I_3) through R_3, given the information of Figure 20-9 and $R_1 + R_3 = 5.6$ kΩ.

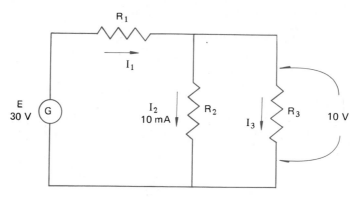

FIGURE 20–9
Circuit for Exercise 20–6, problem 10.

11. Graphically determine the value of load resistance that will cause maximum power to be developed in the load of Figure 20-10.

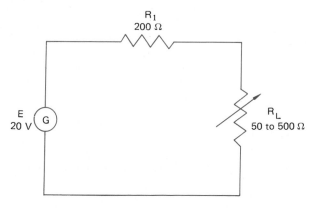

FIGURE 20–10
Circuit for Exercise 20–6, problem 11.

12. Repeat problem 11 when $R_1 = 1 \text{ k}\Omega$ and R_L varies between 500Ω and $3 \text{ k}\Omega$.

SELECTED TERMS

discriminant   The radicand in the quadratic equation. The number of real roots depends on the sign of the radicand.

quadratic equation   An equation containing the square of the independent variable but no higher order terms.

vertex   The turning point of the graph of a quadratic equation.

EXPONENTS, RADICALS, AND EQUATIONS

In this chapter the laws of exponents are applied to literal expressions and to expressions with zero and negative integers as exponents. The laws for radicals are used to simplify expressions with radicals. The chapter concludes with the solution of radical equations.

21-1 LAWS OF EXPONENTS

The fundamental laws for working with exponents along with an example of the application of each are given below. Exponents follow the five stated laws:

RULE 21-1. FIVE LAWS OF EXPONENTS

1. $x^m x^n = x^{m+n}$ (21-1)
   Example: $x^4 x^3 = x^{4+3} = x^7$

2. $(x^m)^n = x^{mn}$ (21-2)
   Example: $(x^7)^3 = x^{7\times3} = x^{21}$

3. $(xy)^m = x^m y^m$ (21-3)
   Example: $(xy)^3 = x^3 y^3$

4. $\left(\dfrac{x}{y}\right)^m = \dfrac{x^m}{y^m}, \ y \neq 0$ (21-4)

   Example: $\left(\dfrac{x}{y}\right)^3 = \dfrac{x^3}{y^3}$

5. $\dfrac{x^m}{x^n} = x^{m-n}, \ x \neq 0$ (21-5)

   Example: $\dfrac{x^5}{x^3} = x^{5-3} = x^2$

EXERCISE 21-1
Use the five laws of exponents to simplify each of the following.

1. $a^3 a^5$ 2. $(x^2)^3$ 3. $(2b)^3$

4. $(x/y)^2$ 5. $\dfrac{c^3}{c^2}$ 6. $b^x b^{5x}$

7. $-(7c)^2$ 8. $c^{3a} c^{-2a}$ 9. $(y^{2x})^2$

10. $\dfrac{8x^3}{x^2}$ 11. $\left(\dfrac{2x}{3a}\right)^3$ 12. $(x^m y^n)^2$

13. $(a^b)^c$ 14. $\dfrac{24x^5 y^4}{-6x^3 y^3}$ 15. $-2x(xy)^3$

16. $(x^{3a})^a$ 17. $y^{a+b}y^{2-b}$ 18. $(2ab^2)^3$

19. $\left(\dfrac{2x^5y}{x^2}\right)^3$ 20. $\left(\dfrac{2a^2}{b^3}\right)^3$ 21. $\left(\dfrac{5S^2}{T}\right)^2$

21-2 ZERO AND NEGATIVE INTEGERS AS EXPONENTS

Zero Exponent

When evaluating the expression a^m/a^m, the solution is 1. Applying the laws of exponents to this expression results in $a^m/a^m = a^{m-m} = a^0$. However, we already know that $a^m/a^m = 1$, so a^0 must also be 1. Rule 21-2 states this.

> **RULE 21-2. ZERO EXPONENT**
>
> Any number, except zero, raised to the zero power is 1. Thus $a^0 = 1$ $(a \neq 0)$.

Negative Exponents

What does the expression a^{-3} mean? Let's investigate the meaning of this result: $a^4/a^7 = a^{4-7} = a^{-3}$. However, a^4/a^7 and a^7/a^4 are reciprocals. $a^4/a^7 = a^{-3}$, and $a^7/a^4 = a^3$. Then a^{-3} and a^3 are also reciprocals. Therefore, $a^3 = 1/a^{-3}$, and $a^{-3} = 1/a^3$. We see that a^{-3} is the same as $1/a^3$.

> **RULE 21-3. NEGATIVE EXPONENTS**
>
> Numbers written with negative exponents may be written with positive exponents by transferring the factor from one term of the fraction to the other term.
> *Examples:* a^{-m} may be written as $1/a^m$
> *and* $1/b^{-n}$ may be written as b^n.

EXAMPLE 21-1   Express $\dfrac{3}{2x^{-2}}$ with positive exponents.

Solution
$$\frac{3}{2x^{-2}} = \frac{3x^2}{2}$$

EXAMPLE 21-2 Express $\dfrac{x^{-5}}{3^{-1}x}$ with positive exponents and simplify.

Solution
$$\frac{x^{-5}}{3^{-1}x} = \frac{3}{x^5 \cdot x} = \frac{3}{x^6}$$

EXERCISE 21-2
Write the value of each of the following.

1. 5^0
2. $3^0(2)$
3. $4a^0$
4. $(4a)^0$
5. $(-4)^0$
6. $6/(3x^0)$
7. $(2 + 2a^0)/4$
8. 2^{-1}
9. 2^{-2}
10. $(-3)^{-2}$
11. $1/3^{-2}$
12. $5/3^{-1}$
13. $32(4^{-2})$
14. $1/(2 + a^0)^{-2}$
15. $1/(2^0 + a^0)$

Express each of the following with positive exponents.

16. a^{-3}
17. xy^{-2}
18. $-5b^{-2}$
19. $\dfrac{2a^{-3}}{b}$
20. $\dfrac{a^{-1}b^{-1}}{c^{-1}}$
21. $\dfrac{3a}{b^{-2}}$
22. $(x^{-1}y)^{-2}$
23. $(x^2y^{-3})^{-1}$
24. $(x^{-2}/y^{-3})^{-2}$
25. $(x/y)^{-2}$
26. $\left(\dfrac{2ab}{a^2}\right)^{-2}$
27. $(3b^2c)^{-3}$

21-3 FRACTIONAL EXPONENTS

Fractional exponents follow the laws of exponents where the numerator indicates the power of the base and the denominator indicates the root of the base. Thus, $a^{m/n} = \sqrt[n]{a^m}$, read as "the nth root of a to the mth power."

RULE 21-4. FRACTIONAL EXPONENTS

Fractional exponents follow the laws of integer exponents:

1. The numerator of the fraction is the power of the base.
2. The denominator of the fraction is the root of the base.

Fractional exponents are used to simplify radicals, as is shown in the following examples. They are the basis of the system of logarithms studied in Chapter 22.

EXAMPLE 21-3   Write $8^{2/3}$ with a radical and evaluate.

Solution

$$8^{2/3} = \sqrt[3]{8^2}$$

Square 8:

$$8^{2/3} = \sqrt[3]{64}$$

Extract root:

$$\therefore \quad 8^{2/3} = 4$$

EXAMPLE 21-4   Evaluate $\sqrt[4]{3^6}$ with your calculator.

Solution   Express $\sqrt[4]{3^6}$ with fractional exponents:

$$3^{6/4}$$

Express the fraction as a decimal:

$$3^{6/4} = 3^{1.5}$$

Evaluate $3^{1.5}$ using $\boxed{y^x}$ on your calculator:

$$3\,\boxed{y^x}\,1.5 \Rightarrow 5.196$$

$$\therefore \quad \sqrt[4]{3^6} = 5.196$$

EXAMPLE 21-5   Simplify $\left(\dfrac{a^2 b^3}{a^3}\right)^{1/6}$.

Solution   Distribute the 1/6 to the other exponents:

$$\left(\frac{a^2 b^3}{a^3}\right)^{1/6} = \frac{a^{2/6} b^{3/6}}{a^{3/6}}$$

Clear the denominator:

$$a^{2/6}a^{-3/6}b^{3/6}$$

Simplify:

$$a^{-1/6}b^{1/2} = b^{1/2}/a^{1/6}$$

$$\therefore \quad \left(\frac{a^2b^3}{a^3}\right)^{1/6} = b^{1/2}/a^{1/6}$$

EXAMPLE 21-6 Simplify $(\sqrt[3]{x^2})(\sqrt[2]{x})^3$.

Solution Express with fractional exponents:

$$x^{2/3}(x^{1/2})^3 = x^{2/3}x^{3/2}$$

Add exponents:

$$x^{4/6}x^{9/6} = x^{13/6}$$

Simplify:

$$x^{13/6} = x^2 \cdot x^{1/6} = x^2\sqrt[6]{x}$$

$$\therefore \quad (\sqrt[3]{x^2})(\sqrt[2]{x})^3 = x^2\sqrt[6]{x}$$

EXERCISE 21-3

Calculator Drill
Evaluate each of the following.

1. $25^{1/2}$	2. $36^{1/2}$	3. $144^{1/2}$
4. $27^{1/3}$	5. $64^{1/3}$	6. $32^{1/5}$
7. $4^{3/2}$	8. $125^{2/3}$	9. $32^{2/5}$
10. $48^{5/6}$	11. $96^{3/8}$	12. $50^{7/5}$
13. $(\frac{1}{16})^{1/2}$	14. $(\frac{4}{9})^{3/2}$	15. $(0.062)^{1/3}$
16. $(0.125)^{3/2}$	17. $(6^2)^{1/4}$	18. $(5^4)^{3/4}$
19. $(7^3)^{3/10}$	20. $(3^5)^{2/5}$	21. $(4)^{5/6}$

452

Written Exercise

Express each of the following as a radical.

22. $5^{1/2}$ 23. $a^{1/3}$ 24. $b^{2/3}$

25. $x^{5/2}$ 26. $5a^{1/2}$ 27. $7a^{4/3}$

28. $-(2a^3)^{1/2}$ 29. $-(7b^5)^{1/3}$ 30. $(5c^2)^{3/2}$

Express the following without a radical sign.

31. $\sqrt{2}$ 32. $\sqrt{5}$ 33. $-2\sqrt{a}$

34. $\sqrt[3]{b^2}$ 35. $-5b\sqrt[5]{c^2d}$ 36. $-7c\sqrt{a+b}$

37. $\sqrt[3]{T+s}$ 38. $\sqrt[4]{4\pi+\theta}$ 39. $-\sqrt[5]{R-t}$

Simplify.

40. $\left(\dfrac{a^{1/3}}{b^{2/3}}\right)^6$ 41. $(a^{2/3})^{1/2}$ 42. $\left(\dfrac{a^{1/2}b^{4/3}}{a^{-2}b}\right)^3$

43. $\sqrt[7]{x^4}(\sqrt[7]{x})^3$ 44. $\dfrac{(\sqrt[3]{y})^5}{\sqrt[3]{y^2}}$ 45. $\dfrac{\sqrt[3]{a^4}}{\sqrt[4]{a^3}}$

21-4 LAWS OF RADICALS

The following five basic laws for radicals are used when operating on expressions containing radicals.

RULE 21-5. FIVE LAWS FOR RADICALS

1. $(\sqrt[n]{x})^n = x$ (21-6)

   Example: $(\sqrt[5]{x})^5 = (x^{1/5})^5 = x^{5/5} = x$

2. $\sqrt[n]{xy} = \sqrt[n]{x}\sqrt[n]{y}$ (21-7)

   Example: $\sqrt[3]{48} = \sqrt[3]{8 \cdot 6} = \sqrt[3]{8}\,\sqrt[3]{6} = 2\sqrt[3]{6}$

3. $\sqrt[n]{\dfrac{x}{y}} = \dfrac{\sqrt[n]{x}}{\sqrt[n]{y}}$ (21-8)

$$\textit{Example:} \quad \sqrt[3]{\frac{27}{8}} = \frac{\sqrt[3]{27}}{\sqrt[3]{8}} = \frac{3}{2}$$

4. $\sqrt[m]{\sqrt[n]{x}} = \sqrt[mn]{x}$ (21-9)

$$\textit{Example:} \quad \sqrt[5]{\sqrt[3]{x}} = \sqrt[15]{x} = x^{1/15}$$

5. $\sqrt[m]{x^n} = x^{n/m}$ (21-10)

$\textit{Example:} \; \sqrt[3]{x^5} = x^{5/3}$

EXERCISE 21-4

Apply the indicated law for radicals to the following.

Equation 21-6:

1. $(\sqrt[3]{7})^3$ 2. $(\sqrt[3]{bc})^3$ 3. $-(\sqrt[3]{6^3})^3$

4. $-(\sqrt{a^3})^2$ 5. $(\sqrt[7]{x^{14}})^7$ 6. $-(\sqrt[3]{y^9})^3$

Equation 21-7:

7. $\sqrt[3]{ab}$ 8. $\sqrt[5]{5x}$ 9. $-\sqrt[4]{cd}$

10. $-\sqrt{13xy}$ 11. $\sqrt[7]{4mn}$ 12. $-\sqrt[3]{11bc}$

Equation 21-8:

13. $\sqrt{\frac{2}{3}}$ 14. $\sqrt{\frac{7}{13}}$ 15. $\sqrt{\frac{a}{b}}$

16. $-\sqrt{\frac{2b}{c}}$ 17. $-\sqrt{\frac{3x}{5y}}$ 18. $-\sqrt{\frac{ax}{by}}$

Equation 21-9:

19. $\sqrt{\sqrt[3]{7}}$ 20. $\sqrt{\sqrt[5]{5}}$ 21. $\sqrt[3]{\sqrt{a}}$

22. $\sqrt[3]{\sqrt[4]{ax}}$ 23. $-\sqrt[5]{\sqrt{aby}}$ 24. $-\sqrt[3]{\sqrt[3]{3x}}$

454

Equation 21-10:

25. $\sqrt[3]{3^2}$ 26. $\sqrt{2^3}$ 27. $\sqrt{6^5}$

28. $\sqrt[5]{c^3}$ 29. $-\sqrt[9]{a^7}$ 30. $-\sqrt[7]{b^3}$

21-5 SIMPLIFYING RADICALS

A radical may be simplified by the following means: (1) removing factors from the radicand, (2) lowering the index of the radical, or (3) rationalizing the denominator. Each of these means of simplifying a radical is investigated separately.

Simplifying by Removing Factors

Remove factors from the radicand by expressing the radicand as a product of factors using Equation 21-7. Select the factors so that one factor contains all the perfect nth powers, where n is the index of the radical. This is demonstrated in the following examples.

EXAMPLE 21-7   Simplify $\sqrt{27a^3b}$.

     Solution   Factor into two radicals, one containing perfect squares:

$$\sqrt{27a^3b} = \sqrt{9a^2}\,\sqrt{3ab}$$

$$\therefore \quad \sqrt{27a^3b} = 3a\sqrt{3ab}$$

EXAMPLE 21-8   Simplify $\sqrt[3]{-x^7}$.

     Solution   Factor into two radicals, one containing perfect cubes:

$$\sqrt[3]{-x^7} = \sqrt[3]{(-1)^3x^6}\,\sqrt[3]{x}$$

$$\therefore \quad \sqrt[3]{-x^7} = -x^2\sqrt[3]{x}$$

Simplifying by Lowering the Index

Lower the index by first expressing the radicand as a power of some quantity, and then apply Equation 21-10. This procedure is shown in Example 21-9.

EXAMPLE 21-9 Simplify $\sqrt[6]{27}$.

Solution Express 27 as a power of 3:

$$\sqrt[6]{27} = \sqrt[6]{3^3}$$
$$= 3^{3/6}$$
$$= 3^{1/2}$$
$$\therefore \quad \sqrt[6]{27} = \sqrt{3}$$

Simplifying by Rationalizing the Denominator

The process of removing the radical from the denominator of a fraction is called *rationalizing the denominator*. When the denominator is rationalized, it is changed from an irrational to a rational form. The procedure for rationalizing the denominator is given in Rule 21-6 and demonstrated in the following examples.

> **RULE 21-6. RATIONALIZING THE DENOMINATOR**
>
> To rationalize the denominator, multiply the numerator and denominator of the radicand by a factor that will make the denominator a perfect nth power, where n is the index of the radical.

EXAMPLE 21-10 Rationalize $\sqrt{\dfrac{1}{3}}$.

Solution Multiply numerator and denominator by 3, thus making the denominator a perfect square:

$$\sqrt{\frac{1 \cdot 3}{3 \cdot 3}} = \sqrt{\frac{3}{9}}$$

Apply Equation 21-8:

$$\sqrt{\frac{3}{9}} = \frac{\sqrt{3}}{\sqrt{9}} = \frac{\sqrt{3}}{3} = \frac{1}{3}\sqrt{3}$$

$$\therefore \qquad \sqrt{\frac{1}{3}} = \frac{1}{3}\sqrt{3}$$

EXAMPLE 21-11

Rationalize $\sqrt[3]{\dfrac{3}{2x}}$.

Solution Multiply numerator and denominator by $4x^2$ to make the denominator a perfect cube:

$$\sqrt[3]{\frac{3}{2x}} = \sqrt[3]{\frac{12x^2}{8x^3}}$$

Apply Equation 21-8:

$$\sqrt[3]{\frac{3}{2x}} = \frac{\sqrt[3]{12x^2}}{\sqrt[3]{8x^3}}$$

$$\therefore \qquad \sqrt[3]{\frac{3}{2x}} = \frac{1}{2x}\sqrt[3]{12x^2}$$

To rationalize the denominator when the radicals are of different *orders,* multiply the numerator and denominator by a factor that will make the denominator a rational form.

EXAMPLE 21-12

Rationalize $\dfrac{\sqrt[3]{3}}{\sqrt{2}}$.

Solution Multiply numerator and denominator by $\sqrt{2}$:

$$\frac{\sqrt[3]{3}}{\sqrt{2}} = \frac{\sqrt[3]{3}\sqrt{2}}{\sqrt{2}\sqrt{2}} = \frac{\sqrt[3]{3}\sqrt{2}}{2}$$

Combine the radicals; express them using fractional exponents:

$$\frac{\sqrt[3]{3}\sqrt{2}}{2} = \frac{3^{1/3} \cdot 2^{1/2}}{2}$$

Express the exponents with common denominators:

$$\frac{3^{2/6} \cdot 2^{3/6}}{2} = \frac{(3^2 \cdot 2^3)^{1/6}}{2}$$

Simplify and write as a radical:

$$\therefore \quad \frac{\sqrt[3]{3}}{\sqrt{2}} = \frac{\sqrt[6]{72}}{2}$$

EXERCISE 21-5

Simplify each of the following radicals, expressing the answer as a radical.

1. $\sqrt{64x^3y^5}$ 2. $\sqrt[3]{8a^2y^6}$

3. $\sqrt[3]{54x^{11}y^8}$ 4. $\sqrt{32x^4y^3}$

5. $-\sqrt{18a^6b^8}$ 6. $\sqrt{\dfrac{16x^7}{a^2y^6}}$

7. $\sqrt{\dfrac{12a^5}{9b^{10}}}$ 8. $\sqrt[8]{16}$

9. $\sqrt[9]{64}$ 10. $\sqrt[6]{125}$

11. $\sqrt[10]{32}$ 12. $\sqrt[6]{25x^4}$

13. $\sqrt[8]{81a^2}$ 14. $\sqrt[10]{x^4y^8}$

15. $\sqrt{\dfrac{3}{5}}$ 16. $\sqrt{\dfrac{3}{7}}$

17. $\sqrt{\dfrac{5}{3x^2}}$ 18. $\dfrac{\sqrt[3]{4}}{\sqrt{2}}$

19. $\dfrac{\sqrt[3]{2a}}{\sqrt{3a}}$

20. $\dfrac{-\sqrt{6x^2}}{\sqrt[3]{9}}$

21. $\sqrt{\dfrac{1 - \frac{1}{2}}{2}}$

22. $\sqrt{2 - \left(\frac{1}{5}\right)^2}$

23. $\sqrt{1 - \left(\dfrac{\sqrt{3}}{2}\right)^2}$

24. $\sqrt{1 - \left(\frac{3}{5}\right)^2}$

25. $\sqrt{4y^2 + 36}$

26. $\sqrt[3]{\dfrac{16x^7}{y^4}}$

27. $\sqrt{\left(\dfrac{b^{2/3}c^{1/2}}{2^{-1}a^{-2}}\right)^6}$

28. $\sqrt[3]{\left(\dfrac{a^2b}{125a^{-3}}\right)^{-1}}$

29. $\sqrt{2x + 6 + \dfrac{9}{2x}}$

30. $\sqrt{\sqrt[5]{32} \cdot \sqrt{4}}$

31. $\sqrt{\dfrac{a}{8} + \dfrac{-3}{y^3}}$

32. $\sqrt{\dfrac{4b^2}{9} - \dfrac{16}{x^2}}$

21-6 RADICAL EQUATIONS

An equation in which a variable appears as part of the radicand (within the radical) or has a fractional exponent is called a *radical equation.* Thus, $\sqrt{y} = 2$ and $\sqrt{x + 1} = 5$ are radical equations. The equation $x + \sqrt{3} = 0$ is not a radical equation because x is not part of the radicand.

RULE 21-7. SOLVING RADICAL EQUATIONS

To solve a radical equation:

1. Isolate the radical term in one member of the equation.
2. Raise each member of the equation to a power equal to the index of the radical.
3. Solve the resulting equation for the root.
4. Check the root, as the preceding process may introduce roots into the derived equation that the original equation did not have.

EXAMPLE
21-13

Solve $5 + \sqrt{x} = 10$ for x.

Solution Use Rule 21-7 to solve.

Step 1: Isolate $\sqrt{x}$ in the left member:

$$\sqrt{x} = 5$$

Step 2: Square both members:

$$(\sqrt{x})^2 = 5^2$$

Step 3: Solve:

$$x = 25$$

Step 4: Check.

Check Does $5 + \sqrt{x} = 10$ when $x = 25$?

$$5 + \sqrt{25} = 10$$
$$5 + 5 \quad = 10 \quad \text{Yes.}$$

$$\therefore \quad x = 25 \text{ is the root.}$$

EXAMPLE
21-14

Solve $2\sqrt{y-2} + 4 = 8$.

Solution Use Rule 21-7.

Step 1: Isolate $\sqrt{y-2}$ in the left member:

$$\sqrt{y-2} = \frac{8-4}{2} = 2$$

Step 2: Square both members:

$$y - 2 = 4$$

Step 3: $$y - 6$$

Check Does $2\sqrt{y-2} + 4 = 8$ when $y = 6$?

$$2\sqrt{6-2} + 4 = 8$$
$$2\sqrt{4} + 4 = 8$$
$$4 + 4 = 8$$
$$8 = 8 \quad \text{Yes.}$$

$$\therefore \quad y = 6 \text{ is the solution.}$$

EXAMPLE
21-15

Solve $-4 = x - \sqrt{3x + 10}$ for x.

Solution Use Rule 21-7.

Step 1: Isolate $\sqrt{3x + 10}$ in the left member:

$$\sqrt{3x + 10} = x + 4$$

Step 2: Square both members:

$$(\sqrt{3x + 10})^2 = (x + 4)^2$$
$$3x + 10 = x^2 + 8x + 16$$

Step 3: Combine like terms and equate to zero:

$$x^2 + 5x + 6 = 0$$

Solve the quadratic by factoring:

$$(x + 3)(x + 2) = 0$$
$$x + 3 = 0 \qquad\qquad x + 2 = 0$$
$$x = -3 \qquad\qquad x = -2$$

Step 4: There are two possible roots; check both.

Check Does $-4 = x - \sqrt{3x + 10}$ when $x = -3$?

$$-4 = -3 - \sqrt{-9 + 10}$$
$$-4 = -4 \quad \text{Yes.}$$

Does $-4 = x - \sqrt{3x + 10}$ when $x = -2$?

$$-4 = -2 - \sqrt{-6 + 20}$$
$$= -2 - 2$$
$$-4 = -4 \quad \text{Yes.}$$

$$\therefore \quad x = -2 \text{ and } x = -3 \text{ are both roots.}$$

Solve and check each of the following.

1. $\sqrt{x} = 7$

2. $\sqrt{x} = 5$

3. $\sqrt{4x} = 8$

4. $\sqrt{3a} = 6$

5. $\sqrt{6x + 1} = 5$

6. $\sqrt{b - 1} = 2$

7. $\sqrt{a + 2} = 8$

8. $5 = \sqrt{1 + 4x}$

9. $2\sqrt{x + 3} = 6$

10. $4\sqrt{2a} = 8$

11. $5\sqrt{3y} = 15$

12. $3\sqrt{x + 1} = 9$

13. $2\sqrt{a} - 1 = 2$

14. $\sqrt{2y} + 3 = 1$

15. $\sqrt{6a} - 2 = 5$

16. $3 - \sqrt{2x - 7} = 0$

17. $\sqrt{5c - 2} + 3 = 6$

18. $\sqrt{x^2 - 8} + 4 = x$

19. $\sqrt{5 + a^2} - 5 = -a$

20. $x + 3\sqrt{x} = 10$

21. $\sqrt{y + 4} = 2 + y$

22. $4 + \sqrt{x + 2} = x$

23. $3 - \sqrt{7x - 3} = 2x$

24. $\sqrt{\dfrac{6 + 2x}{5}} = 4$

25. $\sqrt{\dfrac{4y}{3}} - 6 = 2$

26. Solve $4a = c\sqrt{3}$ for c.

27. Solve $t = \sqrt{\dfrac{2S}{g}}$ for g.

28. Solve $t = \pi\sqrt{\dfrac{l}{g}}$ for l.

29. Solve $r = \sqrt[3]{\dfrac{3V}{4\pi}}$ for V.

30. Solve $E = \sqrt{V^2 + 2IR}$ for R.

SELECTED TERMS

base   A number raised to a power; for example, in x^2, x is the base.

exponent   The power to which a number or base is raised; for example, in z^n, n is the exponent.

fractional exponent   An exponent that is a fraction; an exponent of 1/n indicates the nth root.

LOGARITHMIC AND EXPONENTIAL FUNCTIONS

In the preceding chapters, functions involving addition, subtraction, multiplication, division, and raising variables to constant powers have been discussed. In this chapter, we look at raising constants to *variable* powers. We have examined $y = x^2$; now we are going to examine $y = 10^x$.

22-1 COMMON LOGARITHMS

Suppose we know that $a = 10^b$, and we want to solve for b. What do we do? Read $a = 10^b$ as *"b is the exponent to which 10 must be raised to yield a."* In mathematics, we define *common logarithm* so that the following statement means the same as the preceding: *b is the common logarithm of a.* This can be written in the following manner:

If $a = 10^b$, then $b = \log(a)$

The symbol log (a) is read "log of a" and stands for the common logarithm function of a. Notice that the parentheses here indicate "function of" and not multiplication. The parentheses are frequently omitted.

The common logarithm of a number is defined as the exponent to which 10 must be raised to yield the number. In the following sections you will see that logarithms follow the rules of exponents.

EXAMPLE 22-1 Find the common logarithm of 1, 10, and 100.

Observation $1 = 10^0$, $10 = 10^1$, and $100 = 10^2$.

Solution Use the definition of common logarithms and the observation:

$$\log(1) = \log(10^0)$$
$$= 0$$
$$\log(10) = \log(10^1)$$
$$= 1$$
$$\log(100) = \log(10^2)$$
$$= 2$$

∴ Log (1) = 0, log (10) = 1, and log (100) = 2.

Now, turn to your calculator and your owner's guide. Locate the common logarithm function in the guide and the function key $\boxed{\log\ x}$ or $\boxed{\log}$ on your calculator. This function will calculate the common logarithm of any positive number for you. Use your calculator to work the following examples.

EXAMPLE 22-2   Find the common logarithm of 5, 6.5, and 2.98.

Observation   When writing logarithms, we generally write four digits to the right of the decimal point. This corresponds to three place accuracy in the number.

Solution   Use your calculator:

$$5 \boxed{\text{log}} \Rightarrow 0.6990$$

$$6.5 \boxed{\text{log}} \Rightarrow 0.8129$$

$$2.98 \boxed{\text{log}} \Rightarrow 0.4742$$

EXAMPLE 22-3   Find the common logarithm of $10^{0.5}$.

Solution 1   Use the definition of common logarithm:

$$\log (10^{0.5}) = 0.5$$

Observation   Recall that $10^{0.5}$ means the square root of 10.

Solution 2   Use your calculator:

$$\boxed{\sqrt{x}} \quad \begin{aligned} \log (10^{0.5}) &= \log (\sqrt{10}) \\ \log (10^{0.5}) &= \log (3.162) \\ \log (10^{0.5}) &= 0.5 \end{aligned}$$

$$\therefore \quad \text{Log } (10^{0.5}) = 0.5 \text{ by definition and by calculator.}$$

EXERCISE 22-1

Use the definition to find the common logarithms of the following numbers.

1. 10^3 2. 10^1 3. 10^4

4. $10^{1.5}$ 5. $10^{2.5}$ 6. $10^{0.25}$

Calculator Drill

Use your calculator to find the common logarithms of the following numbers.

7. 2   8. 4   9. 8

10. 16   11. 32   12. 2.1

13. 3.1	14. 4.1	15. 5.1
16. 6.1	17. 7.1	18. 6.2
19. 3.41	20. 3.78	21. 5.92
22. 1.05	23. 2.95	24. 8.01
25. 7.62	26. 9.15	27. 4.61

22-2 COMMON LOGARITHMS AND SCIENTIFIC NOTATION

In the preceding section we learned the definition of common logarithms. We also learned how to use our calculators to find the common logarithm of a number. We are going to perform an experiment with our calculators on a series of numbers in scientific notation.

EXAMPLE 22-4   Find the common logarithms of 3×10^0, 3×10^1, 3×10^2, and 3×10^9.

Solution   Use your calculator:

$$3 \boxed{EE}\, 0 \boxed{log} \Rightarrow 0.4771$$

$$3 \boxed{EE}\, 1 \boxed{log} \Rightarrow 1.4771$$

$$3 \boxed{EE}\, 2 \boxed{log} \Rightarrow 2.4771$$

$$3 \boxed{EE}\, 9 \boxed{log} \Rightarrow 9.4771$$

Observation   Notice that the fractional part of each answer is the same number, and that the integer part of each answer is the same as the exponent of 10.

In Example 22-4, notice that common logarithms of the numbers 3×10^0, 3×10^1, 3×10^2, and 3×10^9 all had the same fractional part. It is a fact that the fractional part of a logarithm depends only on the digits of a number and not on the location of the decimal point nor on the exponent of 10. The fractional part of a logarithm is called the *mantissa*. Notice also that the integer part of the logarithm depends only on the exponent of 10. The integer part of the logarithm is called the *characteristic*. If we have a table of common logarithms for numbers between 1 and 10, or if we have a calculator that will find the common logarithm for a number, we can use Rule 22-1 to find the common logarithm of any positive number.

RULE 22-1. FINDING THE COMMON LOGARITHM OF A NUMBER

To find the logarithm of a number:

1. Write the number in scientific notation.
2. Determine the mantissa as the common logarithm of the decimal coefficient. (This will be a fraction between zero and one.)
3. Determine the characteristic as the exponent of 10.
4. Form the common logarithm by adding the characteristic and the mantissa.

EXAMPLE 22-5 Find the mantissa, the characteristic, and the common logarithm of 654, 7.83, and 0.0491.

Solution Use Rule 22-1 and your calculator.

Step 1: Write the number in scientific notation:

$$\text{SN:} \qquad 654 = 6.54 \times 10^2$$

Step 2: Determine the mantissa:

$$\boxed{\text{log}} \qquad \text{mantissa} = \log (6.54) = 0.8156$$

Step 3: Determine the characteristic:

$$\text{characteristic} = 2$$

Step 4: Form the common logarithm:

$$\therefore \qquad \log (654) = 2.8156$$

Solution 2 Use Rule 22-1 and your calculator.

Step 1: Write the number in scientific notation:

$$\text{SN:} \qquad 7.83 = 7.83 \times 10^0$$

Step 2: Determine the mantissa:

$$\boxed{\text{log}} \qquad \text{mantissa} = \log (7.83) = 0.8938$$

Step 3: Determine the characteristic:

$$\text{characteristic} = 0$$

Step 4: Form the common logarithm:

$$\therefore \quad \log{(7.83)} = 0.8938$$

Solution 3 Use Rule 22-1 and your calculator.
Step 1: Write the number in scientific notation:

$$\text{SN:} \quad 0.0491 = 4.91 \times 10^{-2}$$

Step 2: Determine the mantissa:

$$\boxed{\text{log}} \quad \text{mantissa} = \log{(4.91)} = 0.6911$$

Step 3: Determine the characteristic:

$$\text{characteristic} = -2$$

Step 4: Form the common logarithm:

$$\therefore \quad \log{(0.0491)} = -2 + 0.6911 = -1.3089$$

Summary

$$\log{(654)} = 2 + 0.8156 = 2.8156$$
$$\log{(7.83)} = 0 + 0.8938 = 0.8938$$
$$\log{(0.0491)} = -2 + 0.6911 = -1.3089$$

Although the mantissa is positive, the characteristic may be positive or negative. The characteristic is always negative for fractions. In the case of fractions, we combine the characteristic and the mantissa by subtraction. The integer and fractional parts of the resulting logarithm are both negative. This is the answer your calculator will give.

The graph of Figure 22-1 is the plot of the common logarithm function, which reveals several properties of the common logarithm function. For $x > 1$, $\log{(x)} > 0$; for $1 > x > 0$, $\log{(x)} < 0$; and for $x = 1$, $\log{(x)} = 0$.

EXERCISE 22-2
Find the characteristics, mantissas, and common logarithms of the following numbers.

1. (a) 7   (b) 70   (c) 700   (d) 7000
2. (a) 2   (b) 20   (c) 200   (d) 2000
3. (a) 0.075   (b) 0.75   (c) 7.5   (d) 75

4. (a) 0.008   (b) 0.8   (c) 80   (d) 8000
5. (a) 1.62   (b) 16.2   (c) 162   (d) 1620
6. (a) 99.1   (b) 9.91   (c) 0.991   (d) 0.0991
7. (a) 5.55   (b) 55.5   (c) 555   (d) 5550
8. (a) 3.61   (b) 0.361   (c) 0.0361   (d) 0.003 61
9. (a) 0.001 05 (b) 0.0105   (c) 0.105   (d) 1.05

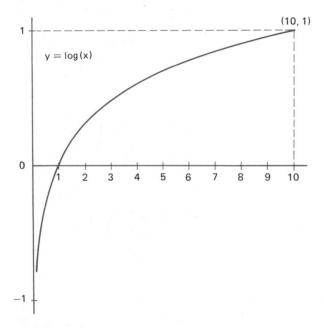

FIGURE 22–1
Plot of the common logarithm function.

22-3 ANTILOGARITHMS

The common logarithm of a number is the exponent to which 10 must be raised to give the original number. Knowing this definition, how can we find the original number if we know the common logarithm? By raising 10 to the common logarithm power! This operation is the definition of the *common antilogarithm* function, which is the *inverse* of the common logarithm function. All of this can be written in symbols:

$$\text{If } b = \log(a), \text{ then } a = \text{antilog}(b) = 10^b$$

Check your calculator and your owner's guide to find out how to evaluate the antilogarithm of a number. Your calculator may use $\boxed{10^x}$ or $\boxed{y^x}$ where you supply the 10 for y. Or your calculator may use two key strokes, as in $\boxed{\text{INV}}$ $\boxed{\log}$ or $\boxed{f^{-1}}$ $\boxed{\log}$.

EXAMPLE 22-6   If the common logarithm of A is 2, what is the value of A?

Solution   Use the antilogarithm to find A:

$$2 = \log(A)$$

Apply antilog; see Figure 22-2:

$$A = \text{antilog}(2) = 10^2$$
$$\therefore \quad A = 100$$

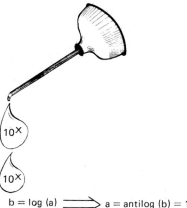

$$b = \log(a) \implies a = \text{antilog}(b) = 10^b$$

FIGURE 22–2
Effects of antilog.

EXAMPLE 22-7   If the common logarithm of Z is 1.3324, what is the value of Z?

Solution   Use the antilogarithm and your calculator:

$$1.3324 = \log(Z)$$

Apply antilog:

$$Z = \text{antilog}(1.3324) = 10^{1.3324}$$
$$1.3324 \boxed{10^x} \Rightarrow 21.5$$

$$\therefore \quad Z = 21.5$$

Calculator Drill

Find the antilogarithm of the following numbers.

1. 0.3522
2. 0.9706
3. 0.1592
4. 0.7041
5. 0.5223
6. −0.3871
7. −1.0526
8. 1.3952
9. 1.4671
10. 2.8352
11. 0.3010, 1.3010, 2.3010
12. 0.4771, 1.4771, 2.4771
13. 0.6021, 2.6021, 4.6021
14. 0.6990, 2.6990, 4.6990
15. −0.1549, −1.1549, −2.1549

22-4 LOGARITHMS, PRODUCTS, AND QUOTIENTS

In this section we will explore two important properties of logarithms. They will each be stated as rules. These rules will be useful in solving logarithmic equations.

RULE 22-2. LOGARITHM OF A PRODUCT

The logarithm of a product is equal to the sum of the logarithms of the factors.

$$\log (AB) = \log (A) + \log (B)$$

EXAMPLE 22-8
Solution

Check the equation $\log (20 \cdot 15) = \log (20) + \log (15)$. Evaluate the left and right members separately:

Left member:

$$\boxed{\times} \qquad \log (20 \cdot 15) = \log (300)$$
$$\boxed{\log} \qquad\qquad\qquad\quad = 2.4771$$

Right member:

$$\boxed{\log} \qquad \log (20) + \log (15) = 1.3010 + 1.1761$$
$$\boxed{+} \qquad\qquad\qquad\qquad\quad = 2.4771$$

$$\therefore \qquad \log (20 \cdot 15) = \log (20) + \log (15)$$

EXAMPLE 22-9 Solve $P = 3 \times 6$ using logarithms and Rule 22-2.

Solution $P = 3 \times 6$

Take the log of each member.

$$\log (P) = \log (3 \times 6)$$

Apply Rule 22-2:

$$\log (P) = \log (3) + \log (6)$$

| log | $\log (P) = 0.4771 + 0.7782$
| + | $\log (P) = 1.2553$

Take the antilog of each member:

| 10ˣ | $P = \text{antilog} (1.2553) = 10^{1.2553}$

$$\therefore \quad P = 18$$

RULE 22-3. LOGARITHM OF A QUOTIENT

The logarithm of a quotient is equal to the difference between the logarithm of the dividend and the logarithm of the divisor.

$$\log (A/B) = \log (A) - \log (B)$$

EXAMPLE 22-10

Check the equation $\log \left(\frac{75}{15}\right) = \log 75 - \log 15$.

Solution   Evaluate each member of the equation separately. Left member:

| ÷ | $\log \left(\frac{75}{15}\right) = \log (5)$
| log | $= 0.6990$

Right member:

| log | $\log (75) - \log (15) = 1.8751 - 1.1761$
| − | $= 0.6990$

$$\therefore \quad \log \left(\frac{75}{15}\right) = \log (75) - \log (15)$$

EXAMPLE
22-11
Solve $Q = \frac{20}{4}$ using logarithms and Rule 22-3.

Solution

$$Q = \frac{20}{4}$$

Take the log of each member:

$$\log (Q) = \log \left(\frac{20}{4}\right)$$

Apply Rule 22-3:

$$\boxed{\log}\quad \begin{aligned} \log (Q) &= \log (20) - \log (4) \\ \log (Q) &= 1.3010 - 0.6021 \\ \log (Q) &= 0.6989 \end{aligned}$$
$$\boxed{-}$$

Take the antilog of each member:

$$\boxed{10^x}\qquad Q = \text{antilog}\ (0.6989) = 10^{0.6989}$$

$$\therefore\qquad Q = 5$$

EXERCISE 22-4

Check the following equations as in Examples 22-8 and 22-10:

1. $\log (3 \times 5) = \log (3) + \log (5)$
2. $\log (2 \times 7) = \log (2) + \log (7)$
3. $\log (17 \times 3.5) = \log (17) + \log (3.5)$
4. $\log \left(\frac{625}{25}\right) = \log (625) - \log (25)$
5. $\log \left(\frac{18}{2}\right) = \log (18) - \log (2)$

Determine whether the following statements are true or false:

6. $\log (75) = \log (5) + \log (15)$
7. $\log (14) = \log (7 + 2)$
8. $\log (7) = \log (35) - 5$
9. $\log (20) = \log (4) \log (5)$
10. $\log (9) = \log (72) - \log (8)$

Use logarithms to solve the following as in Examples 22-9 and 22-11:

11. $P = 7 \times 4$
12. $P = 23.45 \times 0.0574$
13. $Q = \frac{16}{4}$
14. $Q = 0.213/0.0790$
15. $Q = 1218/203$

22-5 LOGARITHMS, POWERS, AND RADICALS

In the preceding section we learned that through logarithms multiplication becomes addition and division becomes subtraction. In a similar fashion, raising to powers becomes multiplication.

RULE 22-4. LOGARITHM OF A POWER

The logarithm of a power is equal to the product of the exponent and the logarithm of the base.

$$\log (a^n) = n \log (a)$$

EXAMPLE
22-12

Check the equation $\log (5^4) = 4 \log (5)$.

Solution   Evaluate each member of the equation.
Left member:

$\boxed{y^x}$ $\log (5^4) = \log (625)$
$\boxed{\log}$ $\qquad\quad = 2.7959$

Right member:

$\boxed{\log}$ $4 \log (5) = 4 (0.6990)$
$\boxed{\times}$ $\qquad\quad = 2.7959$

$\therefore \qquad \log (5^4) = 4 \log (5)$

Square roots can be indicated with a radical sign and with an exponent of one-half. This combined with Rule 22-4 gives a new rule for the logarithm of the square root of a number.

RULE 22-5. LOGARITHM OF THE SQUARE ROOT OF A NUMBER

The logarithm of the square root of a number is equal to one-half the logarithm of the number.

$$\log (\sqrt{A}) = \frac{1}{2} \log (A)$$

EXAMPLE 22-13

Check the equation $\log (\sqrt{49}) = \frac{1}{2} \log (49)$.

Solution Evaluate each member of the equation.
Left member:

$$\boxed{\sqrt{x}}$$
$$\boxed{\log}$$

$$\log (\sqrt{49}) = \log (7)$$
$$= 0.8451$$

Right member:

$$\boxed{\log}$$
$$\boxed{\div}$$

$$\frac{1}{2} \log (49) = \frac{1}{2} (1.6902)$$
$$= 0.8451$$

$$\therefore \quad \log (\sqrt{49}) = \frac{1}{2} \log (49)$$

EXAMPLE 22-14

Calculate $\log (\sqrt{72.5})$ without using square root.

Solution Use Rule 22-5.

$$\log (\sqrt{72.5}) = \frac{1}{2} \log (72.5)$$
$$= \frac{1}{2} (1.8603)$$
$$= 0.9302$$
$$\therefore \quad \log (\sqrt{72.5}) = 0.9302$$

In the following example we use Rule 22-5 to evaluate the square root of a number. The same steps can be used with Rule 22-4 to evaluate any power of a number.

EXAMPLE 22-15

Calculate $\sqrt{172}$ without using square root.

Solution   Use Rule 22-5 and logarithms:

$$\boxed{\text{log}} \qquad \log(\sqrt{172}) = \tfrac{1}{2} \log (172)$$
$$\boxed{\div} \qquad\qquad\qquad = \tfrac{1}{2} (2.2355)$$
$$= 1.1178$$

Apply antilog:

$$\boxed{10^x} \qquad \sqrt{172} = \text{antilog} (1.1178) = 10^{1.1178}$$

$$\therefore \qquad \sqrt{172} = 13.1$$

**EXAMPLE
22-16**

Calculate $17^{3.49}$

Solution   Use Rule 22-4:

$$\boxed{\text{log}} \qquad \log (17^{3.49}) = 3.49 \log (17)$$
$$\boxed{\times} \qquad\qquad\qquad = 3.49(1.2304)$$
$$= 4.2941$$

Apply antilog:

$$\boxed{10^x} \qquad 17^{3.49} = \text{antilog} (4.2941) = 10^{4.2941}$$

$$\therefore \qquad 17^{3.49} = 19\ 700$$

**EXAMPLE
22-17**

Solve $N = 4^3$ using logarithms and Rule 22-4.

Solution   Take the log of each member:

$$\log (N) = \log (4^3)$$

Apply Rule 22-4:

$$\boxed{\text{log}} \qquad \log (N) = 3 \log (4)$$
$$\boxed{\times} \qquad\qquad\quad = 3(0.6021)$$
$$= 1.8063$$

Take the antilog of each member:

$\boxed{10^x}$ $N = \text{antilog } (1.8063) = 10^{1.8063}$

$\therefore$ $N = 64$

EXERCISE 22-5
Check the following equations as in Examples 22-12 and 22-13.

1. $\log (3^2) = 2 \log (3)$
2. $\log (12^2) = 2 \log (12)$
3. $\log (29^2) = 2 \log (29)$
4. $\log (5^3) = 3 \log (5)$
5. $\log (\sqrt{81}) = \frac{1}{2} \log (81)$
6. $\log (\sqrt{961}) = \frac{1}{2} \log (961)$
7. $\log (\sqrt{2^3}) = \frac{3}{2} \log (2)$

Use Rules 22-4 and 22-5 in evaluating the following.

8. $\sqrt{39}$	9. $4^{7.5}$	10. $\sqrt{409}$
11. $\sqrt{3.5}$	12. $\sqrt{11}$	13. $(5.2)^{0.28}$
14. $(1.01)^{120}$	15. $(87)^{0.314}$	16. $\sqrt[3]{6.01 \times 10^{-5}}$

22-6 NATURAL LOGARITHMS

In electronics there are two important logarithm functions. The first is the common logarithm based on the number 10. The second is the natural logarithm based on the number e. The value of e to five digits is 2.7183.

To distinguish these two logarithm functions, we will continue to use log ( ) for common logarithms. And we introduce a new symbol ln ( ) for natural logarithms. The following is the definition of the natural logarithm function:

If $A = e^B$, then $B = $ natural logarithm of $A = \ln (A)$

The symbol ln (A) is read "ell-en of a" or "lin of a." It may also be read "natural log of a." As with common logarithms, the parentheses are frequently omitted. Remember that they are used here to denote "function," not multiplication.

Check your calculator and operating instructions for natural logarithms. Locate the function key $\boxed{\ln x}$ or $\boxed{\ln}$. This function will calculate the natural logarithm of any positive number for you.

EXAMPLE 22-18

Find the natural logarithms of 5, 10, and 15.

Solution Use your calculator:

$$5 \; \boxed{\ln} \Rightarrow 1.6094$$
$$10 \; \boxed{\ln} \Rightarrow 2.3026$$
$$15 \; \boxed{\ln} \Rightarrow 2.7081$$

Figure 22-3 presents a graph of both logarithm functions. Notice that both graphs pass through the point (1, 0). Also, both have the same *general* behavior. There is no graph for $x \leqslant 0$. For $0 < x < 1$, both graphs are below

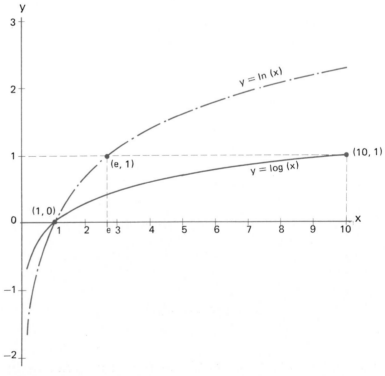

FIGURE 22–3
Comparison of log (x) and ln (x).

the x-axis. This means both functions are negative in this region. For $x > 1$, both graphs are shown above the x-axis and both functions are positive. Finally, as x increases both graphs increase.

Antilogarithm

Recall the definition of the natural logarithm of a number as the exponent to which e must be raised to give the number. Knowing the natural logarithm of a number, how can we find the number? By raising e to the natural logarithm power. This is the definition of the *natural antilogarithm* function, which is the *inverse* of the natural logarithm function. The following symbols express this definition:

$$\text{If } a = \ln (b), \text{ then } b = \text{antiln} (a) = e^a$$

The antiln function is also known as the *exponential function*. The graph of the exponential function is shown in Figure 22-4.

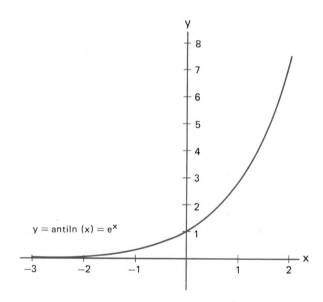

FIGURE 22–4
Graph of the exponential function, e^x.

Check your calculator and owner's guide on how to calculate antiln (x). It may use $\boxed{e^x}$ or a two-key-stroke sequence, as $\boxed{f^{-1}}$ $\boxed{\ln}$ or $\boxed{\text{INV}}$ $\boxed{\ln}$. Figure 22-5 shows the effects of the antiln function.

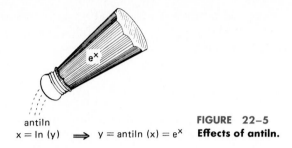

antiln
$$x = \ln{(y)} \implies y = \text{antiln}(x) = e^x$$

FIGURE 22-5
Effects of antiln.

**EXAMPLE
22-19**   Find y if $\ln{(y)} = 1.9782$.

Solution   Use antiln:

$$\ln{(y)} = 1.9782$$
$$y = \text{antiln}(1.9782) = e^{1.9782}$$
$\boxed{e^x}$ $y = 7.23$

Check   Does $\ln{(y)} = 1.9782$ when $y = 7.23$?

$\boxed{\ln}$ $\ln{(7.23)} = 1.9782$ Yes!
$\therefore$ $y = 7.23$

EXERCISE 22-6
Use the definition to find the natural logarithm of the following terms.

1. $e^{1.2}$ 2. $e^{3.0}$ 3. $e^{0.75}$ 4. $e^{0.12}$ 5. $e^{-4.1}$
6. $e^{-0.51}$ 7. e^{a} 8. e^{-Rt} 9. e^{-z} 10. e^{2n}

Calculator Drill
Calculate the natural logarithms of the following numbers.

11. 1.25	12. 3.16
13. 4.0	14. 5.0
15. 6.0	16. 7.0
17. 0.156	18. 0.029
19. 0.015	20. 18.4
21. 927	22. 7040
23. 143.2	24. 40.94
25. 6.35×10^4	26. 1045

27. 0.831	28. 0.0392
29. 79.3	30. 2001

Find the natural antilogarithms of the following numbers.

31. 1.0	32. 0.0
33. 2.3026	34. 2.9444
35. 2.1983	36. -5.2785
37. -3.6045	38. -1.3863
39. 0.5241	40. -99.0919
41. -0.0590	42. 4.7205
43. -7.0646	44. 0.8751
45. -4.6702	46. -3.0011
47. 14.3285	48. -4.4273
49. -16.2345	50. 0.5623

22-7 CHANGING BASE

Sometimes the question arises of converting a logarithm from one base to another. With the aid of a calculator this is a simple task when the value of the logarithm is known. Take the antilogarithm and then the second logarithm of the number. A more direct method is provided by the following rule:

RULE 22-6. CHANGING BASE

1. To change a common logarithm to a natural logarithm, multiply by 2.3026, ln of 10:

$$\ln (x) = 2.3026 \log (x)$$

2. To change a natural logarithm to a common logarithm, divide by 2.3026, ln of 10:

$$\log (x) = \ln (x)/2.3026$$

EXAMPLE
22-20

Check Rule 22-6 when $x = 15$.

Solution Substitute for x; then evaluate each member of the equation:

$$\ln (15) = 2.3026 \log (15)$$

Left member:

$$\boxed{\ln x} \qquad \ln (15) = 2.7081$$

Right member:

$$\boxed{\log x} \qquad 2.3026 \log (15) = 2.3026 \times 1.1761$$
$$\boxed{\times} \qquad\qquad\qquad\qquad = 2.7081$$

$$\therefore \qquad\qquad \ln (15) = 2.3026 \log (15)$$

The number 2.3026 may be remembered as $\ln (10)$ or $1/\log (e)$. In fact, for any number n, where $n \ne 1$, $\ln (n)/\log (n) = 2.3026$.

> **EXERCISE 22-7**
> Use common logarithms to calculate the following.
>
> | | | |
> |---|---|---|
> | 1. $\ln (4)$ | 2. $\ln (5.1)$ | 3. $\ln (0.49)$ |
> | 4. $\ln (8.2)$ | 5. $\ln (0.79)$ | 6. $\ln (10.9)$ |
>
> Use natural logarithms to calculate the following.
>
> | | | |
> |---|---|---|
> | 7. $\log (60)$ | 8. $\log (21)$ | 9. $\log (5.7)$ |
> | 10. $\log (0.015)$ | 11. $\log (0.29)$ | 12. $\log (0.091)$ |

22-8 FURTHER PROPERTIES OF NATURAL LOGARITHMS

As we noted earlier, the natural logarithm function and the common logarithm function have the same general behavior. In fact, the rules for logarithms and products, quotients, powers, and radicals are the same for both logarithm functions.

EXAMPLE 22-21

Check Rule 22-2 for natural logarithms: $\ln (AB) = \ln (A) + \ln (B)$ when $A = 16$ and $B = 3.5$.

Solution　Substitute for A and B; then evaluate each member of the equation:

$$\ln (16 \times 3.5) = \ln (16) + \ln (3.5)$$

Left member:

$$\boxed{\times} \quad \ln (16 \times 3.5) = \ln (56)$$
$$\boxed{\text{In}} \quad\quad\quad\quad\quad = 4.0254$$

Right member:

$$\boxed{\text{In}} \quad \ln (16) + \ln (3.5) = 2.7726 + 1.2528$$
$$\boxed{+} \quad\quad\quad\quad\quad\quad = 4.0254$$

$$\therefore \quad \ln(16 \times 3.5) = \ln (16) + \ln (3.5)$$

EXAMPLE 22-22

Check Rule 22-3 for natural logarithms: $\ln (A/B) = \ln (A) - \ln (B)$ when $A = 54$ and $B = 60$.

Solution　Substitute for A and B; then evaluate each member of the equation:

$$\ln \left(\tfrac{54}{60}\right) = \ln (54) - \ln (60)$$

Left member:

$$\boxed{\div} \quad \ln \left(\tfrac{54}{60}\right) = \ln (0.9)$$
$$\boxed{\text{In}} \quad\quad\quad\quad = -0.1054$$

Right member (use chain operations):

$$\boxed{\text{In}} \quad \ln (54) - \ln (60) = 3.9890 - 4.0943$$
$$\boxed{-} \quad\quad\quad\quad\quad\quad = -0.1054$$

$$\therefore \quad \ln \left(\tfrac{54}{60}\right) = \ln (54) - \ln (60)$$

EXAMPLE 22-23

Check Rule 22-4 for natural logarithms: $\ln(a^n) = n \ln(a)$ when $a = 1.67$ and $n = 3$.

Solution   Substitute for a and n and then evaluate each member of the equation:

$$\ln(1.67^3) = 3 \ln(1.67)$$

Left member (perform chain operations):

$$\boxed{y^x} \qquad \ln(1.67^3) = \ln(4.66)$$
$$\boxed{\ln} \qquad\qquad\qquad\quad = 1.5385$$

Right member:

$$\boxed{\ln} \qquad 3 \ln(1.67) = 3(0.5128)$$
$$\boxed{\times} \qquad\qquad\qquad\quad = 1.5385$$

$$\therefore \qquad \ln(1.67^3) = 3 \ln(1.67)$$

EXAMPLE 22-24

Check Rule 22-5 for natural logarithms: $\ln(\sqrt{A}) = \frac{1}{2} \ln(A)$ when $A = 79.6$ (use chain operations).

Solution   Substitute for A and then evaluate each member of the equation:

$$\ln(\sqrt{79.6}) = \frac{1}{2} \ln(79.6)$$

Left member:

$$79.6 \boxed{\sqrt{x}} \boxed{\ln} \Rightarrow 2.1885$$

Right member:

$$79.6 \boxed{\ln} 2 \boxed{\div} \Rightarrow 2.1885$$

$$\therefore \qquad \ln(\sqrt{79.6}) = \frac{1}{2} \ln(79.6)$$

EXERCISE 22-8
Check the following equations:

1. $\ln(17 \cdot 3) = \ln(17) + \ln(3)$
2. $\ln(9 \cdot 5) = \ln(9) + \ln(5)$
3. $\ln\left(\frac{65}{4}\right) = \ln(65) - \ln(4)$

4. $\ln\left(\dfrac{0.695}{0.421}\right) = \ln(0.695) - \ln(0.421)$

5. $\ln(15^2) = 2\ln(15)$
6. $\ln(0.71^2) = 2\ln(0.71)$
7. $\ln(\sqrt{64}) = \frac{1}{2}\ln(64)$

8. $\ln(\sqrt{3.06}) = \frac{1}{2}\ln(3.06)$

Use natural logarithms, Rules 22-4 and 22-5, and Example 22-15 to calculate:

9. $\sqrt{5.41}$ 10. $\sqrt{0.0792}$ 11. 17.1^2 12. $(6.45)^{0.295}$

13. $(18.7)^{1/3}$ 14. $(1.51)^{17}$ 15. $\sqrt{71.3}$ 16. $(1.01)^{144}$

22-9 LOGARITHMIC EQUATIONS

An equation involving the logarithm of the unknown is referred to as a **logarithmic equation.** Thus, $16 = 4\log(x)$ is a logarithmic equation. On the other hand, $25x = x\log(14)$ is not a logarithmic equation. The key to solving a

TABLE 22-1

PROPERTIES OF LOGARITHMS		
Property	**Common**	**Natural**
Base	10	$e = 2.7183$
Logarithm of 1	$\log(1) = 0$	$\ln(1) = 0$
Logarithm of e	$\log(e) = 0.4343$	$\ln(e) = 1$
Logarithm of 10	$\log(10) = 1$	$\ln(10) = 2.3026$
Product	$\log(A \cdot B) = \log(A) + \log(B)$	$\ln(A \cdot B) = \ln(A) + \ln(B)$
Quotient	$\log(A/B) = \log(A) - \log(B)$	$\ln(A/B) = \ln(A) - \ln(B)$
Power	$\log(A^n) = n\log(A)$	$\ln(A^n) = n\ln(A)$
Radical	$\log(\sqrt{A}) = \frac{1}{2}\log(A)$	$\ln(\sqrt{A}) = \frac{1}{2}\ln(A)$
Antilogarithm	$\text{antilog}(x) = 10^x$	$\text{antiln}(x) = e^x$
Change of base	$\log(x) = \ln(x)/2.3026$	$\ln(x) = 2.3026\log(x)$

logarithmic equation is to treat the logarithm of the unknown as a new variable. Solve for this new variable and apply the antilogarithm to find the unknown.

In Table 22-1 you will find a summary of the properties of logarithms that we have studied. This table will help in solving logarithmic equations.

EXAMPLE 22-25

Solve $16 = 4 \log (x) + 8$.

Solution   First solve for $\log (x)$:

$$8 = 4 \log (x)$$
$$\log (x) = 2$$

Apply antilog to find x:

$$x = \text{antilog} (2) = 10^2$$
$$\therefore \quad x = 100$$

EXAMPLE 22-26

Solve $4.5 + \ln (x^2) = \ln (x)$.

Solution   Use Table 22-1 to simplify $\ln (x^2)$:

$$4.5 + 2 \ln (x) = \ln (x)$$

Solve for $\ln (x)$:

$$\ln (x) = -4.5$$

Apply antiln to find x:

$$x = \text{antiln} (-4.5) = e^{-4.5}$$
$$\boxed{e^x} \qquad = 1.11 \times 10^{-2}$$
$$\therefore \quad x = 1.11 \times 10^{-2}$$

EXAMPLE 22-27

Solve $\ln (x) - 1 = \log (x)$.

Solution   Convert $\ln (x)$ to $\log (x)$; then proceed:

$$2.3026 \log (x) - 1 = \log (x)$$
$$1.3026 \log (x) = 1$$

$$\log{(x)} = 1/1.3026$$
$$\boxed{1/x} \qquad = 0.7677$$

Apply antilog to find x:

$$x = \text{antilog}\,(0.7677) = 10^{0.7677}$$
$$\boxed{10^x} \qquad = 5.86$$
$$\therefore \qquad x = 5.86$$

EXERCISE 22-9
Solve the following equations.

1. $1.05 = \log{(x^2)}$
2. $17 + 2 \log{(x)} = 6$
3. $5 \log{(x)} - 3 = 2$
4. $7 \ln{(x)} = 2 \ln{(x^3)} + 2.71$
5. $4 \ln{(\sqrt{x})} = -1.53$
6. $\ln{(x/6)} = 0.15$
7. $\log{(4x)} + \log{(x)} = 2.459$
8. $4 \log{(x)} - 3 = \ln{(x^3)}$
9. $5 \log{(\sqrt{x})} = \ln{(x)} + 1.59$
10. $\log{(x/4)} + \log{(x/7)} = 0.361$
11. $\ln{(17/V)} = 4.5$
12. $\ln{(13z)} + \ln{(z^2)} = 9.021$
13. $\ln{(V^2)} - 3.29 = \ln{(V)}$
14. $\ln{(Y^3)} - 2 \ln{(Y)} = 2.3026$
15. $2.3026 \log{(Z)} = 1$
16. $\ln{(t^2)} + \ln{(1/t)} = 4.39$
17. The strength of signal to the strength of the noise measured in decibels is defined by the equation $N_{dB} = 10 \log{(P_S/P_N)}$. Solve for P_N.
18. In a particular circuit, the signal-to-noise ratio was found to be 15. How many decibels is this? (See problem 17.)
19. The number of octaves between two frequencies is given by the formula $N = \log{(f_2/f_1)}/\log{(2)}$. Solve for f_2.
20. Find the frequency 4.5 octaves above 60 Hz (see problem 19).

22-10 EXPONENTIAL EQUATIONS

An equation with the unknown appearing in an exponent is called an *exponential equation*. Thus, $90 = 120\, e^{-0.01/t}$ is an exponential equation. Exponential equations may be solved by first solving for the term containing the unknown exponent, and then using logarithms to find the unknown.

EXAMPLE 22-28

Solve $90 = 120 \, e^{-0.01/t}$ for t.

Solution   Solve for $e^{-0.01/t}$; then use natural logarithms:

$$90 = 120 \, e^{-0.01/t}$$

$$e^{-0.01/t} = \frac{90}{120} = 0.75$$

Take the natural logarithm of both members of the equation:

$$-0.01/t = \ln(0.75)$$

Solve for t:

$$t = -0.01/\ln(0.75)$$
$$= -0.01/(-0.2877)$$

$$\therefore \quad t = 0.0348$$

EXAMPLE 22-29

Solve $10^{2x} - 4 = 16$ for x.

Solution   Solve for 10^{2x} ; then use common logarithms:

$$10^{2x} = 20$$

Take the common log of both members of the equation:

$$2x = \log(20)$$
$$x = \log(20)/2$$
$$= 1.3010/2$$
$$= 0.6505$$

$$\therefore \quad x = 0.651$$

EXAMPLE 22-30

Solve $f_2 - f_1 2^n = 0$ for n.

Solution Solve for 2^n; then use common logarithms to find n:

$$f_1 2^n = f_2$$
$$2^n = f_2/f_1$$
$$\log(2^n) = \log(f_2/f_1)$$

Simplify the left member:

$$n \log(2) = \log(f_2/f_2)$$
$$n = \log(f_2/f_1)/\log(2)$$

Observation A similar solution could have been arrived at using natural logarithms. Both equations would give the same numerical value for n.

EXAMPLE 22-31

Evaluate n in the following equations when $f_1 = 330$ Hz and $f_2 = 2700$ Hz.

Solution Use common logarithms
$$n = \log(f_2/f_1)/\log(2)$$
$$= \log(2700/300)/\log(2)$$
$$= \log(8.18)/\log(2)$$
$$= 0.9128/0.3010$$
$$n = 3.03$$

Use natural logarithms
$$n = \ln(f_2/f_1)/\ln(2)$$
$$= \ln(2700/330)/\ln(2)$$
$$= \ln(8.18)/\ln(2)$$
$$= 2.1019/0.6931$$
$$n = 3.03$$

$\therefore$ The two equations give the same answers:
$$n = 3.03$$

EXERCISE 22-10

Solve the following equations.

1. $14 = 10^x$

2. $3.5 = 10^x - 1$

3. $81 = 0.9(10^y)$

4. $0.92 = 3(10^a) + 0.14$

5. $3(10^s) = 11 - 2(10^s)$ 6. $18 = e^T$

7. $e^x - 4 = 0$ 8. $1.52\, e^{0.1/y} = 40.3$

9. $3.4\, e^x = 0.015 + 1.5\, e^x$ 10. $75\, e^x - 125 = 13 + 6\, e^x$

11. $e^{-4/t} = 0.8$ 12. $e^{-15/t} = 0.5$

13. $e^{-1/(6t)} = 0.95$ 14. $1 - e^{-7.4/t} = 0.26$

15. $1 + e^{8.6/t} = 3.54$ 16. $2^x = 32$

17. $3^y = 243$ 18. $5^t = 17.3$

19. $21^{1/a} = 0.21$ 20. $15^{-1/b} = 0.95$

22-11 SEMILOG AND LOG–LOG PLOTS

Graphs are tools used to present mathematical data. As such, there is no one *right* way to draw a graph. The scales are dictated by use of the graph. Two scales useful in electronics are *semilog* and *log–log*. A semilog plot has one

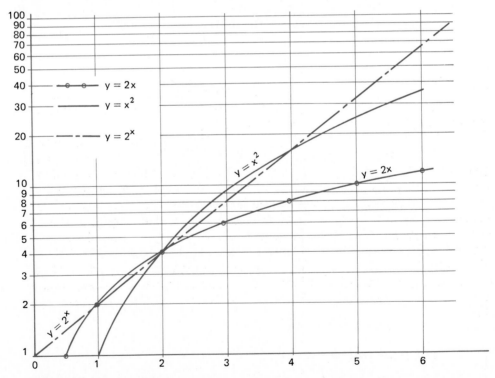

FIGURE 22–6
Examples of semilog plots.

CHAPTER 22 LOGARITHMIC AND EXPONENTIAL FUNCTIONS

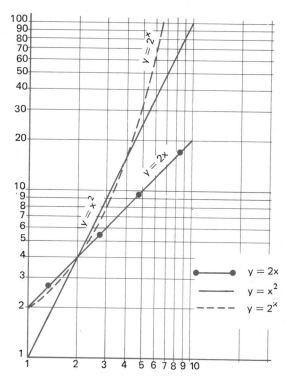

FIGURE 22-7
Examples of log–log plots.

regular (linear) scale and one with logarithmic spacing. Figure 22-6 is an example of a semilog plot. In a log–log plot both scales have logarithmic spacing. Figure 22-7 is an example of a log–log plot.

In Figure 22-6 there are three curves. On rectangular (linear) graph paper, $y = 2x$ would be a straight line, but on semilog paper it is a curve. The graph of $y = x^2$ is a curve on both types of paper, but the curves do not look the same. The graph of $y = 2^x$ is a straight line on semilog paper. These last two curves show the usefulness of semilog plots. First, they compress large values (and expand values close to zero) so that more data may be presented on a single graph. Second, semilog graphs convert exponential curves into straight lines.

In Figure 22-7 the same three curves are plotted on log–log paper. Here, $y = 2x$ and $y = x^2$ are straight lines, but $y = 2^x$ is a curve. Log–log plots are used to present wide ranges of data, especially data that have the form $y = ax^n$.

EXERCISE 22-11

1. Create a semilog plot of $y = e^x$.
2. Create a semilog plot of $y = 10^x$.

3. Create a log–log plot of $y = x^3$.
4. Create a log–log plot of $y = x^{1.5}$.
5. On log–log paper, plot $y = \sqrt{x}$ and $y = \ln(x)$.
6. Plot the following points on rectangular, semilog, and log–log paper:

x	0.1	0.2	0.3	0.4	0.5	0.6	0.7
y	0.1	0.203	0.309	0.423	0.546	0.684	0.842

x	0.8	0.9	1.0	1.1	1.2	1.3	1.4
y	1.029	1.26	1.56	1.96	2.57	3.60	5.80

7. Plot the following data on semilog and log–log paper. Do they plot as a straight line?

x	0.5	1	1.5	2	2.5	3
y	0.5	2	4.5	8	12.5	18

8. Plot the following data on semilog and log–log paper. Do they plot as a straight line?

x	1	1.5	2	2.5	3	3.5	4	4.5	5
y	1.07	1.10	1.14	1.18	1.21	1.27	1.31	1.36	1.40

x	5.5	6	6.5	7	7.5	8	8.5	9	9.5	10
y	1.45	1.50	1.55	1.61	1.66	1.72	1.78	1.84	1.90	1.97

22-12 NOMOGRAPHS

In electronics and especially communications, it is often convenient to solve problems graphically. There are many graphical aides to solutions presented in handbooks and periodicals. These graphical aides are called *nomographs*. Figure 22-8 presents a simple nomograph consisting of a single straight line with different gradations on either side. To use this nomograph, find the power ratio on the top; directly under it is the corresponding number of decibels (dB). The reverse process will convert decibels to power ratio. For example, a power ratio of 3 corresponds to 4.8 decibels.

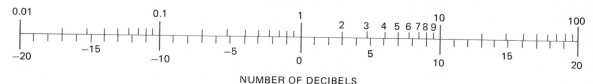

FIGURE 22-8
Nomograph for converting from power ratio to number of decibels.

A second type of nomograph relates three variables. The formula for inductive reactance is $X_L = 2\pi fL$. A nomograph relating X_L, f, and L is presented in Figure 22-9. When two of the parameters are known, the third is found by placing a straight edge to join the two known values. The value of the third parameter is read from the appropriate scale under the straight edge.

EXAMPLE 22-32

Find the frequency if the inductance is 400 μH and the reactance is 2 kΩ.

Solution Use the nomograph in Figure 22-9. Place a mark at 400 μH and a second at 2 kΩ. Draw a straight line through these points. Read the frequency as 800 kHz.

Check Check the graphical solution:

$$X_L = 2\pi fL$$

Solve for f:

$$f = X_L/(2\pi L)$$

Substitute:

$$f = 2(10^3)/(2\pi 400 \times 10^{-6})$$
$$= 796\ 000\ \text{Hz}$$
$$= 796\ \text{kHz} \approx 800\ \text{kHz}$$

$\therefore$ The graphical solution checks.

EXERCISE 22-12
Use the nomograph in Figure 22-8 to convert the following.

1. 6 dB to power ratio
2. 15 dB to power ratio
3. 0.02 to number of decibels
4. 20 to number of decibels

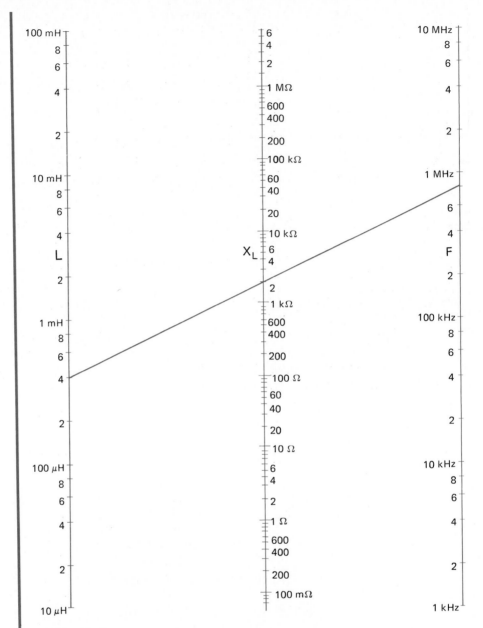

FIGURE 22–9
Inductive reactance versus frequency nomograph.

Use the nomograph in Figure 22-9 to find the third parameter.

5. $L = 200 \ \mu\text{H}$ $f = 1 \ \text{MHz}$ find X_L
6. $L = 6 \ \text{mH}$ $f = 50 \ \text{kHz}$ find X_L
7. $X_L = 60 \ \Omega$ $f = 10 \ \text{kHz}$ find L
8. $X_L = 10 \ \text{k}\Omega$ $f = 10 \ \text{MHz}$ find L
9. $X_L = 1 \ \text{M}\Omega$ $L = 100 \ \text{mH}$ find f
10. $L = 30 \ \text{mH}$ $X_L = 300 \ \Omega$ find f

SELECTED TERMS

antilogarithm   The inverse of the logarithm function: when $a = \log(b)$ then $b = \text{antilog}\ (a)$.

characteristic   The integer part of a logarithm.

common logarithm   Base ten logarithm.

logarithm   The exponent to which the base must be raised to equal the original number.

mantissa   The fractional part of a logarithm.

natural logarithm   Base e (2.71828 . . . .) logarithm.

23

APPLICATIONS OF LOGARITHMIC AND EXPONENTIAL EQUATIONS TO ELECTRONIC CONCEPTS

This chapter applies the techniques of logarithmic and exponential equations to two areas of electronics. The first area is gain and loss in an electronic system. The second is rise time in *RC* and *RL* series circuits. The topics have been arranged to progress from the decibel, through system calculations, to *RC* and *RL* transient behavior.

23-1 THE DECIBEL

Power in Decibels

The human ear hears in a *logarithmic* manner. That is, an apparent doubling in the loudness of the sound heard by the ear requires approximately a ten times increase in the acoustic power source. For example, to double the loudness of sound caused by an audio amplifier operating at 4 watts, the power must be increased to 40 watts.

Because of its logarithmic nature, the *decibel* (dB) has been adopted as the practical unit for measuring what the ear hears. Over the years the decibel has been applied to indicate the performance of all types of electronic circuits and devices, including amplifiers (audio frequency as well as radio frequency), filters, antennas, and microphones.

The decibel is a logarithmic expression that compares two power levels. Expressed mathematically,

$$N_{dB} = 10 \log \left(\frac{P_{out}}{P_{in}} \right) \quad dB \qquad (23\text{-}1)$$

where N_{dB} = gain or loss in decibels
P_{in} = input power
P_{out} = output power

EXAMPLE 23-1   Determine the power gain of an amplifier (N_{dB}) when an input power of 0.5 W produces an output power of 50 W.

Solution   Use Equation 23-1 and substitute:

$$N_{dB} = 10 \log (P_{out}/P_{in})$$
$$= 10 \log (50/0.5)$$

$$\therefore \quad N_{dB} = 20 \text{ dB}$$

Expressing Gain and Loss in Decibels

In the preceding example the larger power of 50 W was placed over the smaller power of 0.5 W. This caused the log of the ratio to be positive. However, when this ratio is inverted, the log of the ratio is negative. A gain is indicated with a plus sign: $P_{out}/P_{in} > 1 \Rightarrow \log (P_{out}/P_{in}) > 0 \ (+)$. A loss is indicated with a minus sign: $P_{out}/P_{in} < 1 \Rightarrow \log (P_{out}/P_{in}) < 0 \ (-)$.

EXAMPLE 23-2   Use an appropriate sign to express each of the stated conditions as a gain or a loss: gain of 32 dB and loss of 17 dB.

Solution

$$\text{Gain of 32 dB} = +32 \text{ dB}$$
$$\text{Loss of 17 dB} = -17 \text{ dB}$$

EXAMPLE 23-3   The attenuator network pictured in Figure 23-1 has a loss of 6 dB (-6 dB). Determine the input power (P_{in}) if the output power (P_{out}) is 300 mW.

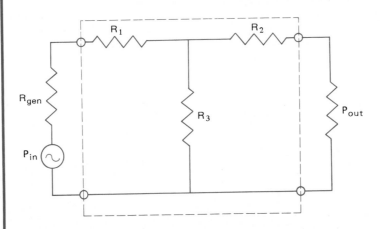

FIGURE 23-1
Circuit for Example 23-3 with the attenuator network pictured within the dashed lines.

Solution   Substitute into Equation 23-1 and solve for P_{in}:

$$N_{dB} = 10 \log \left(\frac{P_{out}}{P_{in}} \right)$$

$$-6 = 10 \log\left(\frac{300 \times 10^{-3}}{P_{in}}\right)$$

$$-0.6 = \log\left(\frac{300 \times 10^{-3}}{P_{in}}\right)$$

Take the antilog of each member:

$$0.25 = \frac{300 \times 10^{-3}}{P_{in}}$$

$$P_{in} = \frac{300 \times 10^{-3}}{0.25}$$

$$\therefore \quad P_{in} = 1.20 \text{ W}$$

EXAMPLE 23-4 Determine the voltage (V_o) developed across the load (1.0 kΩ) of the $+24$-dB amplifier shown in Figure 23-2.

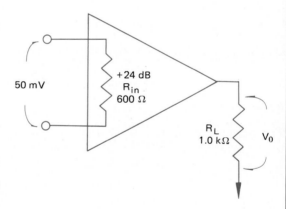

FIGURE 23–2

Circuit for Example 23–4. The amplifier has a 24-dB gain.

Solution Determine P_{in}:

$$P_{in} = E^2/R_{in}$$

$$= \frac{(50 \times 10^{-3})^2}{600}$$

$$P_{in} = 4.167 \ \mu\text{W}$$

Substitute into Equation 23-1 and solve for P_{out}:

$$N_{dB} = 10 \log \left(\frac{P_{out}}{P_{in}} \right)$$

$$24 = 10 \log \left(\frac{P_{out}}{4.167 \times 10^{-6}} \right)$$

$$2.4 = \log \left(\frac{P_{out}}{4.167 \times 10^{-6}} \right)$$

Take the antilog of each member:

$$251.2 = \frac{P_{out}}{4.167 \times 10^{-6}}$$
$$P_{out} = 1.047 \text{ mW}$$

Solve for V_o:

$$P_{out} = V_o^2/R_L$$
$$V_o = \sqrt{P_{out}R_L}$$
$$= \sqrt{(1.047 \times 10^{-3})(1 \times 10^3)}$$

$$\therefore \quad V_O = 1.02 \text{ V}$$

Power Gain or Loss in Decibels from Voltage Ratio

The power gain or loss in decibels may be computed from a voltage ratio if the measurements are made across equal resistances. Equation 23-1 is expressed as Equation 23-2 when $R_{in} = R_{out}$.

$$N_{dB} = 10 \log \left(\frac{P_{out}}{P_{in}} \right)$$

However,

$$P_{out} = E_{out}^2/R_{out} \text{ and } P_{in} = E_{in}^2/R_{in}$$

Thus,

$$N_{dB} = 10 \log \left(\frac{E_{out}^2/R_{out}}{E_{in}^2/R_{in}} \right)$$

Substitute $R_{out} = R_{in}$ and simplify:

$$N_{dB} = 10 \log \left(\frac{E_{out}^2/R_{out}}{E_{in}^2/R_{out}} \right)$$

$$= 10 \log \left[\left(\frac{E_{out}}{E_{in}} \right)^2 \right]$$

$$= 2(10) \log \left(\frac{E_{out}}{E_{in}} \right)$$

$$N_{dB} = 20 \log \left(\frac{E_{out}}{E_{in}} \right) \qquad dB \qquad\qquad (23\text{-}2)$$

EXAMPLE 23-5   Determine the decibel loss in a 50-Ω coaxial cable that has an input of 12 V. The output is 4 V across a 50-Ω termination.

Solution   Use Equation 23-2 because each resistance is 50 Ω:

$$N_{dB} = 20 \log (E_{out}/E_{in})$$
$$= 20 \log (4/12)$$

$$\therefore \qquad N_{dB} = -9.54 \ dB$$

EXERCISE 23-1

1. An audio amplifier develops 20 W in the output load from an input of 75 mW. Determine the power gain of the amplifier in decibels.
2. An RF linear amplifier requires 2.5-W input to develop 54-W output. Determine the power gain of the amplifier in decibels.
3. An equalizer has an insertion loss of 6 dB when placed in an audio line. Determine the output power when the input power is 1.5 W.
4. A low-pass filter has an attenuation of 40 dB at 50 MHz. Determine the output power at 50 MHz when the input power is 5 W.
5. The input to a preamp is 10 mV and the output is 2 V. Assuming that each voltage is developed across the same amount of resistance, determine the power gain of the preamp.
6. Determine the input voltage needed to develop 37.5 V across a resistance if the power gain of an amplifier is +37 dB. Both input and output resistances are equal.

7. Determine the amount of input voltage needed to develop 650 mW in a 16-Ω load when the power gain of the amplifier is +27 dB. The input resistance is 10 kΩ.

8. The input resistance to an amplifier is 1.2 kΩ, and the output resistance is 200 Ω. If the amplifier has a +70-dB gain, determine what input voltage is needed to develop 28 V in the output.

9. A ceramic cartridge develops 180 mV across 15-kΩ input resistance of an amplifier. The output power developed across an 8-Ω load is 24 W. Determine the decibel power gain of the amplifier.

10. Determine the voltage dropped across a 4-Ω load when a microphone develops 6 mV across 600 Ω. The gain of the amplifier is +36 dB.

23-2 SYSTEM CALCULATIONS

In the preceding section you learned that decibels (dB) are used to compare two power levels. When working with electronic systems, it is sometimes necessary to know the actual power or voltage of the system in relation to a stated reference. We will consider two ways of using decibels to state absolute values. One term, the decibel milliwatt (dBm), represents the actual power. The other term, the decibel millivolt (dBmV), represents the actual voltage.

Decibel Milliwatt (dBm)

Audio communication systems having input and output resistances of 600 Ω use a reference power of 1 mW as 0 dB. The unit decibel milliwatt (dBm) is used to indicate the actual power of an amplifier, an attenuator, or an entire system. The gain or loss in decibels based on 1 mW is computed by Equation 23-3.

$$N_{dBm} = 10 \log \left(\frac{P}{1 \text{ mW}} \right) \qquad \text{dBm} \qquad \qquad (23\text{-}3)$$

EXAMPLE 23-6 Determine the power represented by +20 dBm.
  Solution Substitute into Equation 23-3 and solve for P:

$$20 = 10 \log \left(\frac{P}{1 \text{ mW}} \right)$$

$$2 = \log \left(\frac{P}{1 \times 10^{-3}} \right)$$

Take the antilog:

$$100 = \frac{P}{1 \times 10^{-3}}$$
$$P = 100 \times 10^{-3}$$

$$\therefore \quad P = 100 \text{ mW}$$

From this example we see that a $+20$ dBm represents 100 mW. Because the decibel is logarithmic, the actual power present in an audio communication system may be determined by first algebraically adding the powers represented in decibel milliwatts and then solving Equation 23-3 for P.

A quick estimate of the power represented by N_{dBm} may be made with the information of Table 23-1.

TABLE 23-1

SIGNAL POWER COMPARED TO DECIBELS MILLIWATT

dBm	P/1 mW	Signal Power
+40	10 000	10 W
+30	1 000	1 W
+20	100	100 mW
+10	10	10 mW
+6	4	4 mW
+3	2	2 mW
0	1	1 mW
−3	0.5	0.5 mW
−6	0.25	0.25 mW
−10	0.10	100 μW
−20	0.01	10 μW
−30	0.001	1 μW
−40	0.000 1	0.1 μW

EXAMPLE 23-7 Determine the actual power in the audio system pictured in Figure 23-3.

Solution Algebraically add the gains and losses:

$$N_{dBm} = +15 - 6 + 18$$
$$N_{dBm} = +27 \text{ dBm}$$

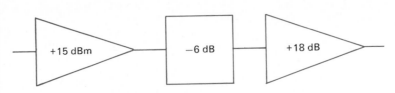

FIGURE 23-3
Audio system for Example 23–7.

Solve Equation 23-3 for P:

$$N_{\text{dBm}} = 10 \log \left(\frac{P}{1 \text{ mW}} \right)$$

$$27 = 10 \log \left(\frac{P}{1 \times 10^{-3}} \right)$$

$$2.7 = \log \left(\frac{P}{1 \times 10^{-3}} \right)$$

$$501.2 = \frac{P}{1 \times 10^{-3}}$$

$$P = 0.501 \text{ W}$$

$\therefore$ The output power is 0.501 W for the 1 mW of input.

Decibel Millivolt (dBmV)

Another special application of the decibel is in community antenna television (CATV) for the measurement of the signal intensity. In this system, a signal of 1 mV across 75 Ω is the reference level that corresponds to 0 dB. The unit decibel millivolt (dBmV) is used to indicate the actual voltage of an amplifier, an attenuator, or an entire system. The actual gain or loss in dBmV is computed by Equation 23-4.

$$N_{\text{dBmV}} = 20 \log \left(\frac{V}{1 \text{ mV}} \right) \text{ dBmV} \tag{23-4}$$

EXAMPLE 23-8 Express a signal strength of 100 μV in dBmV.

Solution Use Equation 23-4:

$$N_{\text{dBmV}} = 20 \log \left(\frac{V}{1 \text{ mV}} \right)$$

$$= 20 \log \left(\frac{100 \times 10^{-6}}{1 \times 10^{-3}} \right)$$

$$N_{\text{dBmV}} = -20 \text{ dBmV}$$

Table 23-2 provides information to make a quick estimate of the actual voltage represented by N_{dBmV}.

TABLE 23-2

SIGNAL VOLTAGE COMPARED TO DECIBELS MILLIVOLT

dBmV	V/1 mV	Signal Voltage	dBmV	V/1 mB	Signal Voltage
+60	1 000	1 V	−3	0.7	0.7 mV
+40	100	100 mV	−6	0.5	0.5 mV
+20	10	10 mV	−20	0.1	100 μV
+6	2	2 mV	−40	0.01	10 μV
+3	1.4	1.4 mV	−60	0.001	1 μV
0	1	1 mV			

EXAMPLE 23-9 Determine the actual signal level across a resistance of 75 Ω for a power of +25 dBmV.

Solution Substitute the given information into Equation 23-4:

$$25 = 20 \log \left(\frac{V}{1 \text{ mV}} \right)$$

$$1.25 = \log \left(\frac{V}{1 \times 10^{-3}} \right)$$

Take the antilog of each member:

$$17.8 = \frac{V}{1 \times 10^{-3}}$$

$$V = 17.8 \text{ mV}$$

EXAMPLE 23-10 Determine the actual signal level in millivolts (mV) across the connecting terminals of the television set of

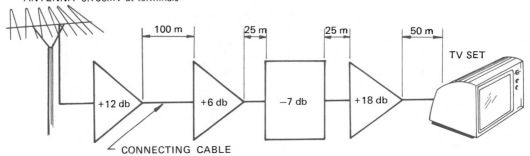

ANTENNA 0.700mV at terminals

100 m 25 m 25 m 50 m

TV SET

+12 db   +6 db   −7 db   +18 db

CONNECTING CABLE

FIGURE   23–4
Simplified distribution system for Example 23–10.

Figure 23-4. The signal at the antenna terminals is 700 μV, and the connecting cable has an attenuation (loss) of 9 dB/100 m. The resistance of the system is the same throughout.

Solution   State the incoming signal of 700 μV in dBmV:

$$N_{\text{dBmV}} = 20 \log \left(\frac{700 \times 10^{-6}}{1 \times 10^{-3}} \right)$$

$$= -3 \text{ dBmV}$$

Compute the cable loss at 9 dB/100 m:

$$100 \text{ m} + 25 \text{ m} + 25 \text{ m} + 50 \text{ m} = 200 \text{ m}$$
$$200 \text{ m} \, (-9 \text{ dB/100 m}) = -18 \text{ dB}$$

Compute the system gain in decibel millivolts:

$$-3 \text{ dBmV} - 18 \text{ dB} + 12 \text{ dB} + 6 \text{ dB}$$
$$- 7 \text{ dB} + 18 \text{ dB} = +8 \text{ dBmV}$$

Compute the signal level across the connecting terminals of the television. Use Equation 23-4:

$$N_{\text{dBmV}} = 20 \log \left(\frac{V}{1 \text{ mV}} \right)$$

$$8 = 20 \log \left(\frac{V}{1 \times 10^{-3}} \right)$$

$$0.40 = \log\left(\frac{V}{1 \times 10^{-3}}\right)$$

$$2.51 = \frac{V}{1 \times 10^{-3}}$$

$$V = 2.51 \times 10^{-3}$$

$$\therefore \quad V = 2.51 \text{ mV}$$

Summary

- The actual signal level may be determined in an electronic system by referencing decibel (dB) gains and decibel (dB) losses in the system to a definite amount of voltage or power.
- The overall gain or loss in a system may be determined by algebraically adding the gains and losses.
- Gains or losses in decibel milliwatts (dBm) and decibel millivolts (dBmV) are not used together to describe the same system.

EXERCISE 23-2

1. Express each of the following power levels in decibel milliwatts.

   (a) 15 mW   (b) 200 mW   (c) 18 μW
   (d) 2.2 W   (e) 610 μW   (f) 87 mW

2. What is the power level in watts (W) represented by the following?

   (a) +17 dBm   (b) −7 dBm   (c) +30.5 dBm
   (d) −5.6 dBm   (e) −18 dBm   (f) +42 dBm

3. Express each of the following voltage levels in decibel millivolts.

   (a) 0.32 V   (b) 150 μV   (c) 3.8 V
   (d) 72 mV   (e) 840 mV   (f) 27 μV

4. What is the signal level in millivolts represented by the following?

   (a) +23 dBmV   (b) +2.8 dBmV   (c) −6.9 dBmV
   (d) −10.8 dBmV   (e) +11 dBmV   (f) −1.4 dBmV

5. A certain 75-Ω antenna develops a signal of 4700 μV. Express this signal level in dBmV.

6. Determine the specifications in decibel milliwatts of an amplifier that has an output power of 8 W.

7. Determine the specifications in decibel millivolts of a CATV system that provides an output of 8.7 mV of signal to the subscriber's set.

8. A certain amplifier has an output of 290 mV across 600 Ω. Express this signal level in dBm.

9. Determine the actual voltage level across a 600-Ω load in an audio system that has a preamp with an output power of +28 dBm connected through a matching transformer of −4 dB to an amplifier with a gain of +18 dB. The output of the amplifier is connected through a −6-dB matching pad to the 600-Ω load.

10. Specify the gain needed in decibels to provide 1500 μV of signal to the 75 Ω input of an FM receiver if the signal at the antenna is 100 μV. The 75 Ω antenna is connected to the receiver by 50 m of 75-Ω coaxial cable having a loss of 4 dB/100 m.

23-3 *RC* AND *RL* TRANSIENT BEHAVIOR

RL Transient Behavior

When the switch of Figure 23-5 is closed, the current through the resistance goes instantly from 0 to 2 A. The graph of current as a function of time verifies this instantaneous change in current. However, when an inductor (coil) is

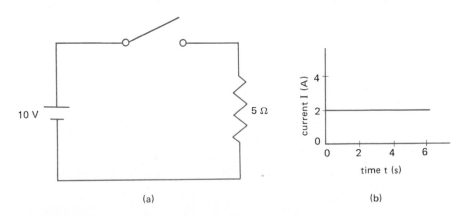

(a)   (b)

FIGURE   23–5
(a) The current in a resistive circuit changes from 0 to 2 A when the switch is closed; (b) the graph of *I* as a function of time.

CHAPTER 23   APPLIC. OF LOG. AND EXPONENTIAL EQS. TO ELECTRONIC CONCEPTS

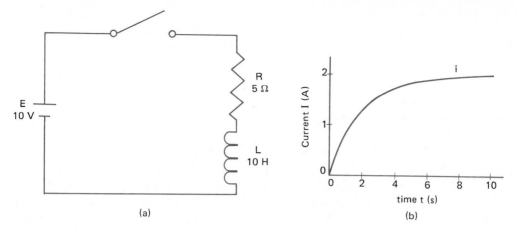

FIGURE 23–6
Current in an *RL* circuit: circuit (a) has a transient behavior as shown in (b).

added in series with the resistance of Figure 23-5, the current no longer changes instantaneously. Figure 23-6 shows how the inductance (measured in henries, H) of the inductor changes the characteristic of the current. We see that it takes some time for the current to come to a *steady state*. The increase in the circuit current is a *transient behavior*. The graph of Figure 23-6 is not linear, but is an exponential function. During the transient time, the current in the circuit is determined with Equation 23-5.

$$i = \frac{E}{R}[1 - e^{-t(L/R)}] \qquad\qquad (23\text{-}5)$$

where i = circuit instantaneous current in amperes at time t
      t = time in seconds, after switch is closed
      e = base of the natural logarithms
      E = source voltage in volts
      R = circuit resistance in ohms
      L = circuit inductance in henries

EXAMPLE 23-11
Determine the instantaneous current in the circuit of Figure 23-6 two seconds after the switch is closed.

Solution   Substitute $R = 5\ \Omega$, $L = 10$ H, $t = 2$ s, and $E = 10$ V into Equation 23-5:

$$i = \frac{E}{R}[1 - e^{-t/(L/R)}]$$

$$= \frac{10}{5} [1 - e^{-2/(10/5)}]$$
$$= 2(1 - 0.3679)$$

$$\therefore \quad i = 1.26 \text{ A}$$

**EXAMPLE
23-12**

In the circuit of Figure 23-6, determine the instantaneous current 3.5 s after the switch is closed.

Solution   Substitute into Equation 23-5:

$$i = \frac{10}{5} [1 - e^{-3.5/(10/5)}]$$
$$= 2(1 - 0.1738)$$

$$\therefore \quad i = 1.65 \text{ A}$$

Table 23-3 is a summary of the exponential equations for the *charging* of an *RL* series circuit. The equations in Table 23-3 refer to the circuit in Figure 23-7.

**EXAMPLE
23-13**

Determine the inductance in henries (H) of the inductor of Figure 23-7 if the instantaneous current is 500 mA 3 seconds after the closing of the switch. $E = 12$ V and $R = 8 \ \Omega$.

Solution   Let $x = -t/(L/R)$ in Equation 23-5:

$$i = \frac{E}{R} (1 - e^x)$$

Solve for x:

$$\frac{iR}{E} = 1 - e^x$$

$$e^x = 1 - \frac{iR}{E}$$

TABLE 23-3

THE *RL* EXPONENTIAL EQUATIONS FOR SERIES CIRCUITS UNDER CHARGE

Instantaneous
circuit current

$$i = \frac{E}{R}(1 - e^{-t/(L/R)})$$

(23-5)

Instantaneous
voltage across
inductance

$$v_L = Ee^{-t/(L/R)}$$

(23-6)

Instantaneous
voltage across
resistance

$$v_R = E(1 - e^{-t/(L/R)})$$

(23-7)

where i = circuit instantaneous current in amperes at time t
  t = time in seconds, after switch is closed
  e = base of the natural logarithms
  E = source voltage in volts
  R = circuit resistance in ohms
  L = circuit inductance in henries

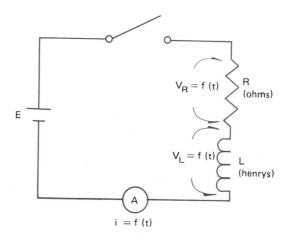

FIGURE 23-7
Series *RL* circuit for Equations 23-5
through 23-7.

$$\ln(e^x) = \ln\left(1 - \frac{iR}{E}\right)$$

$$x = \ln\left(1 - \frac{0.5(8)}{12}\right)$$

$$x = -0.4055$$

Substitute $-t/(L/R)$ for x:

$$-t/(L/R) = -0.4055$$

Solve for L:

$$t = 0.4055\,(L/R)$$

$$\frac{tR}{0.4055} = L$$

$$L = \frac{3(8)}{0.4055}$$

$$\therefore \quad L = 59.2\ H$$

EXERCISE 23-3

1. Determine the instantaneous current in the circuit of Figure 23-7 two seconds after the switch is closed. $R = 2\ \Omega$, $L = 5\ H$, and $E = 30\ V$.
2. Repeat problem 1 for $R = 1\ \Omega$.
3. For the circuit of Figure 23-7:
   (a) Find the instantaneous current (i) in the circuit 5 ms after the switch is closed. $R = 200\ \Omega$, $L = 500\ mH$, and $E = 50\ V$.
   (b) Repeat part (a) for v_L.
   (c) Repeat part (a) for v_R.
4. An inductor of 12 H is connected in series with a 90-V dc source. What is the instantaneous current in the inductor 0.2 s after the source is connected? The internal resistance of the inductor is 50 Ω.

Determine the values missing in the following table. Use Equations 23-5 through 23-7 and Figure 23-7.

	E	R	L	t	i	v_R	v_L
5.	18 V	56 Ω	1 H	35 ms	?	?	?
6.	36 V	150 Ω	250 mH	800 μs	?	?	?
7.	10 V	8 Ω	?	2.5 s	?	?	3 V
8.	20 V	4 Ω	?	1 s	?	12 V	?

RC Transient Behavior

When a capacitance is connected in series with a resistance and a dc source of voltage (as shown in Figure 23-8), the instantaneous current is not linear, but is an exponential function. The *transient behavior* of the circuit of Figure 23-8 is described mathematically by Equations 23-8 through 23-10 in Table 23-4.

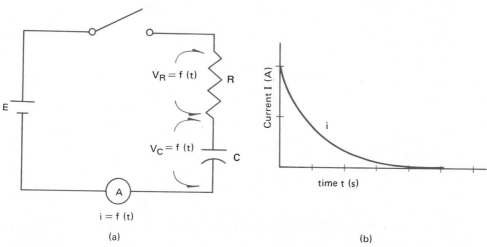

FIGURE 23-8
Current in an *RC* circuit: circuit (a) has a transient behavior as shown in (b).

EXAMPLE 23-14

In the circuit of Figure 23-8(a), determine the instantaneous voltage across the capacitor 10 s after the switch is closed. $R = 100$ kΩ, $C = 50$ μF, and $E = 100$ V.

Solution Use Equation 23-9 from Table 23-4:

$$v_C = E(1 - e^{-t/RC})$$
$$= 100[1 - e^{-10/(100 \times 10^3 \times 50 \times 10^{-6})}]$$

$$\therefore \quad v_C = 86.5 \text{ V}$$

EXAMPLE 23-15

Determine how long it will take for the voltage across the capacitor of Figure 23-8(a) to reach 90% of the

---

TABLE 23-4

THE *RC* EXPONENTIAL EQUATIONS FOR SERIES CIRCUITS UNDER CHARGE

Instantaneous circuit current

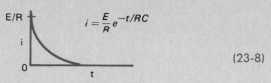

$$i = \frac{E}{R}e^{-t/RC}$$

(23-8)

Instantaneous voltage across capacitance

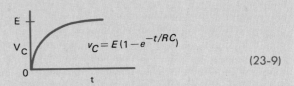

$$v_C = E(1 - e^{-t/RC})$$

(23-9)

Instantaneous voltage across resistance

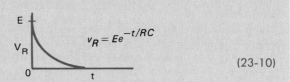

$$v_R = Ee^{-t/RC}$$

(23-10)

where i = circuit instantaneous current in amperes at time t
t = time in seconds, after switch is closed
e = base of the natural logarithms
E = source voltage in volts
R = circuit resistance in ohms
C = circuit capacitance in farads
v_C = instantaneous voltage across the capacitance
v_R = instantaneous voltage across the resistance

source voltage once the switch is closed. $R = 15 \text{ k}\Omega$, $C = 10 \text{ }\mu\text{F}$, and $E = 30 \text{ V}$.

Solution Let $x = -t/RC$ in Equation 23-9:

$$\begin{cases} v_C = E(1 - e^x) \\ v_C = 0.9E \end{cases}$$

Solve for x by eliminating v_C:

$$0.9E = E(1 - e^x)$$
$$e^x = 1 - 0.9$$
$$x = \ln(0.1)$$
$$x = -2.303$$

Substitute $-t/RC$ for x:

$$-t/RC = -2.303$$

Solve for t:

$$t = 2.303RC$$
$$= 2.303 \times 15 \times 10^3 \times 10 \times 10^{-6}$$

$$\therefore \qquad t = 0.346 \text{ s}$$

Summary

Before going on to the next exercise, take a moment to look over Tables 23-3 and 23-4. Compare the equation forms to their exponential curves. Here you will see that quantities (voltage or current) that *rise,* as in Equations 23-5, 23-7, and 23-9, have a general equation form of

$$y = Y_{mx}(1 - e^{-t/\tau}) \tag{23-11}$$

Quantities that *fall,* as in Equations 23-6, 23-8, and 23-10, have a general equation form of

$$y = Y_{mx}e^{-t/\tau} \tag{23-12}$$

In each of these general equations, y represents the instantaneous quantity (i if current or e if voltage). Y_{mx} represents the maximum value of source voltage (E) or source current ($I = E/R$). The t represents time in seconds after the start of charge or discharge. The Greek letter tau, τ, is used to represent RC ($\tau = RC$) in a capacitive circuit or L/R ($\tau = L/R$) in an inductive circuit. By memorizing these general forms, you will be able to recall the appropriate equation for rise or fall of a given quantity in a series RC or RL circuit.

EXERCISE 23-4

1. A 10-μF capacitor is charged through a 47-kΩ resistor by a 36-V source. Determine the instantaneous current in the circuit after 2 s.
2. Determine the voltage across the capacitor (v_C) of problem 1 after 0.8 s.
3. Determine the voltage across the resistor of problem 1 after 1.5 s.

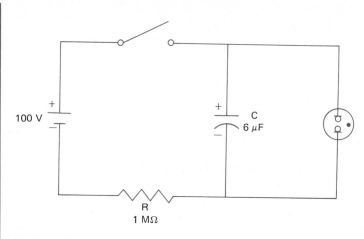

FIGURE 23-9
Circuit for Exercise 23-4, problem 4.

4. How long will it be after the switch of Figure 23-9 is closed before the neon lamp will flash? The lamp flashes when there is 80 V across it.

Determine the values missing in the following table. Use Equations 23-8 through 23-10 and Figure 23-8.

	E	R	C	t	i	v_R	v_C
5.	25 V	10 kΩ	470 μF	?	?	18 V	?
6.	10 V	68 kΩ	?	30 ms	?	?	4 V
7.	100 V	100 kΩ	0.01 μF	1.8 ms	?	?	?
8.	36 V	820 kΩ	1 μF	?	23.9 μA	?	?

SELECTED TERMS

amplifier A device used to increase the magnitude of an electric signal.

attenuator A resistive network that reduces the magnitude of an electric signal.

decibel A logarithmic expression that compares two power levels. The unit used to express gain, loss, and relative power levels.

instantaneous value The value of current or voltage at a particular instant of time.

transient A changing action occurring in an electric circuit during the time between the initial application of power and the settling to a steady-state condition.

24

ANGLES AND TRIANGLES

This chapter deals with geometric concepts. Many of the ideas presented will be reinforced by drawings. You will make better progress if you make your own drawings for these ideas.

24-1 POINTS, LINES, AND ANGLES

The first geometric concept that we will consider is the *point*. We have already used points when we were constructing graphs. In this chapter, capital letters are used to designate points. Thus, A and O represent points in Figure 24-1.

The second geometric concept to be considered is the *line segment*. A line segment is the portion of a straight line connecting two points. Figure 24-2 presents a line segment joining the points A and B. The symbol $\overline{AB}$ is used to denote the line segment connecting points A and B.

Building on the concepts of points and line segments, we can define the concept of an angle. An *angle* is the geometric figure formed when one end point of a line segment is held fixed while the second end point is moved to a new location without changing the length of the line segment. Thus, in Figure 24-3, the figure represented by $\overline{OA}$ and $\overline{OA'}$ is the angle "swept out" by $\overline{OA}$ as A moves to A', while O is held fixed. The symbol $\angle AOA'$ is used for the angle AOA'. The fixed end point, O, is called the *vertex* of the angle.

Like other mathematical quantities, angles have size and direction. One unit of size is the *revolution*. The angle swept out by moving one end point all the way around the fixed end point and back to its original position is one revolution in size. This is illustrated in Figure 24-4. The abbreviation for revolution is *rev*.

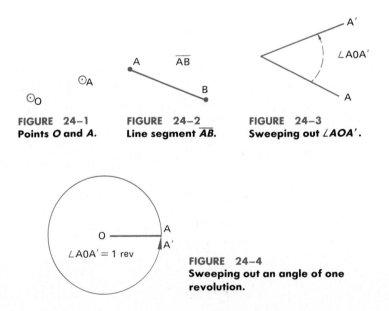

FIGURE 24–1
Points *O* and *A.*

FIGURE 24–2
Line segment $\overline{AB}$.

FIGURE 24–3
Sweeping out $\angle AOA'$.

FIGURE 24–4
Sweeping out an angle of one revolution.

EXAMPLE 24-1 How many revolutions are swept out by a second hand in 2 minutes 45 seconds?

Solution Each minute the second hand sweeps out an angle of one revolution. Let n represent the number of revolutions.

$$n = (2 \text{ min } 45 \text{ s})(1 \text{ rev/min})$$

Converting seconds to minutes:

$$n = (2.75 \text{ min}) \left(\frac{1 \text{ rev}}{\text{min}} \right) = 2.75 \text{ rev}$$

∴ The second hand sweeps out 2.75 rev in 2 min 45 s.

EXAMPLE 24-2 Engine speed is measured in revolutions per minute (rev/min). If an automobile engine is operated at 5500 rev/min for 6 min, how many revolutions will have been swept out by the crankshaft?

Solution Let n be the number of revolutions:

$$n = 5500 \times 6 \left(\frac{\text{rev}}{\text{min}} \right) (\text{min})$$

∴ $n = 33\ 000 \text{ rev}$

Often we want to measure fractions of revolutions. For this a common unit of measure is the degree. One revolution is equal to 360 *degrees,* or 360°, where the superscript ° stands for degree. The following example explores the relation between revolutions and degrees.

EXAMPLE 24-3 What angle size in degrees is swept out by a second hand in 20 s?

Solution Let n equal the number of revolutions:

$$n = (20 \text{ s})(1 \text{ rev/min})(1 \text{ min/60 s})$$
$$= \tfrac{20}{60} \text{ rev}$$
$$n = \tfrac{1}{3} \text{ rev}$$

Multiply by 360 to convert to degrees:

$$n = \tfrac{1}{3} \text{rev} \, \frac{360°}{1 \text{ rev}}$$

$$\therefore \quad n = 120°$$

A second unit of size for measuring fractions of revolutions is the *radian*. The distance around a circle is 2π times the length of the radius (see Figure 24-5). In like manner, the angle swept out in one revolution is 2π radians, or $2\pi^r$ where the superscript ^{r} stands for radians. While the abbreviation for radians is rad, it is not proper to use any unit indication with angles measured in radians. We introduce the superscript ^{r} in this chapter as a convenience for indicating radians unless the angle contains the symbol π. Later, when the student can tell from the context of a problem that a number without unit is intended to be an angle in radians, we will drop this practice.

FIGURE 24–5
Geometric properties of radians. (a) The circumference of a circle is related to the radians in one revolution. (b) When the arc length *BB'* is equal to the sides *OB* and *OB'*, the angle is one radian.

Although degrees are *"nicer"* numbers than radians, advanced mathematics and engineering frequently use radians. So we will practice with both units. Table 24-1 shows conversion factors for revolutions, degrees, and radians. Remember, the number π is approximately 3.14159.

EXAMPLE 24-4   How many degrees are in one radian?
Solution   Start with 1 rev = 360° and 1 rev = $2\pi^r$; solve for 1^r:

$$2\pi = 360$$
$$1 = 360/2\pi$$
$$\therefore \quad 1^r \approx 57.3°$$

TABLE 24-1

CONVERSION FACTORS FOR REVOLUTIONS, DEGREES, AND RADIANS			
Revolutions	*Radians (Using π)*	*Radians (Decimal)*	*Degrees*
1	2π	6.28	360
0.5	π	3.14	180
0.159	1	1	57.3
0.002 78	$\pi/180$	0.0175	1

EXAMPLE 24-5   Convert 75.2° to radians.

Solution   Look up 1° in Table 24-1; multiply 75.2° by the corresponding value under radians (0.0175):

$$75.2 = 75.2(0.0175)$$
$$\therefore \quad 75.2° = 1.316^r$$

EXAMPLE 24-6   Convert 5.25^r to revolutions.

Solution   Look up 1^r in Table 24-1 and multiply 5.25^r by the corresponding value under revolutions (0.159):

$$5.25 = 5.25(0.159)$$
$$\therefore \quad 5.25^r = 0.835 \text{ rev}$$

Observation   Some calculators have key strokes that will convert degrees to radians and radians to degrees. Check your owner's guide to see if your calculator has this feature.

EXERCISE 24-1

1. How many revolutions does a second hand sweep out in 1 h?
2. The earth spins on its axis at the rate of 1 rev/day. How many revolutions does it sweep out in 60 h?
3. What angle in revolutions is swept out by a second hand during 48 s? In degrees?
4. How many revolutions does a turntable make while playing a 3-min 15-s recording on a 45-RPM record.

Convert the following angles to degrees. (Remember that angles containing π are in radians.)

5. $\frac{1}{4}$ rev   6. 2.5ʳ   7. 1.76π 8. 0.15 rev

9. 1.24 rev   10. 0.81 rev   11. 0.45ʳ   12. 5.75ʳ

Convert the following angles to revolutions.

13. 300°   14. 544°   15. 4.8π 16. 172π

17. 29.6ʳ   18. 31.4ʳ   19. 5.27π 20. 1700°

Convert the following angles to decimal radians.

21. 114.6°   22. 29.3°   23. 0.75 rev   24. 1.54 rev

25. 175°   26. 1.4π 27. 0.75π 28. $\pi/3$

24-2 SPECIAL ANGLES

Several angles are so important that they have their own names. The first special angle is 180° or $\frac{1}{2}$ rev. This angle, which is a straight line, is called a *straight angle*. Angle β in Figure 24-6 is a straight angle. Two angles that sum to a straight angle are called *supplementary angles*. In Figure 24-7, $\angle\theta$ and $\angle\phi$ are supplementary.

The second special angle is 90° or $\frac{1}{4}$ rev. This angle is called a *right angle*. The special symbol └ is used to indicate a right angle, as in Figure 24-8. Two angles that sum to a right angle are called *complementary angles*. In Figure 24-9, $\angle\alpha$ and $\angle\beta$ are complementary.

Other angles may be classified in terms of straight angles and right angles. If an angle is between a straight angle and a right angle, it is called an *obtuse angle*. If an angle is between zero and a right angle, it is called an *acute*

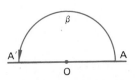

FIGURE 24-6
An angle of 180° is a straight angle.

FIGURE 24-7
Two supplementary angles, θ and ϕ, sum to a straight angle.

SYMBOL FOR
RIGHT ANGLE

FIGURE 24-8
An angle of 90° is a right angle.

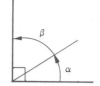

FIGURE 24–9
**The complementary angles α and β
sum to a right angle.**

angle. More special angles are presented in Table 24-2. The size of each angle is given in revolutions, degrees, and radians.

TABLE 24-2

FRACTIONAL PARTS OF A REVOLUTION			
Revolutions	**Radians (Using π)**	**Radians (Decimal)**	**Degrees**
0	0	0	0
$\frac{1}{12}$	π/6	0.524	30
$\frac{1}{8}$	π/4	0.785	45
$\frac{1}{6}$	π/3	1.05	60
$\frac{1}{4}$	π/2	1.57	90
$\frac{1}{2}$	π	3.14	180
$\frac{3}{4}$	3π/2	4.71	270
1	2π	6.28	360

Most devices used to measure angles are similar in operation to the protractor pictured in Figure 24-10. Almost all such devices measure angles in degrees.

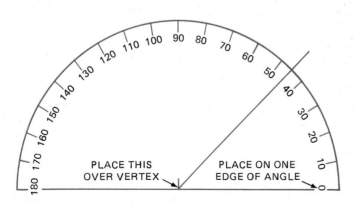

PLACE THIS
OVER VERTEX

PLACE ON ONE
EDGE OF ANGLE

FIGURE 24–10
Protractor measuring an angle.

EXERCISE 24-2

Decide whether the following pairs of angles are supplementary, complementary, or neither.

1. 33°, 57°	2. 0.374 rev, 0.125 rev
3. 135°, 45°	4. 75°, 85°
5. $\pi/3$, $\pi/6$	6. 0.125 rev, 0.125 rev
7. $\frac{3}{8}$ rev, $\frac{1}{4}$ rev	8. 1.07ʳ, 0.50ʳ
9. 2.00ʳ, 1.14ʳ	10. 0.65π, 0.45π

Find the supplement of the following angles.

11. 144°	12. 37.5°	13. 80.7°
14. 105°	15. 90°	16. 120°
17. 3.01ʳ	18. 1.57ʳ	19. 2.61ʳ

Find the complement of the following angles.

20. 38°	21. 57.3°	22. 22.1°
23. 45°	24. 83.4°	25. 49.9°
26. 0.785ʳ	27. 1.07ʳ	28. 0.52ʳ

24-3 TRIANGLES

The fourth geometric concept is the triangle. A triangle is a figure formed by
joining three line segments end point to end point as in Figure 24-11. A triangle has three angles, three vertices, and three sides. In Figure 24-11(a), the
vertices are labeled A, B, and C. The angles are labeled α, β, and γ. The sides
are not labeled in Figure 24-11(a). We can use the line segment notation to

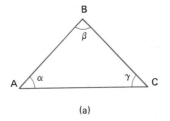

(a)

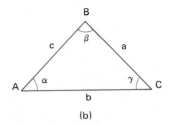

(b)

FIGURE 24–11
**Labeling of the parts of a triangle. The first three letters of the Greek
alphabet are α, β, and γ. We use α for the angle opposite side a, β for
the angle opposite side b, and γ for the angle opposite side c.**

refer to them as $\overline{AB}$, $\overline{BC}$, and $\overline{AC}$. This notation is very cumbersome, so we will use the notation of Figure 24-11(b), where the sides are labeled as *a, b,* and *c*. Notice that side *a* is opposite vertex *A*, side *b* is opposite vertex *B*, and side *c* is opposite vertex *C*.

A property of triangles that can help you to check your work is stated as Rule 24-1.

RULE 24-1. RELATIVE SIZE OF THE SIDES OF A TRIANGLE

The longest side is opposite the largest angle.
The shortest side is opposite the smallest angle.

EXAMPLE 24-7   Examine the triangle in Figure 24-12 to determine the smallest angle and the shortest side.

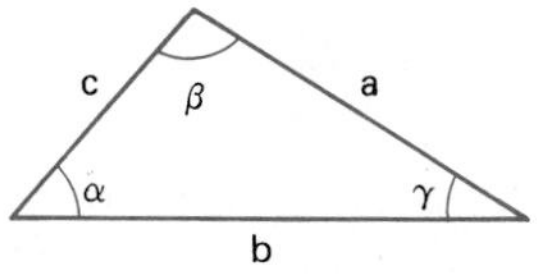

FIGURE   24–12
Triangle for Example 24–7.

Solution   By inspection, the smallest angle is $\angle\gamma$, and the shortest side is *c*.

A remarkable property of triangles is that the sum of the angles is a straight angle. This fact can be used to find the size of the third angle if two angles are known. The procedure for finding the third angle is presented as Rule 24-2.

RULE 24-2. FINDING THE THIRD ANGLE OF A TRIANGLE

To find the third angle of a triangle:

1. Form the sum of the first two angles.
2. The third angle is the supplement of the sum.

EXAMPLE 24-8   Two angles of a triangle are 35° and 47°; what is the third angle?

Solution   Use Rule 24-2:
Step 1:   Form the sum:

$$35° + 47° = 82°$$

Step 2:   The supplement of 82° is 180° − 82°, which is 98°.

∴   The third angle is 98°.

EXAMPLE 24-9   In a triangle, $\angle\alpha = 1.2^r$ and $\angle\beta = 0.9^r$; find $\angle\gamma$.
Solution   Use Rule 24-2.
Step 1:   Form the sum:

$$1.2 + 0.9 = 2.1$$

Step 2:   Find the supplement:

$$\angle\gamma = 3.14 - 2.1$$

$$\therefore \quad \angle\gamma = 1.04^r$$

EXERCISE 24-3
Using the notation of Figure 24-11(b), find the missing angle and determine which is the longest side.

1. $\angle\alpha = 57°, \angle\beta = 22°$ 2. $\angle\beta = 51°, \angle\gamma = 78°$

3. $\angle\alpha = 101°, \angle\gamma = 30°$ 4. $\angle\alpha = 38°, \angle\beta = 90°$

5. $\angle\alpha = 29.5°, \angle\beta = 50.5°$ 6. $\angle\alpha = 115°, \angle\gamma = 15.4°$

7. $\angle\alpha = 1.5^r, \angle\beta = 0.5^r$ 8. $\angle\alpha = 1.57^r, \angle\beta = 0.52^r$

9. $\angle\beta = 1^r, \angle\gamma = 1^r$ 10. $\angle\alpha = 1.21^r, \angle\beta = 1^r$

11. $\angle\alpha = \pi/2, \angle\gamma = \pi/4$ 12. $\angle\beta = \pi/3, \angle\gamma = \pi/6$

13. $\angle\alpha = \pi/3, \angle\beta = \pi/3$ 14. $\angle\alpha = \pi/2, \angle\gamma = \pi/6$

15. $\angle\alpha = 57.3°, \angle\beta = 57.3°$ 16. $\angle\beta = 90°, \angle\gamma = 85°$

17. $\angle\alpha = 79.5°, \angle\gamma = 79.6°$ 18. $\angle\alpha = 70°, \angle\beta = 42°$

19. $\angle\beta = 45°, \angle\gamma = 45°$ 20. $\angle\alpha = 15.2°, \angle\gamma = 15.2°$

24-4 RIGHT TRIANGLES AND THE PYTHAGOREAN THEOREM

Triangles containing a right angle make up an important class of triangles. A triangle containing a right angle is called a *right triangle.* The side opposite the right angle is called the *hypotenuse,* as shown in Figure 24-13.

The ancient Greek mathematician Pythagoras is given credit for discovering that, if the lengths of any two sides of a right triangle are known, the length of the third side can be easily determined. This relation, known as the Pythagorean theorem, is stated as Rule 24-3.

FIGURE 24–13
Parts of a right triangle. The side opposite the right angle is called the hypotenuse.

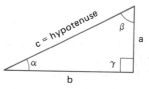

RULE 24-3. PYTHAGOREAN THEOREM

In every right triangle, the square of the hypotenuse equals the sum of the squares of the other two sides. In Figure 24-13,

$$c^2 = a^2 + b^2$$

EXAMPLE 24-10

Find the length of the hypotenuse in a right triangle if the other two sides are 3 m and 4 m.

Solution Use Rule 24-3. Substitute 3 for a and 4 for b, then solve for c:

$$c^2 = a^2 + b^2$$
$$= 3^2 + 4^2$$
$$= 9 + 16$$
$$\boxed{\sqrt{x}} \quad c^2 = 25$$
$$\therefore \quad c = 5 \text{ m}$$

EXAMPLE 24-11

If the hypotenuse is 27 m and a second side is 17 m, how long is the third side in a right triangle?

Solution   Use Rule 24-3. Substitute 27 for c and 17 for a; solve for b:

$$c^2 = a^2 + b^2$$
$$27^2 = 17^2 + b^2$$
$$b^2 = 729 - 289$$
$$\boxed{\sqrt{x}} \quad b^2 = 440$$
$$\therefore \quad b = 21 \text{ m}$$

The angle γ is a right angle in the right triangle of Figure 24-13. We know that the sum of $\angle\alpha$ and $\angle\beta$ is the supplement of $\angle\gamma$. Thus, the sum of $\angle\alpha$ and $\angle\beta$ is a right angle. And $\angle\alpha$ and $\angle\beta$ are complements. This important fact is restated as Rule 24-4.

RULE 24-4. SUM OF TWO ACUTE ANGLES IN A RIGHT TRIANGLE

The sum of the two acute angles in a right triangle is a right angle. In Figure 24-13, $\angle\alpha$ and $\angle\beta$ are complements, and

$$\alpha + \beta = 90°$$

EXAMPLE 24-12   Find $\angle\beta$ when $\angle\alpha = 37°$ in a right triangle.

Solution   Use Rule 24-4:

$$\alpha + \beta = 90$$
$$37 + \beta = 90$$
$$\therefore \quad \beta = 53°$$

EXERCISE 24-4
Find the missing sides and angles in the following triangles. Use the notation of Figure 24-13.

1.	$a = 1.5$ m	$b = 2.0$ m	$\alpha = 36.9°$
2.	$a = 13$ m	$b = 13$ m	$\alpha = 45°$
3.	$c = 39$ m	$a = 16$ m	$\beta = 65.8°$
4.	$c = 109$ m	$b = 74$ m	$\beta = 43.5°$
5.	$a = 78.5$ m	$b = 44.2$ m	$\beta = 29.4°$

6. $a = 0.152$ m $\quad c = 0.305$ m $\quad \alpha = 29.9°$
7. $b = 2.59$ m $\quad c = 4.00$ m $\quad \alpha = 49.6°$
8. $a = 7$ m $\quad\quad b = 24$ m $\quad\quad \alpha = 16.3°$
9. $a = 60$ m $\quad\quad c = 61$ m $\quad\quad \alpha = 79.6°$
10. $b = 12$ m $\quad\quad c = 13$ m $\quad\quad \beta = 67.4°$

24-5 SIMILAR TRIANGLES; TRIGONOMETRIC FUNCTIONS

Similar Triangles

The two triangles in Figure 24-14 have the same proportions. The ratios of the corresponding sides, a'/a, b'/b, and c'/c, are all equal. When these three ratios are equal, the two triangles are said to be *similar*. If two triangles are similar, their angles are the same size. Thus, in Figure 24-14, $\angle\alpha = \angle\alpha'$, $\angle\beta = \angle\beta'$, and $\angle\gamma = \angle\gamma'$.

In two similar triangles, the ratio of two sides in one triangle is equal to the ratio of the corresponding sides in the second triangle. Thus, in Figure 24-14, $a/b = a'/b'$, $a/c = a'/c'$, and $b/c = b'/c'$. These ratios become the function of one of the acute angles in a right triangle. This is because in a right triangle we know that one angle is a right angle, and knowing a second angle gives us enough information to determine the third angle. Knowing all three angles, we know the proportions of the sides. Thus, the ratios a/c, b/c, and a/b can be considered functions of the angle.

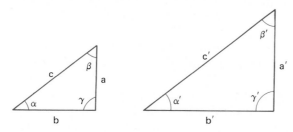

FIGURE 24–14
Similar Triangles have equal angles: $\angle\alpha = \angle\alpha'$, $\angle\beta = \angle\beta'$, and $\angle\gamma = \angle\gamma'$.

Trigonometric Functions

In a right triangle we define the *sine function* of an angle as the *ratio* of the length of the **side opposite** the angle **to** the length of the **hypotenuse.** It is customary to use the symbol sin ( ) to indicate the sine function. Thus, in Figure 24-15, sin (α) = side opposite/hypotenuse = a/c. Sin (α) is read "sine of alpha." Remember that the parentheses indicate function, not multiplication.

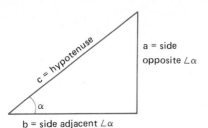

a = side
opposite ∠α

b = side adjacent ∠α

FIGURE 24–15
**Right triangle used in definition of
the trigonometric functions.**

Check your calculator and your owner's guide for how to calculate the sine function. Look for a key like $\boxed{\sin x}$ or $\boxed{\text{SIN}}$. Pay special attention to how you tell your calculator whether the angle is in degrees or in radians. Two systems are in common use. The first is a positional switch with which you can select degrees or radians. The second is an indicator light and key stroke ($\boxed{\text{rad}}$ or $\boxed{\text{DRG}}$). The indicator light tells you in which system the calculator is working. The key stroke allows you to change the system. Are there any restrictions on the size of angle your calculator will accept? Consult your owner's guide.

In a right triangle we define the *cosine function* of an angle as the *ratio* of the length of the **side adjacent** to the angle **to** the length of the **hypotenuse.** It is customary to use the symbol cos ( ) to indicate the cosine function. Thus, in Figure 24-15, cos (α) = side adjacent/hypotenuse = b/c. Cos (α) is read "cosine of alpha." Again, the parentheses indicate function and not multiplication.

Check your calculator and your owner's guide for how to calculate the cosine function. Look for a key like $\boxed{\cos x}$ or $\boxed{\text{COS}}$. How do you select degrees or radians? Are there any constraints on the size of the angle?

In a right triangle we define the *tangent function* of an angle as the *ratio* of the length of the **side opposite** the angle **to** the length of the **side adjacent** to the angle. The common symbol for the tangent function is tan ( ). Thus, in Figure 24-15, tan (α) = side opposite/side adjacent = a/b. Tan (α) is read "tangent alpha." The parentheses are used to indicate function, not multiplication.

Check your calculator and your owner's guide for how to calculate the tangent function. Look for a key like $\boxed{\tan x}$ or $\boxed{\text{TAN}}$. How do you select degrees or radians? What limits are placed on the size of the angle?

The definitions of the three functions—sine, cosine, and tangent—are summarized in Table 24-3. Refer also to Figure 24-15.

TABLE 24-3

DEFINITION OF THE TRIGONOMETRIC FUNCTIONS

Function	Symbol	Definition	Figure 24-15
sine (α)	sin (α)	Side opposite/hypotenuse	a/c
cosine (α)	cos (α)	Side adjacent/hypotenuse	b/c
tangent (α)	tan (α)	Side opposite/side adjacent	a/b

EXAMPLE 24-13

Use your calculator to evaluate the sine, cosine, and tangent functions of 35°.

Solution Set your calculator to work in degrees:

SIN   sin (35°) = 0.574
COS   cos (35°) = 0.819
TAN   tan (35°) = 0.700

EXAMPLE 24-14

Use your calculator to evaluate the sine, cosine, and tangent functions of 1^r.

Solution Set your calculator to work in radians:

SIN   sin (1^r) = 0.842
COS   cos (1^r) = 0.540
TAN   tan (1^r) = 1.56

It is helpful to learn the brief list of trigonometric functions in Table 24-4. The list will aid you in finding approximate solutions to problems and in detecting errors in key strokes while using your calculator.

TABLE 24-4

A BRIEF TABLE OF TRIGONOMETRIC FUNCTIONS

Angle (Degrees)	Angle (Radians)	Sine	Cosine	Tangent
0	0.000	0.000	1.000	0.000
30	0.524	0.500	0.866	0.577
45	0.785	0.707	0.707	1.000
60	1.047	0.866	0.500	1.732
90	1.571	1.000	0.000	—

EXERCISE 24-5

Calculator Drill

Evaluate the sine, cosine, and tangent functions of the following angles in degrees.

1. 1°	2. 25°	3. 79°	4. 15°
5. 89°	6. 0°	7. 51.5°	8. 70.9°
9. 11.4°	10. 17.3°	11. 0.05°	12. 5.37°
13. 81.6°	14. 77.7°	15. 49.2°	16. 63.9°
17. 50.2°	18. 71.4°	19. 21.3°	20. 38.4°

Evaluate the sine, cosine, and tangent of the following angles in radians.

21. 1.5^r	22. 0.25^r	23. 0.79^r	24. 0.017^r
25. 0^r	26. 0.5^r	27. 0.1^r	28. 1.2^r
29. 1.45^r	30. 0.67^r	31. 0.159^r	32. 0.318^r
33. 1.17^r	34. 0.200^r	35. 1.57^r	36. 1.23^r
37. 0.972^r	38. 0.808^r	39. 0.355^r	40. 0.621^r

24-6 USING THE TRIGONOMETRIC FUNCTIONS TO SOLVE RIGHT TRIANGLES

In the previous section we learned about three functions of angles in right triangles. These functions, sine, cosine, and tangent, are known as trigonometric functions because they are used in measuring triangles. With the use of these functions we can find the missing parts of a right triangle if we know one angle and one side.

EXAMPLE 24-15

In the right triangle in Figure 24-16, $\angle\alpha = 25.7°$ and $c = 15.5$ m. Find side *a*, side *b*, and $\angle\beta$.

Solution   Solve for side *a*:

Use the definition of the trigonometric functions to write equations involving the known parts and one unknown side. Which function involves an angle, the hypotenuse, and the side opposite the angle? The sine function.

$$\sin(\alpha) = a/c$$

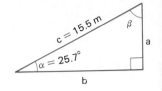

FIGURE 24–16
Right triangle for Example 24–15.

Solve for the unknown part, a:

$$a = c \sin (\alpha)$$

Substitute:

[SIN]
[×]

$a = 15.5 \sin (25.7°)$
$a = 15.5(0.434)$
$a = 6.72$

$\therefore$ $a = 6.72$ m

Solve for side b:
Which function involves an angle, the hypotenuse, and the side adjacent the angle? The cosine function.

$$\cos (\alpha) = b/c$$

Solve for the unknown b:

$$b = c \cos (\alpha)$$

Substitute:

[COS]
[×]

$b = 15.5 \cos (25.7°)$
$b = 15.5(0.901)$
$b = 14.0$

$\therefore$ $b = 14.0$ m

Solve for $\angle\beta$:
How do we find the missing angle β? It is the complement of $\angle\alpha$ (25.7°).

$\therefore$ $\angle\beta = 64.3°$

**EXAMPLE
24-16**

In the right triangle in Figure 24-17, $\angle\alpha = 55.2°$ and $b = 74.5$ mm. Find side c, side a, and $\angle\beta$.

FIGURE 24–17
Right triangle for Example 24–16.

Solution Solve for side c:
Use the definitions of the trigonometric functions to write an equation involving an angle, the side adjacent, and the hypotenuse.

$$\cos(\alpha) = b/c$$

Solve for c:

$$c = b/\cos(\alpha)$$

Substitute:

$$c = 74.5/\cos(55.2°)$$
$$c = 74.5/0.571$$
$$c = 131$$

$$\therefore \quad c = 131 \text{ mm}$$

Solve for side a:
Write an equation involving $\angle\alpha$, b, and a:

$$\tan(\alpha) = a/b$$

Solve for a and substitute:

$$a = b\tan(\alpha)$$
$$a = 74.5\tan(55.2°)$$
$$a = 74.5(1.44)$$
$$a = 107$$

$$\therefore \quad a = 107 \text{ mm}$$

Solve for $\angle\beta$:
Find $\angle\beta$ as the complement of $\angle\alpha$ (55.2°).

$$\therefore \quad \angle\beta = 34.8°$$

**EXAMPLE
24-17**

In the right triangle in Figure 24-18, $\angle\beta = 47.8°$ and $b = 2.83$ m. Find side c, side a, and $\angle\alpha$.

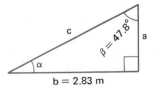

**FIGURE   24–18
Right triangle for Example 24–17.**

$b = 2.83$ m

Solution   Solve for side c:
Write an equation involving $\angle\beta$, b, and c:

$$\sin(\beta) = b/c$$

Solve for c and substitute:

$$c = b/\sin(\beta)$$
$$\boxed{\text{SIN}} \quad c = 2.83/\sin(47.8°)$$
$$\boxed{\div} \quad c = 2.83/(0.741)$$
$$c = 3.82$$

$$\therefore \quad c = 3.82 \text{ m}$$

Solve for side a:
Write an equation involving $\angle\beta$, b, and a:

$$\tan(\beta) = b/a$$

Solve for a and substitute:

$$a = b/\tan(\beta)$$
$$\boxed{\text{TAN}} \quad a = 2.83/\tan(47.8°)$$
$$\boxed{\div} \quad a = 2.83/1.10$$

$$\therefore \quad a = 2.57$$

Solve for $\angle\alpha$:
Find $\angle\alpha$ as the complement of $\angle\beta = 47.8°$:

$$\therefore \quad \angle\alpha = 42.2°$$

EXERCISE 24-6

Find the missing parts of the triangle in Figure 24-19 when you are given the following information.

1. $\angle\alpha = 20°$, $c = 10$ m   2. $\angle\alpha = 59°$, $c = 210$ m
3. $\angle\alpha = 17.5°$, $c = 26.2$ m   4. $\angle\alpha = 83.1°$, $c = 500$ mm
5. $\angle\alpha = 28.4°$, $a = 19.5$ m   6. $\angle\alpha = 52.1°$, $a = 9.05$ m
7. $\angle\alpha = 32.3°$, $a = 2.79$ m   8. $\angle\alpha = 71.5°$, $a = 9.05$ m
9. $\angle\alpha = 14.1°$, $b = 14.1$ m   10. $\angle\alpha = 58.2°$, $b = 351$ mm
11. $\angle\alpha = 45.0°$, $b = 0.70$ km   12. $\angle\alpha = 60°$, $b = 1.86$ km
13. $\angle\beta = 15.3°$, $c = 300$ m   14. $\angle\beta = 79.1°$, $c = 121$ km
15. $\angle\beta = 50.8°$, $c = 875$ mm   16. $\angle\beta = 29.7°$, $c = 515$ m
17. $\angle\beta = 36.2°$, $a = 59.4$ m   18. $\angle\beta = 60.5°$, $a = 2.75$ km
19. $\angle\beta = 75.6°$, $b = 1.07$ km   20. $\angle\beta = 43.3°$, $b = 16.9$ mm

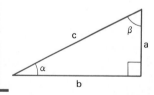

FIGURE 24–19
Right triangle for Exercise 24–6.

24-7 INVERSE TRIGONOMETRIC FUNCTIONS

We have seen that the three trigonometric functions give the value of certain ratios in a right triangle from the knowledge of the size of a particular angle. The inverse functions give the size of the angle from knowledge of the appropriate ratios. Definitions for three inverse trigonometric functions are given below.

In a right triangle we define the *arc sine function* of the ratio of a side to

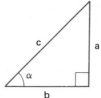

FIGURE 24-20
The inverse trigonometric functions:
$\angle\alpha = \text{Arcsin, } (a/c) = \text{Arccos, } (b/c) = \text{Arctan } (a/b).$

the hypotenuse as *the angle* whose sine is the ratio. There are several symbols in common use for the arc sine: one symbol is $\sin^{-1}(\ )$; a second symbol is Arcsin ( ). Thus, in Figure 24-20, Arcsin $(a/c) = \angle\alpha$.

Check your calculator and owner's guide for how to calculate this function. Your calculator may use two key strokes, such as $\boxed{\text{Arc}}$ $\boxed{\sin}$, $\boxed{\text{INV}}$ $\boxed{\text{SIN}}$, or $\boxed{f^{-1}}$ $\boxed{\sin}$. The angle will be returned in the angular unit you have selected, so check whether the calculator is set for degrees or radians! Finally, the arc sine function will only work for numbers between -1 and 1.

In a right triangle we define the *arc cosine function* of the ratio of a side to the hypotenuse as *the angle* whose cosine is the ratio. We will use the symbol Arccos ( ) to indicate the arc cosine function. Thus, in Figure 24-20, Arccos $(b/c) = \angle\alpha$.

Check your calculator and your owner's guide for how to calculate this function. The same cautions apply to this function that apply to the arc sine function.

In a right triangle, we define the *arc tangent function* of the ratio of the two sides as *the angle* whose tangent is equal to the ratio. We will use the symbol Arctan ( ) to indicate the arc tangent function. Thus, in Figure 24-20, Arctan $(a/b) = \angle\alpha$.

Check your calculator and your owner's guide for how to calculate this function. Notice that there is no limit on the size of the argument for the arc tangent function.

EXAMPLE 24-18   Find the angle in degrees whose sine is 0.244. Then find the same angle in radians.

Solution   Use your calculator; set it to work in degrees:

$$\text{Arcsin } (0.244) = 14.1°$$

Set your calculator to work in radians:

$$\text{Arcsin } (0.244) = 0.246^r$$

EXAMPLE
24-19

Find the angle in degrees whose cosine is 0.398. Then find the same angle in radians.

Solution Use your calculator; set it to work in degrees:

$$\text{Arccos } (0.398) = 66.5°$$

Set your calculator to work in radians:

$$\text{Arccos } (0.398) = 1.16^r$$

EXAMPLE
24-20

Find the angle in degrees whose tangent is 4.87. Then find the same angle in radians.

Solution Use your calculator; set it to work in degrees:

$$\text{Arctan } (4.87) = 78.4°$$

Set your calculator to work in radians:

$$\text{Arctan } (4.87) = 1.37^r$$

EXERCISE 24-7

Calculator Drill

1. Find the arc sine in degrees of the following numbers.

   (a) 0.107   (b) 0.791   (c) 0.255   (d) 0.568
   (e) 0.500   (f) 0.395   (g) 0.642   (h) 0.806
   (i) 0.411   (j) 1.000   (k) 0.043   (l) 0.921

2. Find the arc sine in radians of the numbers in problem 1.
3. Find the arc cosine in degrees of the numbers in problem 1.
4. Find the arc cosine in radians of the numbers in problem 1.
5. Find the arc tangent in degrees of the following numbers.

   (a) 10.0   (b) 0.707   (c) 3.15   (d) 0.444
   (e) 1.57   (f) 1.00   (g) 2.36   (h) 0.290

(i) 6.03   (j) 0.012   (k) 14.3   (l) 7.08

6. Find the arc tangent in radians of the numbers in problem 5.

24-8 SOLVING RIGHT TRIANGLES WHEN TWO SIDES ARE KNOWN

With the aid of the inverse trigonometric functions, we can solve right triangles when two sides are known. The next three examples show how this can be done; it is important that you use your calculator.

EXAMPLE 24-21

In the right triangle in Figure 24-21, the length of the hypotenuse is 200 m and the length of side a is 75 m. Find $\angle\alpha$, $\angle\beta$, and side b.

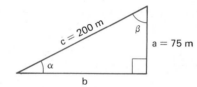

FIGURE 24-21
Right triangle for Example 24-21.

Solution   Find $\angle\alpha$ by writing an equation involving a, c, and $\angle\alpha$:

$$\sin(\alpha) = a/c$$

Apply the arc sine function to both sides to solve for $\angle\alpha$:

$$\text{Arcsin}[\sin(\alpha)] = \text{Arcsin}(a/c)$$
$$\angle\alpha = \text{Arcsin}(a/c)$$

Substitute:

$$\angle\alpha = \text{Arcsin}(75/200)$$
$$\angle\alpha = \text{Arcsin}(0.375)$$

$$\therefore \quad \angle\alpha = 22.0°$$

Find $\angle\beta$ by writing an equation involving a, c, and $\angle\beta$:

$$\cos(\beta) = a/c$$

Apply the arc cosine function to both sides to solve for $\angle\beta$:

$$\text{Arccos}\,[\cos(\beta)] = \text{Arccos}\,(a/c)$$
$$\angle\beta = \text{Arccos}\,(a/c)$$

Substitute:

$\boxed{\div}$
$\boxed{\text{ARC}}\ \boxed{\text{COS}}$ $\qquad \angle\beta = \text{Arccos}\,(75/200)$
$\qquad\qquad\qquad\quad \angle\beta = \text{Arccos}\,(0.375)$

$\therefore \qquad \angle\beta = 68.0°$

Find side b by the Pythagorean theorem, Rule 24-3:

$$a^2 + b^2 = c^2$$

Solve for b and substitute:

$$b = \sqrt{c^2 - a^2}$$
$\boxed{x^2}\ \boxed{-}$ $\qquad b = \sqrt{200^2 - 75^2}$
$\boxed{\sqrt{x}}$ $\qquad\quad b = \sqrt{34\,375}$

$\therefore \qquad b = 185\text{ m}$

**EXAMPLE
24-22**

The length of the hypotenuse is 155 mm and side b is 125 mm. Find $\angle\alpha$, $\angle\beta$, and side a in Figure 24-22.

Solution Find $\angle\alpha$ by writing an equation involving b, c, and $\angle\alpha$:

$$\cos(\alpha) = b/c$$

Apply the arc cosine function to both sides to solve for $\angle\alpha$:

$$\angle\alpha = \text{Arccos}\,(b/c)$$

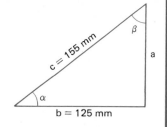

FIGURE 24–22
Right triangle for Example 24–22.

Substitute:

$$\angle\alpha = \text{Arccos } (125/155)$$
$$\angle\alpha = \text{Arccos } (0.806)$$

[ARC] [COS] [÷]

$$\therefore \quad \angle\alpha = 36.2°$$

Find $\angle\beta$ as the complement of $\angle\alpha$:

$$\angle\beta = 90 - 36.2$$
$$\therefore \quad \angle\beta = 53.8°$$

Find side a by writing an equation involving a, b, and $\angle\alpha$:

$$\tan (\alpha) = a/b$$

Solve for a and substitute:

$$a = b \tan (\alpha)$$
$$a = 125 \tan (36.2°)$$
$$a = 125(0.732)$$

[TAN] [×]

$$\therefore \quad a = 91.5 \text{ mm}$$

EXAMPLE 24-23

In the right triangle in Figure 24-23, the length of side a is 155 km, while side b is 125 km. Find $\angle\alpha$, $\angle\beta$, and the hypotenuse c.

Solution   Find $\angle\alpha$ by writing an equation involving a, b, and $\angle\alpha$:

$$\tan (\alpha) = a/b$$

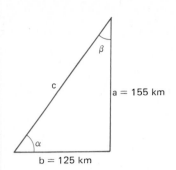

a = 155 km

FIGURE 24–23
Right triangle for Example 24–23.

b = 125 km

Apply the arc tangent function to both sides to solve for $\angle\alpha$:

$$\angle\alpha = \text{Arctan}\ (a/b)$$

Substitute:

$$\angle\alpha = \text{Arctan}\ (155/125)$$
$$\angle\alpha = \text{Arctan}\ (1.24)$$

$$\therefore \quad \angle\alpha = 51.1°$$

Find $\angle\beta$ as the complement of $\angle\alpha$:

$$\angle\beta = 90 - 51.1$$
$$\therefore \quad \angle\beta = 38.9°$$

Find c from a and $\angle\alpha$:

$$\sin\ (\alpha) = a/c$$
$$c = a/\sin\ (\alpha)$$

Substitute:

$$c = 155/\sin\ (51.1°)$$
$$c = 155/0.778$$

$$\therefore \quad c = 199\ \text{km}$$

Find the missing parts in the right triangle in Figure 24-24, given the following information.

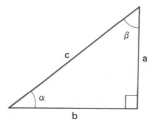

FIGURE 24–24
Right triangle for Exercise 24–8.

1. $a = 15$ m, $c = 25$ m
2. $a = 5$ m, $c = 13$ m
3. $a = 17.5$ mm, $c = 40.2$ mm
4. $a = 57.4$ m, $c = 77.4$ m
5. $b = 24$ m, $c = 25$ m
6. $b = 45$ m, $c = 100$ m
7. $b = 22.1$ km, $c = 38.3$ km
8. $b = 1.47$ km, $c = 2.00$ km
9. $a = 16$ mm, $b = 14$ mm
10. $a = 152$ m, $b = 152$ m
11. $a = 500$ m, $b = 866$ m
12. $a = 757$ mm, $b = 602$ mm

SELECTED TERMS

complementary angles   Two angles which sum to a right angle.

hypotenuse   The side opposite the right angle in a right triangle; also the longest side in a right triangle.

radian   A unit for measuring the size of angles; a complete circle equals 2π (6.283185 . . .) radians.

revolution   A unit for measuring the size of angles; a complete circle equals 1 revolution.

right triangle   A triangle containing a right angle.

25

CIRCULAR FUNCTIONS

In the preceding chapter we learned about functions of angles in right triangles. The angles were limited in size to be between zero and ninety degrees. In this chapter, we will extend these functions to angles of any size.

25-1 ANGLES OF ANY MAGNITUDE

When the concept of angle was introduced, we said that angles have magnitude and direction. It is now time to consider the direction along with the magnitude. Consider a conventional clock with hands. The hands sweep out angles in clockwise (cw) direction when viewed from the front. This is indicated by the arrow ⌢. The opposite direction is called counterclockwise (ccw). It is indicated by the arrow ⌣.

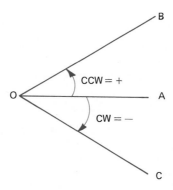

FIGURE 25-1
Positive and negative angles.

In mathematics, we call counterclockwise angles positive and clockwise angles negative. Thus, in Figure 25-1, $\angle AOB$ is positive, while $\angle AOC$ is negative.

Again, consider a clock showing 2 o'clock. Twelve hours later, the clock will also show 2 o'clock. This property is shared with angles. For any angle, the terminal side is in the same position if the angle is changed by one or more complete revolutions. Any angle can be considered as being made up of two parts: **a fraction of a revolution and an integer number of complete revolutions.** For most applications in electronics, we will be concerned with the fractional part.

EXAMPLE 25-1 Write 451° as the sum of two angles: one angle between −180° and 180°, and the other angle a multiple of 360°.

Solution Subtract 360° from 451°:

$$451° - 360° = 91°$$

Note, $-180° < 91° < 180°$

∴ $451° = 360° + 91°$

Observation   We call 451° and 91° *equivalent angles.*

EXAMPLE 25-2   Find an angle between −180° and 180° equivalent to 355°.

Solution   Subtract 360° from 355°:

$$355° - 360° = -5°$$

∴   −5° is equivalent to 355°.

EXAMPLE 25-3   Find an angle between −180° and 180° equivalent to −779°.

Solution   Start by adding 360° to −779°:

$$-779° + 360° = -419°$$

Add a second 360° to −419°:

$$-419° + 360° = -59°$$

Thus,

$$-779° + 2(360°) = -59°$$

∴   −59° is equivalent to −779°.

EXERCISE 25-1

Find angles between −180° and 180° equivalent to the following angles.

1. 365°	2. 270°	3. 545°
4. 600°	5. −270°	6. −405°
7. −659°	8. −288°	9. 382°
10. 417°	11. −515°	12. −323°

25-2 CIRCULAR FUNCTIONS

We are now ready to extend the trigonometric functions to angles of any size through the use of a circle. First, construct a rectangular coordinate system. Then construct a circle of radius ρ with its center at the origin of the coordinate system. See Figure 25-2. We measure $\angle\theta$ from the positive x-axis. Each point, P, on the circle has four numbers associated with it. The first two, x and y, form the rectangular coordinates of the point. The third is the angle θ, and the fourth is the radius ρ, which is the same for each point on the circle.

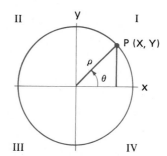

FIGURE 25–2
Diagram for defining the circular functions.

To find the circular functions of any angle, first find the *equivalent angle* between $-180°$ and $180°$, and then construct the angle as in Figure 25-2. The definitions for three circular functions are listed in Table 25-1. For angles in the first quadrant ($0° \leqslant \theta \leqslant 90°$), these definitions coincide with those of the

TABLE 25-1

DEFINITIONS FOR THREE CIRCULAR FUNCTIONS		
Function	*Abbreviation*	*Definition*
sine (θ)	sin (θ)	y/ρ
cosine (θ)	cos (θ)	x/ρ
tangent (θ)	tan (θ)	y/x

TABLE 25-2

SIGNS OF CIRCULAR FUNCTIONS						
Quadrant	*Equivalent Angles*	*x*	*y*	*Sine*	*Cosine*	*Tangent*
I	$0° < \theta < 90°$	+	+	+	+	+
II	$90° < \theta < 180°$	−	+	+	−	−
III	$-180° < \theta < -90°$	−	−	−	−	+
IV	$-90° < \theta < 0°$	+	−	−	+	−

previous chapter. Most calculators will calculate the circular functions for any size of angle.

Because the length of the radius, ρ, is always positive, the signs of the circular functions only depend on the quadrant in which the terminal side of the angle falls. The signs of the functions are summarized in Table 25-2.

EXERCISE 25-2

In which quadrant do the following angles fall?

1. 75°	2. −56°	3. −120°
4. 179°	5. 220°	6. −304°
7. 390°	8. −256°	9. 15°
10. −100°	11. 87°	12. 135°

Calculator Drill

Find the sine of the following angles.

13. 75°	14. −56°	15. −120°
16. 179°	17. 220°	18. −304°
19. 390°	20. −256°	21. 15°
22. −100°	23. 87°	24. 135°

Find the cosine of the following angles.

25. 75°	26. −56°	27. −120°
28. 179°	29. 220°	30. −304°
31. 390°	32. −256°	33. 15°
34. −100°	35. 87°	36. 135°

Find the tangent of the following angles.

37. 75°	38. −56°	39. −120°
40. 179°	41. 220°	42. −304°
43. 390°	44. −256°	45. 15°
46. −100°	47. 87°	48. 135°

25-3 GRAPHS OF THE CIRCULAR FUNCTIONS

Figure 25-3 presents the graph of the sine function. Imagine a point starting on the circle at −180° and moving counterclockwise. The radius of the circle is one unit; therefore, the sine function is equal to the height of the point above (or below) the center of the circle. The graph shows how the sine function varies as the angle changes from −180° to 180°. If the angle were to con-

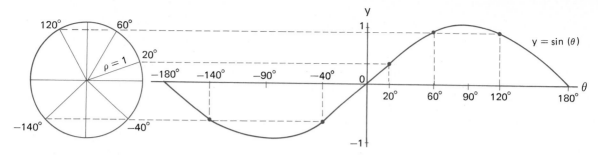

FIGURE 25-3
Circle generating graph of sine function.

tinue to change, the graph would repeat itself. This is called *periodic behavior*. Since the sine function repeats every 360°, the *period* of the sine function is 360°.

Observe that the sine function has a limited range of values. No matter what the angle is, the sine of the angle is between −1 and +1. For any angle θ, −1 ⩽ sin (θ) ⩽ 1. Furthermore, observe from the graph that the sine of an angle, say 90°, has the same magnitude as the sine of minus the angle (−90°), but it is opposite in sign. Thus, sin (−θ) = −sin (θ).

EXAMPLE 25-4 Plot $y = \cos(\theta)$ for $-180° \leq \theta \leq 180°$.
Solution Build a table of values of θ and y:

θ	−180°	−170°	−160°	−140°	−120°	−100°
y	−1.00	−0.985	−0.940	−0.766	−0.500	−0.174

θ	−90°	−80°	−60°	−40°	−20°	−10°
y	0.000	0.174	0.500	0.766	0.940	0.985

θ	0°	10°	20°	40°	60°	80°
y	1.000	0.985	0.940	0.766	0.500	0.174

θ	90°	100°	120°	140°	160°
y	0.000	−0.174	−0.500	−0.766	−0.940

θ	170°	180°
y	−0.985	−1.000

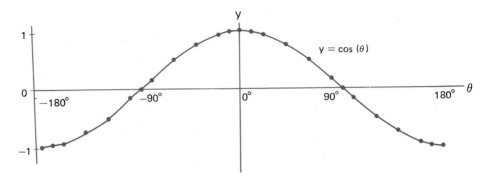

FIGURE 25–4
Graph of the cosine function developed in Example 25–4.

Plot the points from the table (see Figure 25-4). Then draw a smooth curve through the points.

Observation   The graph of the cosine function is similar to sine function in that the range of values is limited from -1 to $+1$. The cosine function is also a periodic function of period $360°$. However, the cosine function is symmetric about $0°$; that is, $\cos(-\theta) = \cos(\theta)$. Finally, the cosine function looks like a shifted sine function. In fact, $\cos(\theta) = \sin(\theta + 90°)$.

EXAMPLE 25-5   Plot $y = \tan(\theta)$ for $-180° \leq \theta \leq 180°$.
Solution   Build a table of values of θ and y. Avoid the exact angles of $-90°$ and $90°$.

θ	$-180°$	$-160°$	$-140°$	$-120°$	$-100°$
y	0.00	0.364	0.839	1.73	5.67

θ	$-91°$	$-89°$	$-80°$	$-60°$	$-40°$
y	57.3	-57.3	-5.67	-1.73	-0.839

θ	$-20°$	$0°$	$20°$	$40°$	$60°$	$80°$	$89°$
y	-0.364	0.00	0.364	0.839	1.73	5.67	57.3

θ	91°	100°	120°	140°	160°	180°
y	−57.3	−5.67	−1.73	−0.839	−0.364	0.00

Plot the points. We cannot draw a smooth curve near −90° and 90°, because the tangent function is *discontinuous* at −90° and 90°. This is indicated in the graph of Figure 25-5 by the dashed vertical lines.

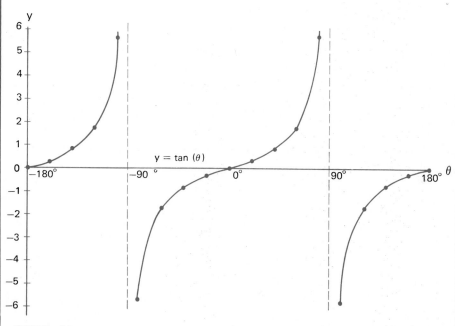

FIGURE 25–5
Graph of the tangent function developed in Example 25–5.

Observation The graph of the tangent function is unlimited in value. The tangent function can be expressed in terms of the sine and cosine: tan (θ) = sin (θ)/cos (θ). It follows that tan (θ) is zero when sin (θ) is zero, and that tan (θ) is undefined when cos (θ) is zero. The graph of the tangent gets closer and closer to the dashed lines as $\angle\theta$ gets closer and closer to −90° or 90°. But it never reaches the dashed lines.

EXERCISE 25-3

Create plots of the following functions for the indicated range of θ.

1. $y = 2 \sin (\theta)$ $-180° \leqslant \theta \leqslant 180°$
2. $y = 3 \cos (\theta)$ $-180° \leqslant \theta \leqslant 180°$
3. $y = \sin (\theta + 90)$ $-270° \leqslant \theta \leqslant 90°$
4. $y = 1/\tan (\theta)$ $0° \leqslant \theta \leqslant 180°$ set $y = 0$ when $\theta = 90$
5. $y = \sin^2 (\theta)$ $0° \leqslant \theta \leqslant 180°$ NOTE: The superscript
6. $y = \cos^2 (\theta)$ $0° \leqslant \theta \leqslant 180°$ 2 indicates the square
7. $y = \sin^2 (\theta) + \cos^2 (\theta)$ $0° \leqslant \theta \leqslant 180°$ of the function thus:

$\sin^2 (\theta) = [\sin (\theta)]^2$.

25-4 INVERSE CIRCULAR FUNCTIONS

The inverse circular functions are extensions of the inverse trigonometric functions. Your calculator will calculate the *principal* value of these functions for you. The sine function takes on all its possible values between $-90°$ and $90°$. The principal value of the arc sine function is an angle between $-90°$ and $90°$. We will capitalize the "a" in arc to indicate principal value. Table 25-3 summarizes the principal values of the inverse circular functions.

TABLE 25-3

PRINCIPAL VALUES OF INVERSE CIRCULAR FUNCTIONS			
Function	*Symbol*	*Range of Argument*	*Range of Principal Value*
Arc sine (x)	Arcsin (x)	$-1 \leqslant x \leqslant 1$	$-90° \leqslant \theta \leqslant 90°$
Arc cosine (x)	Arccos (x)	$-1 \leqslant x \leqslant 1$	$0° \leqslant \theta \leqslant 180°$
Arc tangent (x)	Arctan (x)	No limit	$-90° \leqslant \theta \leqslant 90°$

EXERCISE 25-4

Calculator Drill

Use your calculator to find the principal value of the indicated inverse circular function in degrees:

1. Arcsin (0.707)   2. Arcsin (0.866)   3. Arccos (0.500)
4. Arccos (0.866)   5. Arctan (1.00)   6. Arctan (0.700)
7. Arcsin (−0.500)   8. Arcsin (−0.707)   9. Arcsin (−0.866)
10. Arcsin (−0.342)   11. Arccos (−0.500)   12. Arccos (−0.866)

13. Arccos (-0.707) 14. Arccos (-0.342) 15. Arctan (-1.00)
16. Arctan (-1.54) 17. Arctan (-0.700) 18. Arctan (-0.200)
19. Arctan (-5.67) 20. Arctan (-57.3) 21. Arctan (999)

25-5 THE LAW OF SINES AND THE LAW OF COSINES

In the preceding chapter we learned how to solve for the missing parts of a right triangle using the Pythagorean theorem and trigonometric functions of acute angles (sine, cosine, and tangent). In this section we will learn two new laws that hold in every triangle. We will also learn how to use these laws to solve for the missing parts of a triangle. The first law is the *law of cosines*, which is stated as Rule 25-1.

RULE 25-1. THE LAW OF COSINES

For any triangle labeled as in Figure 25-6, the following three equations hold:

$$a^2 = b^2 + c^2 - 2bc \cos (\alpha)$$

$$b^2 = a^2 + c^2 - 2ac \cos (\beta)$$

$$c^2 = a^2 + b^2 - 2ab \cos (\gamma)$$

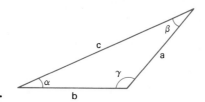

FIGURE 25-6
Triangle for Section 25-5.

If γ is a right angle, then cos (γ) is zero and the third equation of the law of cosines becomes $c^2 = a^2 + b^2$. For this reason the law of cosines is called an extension of the Pythagorean theorem.

The second law states that within a triangle the length of a side divided by the sine of the opposite angle is a constant. This is known as the *law of sines,* and it is stated in the form of two equations in Rule 25-2.

RULE 25-2. THE LAW OF SINES

For any triangle labeled as in Figure 25-6, the following equations hold:

$$\frac{a}{\sin(\alpha)} = \frac{b}{\sin(\beta)} = \frac{c}{\sin(\gamma)}$$

$$\frac{\sin(\alpha)}{a} = \frac{\sin(\beta)}{b} = \frac{\sin(\gamma)}{c}$$

The law of sines is very easy to use when finding the length of a side; however, caution must be used when using the law of sines to find the size of an angle. This is because the sine of an angle is equal to the sine of its supplement. Thus, the Arc sine cannot return a value greater than 90°. There are three additional facts that we can use with the law of sines to solve triangles. The first is that the largest angle is opposite the longest side. The second is that, at most, one angle in a triangle can be larger than 90°. The third is that the sum of the angles of a triangle is 180°.

With the aid of the law of sines and the law of cosines, we can solve three classes of problems involving triangles. The first class of problems is to find the angles of a triangle when the sides are known. This is demonstrated in the following example.

EXAMPLE 25-6 The lengths of the three sides of a triangle are side $a = 37$ m, side $b = 43$ m, and side $c = 56$ m. Find the size of each of the angles. Use the notation of Figure 25-6.

Solution Use the law of cosines (Rule 25-1) to find the angle opposite the longest side, $\angle\gamma$. Thus:

$$c^2 = a^2 + b^2 - 2ab\cos(\gamma)$$

Solve for $\cos(\gamma)$:

$$\cos(\gamma) = \frac{a^2 + b^2 - c^2}{2ab}$$

Substitute for a, b, and c:

$$\cos(\gamma) = \frac{37^2 + 43^2 - 56^2}{2 \times 37 \times 43}$$

$$\cos(\gamma) = 0.0258$$

Use Arccos to find $\angle\gamma$.

$$\therefore \quad \angle\gamma = 88.5°$$

Use the law of sines (Rule 25-2) to find $\angle\beta$. Thus:

$$\frac{\sin(\beta)}{b} = \frac{\sin(\gamma)}{c}$$

Solve for $\sin(\beta)$:

$$\sin(\beta) = \frac{b\sin(\gamma)}{c}$$

Substitute for b, c, and $\angle\gamma$:

$$\sin(\beta) = \frac{43\sin(88.5)}{56}$$
$$\sin(\beta) = 0.768$$

Use Arcsin to find $\angle\beta$:

$$\therefore \quad \angle\beta = 50.1°$$

Use the law of sines to find $\angle\alpha$:

$$\frac{\sin(\alpha)}{a} = \frac{\sin(\gamma)}{c}$$

Substitute and solve for $\sin(\alpha)$:

$$\sin(\alpha) = \frac{37\sin(88.5)}{56} = 0.660$$

Use Arcsin to find $\angle\alpha$:

$$\therefore \quad \angle\alpha = 41.3°$$

Check the solution; the three angles should sum to 180°:

$$41.3° + 50.1° + 88.5° = 179.9°$$

The second class of problems that can be solved with the aid of the law of sines and the law of cosines is finding the missing parts when two sides and the included angle are known, as in the following example.

EXAMPLE 25-7   Two sides and the angle between them are side $a =$ 74.3 m, side $b = 37.8$ m, and $\angle\gamma = 52.3°$. Find side c and the other two angles. Use the notation of Figure 25-6.

Solution   Use the law of cosines to find side c:

$$c^2 = a^2 + b^2 - 2ab \cos (\gamma)$$

Substitute for a, b, and $\angle\gamma$:

$$c^2 = 74.3^2 + 37.8^2 - 2(74.3)(37.8) \cos (52.3)$$
$$\therefore \quad c = 59.3 \text{ m}$$

Use the law of sines to find the angle opposite the short side:

$$\frac{\sin (\beta)}{b} = \frac{\sin (\gamma)}{c}$$

Substitute for b, c, and $\angle\gamma$:

$$\sin (\beta) = \frac{37.8 \sin (52.3)}{59.3}$$

Use Arcsin to find $\angle\beta$:

$$\therefore \quad \angle\beta = 30.3°$$

Use the law of sines to find $\angle\alpha$:

$$\frac{\sin (\alpha)}{a} = \frac{\sin (\gamma)}{c}$$

Substitute for a, c, and $\angle\gamma$:

$$\sin (\alpha) = \frac{74.3 \sin (52.3)}{59.3}$$

Use Arcsin to find $\angle\alpha$:

$$\angle\alpha = 82.5°$$

Check the angles: do they sum to 180?

$$82.5° + 30.3° + 52.3° = 165.1°$$

Observation The sum is wrong; the largest angle, $\angle\alpha$, is too small. Replace $\angle\alpha$ with its supplement. Thus:

$$\angle\alpha = 180° - 82.5° = 97.5°$$

Check the sum of the angles again:

$$97.5° + 30.3° + 52.3° = 180.1°$$

$$\therefore \quad \angle\alpha = 97.5°$$

The third class of problems that can be solved with the aid of the law of sines is finding the missing parts when two angles and one side are known. The following example demonstrates the third class of problems.

EXAMPLE 25-8 One side and two angles in a triangle are side $b = 135$ m, $\angle\alpha = 23.8°$, and $\angle\beta = 44.5°$. Find the other angle and two sides.

Solution Find $\angle\gamma$ as the supplement of $\angle\alpha + \angle\beta$:

$$\angle\gamma = 180 - (23.8 + 44.5)$$

$$\therefore \quad \angle\gamma = 111.7°$$

Use the law of sines to find side a:

$$\frac{a}{\sin(23.8)} = \frac{135}{\sin(44.5)}$$

$$\therefore \quad a = 77.7 \text{ m}$$

Use the law of sines to find side c:

$$\frac{c}{\sin (111.7)} = \frac{135}{\sin (44.5)}$$

$$\therefore \quad c = 179 \text{ m}$$

EXERCISE 25-5
Solve for the missing parts of the triangle in Figure 25-6 given the following information.

1. $a = 15$ m, $b = 20$ m, $c = 25$ m
2. $a = 40$ m, $b = 20$ m, $c = 30$ m
3. $a = 70.2$ m, $b = 106$ m, $c = 74.6$ m
4. $a = 19.6$ km, $b = 44.4$ km, $c = 48.5$ km
5. $a = 35$ mm, $b = 52$ mm, $\angle\gamma = 80°$
6. $a = 156$ m, $c = 76$ m, $\angle\beta = 46°$
7. $b = 79.4$ m, $c = 98.1$ m, $\angle\alpha = 23.3°$
8. $b = 215$ mm, $c = 175$ mm, $\angle\alpha = 31.4°$
9. $a = 40$ m, $\angle\alpha = 40°$, $\angle\beta = 30°$
10. $b = 65$ m, $\angle\alpha = 27°$, $\angle\beta = 45°$
11. $c = 12.5$ m, $\angle\alpha = 10.9°$, $\angle\beta = 53.4°$
12. $c = 58.6$ m, $\angle\alpha = 60°$, $\angle\beta = 60°$

25-6 POLAR COORDINATES

Besides rectangular, semilog, and log-log coordinates, there are many other systems of coordinates. One that is not often used for plots, but which underlies much of electronics, is the system of *polar coordinates.* Construct a skeleton of this new system as in Figure 25-7. Take a piece of paper. Lay out a horizontal and a vertical axis. Mark off 1, 2, and 3 on each axis. Connect the 1s (plus and minus) together with a circle centered at the origin. In like manner, connect the 2s and then the 3s. Label the positive x-axis 0°, the positive y-axis 90°, the negative x-axis 180°, and the negative y-axis 270°. The complete drawing in Figure 25-7 illustrates the basic components of the polar-coordinate system. Notice the origin is called the *pole.*

In polar coordinates, each point is associated with two numbers. The first number is the distance out from the pole. We use the symbol ρ for this number. It is called the *magnitude.* The second number is the direction measured from the positive horizontal axis. We use θ for this number. It is called

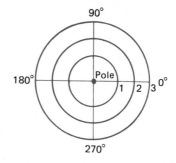

FIGURE 25–7
Skeleton of polar coordinate system.

the *argument*. When writing polar coordinates, we write the numbers in this form, ρ/θ, read *"rho angle theta."*

In rectangular coordinates, each point has only one set of coordinates. In polar coordinates, each point has many sets of coordinates. When a clock indicates 1 o'clock, the time can be 1 A.M. or 1 P.M. This can also be written 0100 hours and 1300 hours. In the same way, the angles $-360°$, $0°$, $360°$, and $720°$ all determine the same direction on polar graph paper. Look at the polar graph paper in Figure 25-8. Notice that the direction $270°$ is also labeled $-90°$. Look at the direction labels around the whole graph paper.

EXAMPLE 25-9   Plot the point $2.5/55°$ on polar graph paper.

Solution   Refer to Figure 25-8. Begin at the pole (P) and move out 2.5 in the $0°$ direction. Follow the arc to the $55°$ direction. Label the point A.

$$\therefore \qquad A \text{ is the point } 2.5/55°.$$

EXAMPLE 25-10

Solution   Determine the coordinates of point B in Figure 25-8.
The argument is between $30°$ and $40°$:

$$\theta = 36.5°$$

The magnitude is between 3 and 3.5:

$$\rho = 3.25$$
$$\therefore \qquad B = 3.25/36.5°$$

Observation   The argument could also have been read as -323.5, and then $B = 3.25/-323.5°$. However, we generally limit the argument to be between $-180°$ and $+180°$. Sometimes we use limits of $0°$ to $360°$.

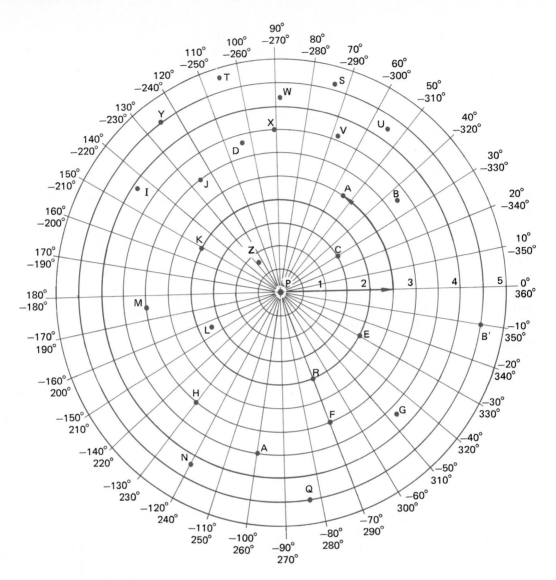

FIGURE 25–8
Polar graph paper.

EXERCISE 25-6
Determine the polar coordinates of the following points in Figure 25-8.

1. *C*	2. *D*	3. *E*	4. *F*
5. *G*	6. *H*	7. *I*	8. *J*
9. *K*	10. *L*	11. *M*	12. *N*

13. Q	14. R	15. S	16. T
17. U	18. V	19. W	20. X
21. Y	22. Z	23. A'	24. B'

Plot the following points on a piece of polar graph paper as in Figure 25-8.

25. $2/75°$	26. $3.7/-90°$	27. $1.2/-45°$
28. $2.1/52°$	29. $0.5/120°$	30. $2.5/-145°$
31. $3.2/-110°$	32. $3.6/135°$	33. $4.0/170°$

25-7 CONVERTING BETWEEN RECTANGULAR AND POLAR COORDINATES

Both rectangular coordinates and polar coordinates represent the location of points on a piece of paper. By inspecting Figure 25-9, we can discover the relationship between (x, y) and ρ/θ. The following rules describe this relationship.

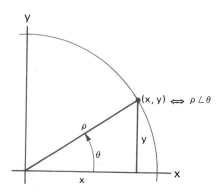

FIGURE 25–9
Relationship between rectangular and polar coordinates.

RULE 25-3. CONVERTING RECTANGULAR TO POLAR COORDINATES

Determine the argument, θ, in two steps:

1. $\theta = \text{Arctan}\ (y/x)$.
2. (a) If (x, y) is in the second quadrant, add 180°.
   (b) If (x, y) is in the third quadrant, subtract 180°.
3. Determine the magnitude, ρ, by

$$\rho = \sqrt{x^2 + y^2}$$

RULE 25-4. CONVERTING POLAR TO RECTANGULAR COORDINATES

Calculate the abscissa, x, and the ordinate, y, by:

1. $x = \rho \cos (\theta)$.
2. $y = \rho \sin (\theta)$.

Some calculators will perform these conversions for you with one or two key strokes ($\boxed{\rightarrow P}$, $\boxed{\rightarrow R}$). Check your owner's guide. If your calculator has these features, pay close attention to the order of entering the numbers and reading the answers. How are you to input both x and y (or ρ and θ)? How are ρ and θ (or x and y) displayed?

EXAMPLE 25-11

Convert $(-3.5, -2.7)$ to polar coordinates.

Solution Use Rule 25-3.

Step 1: $\theta = \text{Arctan} (y/x)$:

$$\theta = \text{Arctan} (-2.7/-3.5)$$
$$= \text{Arctan} (0.771)$$
$$\theta = 37.6°$$

Step 2: $(-3.5, -2.7)$ is in the third quadrant; therefore, subtract 180° from θ.

$$\theta = 37.6° - 180° = -142.4°$$

Step 3: Determine the magnitude:

$$\rho = \sqrt{x^2 + y^2}$$
$$= \sqrt{(3.5)^2 + (-2.7)^2}$$
$$= \sqrt{12.3 + 7.29}$$
$$= \sqrt{19.5}$$
$$\rho = 4.4$$

$\therefore$ $(-3.5, -2.7) \Rightarrow 4.4/\underline{-142.4°}$, as shown in Figure 25-10.

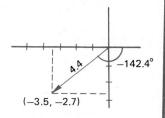

FIGURE 25-10
Diagram for Example 25-11. (−3.5, −2.7)

**EXAMPLE
25-12**

Convert $16.5\underline{/57.2°}$ to rectangular coordinates.

Solution Use Rule 25-4.

Step 1: $x = \rho \cos (\theta)$:

$$x = 16.5 \cos (57.2°)$$
$$= 16.5(0.542)$$
$$x = 8.94$$

Step 2: $y = \rho \sin (\theta)$:

$$y = 16.5 \sin (57.2°)$$
$$= 16.5(0.841)$$
$$y = 13.9$$

$\therefore$ $16.5\underline{/57.2°} \Rightarrow (8.94, 13.9)$, as shown in Figure 25-11.

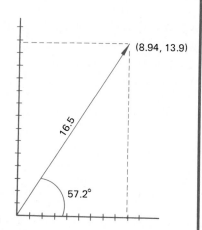

FIGURE 25-11
Diagram for Example 25-12.

EXERCISE 25-7

Calculator Drill
Using Rule 25-3, convert from rectangular to polar coordinates.

1. (15, 0)	2. (0, −3.57)	3. (−8.7, 0)
4. (1.0, 1.0)	5. (5.0, 8.7)	6. (6.34, −4.21)
7. (4.15, −5.09)	8. (174, −108)	9. (−20.2, 153)
10. (−26.9, 12.8)	11. (−0.866, −0.5)	12. (15.6, 28.7)
13. (39.2, −20.1)	14. (−15.7, −30.4)	15. (−4.07, 4.07)
16. (15.1, −2)	17. (−13.4, 5.6)	18. (134, 225)
19. (−1.25, 3.45)	20. (757, −707)	21. (1.86, −1.41)

Using Rule 25-4, convert from polar to rectangular coordinates.

22. $17/\underline{90°}$	23. $16/\underline{180°}$	24. $12.7/\underline{-90°}$
25. $100/\underline{50°}$	26. $25/\underline{30°}$	27. $27/\underline{-75°}$
28. $56.2/\underline{19.3°}$	29. $6.27/\underline{42.3°}$	30. $79.3/\underline{-85.1°}$
31. $0.578/\underline{-50.9°}$	32. $444/\underline{-135°}$	33. $74.6/\underline{156°}$
34. $12.8/\underline{-85°}$	35. $17.6/\underline{492°}$	36. $10.4/\underline{-512°}$
37. $22.3/\underline{1.1^r}$	38. $6.95/\underline{2.8^r}$	39. $30.7/\underline{6.04^r}$
40. $84.7/\underline{5.31^r}$	41. $78.3/\underline{4.95^r}$	42. $517/\underline{3.89^r}$

SELECTED TERMS

equivalent angles   Angles that differ by an integer number of revolutions.

periodic behavior   A behavior or pattern that repeats over and over again, especially in a graph.

polar coordinates   A coordinate system based on a set of concentric circles. The coordinates of a point consist of the distance from a reference point (pole) and the direction from a reference line.

VECTORS AND PHASORS

This chapter introduces the application of the circular functions to electronics. To apply these functions, we need to introduce several new concepts. We will look at physical quantities that have only magnitude, and quantities that have both magnitude and direction. For quantities with both magnitude and direction, we will need the concept of vectors.

26-1 SCALARS AND VECTORS

When length is measured, the quantity that results is called a *scalar quantity*, that is, a number and a unit. Quantities such as 6 cm of wire, a speed of 40 km/h, and a volume of 2 liters are all scalar quantities. Scalar quantities have only magnitude; they do not have direction.

To travel from your house to school, speed alone is not sufficient. Simply driving a car at 40 km/h will not get you there. You need to direct the motion of the car. Directed motion is a *vector quantity*, called *velocity*. In general, quantities expressed with both magnitude and direction are called *vector quantities*.

Vectors are represented by an arrow, as shown in Figure 26-1, in which the magnitude of the vector is the number next to the arrow and the direction of the vector is the direction of the arrow.

It has been traditional to use special notation for vector quantities, such as $\mathbf{V}$ or $\vec{V}$ for velocity. A contemporary method is to use regular symbols to represent vectors and introduce special symbols only for the magnitude of vectors, such as V for velocity and $|V|$ for the magnitude of velocity. We use the contemporary method of notation. Thus, $E = V_1 + V_2$ is a scalar equation for a dc circuit while it is a vector equation for an ac circuit! Whether a symbol represents a scalar or a vector depends on the context of a problem.

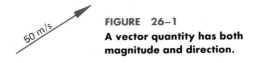

FIGURE 26-1
A vector quantity has both magnitude and direction.

26-2 COMPLEX PLANE

The pilot of a plane flying through the air must keep track of the position of the craft. In order to do this, he must know the compass heading and how the altitude is changing. The position of a plane is seen to be a *three-dimensional* vector (north–south, east–west, up–down).

A concept used with ac circuits is impedance (total opposition to alternating current). Impedance has two components—resistance and reactance. These components do not combine algebraically but combine to form a *two-*

dimensional vector. In electronics we use scalars and two-dimensional vectors.

For notational purposes, we will borrow the concept of the complex plane from mathematics. Figure 26-2 shows the axes labeled R and j and a vector represented by the *complex number $r + jq$*. In complex numbers, the symbol j is used to indicate the component that is plotted in the vertical direction. This is similar to the ordered pair notation used in rectangular coordinates.

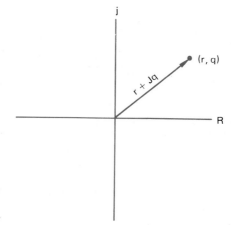

FIGURE 26–2
The complex plane. The complex
number $r + jq$ is an alternate form
of the rectangular coordinate (r, q).
The symbol j is used to delineate
the two components of the number.

26-3 REAL AND IMAGINARY NUMBERS

We are familiar with real numbers. We have been using them throughout the book. What is an imaginary number? We answer this question with another question. What is the square root of minus one ($\sqrt{-1}$)? We define $\sqrt{-1}$ as the number j! This number is outside our everyday experience. It is so strange that we use the word *imaginary* to describe it. Any number of the form jn is called an *imaginary* number.

EXAMPLE 26-1   Express $\sqrt{-25}$ as an imaginary number.
Solution   Write the radicand as the product of -1 and 25:

$$\sqrt{-25} = \sqrt{(-1)25}$$

Write as two radicals:

$$\sqrt{-25} = \sqrt{-1}\,\sqrt{25}$$

Simplify; replace $\sqrt{-1}$ with j:

$$\therefore \quad \sqrt{-25} = j5$$

EXAMPLE 26-2 Express $-\sqrt{-16}$ as an imaginary number.

Solution Factor:

$$-\sqrt{-16} = -\sqrt{(-1)16}$$
$$= -\sqrt{-1}\,\sqrt{16}$$
$$= -j4$$
$$\therefore \quad -\sqrt{-16} = -j4$$

EXERCISE 26-1

Express each number as an imaginary number. Use j in forming the imaginary number.

1. $\sqrt{-9}$ 2. $\sqrt{-64}$ 3. $\sqrt{-36}$

4. $-\sqrt{-4}$ 5. $-\sqrt{-25}$ 6. $-\sqrt{-100}$

7. $\sqrt{-182}$ 8. $\sqrt{-3.76}$ 9. $-\sqrt{-144}$

10. $-\sqrt{-90.8}$ 11. $-\sqrt{-108}$ 12. $-\sqrt{-31.5}$

13. $\sqrt{-0.875}$ 14. $-\sqrt{-1286.5}$ 15. $-\sqrt{-52.82}$

26-4 COMPLEX NUMBERS

A complex number is the sum of a real number and an imaginary number. The *rectangular form* of a complex number is $r + jq$. When $q = 0$, then $r + j0$ is a real number. When $r = 0$ and $q \neq 0$, then $0 + jq$ is an imaginary number. When $r \neq 0$ and $q \neq 0$, then $r + jq$ is a complex number.

The *real part* of the complex number $r + jq$ is r. The *imaginary part* of the complex number $r + jq$ is jq. (NOTE: q is a real number while jq is an imaginary number.) Table 26-1 is a summary of the concepts associated with complex numbers. Observe that the real and the imaginary parts of a complex number can be positive or negative *independent of each other*.

TABLE 26-1

SUMMARY OF COMPLEX NUMBERS AS RELATED TO THE COMPLEX PLANE	
$r + jq$	General form of a complex number
$r + j0$	A real number
$0 + jq, q \neq 0$	An imaginary number
$0 + j0$	Origin in the complex plane
R-axis	Contains all the real numbers
j-axis	Contains all the imaginary numbers
$3 + j4$	A complex number located in the first quadrant
$-3 + j4$	A complex number located in the second quadrant
$-3 - j4$	A complex number located in the third quadrant
$3 - j4$	A complex number located in the fourth quadrant

EXAMPLE 26-3 Write $8 + \sqrt{-16}$ as a complex number.

Solution Write $\sqrt{-16}$ as an imaginary number:

$$\sqrt{-16} = \sqrt{-1}\,\sqrt{16}$$
$$= j4$$

Form the complex number:

$$\therefore \quad 8 + \sqrt{-16} = 8 + j4$$

EXAMPLE 26-4 Classify the following numbers as real, imaginary, or complex.

(a) 7 (b) $0 - j3$ (c) $-\sqrt{-44}$

(d) $5 + j2$ (e) $6 + j0$ (f) $-8 - j4$

Solution

(a) Real (b) Imaginary (c) Imaginary

(d) Complex (e) Real (f) Complex

EXERCISE 26-2

Write each of the following in the form of a complex number.

1. 15 2. $-j7$ 3. $j1.7$

4. $2 - \sqrt{-9}$ 5. $-5 - \sqrt{-16}$ 6. $-3 + \sqrt{-36}$

7. $2 - \sqrt{9}$ 8. -25 9. $17 - \sqrt{-1}$

Classify the following numbers as real, imaginary, or complex.

10. 7
11. $3 - j6$
12. $0 - j0$
13. $-5 + j0$
14. 6
15. $6 - \sqrt{16}$
16. $-7 - \sqrt{-2}$
17. $-\sqrt{18}$
18. $\sqrt{-20}$

Plotting Complex Numbers

The rectangular form of a complex number, $r + jq$, is very similar to the ordered pair notation, (r, q), used to plot in rectangular coordinates. We will take advantage of this similarity to construct a graphic representation of the complex plane as in Figure 26-3. By using the real part of the complex number as the abscissa and the imaginary part of the complex number as the ordinate, we can directly plot the complex number on the complex plane as in Figure 26-4.

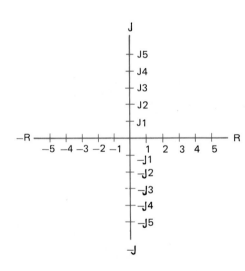

FIGURE 26–3
The coordinates of the horizontal axis (R-axis) are real numbers; the coordinates of the vertical axis (j-axis) are imaginary numbers.

EXERCISE 26-3

Plot the following complex numbers on a complex plane as shown in Figure 26-4.

1. $3 + j2$
2. $4 + j7$
3. $-2 - j5$
4. $6 - j3$
5. $-6 + j4$
6. $0 + j0$
7. $4 - j6$
8. $0 - j8$
9. $1 + j5$

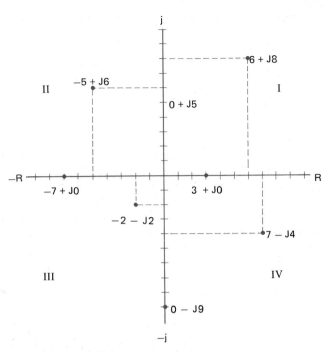

FIGURE 26-4
Points on the complex plane are located with a rectangular coordinate system made up of a real coordinate and an imaginary coordinate.

Record the coordinates of the named points of Figure 26-5.

10. *A*	11. *I*	12. *K*
13. *B*	14. *J*	15. *D*

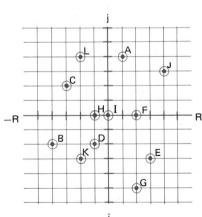

FIGURE 26-5
Points on a complex plane for Exercise 26-3.

16. *E*   17. *H*   18. *C*

19. *F*   20. *G*   21. *L*

26-5 PHASORS

In an ac network, the current and voltage are represented by sine waves having magnitudes and directions that are continually changing. A *rotating vector* is used to represent graphically the changing condition of the ac quantities, current, and voltage.

Figure 26-6 shows how ac sinusoidal quantities are represented by the position of a rotating vector. As the vector rotates it generates an angle. The location of the vector on the plane surface is determined by the magnitude (length) of the vector and by the generated angle. This concept of a rotating vector is shown in Figure 26-6(b). The point at the top of the arrowhead may

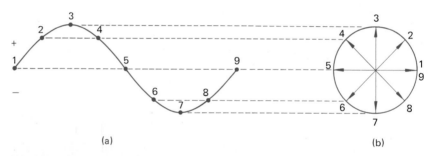

(a)   (b)

FIGURE   26–6
(a) The magnitude of the sine wave (representing the ac voltage or current) is continually changing. (b) A vector with its end fixed at the origin and rotating in a ccw direction represents this varying condition in ac quantities.

be located by the magnitude (the length) and the direction (the angle) of the vector. A rotating vector is represented in general by a function written in *polar form* as

$$\rho \underline{/\alpha(t)}$$

where ρ = vector magnitude, which is constant

$\alpha(t)$ = angular displacement from the reference axis, which is a function of time

572

A *phasor* is a *"stop-action photograph"* of the changing conditions of the ac quantities, voltage, and current, as in Figure 26-7. A phasor, like a rotating vector, is written in polar form as

$$\rho\underline{/\theta}$$

where ρ = phasor magnitude
θ = phase angle, which is equal to $\alpha(t_o)$
t_o = time of the "stop-action photograph"

Phasors are graphed on the complex plane using polar coordinates.

We previously learned that coordinates of the points of the complex plane may be located by a complex number of the rectangular form $r + jq$. Phasors are also used to locate points of the complex plane by a complex number of the polar form $\rho\underline{/\theta}$. We see then that there are two coordinate systems for representing the points of a complex plane: one is a rectangular coordinate system $(r + jq)$ and the other is a polar coordinate system $(\rho\underline{/\theta})$. Because either coordinate system may be used to represent a given point in the plane, the two systems must be equivalent. Thus

$$\rho\underline{/\theta} \Leftrightarrow r + jq$$

Figure 26-8 shows that a point of a complex plane located by a phasor may be described in either the polar or the rectangular form of a complex number.

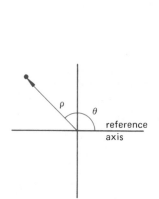

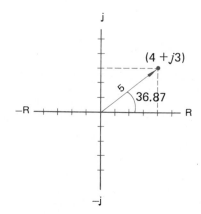

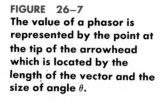

FIGURE 26-7
The value of a phasor is represented by the point at the tip of the arrowhead which is located by the length of the vector and the size of angle θ.

FIGURE 26-8
The point on the complex plane located by the phasor is $4 + j3$ expressed in the rectangular form of a complex number, or $5\angle36.87°$ expressed in the polar form of a complex number.

EXAMPLE 26-5   Express the coordinate of point A of Figure 26-9 in polar form.

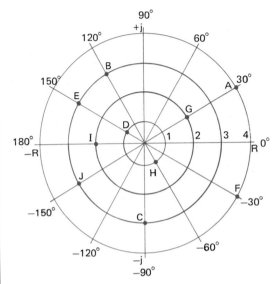

FIGURE 26-9
Complex plane for Exercise 26-4, problems 1-9.

Solution   Draw a phasor from the origin to point A. The magnitude is 4 and the angle is 30.

$$\therefore \quad \text{Point } A \text{ is } 4\underline{/30°}$$

EXERCISE 26-4
Express the coordinates of the following named points of Figure 26-9 in polar form:

1. B	2. F	3. D
4. H	5. C	6. E
7. G	8. I	9. J

Draw each of the following polar phasors on the complex plane of Figure 26-10:

10. $2\underline{/60°}$	11. $3\underline{/-45°}$	12. $4\underline{/120°}$
13. $1\underline{/225°}$	14. $2\underline{/180°}$	15. $3\underline{/-120°}$
16. $1.5\underline{/-135°}$	17. $3.5\underline{/150°}$	18. $2.5\underline{/-60°}$

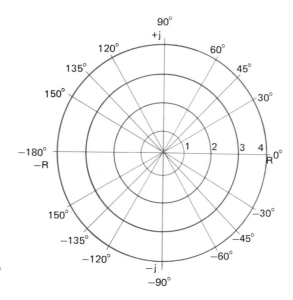

FIGURE 26–10
Complex plane for Exercise 26–4, problems 10–15.

26-6 TRANSFORMING COMPLEX NUMBER FORMS

In working with phasors representing current and voltage, it will be necessary that these complex quantities be expressed in both rectangular and polar form. The process for transforming a number from one form to another will be greatly aided by your calculator.

Rectangular to Polar Transformation

To convert a complex number from the rectangular form to the polar form, apply Rule 26-1.

> RULE 26-1. CONVERTING FROM RECTANGULAR TO POLAR FORM
>
> To change a complex number in rectangular form to polar form $(r + jq \Rightarrow \rho\underline{/\theta})$:
>
> 1. $\rho = \sqrt{r^2 + q^2}$
> 2. $\theta = \text{Arctan } (q/r)$
>   (a) If (r, q) is in quadrant II, then add 180° to θ.
>   (b) If (r, q) is in quadrant III, then subtract 180° from θ.
> 3. Form the complex number in polar form: $\rho\underline{/\theta}$.

Rule 26-1 is **used only** if your calculator does not have a rectangular to polar key $\boxed{\rightarrow P}$. Check your calculator owner's guide to determine how to use this powerful feature of your calculator. It will be assumed that you have this key on your calculator, as the examples will reflect the use of this key stroke.

EXAMPLE 26-6 Express $3 + j4$ in polar form.
Solution (a) Use Rule 26-1:
Step 1:

$$\rho = \sqrt{9 + 16}$$
$$\rho = 5$$

Step 2:

$$\theta = \text{Arctan } \tfrac{4}{3}$$
$$\theta = 53.13°$$
$$\therefore \quad 3 + j4 \Rightarrow 5\underline{/53.13°}$$

Solution (b) Use your $\boxed{\rightarrow P}$ key:

$$\therefore \quad 3 + j4 \boxed{\rightarrow P} 5\underline{/53.13°}$$

EXAMPLE 26-7 Express $-5.2 - j9.5$ in polar form.
Solution Use your calculator:

$$\therefore \quad -5.2 - j9.5 \boxed{\rightarrow P} 10.83\underline{/-118.69°}$$

Polar to Rectangular Transformation

To convert a complex number from the polar form to the rectangular form, apply Rule 26-2.

RULE 26-2. CONVERTING FROM POLAR TO RECTANGULAR FORM

To change a complex number in polar form to rectangular form ($\rho\underline{/\theta} \Rightarrow r + jq$):

1. $r = \rho \cos \theta$.
2. $q = \rho \sin \theta$.
3. Form the complex number in rectangular form: $r + jq$.

Rule 26-2 is **used only** if your calculator lacks the polar to rectangular key $\boxed{\to R}$. Consult your owner's guide for the use of this valuable feature of your calculator. We will use this key exclusively in the remaining chapters of this book.

EXAMPLE 26-8   Express $5/\underline{36.87°}$ in rectangular form.

Solution   (a) Use Rule 26-2.

Step 1:

$$r = 5 \cos (36.87)$$
$$r = -4.00$$

Step 2:

$$q = 5 \sin (36.87)$$
$$q = 3.00$$
$$\therefore \quad 5/\underline{36.87°} \Rightarrow 4 + j3$$

Solution   (b) Use your $\boxed{\to R}$ key:

$$\therefore \quad 5/\underline{36.87°} \boxed{\to R} 4 + j3$$

EXAMPLE 26-9   Express $12.3/\underline{143°}$ in rectangular form.

Solution   Use your calculator:

$$\therefore \quad 12.3/\underline{143°} \boxed{\to R} -9.82 + j7.40$$

EXERCISE 26-5

Calculator Drill
Express the following complex numbers in polar form.

1. $6 + j8$	2. $16 - j12$	3. $-2 + j4$
4. $5 + j3$	5. $-3 - j12$	6. $6 - j7$
7. $3.81 + j5.4$	8. $7.72 - j9.05$	9. $3.2 - j5.52$
10. $11.4 + j17.6$	11. $-14.2 - j19.5$	12. $10.9 - j5.52$
13. $17 - j1.2$	14. $23.1 + j303$	15. $-0.53 - j4.18$
16. $-12.6 + j10.5$	17. $0.225 + j0.903$	18. $3.72 - j10.6$

Express the following complex numbers in rectangular form.

19. $5/1.3^r$ 20. $3/210°$ 21. $7/1^r$

22. $8/45°$ 23. $6/-12°$ 24. $4/-135°$

25. $12/110°$ 26. $15/32°$ 27. $2/-57°$

28. $4.9/-72°$ 29. $7.7/243°$ 30. $10.3/140°$

31. $823/0.78^r$ 32. $30.1/-1.02^r$ 33. $0.19/-2.34^r$

34. $0.587/0.143^r$ 35. $296/4.68^r$ 36. $44.8/-0.28^r$

37. $128/527°$ 38. $42.3/9.56^r$ 39. $0.358/7.73^r$

26-7 RESOLVING SYSTEMS OF PHASORS AND VECTORS

Vectors have the same properties that phasors have—magnitude and direction. Mathematically, vectors and phasors follow the same rules. In this section we chose to use the word "phasor," but the following concepts also apply to vectors.

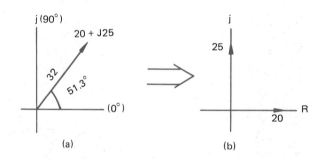

FIGURE 26–11
The polar phasor 32∠51.3° of part (a) is equivalent to the two components shown in part (b).

A phasor can be thought of graphically as having two components: one component along the R-axis and the second component along the j-axis. This is illustrated in Figure 26-11. Mathematically, this is accomplished by polar to rectangular conversion, as in: $1/0°$ $\boxed{\rightarrow R}$ $1 + j0$ and $1/90°$ $\boxed{\rightarrow R}$ $0 + j1$.

The sum of two phasors may be replaced by a single *resultant* phasor in polar form by applying the following guidelines.

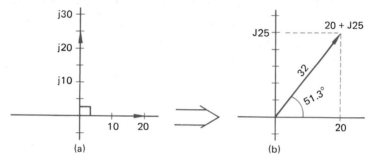

FIGURE 26-12
Resolving two phasors 90° apart by rectangular to polar conversion methods as demonstrated in Example 26-12.

Each of the four cases of Guideline 26-1 are demonstrated in the following series of examples.

**EXAMPLE
26-10**
Resolve the two phasors ($4\underline{/0°}$ and $6\underline{/0°}$) of Figure 26-13(a) into a resultant phasor.

Solution   Use Guideline 26-1 (case 1):

$$4 + 6 = 10$$
$$\therefore \quad 4\underline{/0°} + 6\underline{/0°} = 10\underline{/0°}$$

The resultant is shown as Figure 26-13(b).

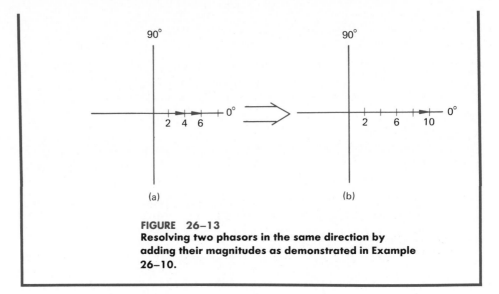

FIGURE 26–13
Resolving two phasors in the same direction by adding their magnitudes as demonstrated in Example 26–10.

**EXAMPLE
26-11**

Resolve the two phasors ($j15$ and $-j5$) of Figure 26-14(a) into a resultant phasor.

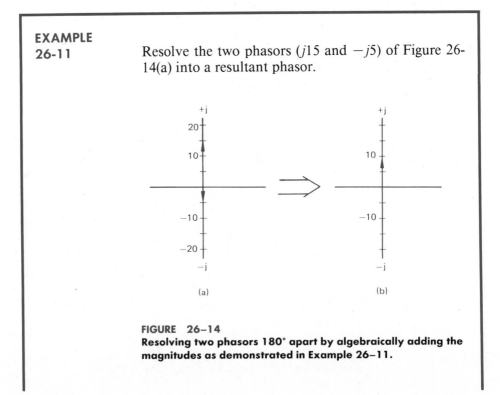

FIGURE 26–14
Resolving two phasors 180° apart by algebraically adding the magnitudes as demonstrated in Example 26–11.

Solution   Use Guideline 26-1 (case 2):

$$\therefore \quad j15 - j5 = j10$$

The resultant is shown as Figure 26-14(b).

**EXAMPLE
26-12**

Resolve the two phasors ($20\underline{/0°}$ and $25\underline{/90°}$) of Figure 26-12(a) into a resultant phasor.

Solution   Use Guideline 26-1 (case 3):

$$20\underline{/0°} \rightarrow 20 + j0$$
$$25\underline{/90°} \rightarrow 0 + j25$$
$$\therefore \quad 20 + j25 \boxed{\rightarrow P}\, 32\underline{/51.3°}$$

The resultant is shown as Figure 26-12(b).

**EXAMPLE
26-13**

Resolve the two phasors ($28.3\underline{/45°}$ and $15\underline{/-30°}$) of Figure 26-15(a) into a resultant phasor.

Solution   Use Guideline 26-1 (case 4):

$$28.3\underline{/45°} \boxed{\rightarrow R}\, 20 + j20$$
$$15\underline{/-30°} \boxed{\rightarrow R}\, 13 - j7.5$$

Plot each phasor in rectangular form as shown in Figure 26-15(b) and algebraically add:

Real component: $13 + 20 = 33$
Imaginary component: $j20 - j7.5 = j12.5$

Plot each component as shown in Figure 26-15(c). Resolve the two 90° phasors into a resultant phasor:

$$\therefore \quad 33 + j12.5 \boxed{\rightarrow P}\, 35.3\underline{/20.8°}$$

The resultant is shown as Figure 26-15(d).

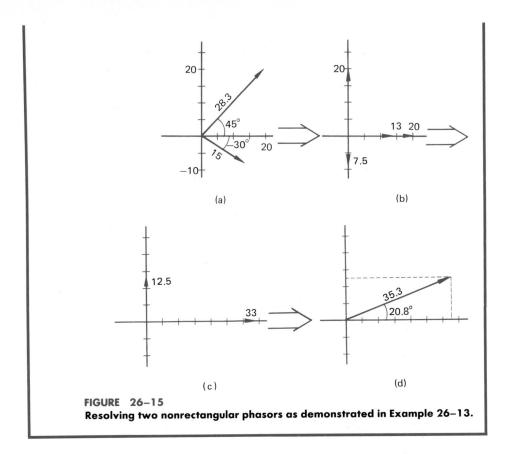

FIGURE 26-15
Resolving two nonrectangular phasors as demonstrated in Example 26-13.

EXERCISE 26-6
Resolve the following systems of phasors into a single resultant phasor by applying Guideline 26-1.

1. $\begin{cases} 5\underline{/0°} \\ 3\underline{/0°} \end{cases}$

2. $\begin{cases} 7\underline{/180°} \\ 12\underline{/0°} \end{cases}$

3. $\begin{cases} 2\underline{/0°} \\ 8\underline{/0°} \end{cases}$

4. $\begin{cases} 6\underline{/-90°} \\ 3\underline{/0°} \end{cases}$

5. $\begin{cases} 5\underline{/90°} \\ 15\underline{/-90°} \end{cases}$

6. $\begin{cases} 4\underline{/-180°} \\ 12\underline{/-90°} \end{cases}$

7. $\begin{cases} 11\underline{/90°} \\ 11\underline{/0°} \end{cases}$

8. $\begin{cases} 9\underline{/90°} \\ 4\underline{/180°} \end{cases}$

9. $\begin{cases} 16\underline{/90°} \\ 14\underline{/90°} \end{cases}$

10. $\begin{cases} 8\underline{/20°} \\ 6\underline{/-40°} \end{cases}$

11. $\begin{cases} 5\underline{/50°} \\ 7\underline{/120°} \end{cases}$

12. $\begin{cases} 4\underline{/-60°} \\ 10\underline{/-160°} \end{cases}$

13. $\begin{cases} 2.5\underline{/-20°} \\ 4.7\underline{/120°} \end{cases}$ 14. $\begin{cases} 7.6\underline{/1.2^r} \\ 2.2\underline{/-2^r} \end{cases}$ 15. $\begin{cases} 6.5\underline{/0.8^r} \\ 8.3\underline{/1.8^r} \end{cases}$

16. $\begin{cases} 15.8\underline{/-3.6^r} \\ 22.7\underline{/0.43^r} \end{cases}$ 17. $\begin{cases} 9.82\underline{/14.3°} \\ 4.36\underline{/-1.4^r} \end{cases}$ 18. $\begin{cases} 32.3\underline{/-5.77^r} \\ -51.7\underline{/-0.835^r} \end{cases}$

SELECTED TERMS

complex number The sum of a real number and an imaginary number.
imaginary number The square root of a negative quantity.
phasor A rotating vector with constant magnitude and angular velocity.
scalar A quantity having only magnitude.
vector A quantity having both magnitude and direction.

27

THE MATHEMATICS OF PHASORS

In the preceding chapter we introduced the three concepts of complex numbers, vectors, and phasors. They can be thought of graphically as arrows or as points. Mathematically, they can be written in the rectangular form or the polar form of complex numbers. Also, complex numbers, vectors, and phasors all follow the same rules.

In this chapter we define phasor operators that are extensions of addition, subtraction, multiplication, and division. Whether quantities are considered as vectors, phasors, or complex numbers, they are combined using the same operators. Thus, the techniques of this chapter can be used with vectors and complex numbers as well as with phasors.

27-1 ADDITION AND SUBTRACTION OF PHASORS

Addition

A phasor may be written in either the polar or rectangular form as a complex number. **To add two phasor quantities, the phasors must be in the rectangular form.** This important concept is summarized in Rule 27-1.

RULE 27-1. ADDING PHASORS

To add phasor quantities, express each in rectangular form and:

1. Add the real parts of the phasors.
2. Add the imaginary parts of the phasors.
3. Form the sum as a phasor written in rectangular form.

The following examples provide you with the opportunity to practice on phasors. It is important that you work along with your calculator. We will be using polar to rectangular $\boxed{\rightarrow R}$ and rectangular to polar $\boxed{\rightarrow P}$ calculator functions throughout this chapter.

EXAMPLE 27-1   Add $3 + j4$ and $5 + j6$.
  Solution   Use Rule 27-1:

$$
\begin{array}{r}
3 + j4 \\
\underline{5 + j6} \\
8 + j10
\end{array}
$$

EXAMPLE 27-2   Add $10\underline{/-53.13°}$ and $-5 - j7$.

Solution   Change $10\underline{/-53.13°}$ into rectangular form:

$$10\underline{/-53.13°} \boxed{\rightarrow \text{R}} 6 - j8$$

Add:

$$\begin{array}{r} 6 - j8 \\ -5 - j7 \\ \hline 1 - j15 \end{array}$$

EXAMPLE 27-3   Add $18.5\underline{/220°}$ and $4.2\underline{/-16°}$. Express the answer in polar form.

Solution   Change each phasor to its rectangular form:

$$18.5\underline{/220°} \boxed{\rightarrow \text{R}} - 14.17 - j11.89$$

$$4.2\underline{/-16°} \boxed{\rightarrow \text{R}} 4.04 - j1.16$$

Add:

$$\begin{array}{r} -14.17 - j11.89 \\ 4.04 - j1.16 \\ \hline -10.13 - j13.05 \end{array}$$

Express in polar form:

$$-10.13 - j13.05 \boxed{\rightarrow \text{P}} 16.52\underline{/-127.8°}$$

Subtraction

For phasors to be subtracted, they must be in the rectangular form. Rule 27-2 summarizes the steps used in subtracting.

RULE 27-2. SUBTRACTING PHASORS

To subtract phasor quantities, express each in rectangular form and:

1. Change the sign of both the real and the imaginary part of the phasor to be subtracted.
2. Add as in Rule 27-1.

EXAMPLE 27-4   Subtract $9 - j4$ from $15 + j12$.

Solution   Change the signs of $9 - j4$:

$$-(9 - j4) = -9 + j4$$

Add:

$$
\begin{array}{r}
15 + j12 \\
-9 + j4 \\
\hline
6 + j16
\end{array}
$$

EXAMPLE 27-5   Subtract $-7 - j3$ from $6 - j5$. Express the difference in polar form.

Solution   Change the signs of $-7 - j3$:

$$-(-7 - j3) = 7 + j3$$

Add:

$$
\begin{array}{r}
6 - j5 \\
7 + j3 \\
\hline
13 - j2
\end{array}
$$

Express in polar form:

$$13 - j2\,\boxed{\rightarrow P}\,13.15\underline{/-8.75°}$$

EXAMPLE 27-6   Subtract $29.3\underline{/142°}$ from $14.3\underline{/-73.5°}$. Express the answer in polar form.

Solution   Change each phasor to its rectangular form:

$$29.3\underline{/142°}\,\boxed{\rightarrow R}\,-23.09 + j18.04$$
$$14.3\underline{/-73.5°}\,\boxed{\rightarrow R}\,4.06 - j13.71$$

Change the signs of $-23.09 + j18.04$:

$$-(-23.09 + j18.04) = 23.09 - j18.04$$

Add:

$$4.06 - j13.71$$
$$\underline{23.09 - j18.04}$$
$$27.15 - j31.75$$

Express in polar form:

$$27.15 - j31.75 \boxed{\rightarrow P} \; 41.78\underline{/-49.47°}$$

EXERCISE 27-1

Perform the indicated addition or subtraction of the given phasor quantities. Express the answer in both rectangular and polar form.

1. $(2 + j5) + (5 - j3)$
2. $(3 + j0) + (-8 - j6)$
3. $(-4 + j7) - (8 + j6)$
4. $(9 - j9) + (0 + j10)$
5. $(12\underline{/30°}) + (5 + j2)$
6. $(16\underline{/47°}) - (-12 + j5)$
7. $(13 - j4) + (5\underline{/60°})$
8. $(-19 + j11) + (18\underline{/-20°})$
9. $(27 + j16) - (10\underline{/-140°})$
10. $(17\underline{/200°}) - (-28 - j12)$
11. $(0.52 + j1.85) + (2.14 - j0.91)$
12. $(-33.7 - j19.5) - (12.5 + j4.14)$
13. $(0.318\underline{/24°}) - (1.5 - j0.85)$
14. $(99 - j10) + (14.1\underline{/-108°})$
15. $(32\underline{/161°}) - (19\underline{/110°})$
16. $(43\underline{/-12°}) + (51\underline{/82°})$
17. $(45\underline{/18°}) + (30\underline{/59°})$
18. $(61\underline{/32°}) - (40\underline{/-72°})$
19. $(127.4\underline{/-2.2^r}) - (9.23\underline{/-0.17^r})$
20. $(0.583\underline{/0.932^r}) + (0.448\underline{/-2.84^r})$

27-2 MULTIPLICATION OF PHASORS

Phasor quantities may be multiplied when they are expressed in either the rectangular or polar form. Table 27-1 may be used to help multiply phasors in rectangular form. **However, the preferred form for multiplying phasors is the polar form.**

TABLE 27-1

DEFINITIONS USED WITH IMAGINARY NUMBERS	
$j = \sqrt{-1}$	$j^3 = -j$
$j^2 = -1$	$j^4 = +1$

Multiplying Phasors in Rectangular Form

Rule 27-3 summarizes the steps used to multiply phasors expressed in rectangular form.

> **RULE 27-3. MULTIPLYING PHASOR QUANTITIES IN RECTANGULAR FORM**
>
> To multiply phasors in rectangular form, multiply the numbers as if they were two binomials by:
>
> 1. Distributing the real part of the first complex number over the second complex number.
> 2. Distributing the imaginary part of the first complex number over the second complex number.
> 3. Replacing j^2 with -1.
> 4. Combining like terms.
> 5. Forming the product as a phasor written in rectangular form.

EXAMPLE 27-7   Multiply $3 + j2$ and $4 - j5$.

Solution   Use Rule 27-3.

Steps 1–2:   Distribute $(3 + j2)$ over $(4 - j5)$:

$$(3 + j2)(4 - j5) = 12 - j15 + j8 - j^2 10$$

Step 3:   Replace j^2 with -1:

$$12 - j15 + j8 + 10$$

Step 4:   Combine like terms:

$$22 - j7$$

$$\therefore \quad (3 + j2)(4 - j5) = 22 - j7$$

TABLE 27-2 SUMMARY OF MULTIPLYING COMPLEX NUMBERS

Pattern		Example
Real number × real number	⟹ real number	$3 \times 5 = 15$
Real number × imaginary number	⟹ imaginary number	$4 \times j2 = j8$
Imaginary number × imaginary number	⟹ real number	$j3 \times j2 = j^26 = -6$
		NOTE: $j^2 = -1$

From the preceding example, we learn that multiplying a real number by an imaginary number results in a product that is an imaginary number, and multiplying two imaginary numbers results in a product that is a real number. These concepts are summarized in Table 27-2.

EXAMPLE 27-8   Multiply $7 - j4$ by $-3 - j5$. Express the answer in polar form.

Solution   Use Rule 27-3.

Steps 1–2:   Distribute $7 - j4$ over $-3 - j5$:

$$(7 - j4)(-3 - j5) = -21 - j35 + j12 + j^220$$

Step 3:   Replace j^2 with -1:

$$-21 - j35 + j12 - 20$$

Step 4:   Combine like terms:

$$-41 - j23$$

Express $-41 - j23$ in polar form:

$$-41 - j23 \boxed{\rightarrow P} \; 47.01 \underline{/-150.7°}$$

$$\therefore \quad (7 - j4)(-3 - j5) = 47 \underline{/-150.7°}$$

Multiplying Phasors in Polar Form

Phasors are easily multiplied when they are expressed in polar form. Rule 27-4 summarizes the steps used to multiply phasors expressed in the polar form.

To multiply phasors in polar form:

1. Multiply the magnitudes.
2. Add the angles.
3. Form the products as a phasor written in polar form.

EXAMPLE 27-9   Multiply $3/\underline{10°}$ and $5/\underline{20°}$.

Solution   Use Rule 27-4.

Step 1:   Multiply the magnitudes:

$$3 \times 5 = 15$$

Step 2:   Add the angles:

$$10° + 20° = 30°$$

Step 3:   Form the product as a polar number:

$$15/\underline{30°}$$

$$\therefore \quad (3/\underline{10°})(5/\underline{20°}) = 15/\underline{30°}$$

EXAMPLE 27-10

Multiply $15.3/\underline{-83°}$ and $26.5/\underline{128°}$.

Solution   Use Rule 27-4.

Step 1:   Multiply the magnitudes:

$$15.3 \times 26.5 = 406$$

Step 2:   Add the angles:

$$-83° + 128° = 45°$$

Step 3:   Form the product as a polar number:

$$406/\underline{45°}$$

$$\therefore \quad (15.3/\underline{-83°})(26.5/\underline{128°}) = 406/\underline{45°}$$

**EXAMPLE
27-11**

Multiply $16.3 - j5.9$ by $22.8\underline{/17.0°}$. Express the answer in polar form.

Solution   Convert $16.3 - j5.9$ into polar form:

$$16.3 - j5.9 \boxed{\rightarrow P} 17.3\underline{/-19.9°}$$

Multiply $17.3\underline{/-19.9°}$ and $22.8\underline{/17.0°}$:

$$(17.3\underline{/-19.9°})(22.8\underline{/17.0°}) = 394.4\underline{/-2.9°}$$

$$\therefore \qquad (16.3 - j5.9)(22.8\underline{/17.0°}) = 394.4\underline{/-2.9°}$$

EXERCISE 27-2

Multiply the following phasors and express the answer in both polar and rectangular form:

1. $(2 + j2)(3 + j3)$
2. $(5 + j4)(6 + j2)$
3. $(7 + j2)(4 + j5)$
4. $(3 + j5)(6 + j3)$
5. $(3 - j4)(5 + j2)$
6. $(-2 + j6)(-2 - j6)$
7. $(5\underline{/15°})(3\underline{/12°})$
8. $(4\underline{/9°})(3\underline{/-20°})$
9. $(6\underline{/-40°})(2\underline{/-25°})$
10. $(7\underline{/14°})(5\underline{/-14°})$
11. $(38\underline{/115°})(17\underline{/-82°})$
12. $(4.5\underline{/-23°})(11.2\underline{/40°})$
13. $(111\underline{/22°})(98\underline{/76°})$
14. $(0.593\underline{/-3.6°})(0.218\underline{/8.3°})$
15. $(70.7\underline{/142°})(31.2\underline{/-105°})$
16. $(44.3\underline{/29.5°})(14.3\underline{/15.7°})$
17. $(29.1\underline{/-13.4°})(12.5\underline{/9.9°})$
18. $(6.08\underline{/-122.5°})(-9.2 - j3.07)$
19. $(0.358 - j1.12)(2.3 + j0.632)$
20. $(70.4 - j16.9)(-42.1 + j33.0)$
21. $(11.2\underline{/1.5^r})(17.6\underline{/2.2^r})$
22. $(0.34\underline{/-121°})(1.04\underline{/0.94^r})$
23. $(122.6\underline{/78.4°})(96.7\underline{/-0.833^r})$
24. $(16.8\underline{/-1.52^r})(7.65 + j10.8)$

27-3 DIVISION OF PHASORS

Division of phasor quantities may be carried out in either the polar or rectangular form. However, division with phasors in polar form is *preferred* because of its simplicity.

Dividing Phasors in Rectangular Form

To divide phasors in rectangular form, the divisor (denominator) must be changed to a real number. To change a divisor from a complex number to a real number, we use the *complex conjugate* of the divisor: A complex conjugate is the complex number with the sign of the imaginary part changed. Thus, $3 + j4$ and $3 - j4$ are complex conjugates of one another.

**EXAMPLE
27-12**

Show that the product of $3 + j4$ and its complex conjugate $3 - j4$ is a real number.

Solution The complex conjugate of $3 + j4$ is $3 - j4$.
Multiply:

$$(3 + j4)(3 - j4) = 9 - j12 + j12 - j^2 16$$
$$= 9 - j^2 16$$
$$= 9 + 16$$
$$= 25$$

$\therefore$ $(3 + j4)(3 - j4) = 25$, which is a real number.

Complex conjugates are used in ac electronics to compute the circuit loads for maximum power transfer.

Rule 27-5 summarizes the steps used to divide phasors represented by complex numbers in rectangular form.

RULE 27-5. DIVIDING PHASORS IN RECTANGULAR FORM

To divide phasors in rectangular form:

1. Multiply the divisor (denominator) and the dividend (numerator) by the complex conjugate of the divisor.
2. Divide the real number and the imaginary number of the dividend by the divisor.
3. Form the quotient as a phasor written in rectangular form.

**EXAMPLE
27-13**

Divide $(15 + j10)$ by $(2 + j1)$.

Solution $\dfrac{15 + j10}{2 + j1}$

Step 1:   Multiply numerator and denominator by $2 - j1$, the complex conjugate of $2 + j1$:

$$\frac{(15 + j10)(2 - j1)}{(2 + j1)(2 - j1)} = \frac{30 - j15 + j20 - j^2 10}{4 - j^2 1}$$

$$= \frac{40 + j5}{5}$$

Step 2: $$= \frac{40}{5} + \frac{j5}{5}$$

Step 3: $$= 8 + j1$$

$$\therefore \quad (15 + j10)/(2 + j1) = 8 + j1$$

**EXAMPLE
27-14**

Divide $4 - j5$ by $3 - j2$.

Solution

$$\frac{4 - j5}{3 - j2}$$

Step 1:   Multiply numerator and denominator by $3 + j2$, the complex conjugate of $3 - j2$:

$$\frac{(4 - j5)(3 + j2)}{(3 - j2)(3 + j2)} = \frac{12 + j8 - j15 - j^2 10}{9 - j^2 4}$$

$$= \frac{12 - j7 + 10}{9 + 4}$$

$$= \frac{22 - j7}{13}$$

Step 2: $$= \frac{22}{13} - \frac{j7}{13}$$

Step 3: $$= 1.7 - j0.54$$

$$\therefore \quad (4 - j5)/(3 - j2) = 1.7 - j0.54$$

Dividing Phasors in Polar Form

Phasors are easily divided when they are written in polar form. Rule 27-6 summarizes the steps used to divide phasors in polar form.

To divide phasors in polar form:

1. Divide the magnitudes.
2. Subtract the angle of the divisor (denominator) from the angle of the dividend (numerator).
3. Form the quotient as a phasor written in polar form.

EXAMPLE 27-15

Divide $15\underline{/20°}$ by $5\underline{/10°}$.

Solution Use Rule 27-6.

Step 1: Divide the magnitudes:

$$\frac{15}{5} = 3$$

Step 2: Subtract 10° from 20°.

$$20° - 10° = 10°$$

Step 3: Form the quotient:

$$3\underline{/10°}$$
$$\therefore \quad (15\underline{/20°})/(5\underline{/10°}) = 3\underline{/10°}$$

EXAMPLE 27-16

Divide $12.5\underline{/-78°}$ by $6.1\underline{/19.5°}$.

Solution

$$(12.5\underline{/-78°})/(6.1\underline{/19.5°}) = 2.05\underline{/-97.5°}$$

EXAMPLE 27-17

Divide $(13.5 - j3.7)$ by $20.9\underline{/42.8°}$.

Solution Change $(13.5 - j3.7)$ into polar form:

$$13.5 - j3.7 \boxed{\rightarrow P} 14.0\underline{/-15.3°}$$

Divide:

$$\therefore \quad (14.0\underline{/-15.3°})/(20.9\underline{/42.8°}) = 0.667\underline{/-58.1°}$$

Write the complex conjugates of the following complex numbers.

1. $5 - j3$
2. $3.2 + j8.1$
3. $12.5 - j4$
4. $-7.3 - j12$
5. $-9.4 + j10.6$
6. $14.3 + j15.7$

Divide the following phasors. Express the answer in both polar and rectangular forms.

7. $(2 + j3)/(3 + j4)$
8. $(5 - j6)/(7 + j1)$
9. $(1 + j2)/(6 - j2)$
10. $(-4 + j8)/(2 - j5)$
11. $(-10 - j5)/(2 + j2)$
12. $(8 + j4)/(2 + j4)$
13. $(18\underline{/45°})/(6\underline{/-15°})$
14. $(33\underline{/-5°})/(3\underline{/-70°})$
15. $(75\underline{/-80°})/(25\underline{/-40°})$
16. $(39\underline{/52°})/(13\underline{/-18°})$
17. $(28\underline{/112°})/(14\underline{/-37°})$
18. $(347\underline{/-161°})/(84\underline{/7.5°})$
19. $(88.9\underline{/-29.2°})/(215\underline{/61.3°})$
20. $(0.913\underline{/102°})/(0.491\underline{/58.8°})$
21. $(18 + j12)/(7\underline{/0.23^r})$
22. $(23\underline{/-0.14^r})/(6 - j5)$
23. $(-0.892 - j1.02)/(2.3\underline{/2.35^r})$
24. $(43.8\underline{/-1.72^r})/(27.2 - j39.5)$

27-4 POWERS AND ROOTS OF PHASORS

A phasor quantity is most easily raised to a power when it is expressed in polar form. Rule 27-7 provides the procedure for raising a phasor to a power.

RULE 27-7. POWER OF A PHASOR

To raise a phasor to a power, express the phasor in polar form and:

1. Raise the magnitude to the specified power.
2. Multiply the angle by the exponent.
3. Form the solution as a phasor written in polar form.

EXAMPLE 27-18   Solve $(5\underline{/12°})^3$.

Solution   Use Rule 27-7.

Step 1: Raise 5 to the third power:

$$\boxed{y^x}\ 5^3 = 125$$

Step 2: Multiply 12° by 3:

$$\boxed{\times}\ 12° \times 3 = 36°$$

Step 3: Form the solution:

$$125\underline{/36°}$$

$$\therefore \quad (5\underline{/12°})^3 = 125\underline{/36°}$$

EXAMPLE 27-19

Solve $(4 + j6)^2$.

Solution Use Rule 27-7. First, express $(4 + j6)$ in polar form:

$$4 + j6\ \boxed{\to P}\ 7.21\underline{/56.3°}$$

Step 1: Raise 7.21 to the second power:

$$\boxed{x^2}\quad 7.21^2 = 52.0$$

Step 2: Multiply 56.3° by 2:

$$\boxed{\times}\quad 56.3 \times 2 = 112.6°$$

Step 3: Form the solution:

$$52.0\underline{/112.6°}$$

$$\therefore \quad (4 + j6)^2 = (7.21\underline{/56.3°})^2 = 52.0\underline{/112.6°}$$

To take the root of a phasor quantity, express the phasor in polar form and apply Rule 27-8.

EXAMPLE 27-20

Solve $\sqrt{25\underline{/10°}}$.

Solution Use Rule 27-8.

Step 1: Take the root of 25:

$$\boxed{\sqrt{x}} \qquad \sqrt{25} = 5$$

Step 2: Divide 10° by the index:

$$\boxed{\div}\, 10°/2 = 5°$$

Step 3: Form the solution:

$$5\underline{/5°}$$

$$\therefore \qquad \sqrt{25\underline{/10°}} = 5\underline{/5°}$$

EXAMPLE 27-21

Solve $\sqrt[3]{7.2 - j10.9}$.

Solution Express $7.2 - j10.9$ as a polar number:

$$7.2 - j10.9 \,\boxed{\rightarrow P}\, 13.06\underline{/-56.55°}$$

Step 1: Take the third root of 13.06:

$$\boxed{y^x} \qquad 13.06^{1/3} = 2.35$$

Step 2: Divide $-56.55°$ by the index of the radical:

$$\boxed{\div} \qquad -56.55°/3 = -18.85°$$

CHAPTER 27   THE MATHEMATICS OF PHASORS

Step 3:
Step 3: Form the solution:

$$2.35\underline{/-18.85°}$$

$$\therefore \quad \sqrt[3]{7.2 - j10.9} = 2.35\underline{/-18.85°}$$

**EXAMPLE
27-22**

Solve $(13.58\underline{/-126°})^{1/2}$. Express the solution in rectangular form.

Solution

$$13.58^{1/2} = 3.69$$
$$-126° \times \tfrac{1}{2} = -63°$$

Form the solution:

$$3.69\underline{/-63°}$$

Change to rectangular form:

$$3.69\underline{/-63°} \boxed{\rightarrow R} 1.68 - j3.29$$

$$\therefore \quad (13.58\underline{/-126°})^{1/2} = 3.69\underline{/-63°} = 1.68 - j3.29$$

EXERCISE 27-4

Solve each of the following. Express the solution in polar form.

1. $(3\underline{/10°})^2$

2. $(3\underline{/4°})^3$

3. $(2\underline{/10°})^4$

4. $(7\underline{/30°})^2$

5. $(12\underline{/-17°})^2$

6. $(3.25\underline{/-20°})^3$

7. $(14.3\underline{/56.2°})^2$

8. $(410\underline{/26.9°})^2$

9. $(16.8 - j4.2)^2$

10. $(-71.4 - j40.8)^3$

11. $\sqrt{49\underline{/40°}}$

12. $\sqrt{121\underline{/-72°}}$

13. $(125\underline{/-96°})^{1/3}$

14. $(196\underline{/46°})^{1/2}$

15. $(2.72\underline{/17.6°})^{1/2}$

16. $\sqrt[3]{84.3\underline{/-118°}}$

17. $\sqrt{-6.13 + j15.2}$

18. $(70.2 + j64.6)^{1/3}$

19. $(-81.0 - j12.5)^{1/2}$

20. $\sqrt{34.7 + j42.9}$

21. $\sqrt[4]{-143 - j391}$

22. $\sqrt{19.8\underline{/-2.87^r}}$

23. $\sqrt{0.843\underline{/0.571^r}}$

24. $(526.7\underline{/2.95^r})^{1/3}$

SELECTED TERMS

complex conjugate   A complex conjugate of a complex number is that complex number with the sign of the imaginary part changed. $2 - j3$ and $2 + j3$ are complex conjugates of one another.

28

ALTERNATING-CURRENT FUNDAMENTALS

This chapter introduces you to alternating current (ac) by exploring the sinusoidal (sine wave) properties of alternating current. The chapter begins by introducing the basic names used to describe the sinusoidal nature of alternating current, and then phasors are used to present a picture of the relationship between the current, the voltage, and the components used in circuits.

28-1 ALTERNATING-CURRENT TERMINOLOGY

Frequency

When a conductor is rotated through a magnetic flux as pictured in Figure 28-1, a voltage is created. The number of times in a second that the conductor is rotated through a complete revolution is called the *frequency* of the voltage. For example, if a conductor is rotated 20 times in 5 seconds, the frequency of the resulting sine wave is 4 cycles per second. The unit for frequency is the hertz (Hz). A frequency of 4 cycles per second is written as 4 Hz.

EXAMPLE 28-1   Express each of the following frequencies in hertz.

      (a) 39 cycles in 3 s
      (b) 100 cycles in 200 ms
      (c) 1500 revolutions in 0.375 s
      (d) 300 rotations in 1.0 min

Solution   (a) $39/3 = 13$ cycles per second $= 13$ Hz
      (b) $100/0.2 = 500$ cycles per second $= 500$ Hz
      (c) $1500/0.375 = 4000$ cycles per second $= 4$ kHz
      (d) $300/60 = 5$ cycles per second $= 5$ Hz

Period

The time it takes for one complete cycle of the conductor through the magnetic flux is called the *period* of the resulting sine wave. Thus, the period of the voltage created by rotating a conductor 60 times a second (60 Hz) through a flux is $\frac{1}{60}$ or 16.67 ms. The relationship between frequency (f) in hertz and period (T) in seconds is expressed as Equation 28-1.

$$T = 1/f \quad \text{s} \tag{28-1}$$

EXAMPLE 28-2   Express the period of each of the following.

(a) 400 Hz   (b) 1.2 MHz   (c) 20 Hz   (d) 5 kHz

Solution   Use Equation 28-1.

(a) $T = 1/400 = 2.50$ ms
(b) $T = 1/(1.2 \times 10^6) = 833$ ns
(c) $T = 1/20 = 50$ ms
(d) $T = 1/(5 \times 10^3) = 200$ μs

Angular Velocity

Each time the conductor of Figure 28-1 rotates one cycle through the flux, it travels 6.28 radians. *Angular velocity* is expressed in radians per second (rad/s). Thus, a wave generated by a conductor traveling 200 radians in 25 seconds would have an angular velocity of 200 rad/25 s or 8 rad/s. Equation 28-2 relates angular velocity (ω) in radians per second to angular displacement (α) in radians and time (t) in seconds.

$$\omega = \alpha/t \qquad \text{rad/s} \tag{28-2}$$

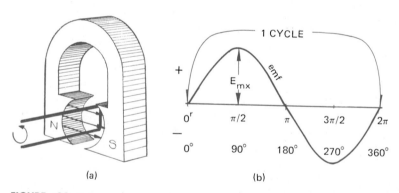

(a)   (b)

FIGURE   28-1

(a) As the conductor is rotated through the magnetic flux, an emf is created. (b) The emf has a sinusoidal wave shape.

EXAMPLE 28-3   Express each of the following as angular velocity in radians per second.

(a) 60 rad in 12 s

(b) 2.4 krad in 0.8 min (0.8 min = 48 s)

(c) 320 rad in 160 s

(d) 45 Mrad in 15 s

Solution   Use Equation 28-2.

(a) $\omega = 60/12 = 5$ rad/s

(b) $\omega = 2400/48 = 50$ rad/s

(c) $\omega = 320/160 = 2$ rad/s

(d) $\omega = (45 \times 10^6)/15 = 3$ Mrad/s

Frequency (f) in hertz and angular velocity (ω) in radians per second are related by Equation 28-3, where 2π (6.28) is the number of radians traveled per cycle. The symbol ω (omega) is used exclusively for angular velocity in electronics.

$$\omega = 2\pi f \quad \text{rad/s} \qquad (28\text{-}3)$$

EXAMPLE 28-4   Express each of the following frequencies as an angular velocity in radians per second (rad/s).

(a) 60 Hz   (b) 400 Hz   (c) 5 kHz   (d) 200 kHz

Solution   Use Equation 28-3.

(a) $\omega = 2\pi 60 = 377$ rad/s

(b) $\omega = 2\pi 400 = 2.51$ krad/s

(c) $\omega = 2\pi 5 \times 10^3 = 31.4$ krad/s

(d) $\omega = 2\pi 200 \times 10^3 = 1.26$ Mrad/s

EXERCISE 28-1

Express the frequency of each of the following waves in hertz.

1. 85 cycles in 5 seconds
2. 200 revolutions in 3 minutes
3. 10 rotations in 2 seconds
4. 120 cycles per second
5. A wave with a period of 28 ms

6. A wave with a period of 8 μs
7. A wave with $\omega = 377$ rad/s
8. A wave with $\omega = 12.57$ krad/s

Determine the period of each of the following waves in seconds.

    9. 60 Hz   10. 15 Hz   11. 12 kHz
  12. 5.4 MHz   13. 25 rev/min   14. 4200 rev/min
  15. 512 rad/s   16. 212 krad/s

Determine the angular velocity of each of the following waves in radians per second.

  17. 3000 rad in 15 s   18. 50 krad in 10 s
  19. 1.2 krev in 4 s   20. 900 Hz
  21. 6 kHz   22. 1.0 MHz
  23. Period of 20 ms   24. Period of 5 μs

28-2 RESISTANCE

Figure 28-2(a) pictures an alternating current passing through the resistance R. The voltage wave of Figure 28-2(a) is the result of the current passing through the resistance. When the curve depicting current is superimposed over the curve depicting voltage, as shown in Figure 28-2(b), you will notice that the two waves start and finish together. Furthermore, both waves reach

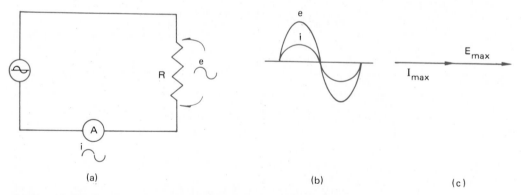

(a)   (b)   (c)

FIGURE 28-2
When an alternating current passes through a resistance (a), the resulting voltage is in phase with the current (b), and the phase angle (the angle between the current and voltage vector) is 0° (c).

their maximum positive amplitude and their maximum negative amplitude together. When the current and voltage pass through corresponding points in their cycle at the same time, the two waves are said to be *in phase*. Figure 28-2(c) is the phasor diagram of the current and voltage in phase. The angle between the current and voltage phasor in a phasor diagram is called the *phase angle*. The Greek letter theta (θ) is used to designate the phase angle in this text.

To summarize, when an ac source of voltage is applied to a resistive load, the current and voltage are in phase ($\angle\theta = 0°$), and Ohm's law ($E = IR$) may be applied.

EXAMPLE 28-5   A 100-Ω resistor is connected to a 200-V ac source.

(a) Determine the current.
(b) Determine the phase angle.
(c) Draw the phasor diagram representing the current and voltage.

Solution   (a) $I = E/R$:

$$I = \frac{200}{100} = 2 \text{ A}$$

(b) Phase angle for a resistive circuit:

$$\angle\theta = 0°$$

(c) See Figure 28-3.

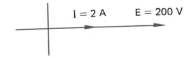

$I = 2$ A $E = 200$ V

FIGURE 28–3
Phasor diagram of current and voltage for Example 28–5.

EXAMPLE 28-6   Determine the value of the resistive load for a circuit having the phasor diagram of Figure 28-4.

Solution

$$R = E/I$$

Substitute $E = 117$ V and $I = 1.20$ A:

$$R = 117/1.20 = 97.5 \ \Omega$$

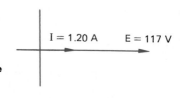

FIGURE 28-4
The phasor diagram for Example
28-6.

I = 1.20 A   E = 117 V

EXERCISE 28-2
Determine the resistive load for each of the phasor diagrams.

1. 5 A   50 V   2. 30 mA   90 V

3. 280 mA   30V   4. 5 V   8 A

5. 510 μA   18 V   6. 300 mA   12.6V

7. 2 V   12.2 A   8. 47 mA   36 V

Determine the source voltage for each of the following.

9. $I = 1.25$ A $R = 50 \, \Omega$ $\theta = 0°$
10. $I = 78$ mA $R = 2.7 \, k\Omega$ $\theta = 0°$
11. $I = 22 \, \mu A$ $R = 1.8 \, M\Omega$ $\theta = 0°$

Determine the circuit current for each of the following.

12. $E = 117$ V $R = 35 \, \Omega$ $\theta = 0°$
13. $E = 2.5$ kV $R = 6.8 \, M\Omega$ $\theta = 0°$
14. $E = 14.7$ V $R = 330 \, \Omega$ $\theta = 0°$
15. $E = 82.0$ V $R = 47 \, \Omega$ $\theta = 0°$

28-3 INDUCTANCE AND INDUCTIVE REACTANCE

Figure 28-5 pictures several inductors (coils) and their schematic symbols.
The letter L is used to designate an inductor in a schematic. The *inductance*
of an inductor is measured in henries (H). The inductance in henries of an
inductor depends upon the physical makeup of the coil. From Equation 28-4
we see that the inductance (L) is dependent on the length (l), cross-sectional
area (A), number of turns of wire (N), and the material contained within the
core (μ).

$$L = \frac{N^2\mu A}{l} \quad H \tag{28-4}$$

$$L = \frac{N^2 \mu A}{l} \quad \text{H}$$

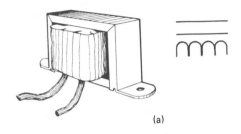

(a)

(b)

FIGURE 28–5
Inductors: (a) iron core for audio-frequency (AF) applications; (b) air core for radio-frequency (RF) applications.

When an alternating current passes through an inductor, the inductor *"reacts"* to the current by producing an opposition to the flow of current. This opposition is the *inductive reactance* of the inductor. The reactance of an inductor (X_L) to alternating current is described by Equation 28-5 and is expressed in ohms.

$$X_L = 2\pi fL \quad \Omega \tag{28-5}$$

where X_L = inductive reactance in ohms (Ω)
$\quad\quad f$ = frequency of the ac sinusoidal current in hertz (Hz)
$\quad\quad L$ = inductance in henries (H)

From Equation 28-5 we see that X_L is a direct function of frequency, and that an inductor has zero reactance when $f = 0$ Hz. Thus, direct current ($f = 0$ Hz) will have no opposition owing to the reactive properties of the inductor. Because of the frequency dependence of the inductive reactance, an inductor may be used to *filter* out unwanted frequencies, while allowing wanted frequencies to pass. This is demonstrated in the circuit of Figure 28-6.

EXAMPLE 28-7   Determine the reactance of the 100-mH RF *choke coil* of Figure 28-6 to a frequency of:

(a) 0 Hz (dc)
(b) 1.0 MHz (RF)

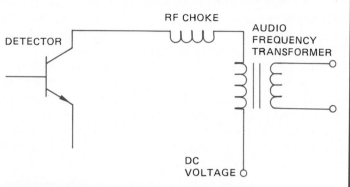

FIGURE 28-6
The indicator in the dc power-supply circuit blocks the passage of radio-frequency current while allowing the dc and audio-frequency currents to pass.

Solution Use Equation 28-5:

(a) Substitute $f = 0$ Hz and $L = 100$ mH:

$$X_L = 2\pi(0)(100 \times 10^{-3}) = 0\ \Omega$$

(b) Substitute $f = 1.0$ MHz and $L = 100$ mH:

$$X_L = 2\pi(1.0 \times 10^6)(100 \times 10^{-3}) = 628\ \text{k}\Omega$$

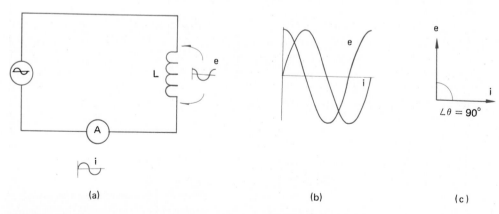

FIGURE 28-7
When ac sinusoidal current passes through an inductor (a), the voltage leads the current (b) by 90°, as shown by the phasor diagram (c).

The inductor derives its opposition to alternating current from its magnetic properties, and these same properties cause the voltage drop across the inductor to lead the current passing through the inductor by 90°. Thus, $\angle\theta = 90°$. Figure 28-7 shows the relationship of the current and voltage waves of the inductor, as well as the phasor diagram of the current and voltage.

To summarize, the application of an ac voltage to an inductive load causes an ac current to flow, which causes the inductor to react. This process causes a *phase shift* between the current and voltage of the inductor, resulting in a phase angle (θ) of 90°. Therefore, in a perfect inductor (one without resistance), voltage leads current by 90°. Ohm's law, in the form $E = IX_L$, is used to compute the circuit conditions.

EXAMPLE 28-8   A 2-H inductor is connected to a 24-V, 60-Hz ac source.

    (a) Determine the circuit current.

    (b) Determine the phase angle.

    (c) Draw the phasor diagram of the current and voltage phasors.

Solution   (a) $I = E/X_L$; $X_L = 2\pi fL$.

Substitute the second equation into the first:

$$I = E/(2\pi fL)$$
$$= 24/[2\pi 60(2)]$$
$$\therefore \quad I = 31.8 \text{ mA}$$

    (b) Phase angle for a perfectly inductive circuit with current as the reference phasor:

$$\angle\theta = 90°$$

    (c) See Figure 28-8.

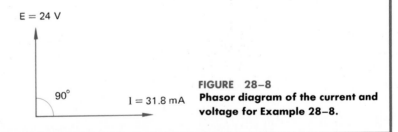

FIGURE 28–8
Phasor diagram of the current and voltage for Example 28–8.

EXAMPLE 28-9 From the phasor diagram of Figure 28-9, determine the inductance of the inductor in henries for a frequency of 2.0 kHz.

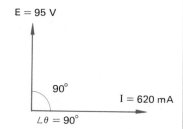

FIGURE 28-9
Phasor diagram for Example 28-9.

Solution

$$E = IX_L$$
$$X_L = E/I$$

Substitute $E = 95$ V and $I = 620$ mA:

$$X_L = 95/(620 \times 10^{-3}) = 153 \ \Omega$$

Solve $X_L = 2\pi f L$ for L:

$$L = X_L/2\pi f$$
$$\therefore \quad L = 153/(2\pi 2 \times 10^3) = 12.2 \text{ mH}$$

EXERCISE 28-3
Determine the inductive reactance X_L of each of the following inductors at (a) 60 Hz, (b) 5 kHz, and (c) 2 MHz.

1. 4 H   2. 10 mH   3. 250 mH
4. 120 μH   5. 12 H   6. 600 mH

Use the information given in each of the following phasor diagrams to determine (a) the inductive reactance and (b) the inductance of the inductor at a frequency of 1 kHz.

7. $E = 100$ V   8. $E = 35$ V
$\quad I = 10$ A $\quad I = 500$ mA

9. $E = 15$ V   10. $E = 75$ V

       $I = 6$ mA $I = 150$ μA

Determine the source voltage for each of the following.

11. $I = 2$ A	$X_L = 100$ Ω	$\theta = 90°$	
12. $I = 220$ mA	$X_L = 1.6$ kΩ	$\theta = 90°$	

Determine the circuit current for each of the following.

13. $E = 28$ V	$L = 2$ H	$f = 100$ Hz	$\theta = 90°$
14. $E = 6.3$ V	$L = 570$ mH	$f = 2.7$ kHz	$\theta = 90°$
15. $E = 132$ V	$L = 800$ μH	$f = 3.15$ MHz	$\theta = 90°$

28-4 CAPACITANCE AND CAPACITIVE REACTANCE

Figure 28-10 shows several capacitors and their schematic symbols. The letter C is used to designate a capacitor in a circuit diagram. The *capacitance* of a capacitor is measured in farads (F). Capacitance depends upon the construction of the capacitor. From Equation 28-6, the capacitance (C) is dependent on the area of the metallic plates (A), the spacing between the plates (d), and

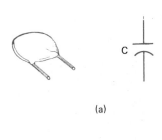

(a)

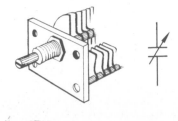

(b)

FIGURE 28–10
(a) Fixed ceramic disc capacitor;
(b) variable capacitor.

the permittivity of the material used as a dielectric (insulator) between the plates (ϵ).

$$C = \epsilon A/d \quad F \tag{28-6}$$

When an ac voltage is placed across a capacitor, the capacitor "reacts" to the flow of current by producing an opposition to the current. This opposition is the *capacitance reactance* of the capacitor to the alternating current. The reactance of a capacitor (X_C) to alternating current is given by Equation 28-7 and is expressed in ohms.

$$X_C = 1/(2\pi fC) \quad \Omega \tag{28-7}$$

where X_C = capacitive reactance in ohms (Ω)
  f = frequency of the current in hertz (Hz) ($\omega = 2\pi f$ rad/s)
  C = capacitance in farads (F)

From Equation 28-7 we see that X_C is an inverse function of frequency and that a capacitor has an extremely large reactance when $f = 0$ Hz. Thus, dc current ($f = 0$ Hz) will be blocked due to the reactive properties of the capacitor. Because of the frequency dependence of the capacitive reactance, a capacitor may be used to *filter* out unwanted frequencies while allowing wanted frequencies to pass. This is demonstrated by the circuit of Figure 28-11.

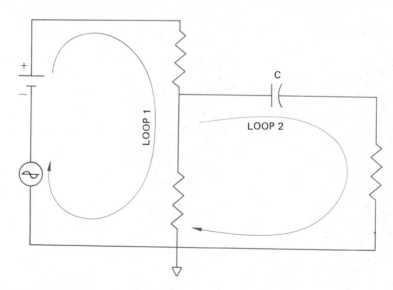

FIGURE 28-11
The current of loop 1 is a mixture of both direct and alternating currents, but the current of loop 2 is only alternating. This is due to the blocking of the direct current by capacitor C.

EXAMPLE 28-10

Determine the reactance of the 1.0-μF capacitor of Figure 28-11 to a frequency of

(a) 0 Hz (dc)
(b) 1.0 MHz (RF)

Solution $X_C = 1/(2\pi fC)$.

(a) Substitute $f = 0$ Hz and $C = 1.0\ \mu$F:

$$X_C = \frac{1}{2\pi(0)(1 \times 10^{-6})} \Rightarrow \infty\ \text{(undefined)}$$

(b) Substitute $f = 1.0$ MHz and $C = 1.0\ \mu$F:

$$X_C = \frac{1}{2\pi(1 \times 10^6 \times 1 \times 10^{-6})} = 0.159\ \Omega$$

The capacitor opposes alternating current because of its inability to instantly take on charge. It is this opposition to the flow of charge that results in the voltage drop across the capacitor lagging the current passing through the capacitor by 90°. Thus, $\angle\theta = -90°$. Figure 28-12 shows the relationship of the current and voltage wave of the capacitor, as well as the phasor diagram of the current and voltage.

To summarize, the application of an ac voltage to a capacitive load causes the capacitor to react ($X_C = 1/\omega C$). In this process the voltage lags behind the current by 90°, which results in a phase angle (θ) of $-90°$. Ohm's

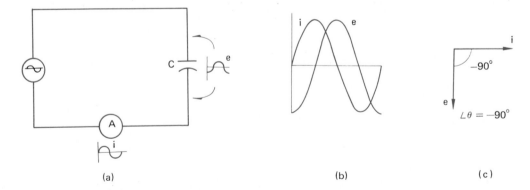

(a)   (b)   (c)

FIGURE 28-12
When alternating sinusoidal current passes through a capacitor (a), the voltage lags the current (b) by 90°, as shown by the phasor diagram (c).

law in the form $E = IX_C$ is used to compute the circuit condition of a perfect capacitive circuit (one without resistance).

EXAMPLE 28-11

A 10-μF capacitor is connected to a 117-V, 60-Hz source.

(a) Determine the circuit current.
(b) Determine the phase angle.
(c) Draw the phasor diagram of the current and voltage phasors.

Solution

(a) $I = E/X_C$, $X_C = 1/(2\pi fC)$.
Substitute the second equation into the first:

$$I = E \Big/ \left(\frac{1}{2\pi fC} \right)$$
$$= E2\pi fC$$
$$= 117(2\pi)(60)(10 \times 10^{-6})$$
$$\therefore \quad I = 441 \text{ mA}$$

(b) Phase angle for a perfectly capacitive circuit with current as the reference phasor:

$$\therefore \angle\theta = -90°$$

(c) See Figure 28-13.

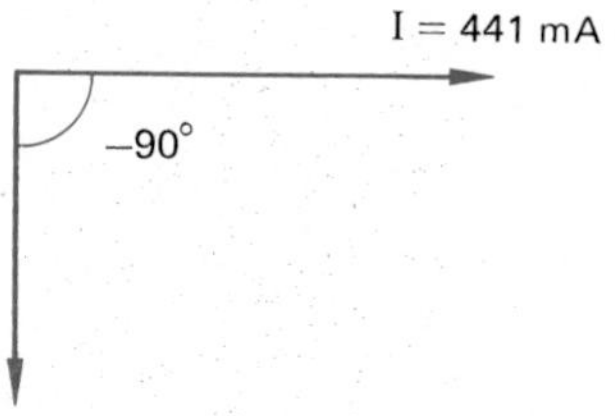

FIGURE 28–13
Phasor diagram of the current and voltage for Example 28–11.
$\angle\theta = -90°$.

I = 441 mA
−90°
E = 117 V

EXAMPLE 28-12

From the phasor diagram of Figure 28-14, determine the capacitance of the capacitor in farads for a frequency of 100 Hz.

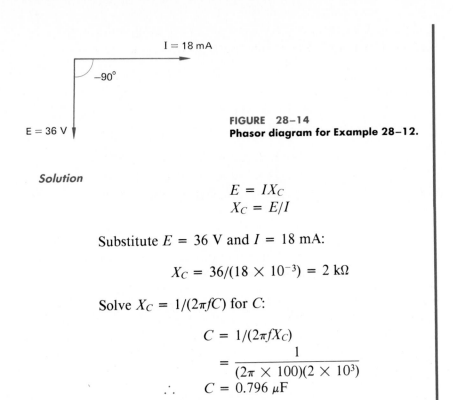

I = 18 mA

−90°

E = 36 V

FIGURE 28–14
Phasor diagram for Example 28–12.

Solution

$$E = IX_C$$
$$X_C = E/I$$

Substitute $E = 36$ V and $I = 18$ mA:

$$X_C = 36/(18 \times 10^{-3}) = 2 \text{ k}\Omega$$

Solve $X_C = 1/(2\pi fC)$ for C:

$$C = 1/(2\pi fX_C)$$
$$= \frac{1}{(2\pi \times 100)(2 \times 10^3)}$$
$$\therefore \quad C = 0.796 \ \mu\text{F}$$

EXERCISE 28-4

Determine the capacitive reactance X_C of each of the following capacitors at (a) 60 Hz, (b) 5 kHz, and (c) 2 MHz.

1. 15 μF
2. 270 pF
3. 0.5 μF
4. 3 μF
5. 1000 μF
6. 82 pF

Use the information given in each of the following phasor diagrams to determine the following: (a) the capacitance reactance and (b) the capacitance of the capacitor at a frequency of 500 Hz.

7. $I = 620$ mA
   $E = 20$ V

8. $I = 500$ μA
   $E = 2.0$ V

9. $I = 4$ A
   $E = 15$ V

10. $I = 12$ mA
    $E = 60$ V

Determine the source voltage for each of the following.

11. $I = 130$ mA $X_C = 42.5\ \Omega$ $\theta = -90°$
12. $I = 4.3$ A $X_C = 10.8\ \Omega$ $\theta = -90°$

Determine the circuit current for each of the following.

13. $E = 523$ V $C = 10\ \mu$F $f = 15$ Hz $\theta = -90°$
14. $E = 27$ V $C = 0.1\ \mu$F $f = 400$ Hz $\theta = -90°$
15. $E = 440$ V $C = 720$ pF $f = 900$ kHz $\theta = -90°$

28-5 VOLTAGE PHASOR FOR SERIES CIRCUITS

In the previous sections we learned that the passage of alternating current through a resistance, an inductance, or a capacitance resulted in a phase angle that was different for each of the devices. In this section, we are concerned with the result of passing alternating current through a series circuit made up of a resistance, a capacitance, and an inductance. Table 28-1 summarizes the concepts of the previous three sections of this chapter.

Current in an AC Series Circuit

In an ac series circuit, the current (I) is the same throughout the circuit. Therefore, current is selected as the reference phasor in ac series circuits. A phasor diagram showing the voltage drops in relation to the circuit may be constructed. The following example develops this idea.

**EXAMPLE
28-13**   Construct the phasor diagram for the circuit of Figure 28-15.

Solution   Determine the voltage drops:

$$|V_R| = IR = 5(8) = 40\text{ V}$$
$$|V_L| = IX_L = 5(6) = 30\text{ V}$$
$$|V_C| = IX_C = 5(12) = 60\text{ V}$$

Use the information of Table 28-1 to determine the phase angle between V and I. Select I as the reference phasor (0°).

TABLE 28-1

SUMMARY OF THE EFFECTS OF ALTERNATING CURRENT ON RESISTANCE, INDUCTANCE, AND CAPACITANCE

Property				Opposition			Ohm's Law
Name	Symbol	Phase (θ)	EI Phasor	Name	Unit	Dependence on Frequency	
Resistance		E and I are in phase: $\theta = 0°$		Resistance	Ohm	None	$E = IR$
Inductance		E leads I by $90°$: $\theta = 90°$		Inductive reactance	Ohm	$X_L = 2\pi f L$	$E = IX_L$
Capacitance		E lags I by $90°$: $\theta = -90°$		Capacitive reactance	Ohm	$X_C = \dfrac{1}{2\pi f C}$	$E = IX_C$

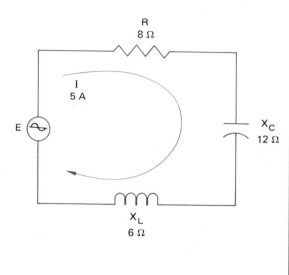

FIGURE 28–15
Series circuit for Example 28–13.

V_R in phase (0°) with I
V_L leads I by 90°
V_C lags I by 90°

Construct the circuit phasor diagram as in Figure 28-16.

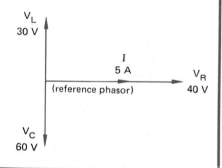

FIGURE 28–16
Phasor diagram for Example 28–13. Voltage leads current in the inductor, while voltage lags current in the capacitor.

Voltage Phasor

From the preceding example we see that in a series circuit the voltage phasor is constructed so that V_R is along the positive R-axis, V_L along the positive j-axis, and V_C along the negative j-axis. Figure 28-17 pictures the voltage phasor diagram for a series ac circuit with resistive, capacitive, and inductive components.

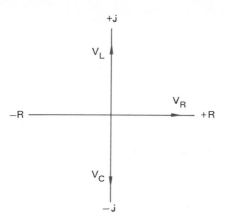

FIGURE 28–17
Generalized voltage phasor diagram for a series LCR circuit.

Kirchhoff's Voltage Law (KVL)

Kirchhoff's voltage law (KVL), $E = V_1 + V_2 + V_3$, etc., is valid for ac series circuits. However, it must be remembered that ac quantities are represented by complex numbers and must be operated on with phasor addition. The following example demonstrates this concept.

EXAMPLE 28-14

Verify Kirchhoff's voltage law for the circuit of Figure 28-18.

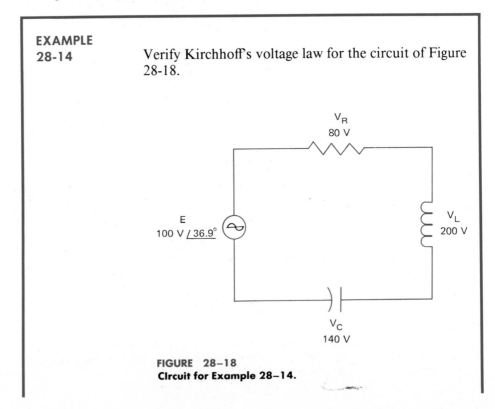

FIGURE 28–18
Circuit for Example 28–14.

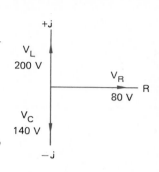

FIGURE 28-19
Voltage phasor diagram for the circuit of Figure 28-18.

Solution   Construct the voltage phasor diagram as shown in Figure 28-19.

Express each voltage phasor as a complex number in rectangular form:

$$V_R = 80 + j0 \text{ V}$$
$$V_L = 0 + j200 \text{ V}$$
$$V_C = 0 - j140 \text{ V}$$

Add the voltage phasors and express the sum in polar form:

$$
\begin{array}{l}
80 + j0 \\
0 + j200 \\
\underline{0 - j140} \\
80 + j60 \quad \boxed{\rightarrow P} \; 100\underline{/36.9°} \text{ V}
\end{array}
$$

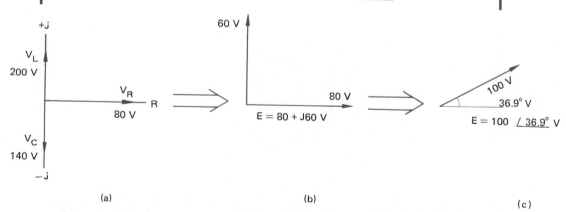

(a)   (b)   (c)

FIGURE 28-20
Solution to Example 28-14: (a) voltage phasor diagram for the voltage drops in the series circuit of Figure 28-18; (b) phasor addition of the voltages; (c) resolution of the rectangular voltage phasors of (b) into a polar phasor.

$$\therefore \quad V_R + V_L + V_C = 100\underline{/36.9°} \text{ V}$$

Kirchhoff's voltage law holds for the circuit of Figure 28-18. Figure 28-20 shows the sequence to the solution of Example 28-14 through phasor diagrams.

EXAMPLE 28-15

A 500-Ω wire-wound resistor is connected in series with a 0.5-μF capacitor and a 1-H inductor. The circuit is attached to a *test oscillator* operating at 300 Hz. The alternating current is measured as 180 mA.

(a) Compute the voltage drop across each component in the circuit.
(b) Construct the voltage phasor diagram.
(c) Determine the source voltage E.

Solution Draw and label a schematic of the circuit as in Figure 28-21.

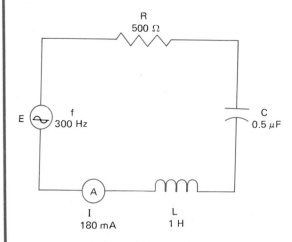

FIGURE 28–21
Circuit for Example 28–15.

(a) Compute the voltage drops:

$$|V_R| = IR = 180 \times 10^{-3} \times 500 = 90 \text{ V}$$
$$|V_L| = IX_L = I(2\pi fL)$$
$$|V_L| = 180 \times 10^{-3} \times 2\pi \times 300 \times 1 = 339.3 \text{ V}$$
$$|V_C| = IX_C = I/(2\pi fC) = \frac{180 \times 10^{-3}}{2\pi \times 300 \times 0.5 \times 10^{-6}}$$
$$|V_C| = 191 \text{ V}$$

(b) Draw the voltage phasor diagram as in Figure 28-22(a).

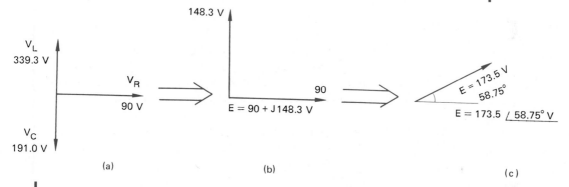

(a)   (b)   (c)

FIGURE 28-22
Voltage phasor diagram for the circuit of Figure 28-21.

(c) Determine the source voltage by first expressing each voltage phasor as a complex number in rectangular form and then adding:

$$V_R = 90 + j0$$
$$V_L = 0 + j339.3$$
$$V_C = 0 - j191.0$$
$$\overline{E = 90 + j148.3 \text{ V}}$$

Convert the source voltage to polar form:

$$\therefore E = 90 + j148.3 \text{ V} \boxed{\rightarrow P} 173.5/58.75° \text{ V}$$

Figure 28-22 shows the sequence to the solution for the source voltage through phasor diagrams.

EXAMPLE
28-16

The current passing through a series circuit consisting of a 1.0-μF capacitor and a 700-Ω resistance is $40\underline{/0°}$ mA. If the circuit is powered by a 60-Hz ac generator, first determine:

(a) The voltage across the capacitor
(b) The voltage across the resistor
(c) The source voltage

then construct:

(d) A voltage phasor diagram
(e) A phasor diagram showing the phase angle between the source voltage and current

Solution (a) Draw and label a schematic of the circuit as in Figure 28-23 and determine V_C :

$$|V_C| = IX_C = I/(\omega C)$$
$$= 40 \times 10^{-3}/[(2\pi 60)(1 \times 10^{-6})]$$
$$|V_C| = 106.1 \text{ V}$$

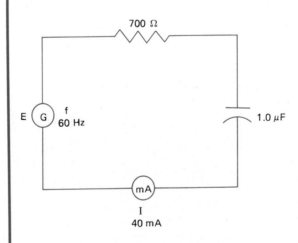

700 Ω

E G f 60 Hz

1.0 μF

mA

I
40 mA

FIGURE 28-23
Schematic for Example 28-16.

(b) Determine V_R :

$$V_R = IR = 40 \times 10^{-3} \times 700 = 28 \text{ V}$$

(c) Determine the source voltage (E) by adding the voltage drops:

$$E = V_R + V_C$$
$$= (28 + j0) + (0 - j106.1)$$
$$E = 28 - j106.1 \;\boxed{\rightarrow \text{P}}\; 109.7\underline{/-75.22°}\ \text{V}$$

(d) Draw the voltage phasor diagram as in Figure 28-24.

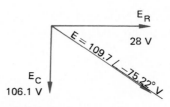

FIGURE 28–24
Voltage phasor diagram for the circuit of Figure 28–23.

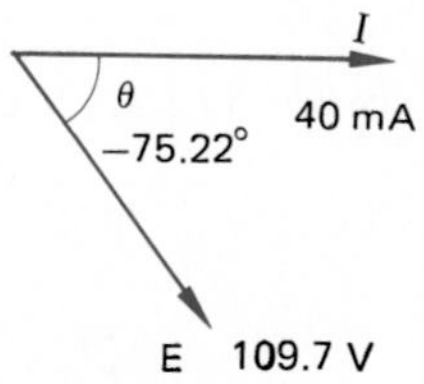

FIGURE 28–25
Phase angle of the current and voltage of the circuit of Figure 28–23.

(e) Draw the phasor diagram showing the phase angle between source voltage and current as in Figure 28-25.

EXERCISE 28-5

Add each of the following voltage drops and express the sum as a phasor in both rectangular and polar form. We recommend that you construct a voltage phasor diagram to aid your solution.

1. $V_R = 50\ \text{V}$ $V_L = 50\ \text{V}$
2. $V_C = 20\ \text{V}$ $V_R = 10\ \text{V}$
3. $V_R = 5\ \text{V}$ $V_L = 2\ \text{V}$
4. $V_C = 18\ \text{V}$ $V_R = 5\ \text{V}$
5. $V_L = 44\ \text{V}$ $V_C = 14\ \text{V}$
6. $V_R = 26\ \text{V}$ $V_C = 9\ \text{V}$ $V_R = 10\ \text{V}$
7. $V_L = 60\ \text{V}$ $V_C = 52\ \text{V}$ $V_L = 12\ \text{V}$
8. $V_R = 7\ \text{V}$ $V_C = 16\ \text{V}$ $V_L = 9\ \text{V}$
9. $V_C = 38\ \text{V}$ $V_R = 75\ \text{V}$ $V_L = 54\ \text{V}$
10. $V_L = 81\ \text{V}$ $V_C = 129\ \text{V}$ $V_R = 30\ \text{V}$

Express each of the following voltages in rectangular form as $V_R + jV_L$ or $V_R - jV_C$.

11. $17\underline{/80°}$ V	12. $36\underline{/25°}$ V
13. $8\underline{/-30°}$ V	14. $15\underline{/-72°}$ V
15. $64\underline{/-275°}$ V	16. $23\underline{/322°}$ V
17. $29\underline{/288°}$ V	18. $40\underline{/-313°}$ V
19. $55\underline{/-420°}$ V	20. $11\underline{/447°}$ V

For each of the following, construct the voltage phasor and compute the source voltage in rectangular and polar form.

21.　　5 V　　3 V

22.　　10 Ω　　$2\underline{/0°}$ A

23.　　8 V　　10 V

24.　　6 Ω　　12 V　　4 Ω　　$I = 2\underline{/0°}$ A

25.　　28 Ω　　21 Ω　　$I = 3\underline{/0°}$ A

26.　　6 V　　9 V　　5 V

In each of the following, draw a schematic of the circuit and determine (a) the voltage drop across each circuit component and (b) the source voltage E in polar form; then draw (c) the voltage phasor diagram and (d) the phasor diagram of the source voltage and circuit current. Label the phase angle.

27. The 200-Hz current passing through a series circuit consisting of 5-kΩ resistance and a 3-H inductor is $200\underline{/0°}$ mA.

28. The 8-kHz current passing through a series circuit consisting of a 100-μH inductor, a 2-μF capacitor, and a 10-Ω resistance is $1.5 \underline{/0°}$ A.

29. The 60-Hz current passing through a series circuit consisting of a 10-H inductor and a 1.0-μF capacitor is $90 \underline{/0°}$ mA.

30. The 400-Hz current passing through a series circuit consisting of a 75-mH inductor, a 15-μF capacitor, and a 33-Ω resistance is $2.2 \underline{/0°}$ A.

28-6 CURRENT PHASOR FOR PARALLEL CIRCUITS

In this section we will investigate the result of passing alternating current through a parallel circuit having branches made up of a resistance, a capacitance, and an inductance. As we have learned, current and voltage are in phase through a resistance, but voltage leads current in an inductor and voltage lags current in a capacitor.

When working with parallel circuits, the voltage is the reference phasor because it is the same across all parts of the parallel circuit. Thus, the current is:

- In phase with voltage across a resistance: $\theta = 0°$
- Leading the voltage across a capacitance: $\theta = 90°$
- Lagging the voltage across an inductance: $\theta = -90°$

These concepts are summarized in Figure 28-26, which shows the current phasor diagram.

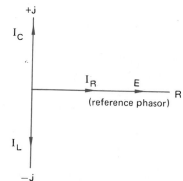

FIGURE 28-26
Generalized current phasor for a parallel LCR circuit when $E \angle 0°$ V is the reference phasor.

Kirchhoff's Current Law (KCL)

Kirchhoff's current law (KCL), $I = I_1 + I_2 + I_3$, etc., is valid for ac parallel circuits. However, ac quantities are phasor quantities represented by complex numbers and, as such, must be operated on with phasor addition. The following example demonstrates this concept.

EXAMPLE
28-17

Verify Kirchhoff's current law for the circuit of Figure 28-27 ($I = I_R + I_L + I_C$).

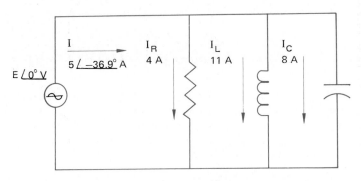

FIGURE 28–27
Circuit for Example 28–17.

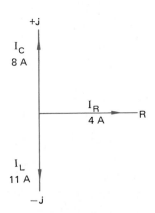

FIGURE 28–28
Current phasor diagram for the circuit of Figure 28–27.

Solution Construct the current phasor diagram as shown in Figure 28-28. Express each current phasor as a complex number in rectangular form:

$$I_R = 4 + j0 \text{ A}$$
$$I_C = 0 + j8 \text{ A}$$
$$I_L = 0 - j11 \text{ A}$$

Add the current phasors and express the sum in polar form:

$$4 + j0$$
$$0 + j8$$
$$\underline{0 - j11}$$
$$4 - j3 \boxed{\rightarrow \text{P}}\ 5\underline{/-36.9°}\ \text{A}$$

∴ The sum of the branch currents $5\underline{/-36.9°}$ A is equal to the source current $5\underline{/-36.9°}$ A. This verifies Kirchhoff's current law.

Figure 28-29 shows the flow of the solution of Example 28-17. Notice the similarity in the sequence of the solution to that of the series circuit of the previous section.

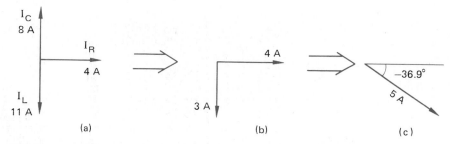

FIGURE 28-29

(a) Current phasor diagram for the current in the branches of the parallel circuit of Figure 28-27; (b) phasor addition of the branch currents; (c) resolution of the rectangular current phasors of (b) into a polar phasor.

EXAMPLE 28-18

A 1-kΩ wire-wound resistor is connected in parallel with a 1-μF capacitor and a 800-mH inductor. The circuit is attached to a 220-V, 400-Hz alternator.

(a) Compute the current flow through each component in the circuit.

(b) Draw the current phasor diagram.

(c) Determine the source current I.

Solution Draw and label a schematic of the circuit as in Figure 28-30.

(a) Compute the branch currents:

$$|I_R| = E/R = 220/(1 \times 10^3) = 220 \text{ mA}$$
$$|I_C| = E/X_C = E(2\pi fC) = 220(2\pi 400 \times 1 \times 10^{-6})$$
$$|I_C| = 553 \text{ mA}$$
$$|I_L| = E/X_L = E/(2\pi fL)$$
$$|I_L| = 220/(2\pi 400 \times 800 \times 10^{-3}) = 109 \text{ mA}$$

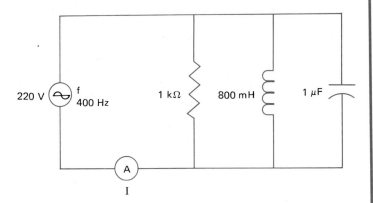

FIGURE 28–30
Circuit for Example 28–18.

(b) Draw the branch-current phasor diagram as in Figure 28-31.

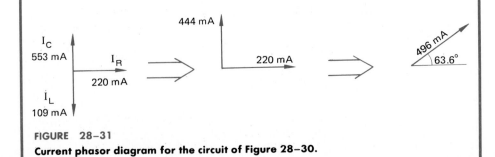

FIGURE 28–31
Current phasor diagram for the circuit of Figure 28–30.

(c) Determine the source current I by first expressing each branch current phasor as a complex number in rectangular form, and then adding:

$$I_R = 220 \times 10^{-3} + j0$$
$$I_C = 0 \qquad\qquad + j553 \times 10^{-3}$$
$$\underline{I_L = 0 \qquad\qquad - j109 \times 10^{-3}}$$
$$I = 220 \times 10^{-3} + j444 \times 10^{-3}\,\text{A}$$

Convert the source current to polar form:

$$\therefore \qquad I = 220 \times 10^{-3} + j444 \times 10^{-3}$$
$$= 496\underline{/63.6°}\,\text{mA}$$

Observation The rectangular form of the current, $220 \times 10^{-3} + j444 \times 10^{-3}$ A, may be expressed as $220 + j444$ mA. Whenever unit prefixes are used with phasor quantities, only one prefix is used, and it applies to both parts of the phasor.

EXERCISE 28-6

Add each of the following branch currents, and express the sum as a phasor in both rectangular and polar form. We recommend that you construct a current phasor diagram to aid your solution.

1. $I_R = 2$ A $I_L = 5$ A
2. $I_C = 7$ A $I_R = 3$ A
3. $I_R = 0.300$ A $I_L = 1.2$ A
4. $I_C = 700\ \mu$A $I_R = 0.52$ mA
5. $I_L = 3.4$ A $I_C = 1.8$ A
6. $I_R = 330$ mA $I_C = 180$ mA $I_R = 75$ mA
7. $I_L = 8.2$ A $I_C = 4.4$ A $I_L = 2.2$ A
8. $I_R = 13$ mA $I_C = 63$ mA $I_L = 43$ mA
9. $I_C = 0.732$ A $I_R = 510$ mA $I_L = 0.270$ A
10. $I_L = 42$ mA $I_C = 87$ mA $I_R = 70$ mA

Express each of the following source currents in rectangular form as $I_R - jI_L$ or $I_R + jI_C$.

11. $12\underline{/72°}$ mA   12. $3.1\underline{/33°}$ A

13. $413\underline{/-19°}$ mA   14. $16\underline{/-63°}$ A

15. $67/\!-\!305°\ \mu A$ 16. $5.4/282°\ mA$

17. $1.6/4.5°\ A$ 18. $51/\!-\!328°\ \mu A$

19. $2.7/443°\ A$ 20. $18/\!-\!395°\ mA$

For each of the following, construct the current phasor diagram and compute the source current in rectangular and polar form.

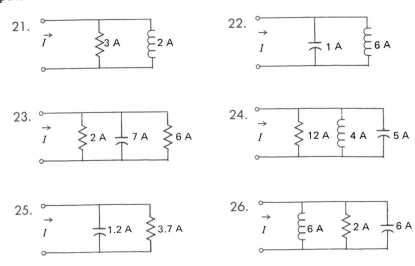

21. I, 3 A, 2 A

22. I, 1 A, 6 A

23. I, 2 A, 7 A, 6 A

24. I, 12 A, 4 A, 5 A

25. I, 1.2 A, 3.7 A

26. I, 6 A, 2 A, 6 A

In each of the following, draw a schematic of the circuit and:

(a) Determine the current through each branch component.
(b) Determine the source current I in polar form.
(c) Draw the branch-current phasor diagram.
(d) Draw the phasor diagram of the source voltage and circuit current, and label the phase angle.

27. The 180-Hz voltage source applied across a parallel circuit consisting of a 4.7-kΩ resistance and a 2.7-H inductor is $100/0°$ V.

28. The 10-kHz voltage source applied across a three-branch parallel circuit consisting of a 120-μH inductor, a 2.2-μF capacitor, and a 12-Ω resistance is $10/0°$ V.

29. The 60-Hz voltage source applied across a parallel circuit consisting of an 8-H inductor and a 2-μF capacitor is $160/0°$ V.

30. The 800-Hz voltage source applied across a three-branch parallel circuit consisting of a 5-mH inductor, a 30-μF capacitor, and a 27-Ω resistance is $12/0°$ V.

SELECTED TERMS

angular velocity   The rate of travel of a rotating vector expressed in radians per second.

capacitive reactance   The opposition to alternating current by a capacitor; measured in units of ohms. Its symbol is X_C.

frequency   The number of complete cycles of alternating current in one second; measured in hertz (Hz).

inductive reactance   The opposition to alternating current by an inductor; measured in units of ohms. Its symbol is X_L.

period   The reciprocal of frequency.

phase angle   The angle between the current and voltage phasor in a phasor diagram.

29

ALTERNATING-CURRENT CIRCUITS

In the previous chapters you were introduced to the rules for series and parallel circuits. This chapter expands upon these rules to introduce you to *impedance* for series circuits and *admittance* for parallel circuits.

29-1 IMPEDANCE OF SERIES AC CIRCUITS

When alternating current passes through a series *RLC* circuit, the intensity of the current is determined by the total opposition offered by the resistance of the resistor and the reactance of the inductor and capacitor. The opposition to the passage of alternating current through a circuit is called *impedance*. Thus, the word impedance may be used to describe the opposition offered by an ac circuit containing only resistance or only capacitance or only inductance. It may also be used to describe ac circuits containing any combination of resistance, capacitance, and inductance. We see, then, that the word impedance is a general label that is used to describe the opposition to the flow of alternating current by any combination of circuit components.

Computing Impedance

Impedance is a *vector quantity* that may be written as a complex number in either the rectangular or polar form. The letter Z is used to represent impedance. Thus, $Z = 5/53.13° \ \Omega = 3 + j4 \ \Omega$ shows an impedance written in both the polar and rectangular form. Ohm's law may be stated in terms of impedance as Equation 29-1.

$$Z = E/I \qquad \Omega \qquad\qquad (29\text{-}1)$$

EXAMPLE 29-1 Determine the impedance of a circuit having a current of $2/0°$ A and a source voltage of $100/30°$ V.

Solution Use Equation 29-1:
Substitute $I = 2/0°$ A and $E = 100/30°$ V:

$$Z = \frac{100/30°}{2/0°}$$

$$\therefore \quad Z = 50/30° \ \Omega$$

EXAMPLE 29-2 Determine the impedance of a circuit having a current of $5/0°$ A and a source voltage of $25/-20°$ V.

Solution Use Equation 29-1:
Substitute $I = 5\underline{/0°}$ A and $E = 25\underline{/-20°}$ V:

$$Z = \frac{25\underline{/-20°}}{5\underline{/0°}}$$

$$\therefore \quad Z = 5\underline{/-20°} \ \Omega$$

EXAMPLE 29-3 Determine the impedance of a circuit having a current of $6\underline{/0°}$A and a source voltage of $36\underline{/0°}$ V.

Solution Use Equation 29-1:
Substitute $I = 6\underline{/0°}$A and $E = 36\underline{/0°}$ V:

$$Z = \frac{36\underline{/0°}}{6\underline{/0°}}$$

$$\therefore \quad Z = 6\underline{/0°} \ \Omega$$

EXAMPLE 29-4 Determine the impedance of a circuit having a current of $5\underline{/0°}$ A and a source voltage of $15\underline{/90°}$ V.

Solution Use Equation 29-1:
Substitute $I = 5\underline{/0°}$ A and $E = 15\underline{/90°}$ V:

$$Z = \frac{15\underline{/90°}}{5\underline{/0°}}$$

$$\therefore \quad Z = 3\underline{/90°} \ \Omega$$

EXAMPLE 29-5 Determine the impedance of a circuit having a current of $4\underline{/0°}$ A and a source voltage of $28\underline{/-90°}$.

Solution Use Equation 29-1:
Substitute $I = 4\underline{/0°}$ A and $E = 28\underline{/-90°}$ V:

$$Z = \frac{28\underline{/-90°}}{4\underline{/0°}}$$

$$\therefore \quad Z = 7\underline{/-90^\circ}\ \Omega$$

Each of the five preceding examples used circuit conditions that were different. By relating the phase angle of the applied voltage to the circuit current, we are able to determine the type of components in each of the series equivalent circuits. This is demonstrated by the following example. Remember that inductive circuits have leading phase angles and capacitive circuits have lagging phase angles.

EXAMPLE 29-6 Determine the circuit components used in each of the preceding five examples.

Solution List the impedance of each circuit, and state the phase relation of the source voltage with respect to the circuit current:

> (1) $50\underline{/30^\circ}\ \Omega$ E leads I by 30°
> (2) $5\underline{/-20^\circ}\ \Omega$ E lags I by 20°
> (3) $6\underline{/0^\circ}\ \Omega$ E and I are in phase
> (4) $3\underline{/90^\circ}\ \Omega$ E leads I by 90°
> (5) $7\underline{/-90^\circ}\ \Omega$ E lags I by 90°

Use the phase angle to identify the circuit components:

> (1) $\theta = 30^\circ$ inductive and resistive
> (2) $\theta = -20^\circ$ capacitive and resistive
> (3) $\theta = 0^\circ$ resistive
> (4) $\theta = 90^\circ$ inductive
> (5) $\theta = -90^\circ$ capacitive

Table 29-1 summarizes the first five examples. Notice that positive phase angles indicate inductive circuits, whereas negative phase angles indicate capacitive circuits.

Impedance Diagram

In an ac series circuit consisting of a resistor, an inductor, and a capacitor, the following are true statements:

$$I = I_R = I_L = I_C \text{ and } E = V_R + j|V_L| - j|V_C|$$

TABLE 29-1

SUMMARY OF EXAMPLES 29-1 THROUGH 29-5

Example	$\dfrac{Z\theta\underline{/\theta}}{\Omega}$	Phase Angle θ	Equivalent Circuit Components	EI Phasor Diagram
29-1	$50\underline{/30°}$	30°	43.3 25 Ω	
29-2	$5\underline{/-20°}$	−20°	4.70 Ω 1.71 Ω	
29-3	$6\underline{/0°}$	0°	6 Ω	
29-4	$3\underline{/90°}$	90°	3 Ω	
29-5	$7\underline{/-90°}$	−90°	7 Ω	

Dividing I of the first equation into each member of the second equation results in

$$\frac{E}{I} = \frac{V_R}{I} + \frac{j|V_L|}{I} - \frac{j|V_C|}{I} \Rightarrow Z = R + jX_L - jX_C \qquad \Omega \qquad (29\text{-}2)$$

When the vector quantities of the right member of Equation 29-2 are graphed, an impedance diagram is formed, as shown in Figure 29-1. It is suggested that the following guidelines be used to aid in the construction of an impedance diagram.

GUIDELINE 29-1 CONSTRUCTING AN IMPEDANCE DIAGRAM

When plotting resistance, inductive reactance, and capacitive reactance on a complex plane:

1. Plot resistance along the positive real axis.
2. Plot inductive reactance along the positive imaginary ($+j$) axis.
3. Plot capacitive reactance along the negative imaginary ($-j$) axis.

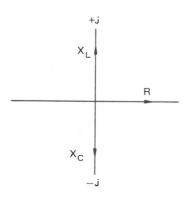

FIGURE 29–1
An impedance diagram has the resistance plotted horizontally, the inductive reactance plotted vertically up, and the capacitive reactance plotted vertically down.

EXAMPLE 29-7 Use the information of Table 29-1 to construct an impedance diagram for the circuit of Example 29-2.

Solution From Table 29-1:

$$R = 4.70 \ \Omega, \quad X_C = 1.71 \ \Omega, \quad Z = 5\underline{/-20°} \ \Omega$$

Construct the impedance diagram as shown in Figure 29-2.

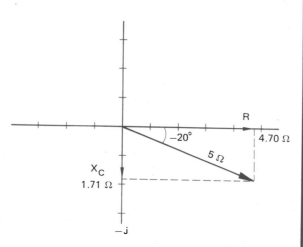

FIGURE 29–2
Impedance diagram for Example 29–7. $Z = 4.70 - j1.71 = \underline{/-20°} \ \Omega$.

Impedance Forms

The impedance of an ac circuit is written in either the rectangular $(R + jX)$ or the polar $(Z\underline{/\theta})$ form of the complex number. The rectangular form is formatted so that the resistance is written first and then the reactance. Thus, $R + jX$ is the general rectangular form of impedance.

EXAMPLE 29-8   Express each of the following series circuit conditions as (a) an impedance diagram, (b) an impedance in rectangular form, and (c) an impedance in polar form.
(1) Resistance of 6 Ω and inductive reactance of 8 Ω
(2) Resistance of 20 Ω and capacitive reactance of 15 Ω
(3) Resistance of 6 Ω and inductive reactance of 20 Ω

Solution   (a) See Figure 29-3 for the impedance diagram of each condition.

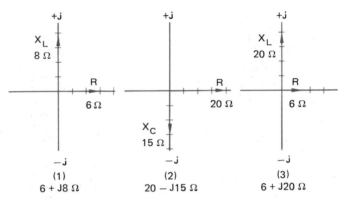

FIGURE   29-3
Impedance diagram for Example 29-8.

(b) Fit each condition into the general form of the rectangular impedance, $R + jX$.
(1) $R = 6\ \Omega,\ X_L = 8\ \Omega$
    $Z = 6 + j8\ \Omega$

(2) $R = 20\ \Omega,\ X_C = 15\ \Omega$
    $Z = 20 - j15\ \Omega$

(3) $R = 6\ \Omega,\ X_L = 20\ \Omega$
    $Z = 6 + j20\ \Omega$

(c) Convert each rectangular form to polar form.
(1) $6 + j8$ →P $10\underline{/53.1°}\ \Omega$

(2) $20 - j15$ →P $25\underline{/-36.9°}\ \Omega$

(3) $6 + j20$ →P $20.88\underline{/73.3°}\ \Omega$

Series Equivalent Circuit

An ac series circuit may have any number of circuit components connected together to form the circuit. The circuit may, however, be expressed at a particular frequency in a simplified form as an equivalent circuit with one resistive component and one reactive component.

   The value of the components of the ac equivalent circuit are determined by applying phasor mathematics to the circuit, as demonstrated in the following example.

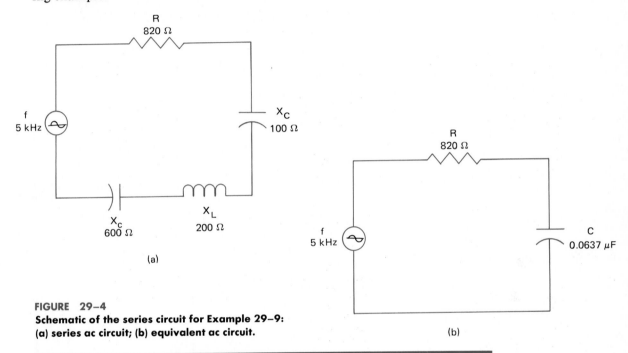

FIGURE 29-4
Schematic of the series circuit for Example 29-9:
(a) series ac circuit; (b) equivalent ac circuit.

EXAMPLE 29-9   The following components are connected in series to form the series circuit shown in Figure 29-4(a):

(1) A resistance of 820 Ω
(2) A capacitor having 100 Ω of reactance
(3) An inductor having 200 Ω of reactance
(4) A capacitor having 600 Ω of reactance

For the circuit:

(a) Determine the total impedance in both rectangular and polar form.

(b) Determine the components of the equivalent series circuit when the operating frequency is 5 kHz.

(c) Draw the impedance diagram for both the series circuit and its equivalent.

Solution (a) First express each circuit component as an impedance in rectangular form $(R + jX)$, and then add to get the total circuit impedance.

(1) 820-Ω resistance $\qquad\qquad$ $820 + j0$
(2) 100-Ω capacitive reactance $\qquad$ $0 - j100$
(3) 200-Ω inductive reactance $\qquad$ $0 + j200$
(4) 600-Ω capacitive reactance $\qquad$ $\underline{0 - j600}$

$$Z_T = 820 - j500 \ \Omega$$

$$\therefore \qquad Z_T = 820 - j500 \ \Omega \ \boxed{\rightarrow\text{P}} \ 960.4\underline{/-31.37°} \ \Omega$$

(b) The components of the series equivalent circuit are expressed as the rectangular form of the impedance $(R - jX_C)$. Thus, $820 - j500 \ \Omega \Rightarrow$ $R = 820 \ \Omega$ and $X_C = 500 \ \Omega$.

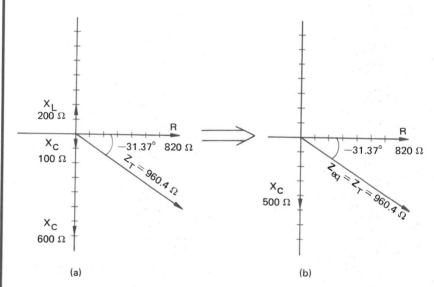

(a) $\qquad\qquad\qquad\qquad\qquad$ (b)

FIGURE 29–5
Impedance diagrams: (a) series circuit for Example 29–9; (b) equivalent series circuit for Example 29–9.

Solving for C when $f = 5$ kHz:

$$X_C = 1/(\omega C)$$
$$C = 1/(\omega X_C)$$
$$= \frac{1}{2\pi \times 5 \times 10^3 \times 500}$$
$$C = 0.0637 \ \mu F$$

∴ The equivalent series circuit components for a frequency of 5 kHz are a resistance of 820 Ω and a capacitor of 0.0637 μF. Figure 29-4(b) shows the schematic of the equivalent circuit. NOTE: It is conventional to express capacitance with μF or pF instead of nF.

(c) The impedance diagram for the series circuit is shown as Figure 29-5(a), and the impedance diagram for the equivalent circuit is shown as Figure 29-5(b).

The preceding example demonstrates that the rectangular form of the circuit impedance represents the components of the series equivalent circuit. If the circuit impedance is in polar form and the equivalent series circuit is wanted, the polar impedance is converted to its rectangular form.

EXAMPLE 29-10

The circuit impedance of a series circuit is $2.6\underline{/74°}$ kΩ. Determine the value of the components of the equivalent series circuit for a frequency of 400 Hz.

Solution Convert from polar form to rectangular form:

$$Z = 2.6\underline{/74°} \ k\Omega \ \boxed{\rightarrow R} \ 0.72 + j2.5 \ k\Omega$$

Use the rectangular form of Z to find the circuit components:

$$0.72 - j2.5 \ k\Omega \Rightarrow R = 720 \ \Omega \text{ and } X_L = 2.5 \ k\Omega$$

Solve for L when $f = 400$ Hz:

$$X_L = \omega L$$
$$L = X_L/\omega = \frac{2.5 \times 10^3}{2\pi 400} = 990 \ \text{mH}$$

$\therefore$ The series equivalent circuit is a resistor of 720 Ω and an inductor of 990 mH.

EXERCISE 29-1

Compute the circuit impedance given the following conditions.

1. $E = 100\underline{/20°}$ V $I = 2.5\underline{/0°}$ A
2. $E = 12\underline{/-90°}$ V $I = 40\underline{/0°}$ mA
3. $E = 18\underline{/0°}$ V $I = 320\underline{/0°}$ mA
4. $E = 320\underline{/33°}$ V $I = 9.4\underline{/0°}$ mA
5. $E = 74\underline{/-8°}$ V $I = 4.3\underline{/0°}$ A
6. $E = 7.5\underline{/-62°}$ V $I = 815\underline{/0°}$ μA

Given the phase angle (θ) between the voltage and the circuit current, determine whether each of the following circuits is resistive, inductive, capacitive, resistive and inductive, or resistive and capacitive. Assume current to be the reference phasor.

7. $\theta = 0°$ 8. $\theta = 90°$ 9. $\theta = 45°$
10. $\theta = -90°$ 11. $\theta = -18°$ 12. $\theta = 60°$

Given the following circuit conditions, express each as an impedance in both the rectangular and polar form of impedance, and construct an impedance diagram for each.

13. A resistance of 10 Ω
14. An inductive reactance of 20 Ω and a resistance of 20 Ω
15. A resistance of 5 Ω and a capacitive reactance of 2 Ω
16. A resistance of 12 Ω and a reactance of $8\underline{/-90°}$ Ω
17. A reactance of $14\underline{/90°}$ Ω
18. A resistance of 25 Ω and a capacitive reactance of 30 Ω

Given the following impedances, determine the resistance and the reactance of the series equivalent circuit for each.

19. $Z = 510\underline{/16°}$ Ω 20. $Z = 4.3\underline{/-50°}$ kΩ
21. $Z = 3.3\underline{/-18°}$ kΩ 22. $Z = 14.3\underline{/5°}$ Ω
23. $Z = 83\underline{/67°}$ Ω 24. $Z = 0.635\underline{/-74°}$ MΩ

644

Given the following circuit conditions, construct an impedance diagram and determine (a) the total impedance in polar form, and (b) the component values for the series equivalent circuit for the stated frequency:

25. $R = 27\ \Omega$ $X_L = 54\ \Omega$ $X_C = 190\ \Omega$ $f = 60$ Hz
26. $X_L = 350\ \Omega$ $R = 95\ \Omega$ $X_L = 65\ \Omega$ $f = 120$ Hz
27. $X_C = 1.5\ \text{k}\Omega$ $X_L = 810\ \Omega$ $R = 500\ \Omega$ $f = 400$ Hz
28. $R = 470\ \Omega$ $X_C = 94\ \Omega$ $X_C = 116\ \Omega$ $f = 800$ Hz
29. $R = 2.2\ \text{k}\Omega$ $X_L = 1.5\ \text{k}\Omega$ $X_C = 3.9\ \text{k}\Omega$ $f = 1.0$ kHz
30. $R = 52\ \Omega$ $X_L = 84\ \Omega$ $X_L = 26\ \Omega$ $f = 60$ Hz

29-2 SOLVING SERIES AC CIRCUITS

An ac series circuit may be made up of several resistors, capacitors, and inductors. When this is the case, each kind of element may be combined into a single *equivalent component* by applying one or more of the following equations.

$$R_T = R_1 + R_2 + R_3 + \cdots + R_n \qquad \Omega \qquad (29\text{-}3)$$

$$L_T = L_1 + L_2 + L_3 + \cdots + L_n \qquad \text{H} \qquad (29\text{-}4)$$

$$C_T = \cfrac{1}{\dfrac{1}{C_1} + \dfrac{1}{C_2} + \dfrac{1}{C_3} + \cdots + \dfrac{1}{C_n}} \qquad \text{F} \qquad (29\text{-}5)$$

EXAMPLE 29-11

Compute the equivalent component for each of the following series-circuit conditions.

(a) Three inductors in series:

$$L_1 = 5\ \text{H} \quad L_2 = 10\ \text{H} \quad L_3 = 2\ \text{H}$$

(b) Three resistors in series:

$$R_1 = 510\ \Omega \quad R_2 = 620\ \Omega \quad R_3 = 390\ \Omega$$

(c) Three capacitors in series:

$$C_1 = 0.22\ \mu\text{F} \quad C_2 = 0.5\ \mu\text{F} \quad C_3 = 0.68\ \mu\text{F}$$

Solution   (a) Use Equation 29-4:

$$L_T = L_1 + L_2 + L_3$$
$$= 5 + 10 + 2$$
$$\therefore \quad L_T = 17 \text{ H}$$

(b) Use Equation 29-3:

$$R_T = R_1 + R_2 + R_3$$
$$= 510 + 620 + 390$$
$$\therefore \quad R_T = 1.52 \text{ k}\Omega$$

(c) Use Equation 29-5:

$$C_T = \cfrac{1}{\cfrac{1}{C_1} + \cfrac{1}{C_2} + \cfrac{1}{C_3}}$$

$$= \cfrac{1}{\cfrac{1}{0.22 \times 10^{-6}} + \cfrac{1}{0.5 \times 10^{-6}} + \cfrac{1}{0.68 \times 10^{-6}}}$$

$$\therefore C_T = 0.125 \ \mu\text{F}$$

Equations 29-3 through 29-5 are used along with the previously learned rules, concepts, and formulas to solve series ac circuits.

EXAMPLE 29-12

A series circuit made up of the following components is connected to a 120-V, 400-Hz alternator.

$$L_1 = 2 \text{ H} \quad R_1 = 1.8 \text{ k}\Omega \quad L_2 = 4 \text{ H}$$
$$C_1 = C_2 = 0.1 \ \mu\text{F} \quad R_2 = 750 \ \Omega$$

Determine:

(a) The equivalent components R_T, L_T, and C_T
(b) The circuit impedance diagram showing the real component and the imaginary component
(c) The impedance in polar form

(d) The circuit current ($|I|$)

(e) The components of the series equivalent circuit

(f) The equivalent series circuit impedance diagram

(g) The current and voltage phasor diagram

Solution   Begin the solution by drawing and labeling a schematic diagram as shown in Figure 29-6.

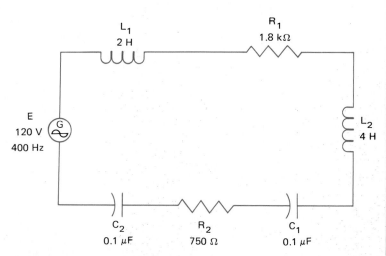

FIGURE   29–6
Schematic diagram for Example 29–12.

(a) Use Equation 29-3 to compute R_T:

$$R_T = 1800 + 750 = 2.55 \text{ k}\Omega$$

Use Equation 29-4 to compute L_T:

$$L_T = 2 + 4 = 6 \text{ H}$$

Use Equation 29-5 to compute C_T:

$$C_T = \cfrac{1}{\cfrac{1}{0.1 \times 10^{-6}} + \cfrac{1}{0.1 \times 10^{-6}}} = 0.05 \text{ }\mu\text{F}$$

(b) Compute the value of the inductive and capacitive reactances:

$$X_{LT} = \omega L_T = 2\pi 400(6) = 15.1 \text{ k}\Omega$$

$$X_{CT} = 1/(\omega C_T) = \frac{1}{2\pi 400(0.05 \times 10^{-6})} = 7.96 \text{ k}\Omega$$

Construct the circuit impedance diagram with R_T = 2.55 kΩ. Figure 29-7(a) is the impedance diagram for the series circuit.

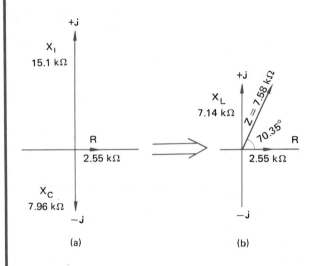

FIGURE 29–7
Impedance diagrams: (a) series circuit of Figure 29–6; (b) series equivalent circuit of Figure 29–8.

(c) Compute the impedance by adding the resistance and reactances of the circuit components:

$$\begin{array}{r} 2550 + j0 \\ 0 + j15\,100 \\ \underline{0 - j7960} \\ Z_T = 2550 + j7140 \ \Omega \\ \therefore \quad Z_T = 2550 + j7140 \ \boxed{\rightarrow\text{P}}\ 7.58\underline{/70.35°} \text{ k}\Omega \end{array}$$

(d) $|I| = |E|/|Z|$

$$\therefore \quad |I| = 120/(7.58 \times 10^3) = 15.8 \text{ mA}$$

(e) The components of the series equivalent circuit

are expressed as the rectangular form of the impedance. Thus,

$$2.55 + j7.14 \text{ k}\Omega \Rightarrow R = 2.55 \text{ k}\Omega \text{ and } X_L = 7.14 \text{ k}\Omega$$

Compute L from X_L:

$$X_L = 2\pi fL$$
$$L = \frac{X_L}{2\pi f} = \frac{7.14 \times 10^3}{2\pi 400} = 2.84 \text{ H}$$

∴ The series-equivalent circuit of $R = 2.55 \text{ k}\Omega$ and $L = 2.84 \text{ H}$ is shown in Figure 29-8.

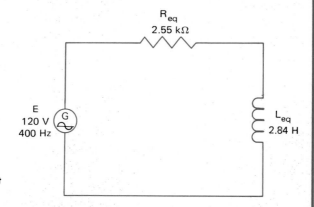

FIGURE 29-8
Series equivalent circuit for Example 29-12.

(f) The series-equivalent-circuit impedance diagram is Figure 29-7(b).

(g) The phase angle between the source voltage (120 V) and the circuit current (15.8 mA) is equal to 70.35°, the angle of the impedance. In series ac circuits, the current phasor is selected as the reference phasor. In this inductive circuit, the voltage phasor leads the current phasor by 70.35°. This result may also be determined by phasor mathematics as:

$$E = IZ$$
$$= (15.8 \times 10^{-3}\underline{/0°})(7.58 \times 10^3\underline{/70.35°})$$
$$E = 120\underline{/70.35°} \text{ V}$$

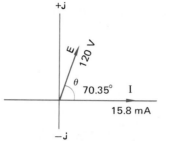

+j

E = 120 V

θ 70.35° I

15.8 mA

−j

Figure 29-9 is the current and voltage phasor diagram with the phase angle $\theta = 70.35°$.

EXAMPLE 29-13

The following components are connected in series to a 3-V, 10-kHz ac source:

$$R_1 = 5.60 \text{ k}\Omega \quad C_1 = 820 \text{ pF} \quad L_1 = 350 \text{ mH}$$
$$C_2 = 1500 \text{ pF}$$

Determine:

(a) The impedance of the circuit
(b) The phase angle between the current and voltage phasors
(c) The circuit current ($|I|$)
(d) The components of the series equivalent circuit

Solution Begin the solution by drawing and labeling a schematic diagram. This is left for you to do.

(a) Compute the impedance by calculating the capacitive reactance and the inductive reactance:

$$C_T = \frac{1}{\dfrac{1}{C_1} + \dfrac{1}{C_2}} = \frac{1}{\dfrac{1}{820 \times 10^{-12}} + \dfrac{1}{1500 \times 10^{-12}}}$$

$$C_T = 530 \text{ pF}$$
$$X_C = 1/(2\pi f C)$$

$$X_C = \frac{1}{(2\pi 10 \times 10^{3})(530 \times 10^{-12})} = 30 \text{ k}\Omega$$

$$X_L = 2\pi fL$$
$$X_L = (2\pi 10 \times 10^3)(350 \times 10^{-3}) = 22 \text{ k}\Omega$$
$$R = 5.6 \text{ k}\Omega$$

Express the impedance as a sum of the resistance and reactance:

$$5600 + j0$$
$$0 - j30\ 000$$
$$\underline{0 + j22\ 000}$$
$$Z = 5600 - j8000 \ \Omega$$
$$\therefore \quad Z = 5600 - j8000 \ \boxed{\rightarrow P}\ 9.77\underline{/{-}55^\circ} \text{ k}\Omega$$

(b) The phase angle (θ) is equal to the angle of the polar form of impedance:

$$\therefore \quad \theta = -55^\circ \quad \text{(voltage lags current} \atop \text{in a capacitive circuit)}$$

(c) Compute the circuit current:

$$|I| = |E|/|Z|$$
$$= 3/(9.77 \times 10^3)$$
$$\therefore \quad |I| = 307 \ \mu\text{A}$$

(d) The components of the series equivalent circuit are expressed as the rectangular form of the impedance:

$$5.6 - j8.0 \text{ k}\Omega \Rightarrow R = 5.6 \text{ k}\Omega \text{ and } X_C = 8.0 \text{ k}\Omega$$

Compute C from X_C:

$$C = 1/(2\pi fX_C)$$
$$= 1/(2\pi \times 10 \times 10^3 \times 8 \times 10^3)$$
$$C = 2000 \text{ pF}$$

$\therefore$ The series-equivalent circuit is a resistance of 5.6 kΩ in series with a capacitor of 2000 pF.

EXERCISE 29-2
Use the given information for each problem to determine (a) the impedance of the circuit (rectangular and polar), (b) the phase angle between the

current and voltage phasors, (c) the circuit current ($|I|$), and (d) the components of the series equivalent circuit. It is suggested that you get in the habit of drawing and labeling complete schematics and constructing impedance diagrams and phasor diagrams of current and voltage. This is usually done as a matter of course.

1. $E = 100$ V, 60 Hz, $R_1 = 1.2$ kΩ, $L_1 = 3$ H, $R_2 = 680$ Ω, $C_1 = 1$ μF

2. $E = 1.5$ V, 5 MHz, $R_1 = 470$ Ω, $L_1 = 4$ μH, $C_1 = 160$ pF, $L_2 = 10$ μH

3. $E = 120$ V, 60 Hz, $R_1 = 51$ Ω, $C_1 = 10$ μF, $L_1 = 400$ mH, $C_2 = 15$ μF

4. $E = 220$ V, 60 Hz, $R_1 = 91$ Ω, $L_1 = 1.8$ H, $R_2 = 120$ Ω, $C_1 = 12$ μF

5. $E = 36$ V, 400 Hz, $C_1 = 5$ μF, $R_1 = 62$ Ω, $C_2 = 8$ μF, $L_1 = 20$ mH

6. $E = 440$ V, 1.0 kHz, $L_1 = 500$ μH, $R_1 = 72$ Ω, $L_2 = 3.3$ mH, $C_1 = 33$ μF

7. A 10-V, 1.0-kHz test oscillator (ac sinusoidal) is attached to a series circuit consisting of a 27-kΩ resistor and a 3.5-H inductor. Determine:
   (a) The inductive reactance of the inductor
   (b) The impedance of the circuit
   (c) The phase angle of the voltage with the current as the reference phasor
   (d) The circuit current $|I|$
   (e) The voltage across the resistor
   (f) The voltage across the inductor
   (g) That $E = V_R + V_L$

8. A series circuit is made up of a 220-Ω resistor and a 25-μF capacitor. The circuit is connected to a 120-V, 60-Hz source. Determine:
   (a) The capacitive reactance
   (b) The impedance of the circuit
   (c) The circuit current $|I|$
   (d) The phase angle between the voltage and the circuit current
   (e) That $E = V_R + V_C$

29-3 ADMITTANCE CONCEPTS

Conductance, the reciprocal of resistance, is used in the solution of dc parallel circuits. *Admittance,* the reciprocal of impedance, is used in the solution of ac parallel circuits. Admittance is the vector quantity used to describe the ease of passage of alternating current. It is measured in siemens (S) and is noted by the letter Y. Thus,

$$Y = 1/Z \quad \text{S} \qquad (29\text{-}6)$$

EXAMPLE 29-14

Express an impedance of $10\underline{/30°}$ as an admittance.

Solution   Use Equation 29-6:

$$Y = 1/(10\underline{/30°})$$
$$\therefore \quad Y = 0.1\underline{/-30°} \text{ S}$$

EXAMPLE 29-15

Express an impedance of $4 - j3$ Ω as an admittance.

Solution   Change Z to polar form:

$$4 - j3 \boxed{\rightarrow\text{P}} 5\underline{/-36.87°} \text{ Ω}$$

Use Equation 29-6:

$$Y = 1/(5\underline{/-36.87°})$$
$$\therefore \quad Y = 0.2\underline{/36.87°} \text{ S}$$

Admittance Diagram

Before constructing the generalized admittance diagram, some additional concepts are needed:

1. The reciprocal of the inductive or capacitive reactance is called *susceptance*, noted by the letter B. Susceptance is a measure of the ability of an inductor or a capacitor to pass alternating current. Thus,

$$B_L = 1/X_L \quad \text{S} \qquad (29\text{-}7)$$
$$B_C = 1/X_C \quad \text{S} \qquad (29\text{-}8)$$

where B_L = inductive susceptance
  B_C = capacitive susceptance

2. In an impedance diagram X_L has an angle of $+90°$ $(+j)$; however, when X_L is reciprocated, the angle becomes $-90°$ $(-j)$. Thus,

$$1/(X_L\underline{/90°}) = B_L\underline{/-90°} = -jB_L \text{ S}$$

and

$$1/(X_C\underline{/-90°}) = B_C\underline{/90°} = +jB_C \text{ S}$$

EXAMPLE 29-16

Express the following reactances as susceptance:

(a) $X_L = 80\underline{/90°}$ (b) $X_C = 4\underline{/-90°}$

Solution (a) Use Equation 29-7:

$$B_L = 1/X_L$$
$$= 1/(80\underline{/90°})$$
$$\therefore \quad B_L = 12.5\underline{/-90°} \text{ mS}$$

(b) Use Equation 29-8:

$$B_C = 1/X_C$$
$$= 1/(4\underline{/-90°})$$
$$\therefore \quad B_C = 0.25\underline{/90°} \text{ S}$$

Admittance may be expressed in either the polar, $Y\underline{/\theta}$, or the rectangular form, $G + jB$, of the complex number. The rectangular form is written with the conductance (G) first and then the susceptance (B).

EXAMPLE 29-17

Express each of the following admittances in its rectangular form.

(a) $92\underline{/-37°} \text{ mS}$ (b) $13\underline{/48°} \text{ }\mu\text{S}$

Solution (a) $92 \times 10^{-3}\underline{/-37°}$ $\boxed{\rightarrow \text{R}}$ $73.5 - j55.4 \text{ mS}$

$$\therefore \quad 92\underline{/-37°} \text{ mS} = 73.5 - j55.4 \text{ mS}$$

The admittance diagram is constructed with the conductance along the reference axis and the susceptance along the j-axis. Figure 29-10 is the general admittance diagram. Notice that the conductance (G) is plotted horizontally, the capacitive susceptance (B_C) is plotted vertically up, and the inductive susceptance (B_L) is plotted vertically down. Some of the concepts presented so far are summarized in Table 29-2.

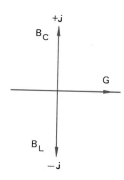

FIGURE 29-10
Admittance diagram for parallel ac circuits.

Computing Admittance

Admittance may be computed from the source voltage and the circuit current. Thus,

$$Y = I/E \quad S \qquad (29\text{-}9)$$

EXAMPLE
29-18

Determine the admittance of a parallel ac circuit having a source voltage of 60 V$\underline{/0°}$ and a circuit current of 180 mA$\underline{/-40°}$. Also construct an admittance diagram.

Solution Use Equation 29-9:

$$Y = \frac{180 \times 10^{-3}\underline{/-40°}}{60\underline{/0°}}$$

$$\therefore \quad Y = 3\underline{/-40°} \ \text{mS} \quad \boxed{\to R} \ 2.30 - j1.93 \ \text{mS}$$

TABLE 29-2

ADMITTANCE SUMMARIZED

Property	Noted by:	Equal to:	Units	Schematic Symbol	Vector Diagram	Polar Form	Rectangular Form
Conductance	G	$1/R$	S			$G\,\underline{/0°}$	$G + j0$
Susceptance (inductive)	B_l	$1/X_l$	S	B_L		$B_l\,\underline{/-90°}$	$0 - jB_l$
Susceptance (capacitive)	B_C	$1/X_C$	S	B_C		$B_C\,\underline{/90°}$	$0 + jB_C$
Admittance	Y	$1/Z$	S	Any one or any combination of the above		$Y\,\underline{/\theta}$	$G + jB$

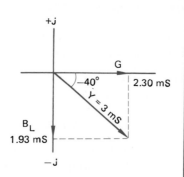

FIGURE 29-11
Admittance diagram for Example 29-18.

Figure 29-11 is the admittance diagram.

EXERCISE 29-3

Express each of the following impedances as an admittance in polar form:

1. $10/\underline{15°}\ \Omega$ 2. $100/\underline{-55°}\ \Omega$ 3. $1000/\underline{82°}\ \Omega$

4. $750/\underline{-30°}\ \Omega$ 5. $56/\underline{-19°}\ \Omega$ 6. $270/\underline{90°}\ \Omega$

7. $120/\underline{-90°}\ \Omega$ 8. $300/\underline{0°}\ \Omega$ 9. $910/\underline{-5°}\ \Omega$

Compute the susceptance of each of the following reactances:

10. $80/\underline{90°}\ \Omega$ 11. $1.5/\underline{-90°}\ k\Omega$ 12. $39/\underline{-90°}\ \Omega$

13. $420/\underline{90°}\ \Omega$ 14. $2.7/\underline{90°}\ k\Omega$ 15. $68/\underline{-90°}\ k\Omega$

16. $143/\underline{90°}\ k\Omega$ 17. $0.510/\underline{-90°}\ M\Omega$ 18. $0.33/\underline{90°}\ k\Omega$

Compute the admittance of each parallel circuit given the source voltage E and circuit current I. Draw an admittance diagram for each circuit.

19. $E = 36/\underline{0°}\ V$ $I = 42/\underline{17°}\ mA$

20. $E = 50/\underline{0°}\ V$ $I = 7.2/\underline{48°}\ mA$

21. $E = 100/\underline{0°}\ V$ $I = 808/\underline{-61°}\ \mu A$

22. $E = 27/\underline{0°}\ V$ $I = 0.618/\underline{-24°}\ A$

23. $E = 18/\underline{0°}\ V$ $I = 1.43/\underline{15°}\ A$

24. $E = 80/\underline{0°}\ V$ $I = 917/\underline{-84°}\ mA$

29-4 ADMITTANCE OF PARALLEL AC CIRCUITS

Forming Equivalent Components

When several capacitors are connected in parallel, the capacity of a single equivalent capacitor may be found by Equation 29-10.

$$C_T = C_1 + C_2 + C_3 + \cdots + C_n \quad \text{F} \qquad (29\text{-}10)$$

Several inductors connected in parallel may be combined into a single equivalent inductor by applying Equation 29-11.

$$L_T = \cfrac{1}{\dfrac{1}{L_1} + \dfrac{1}{L_2} + \dfrac{1}{L_3} + \cdots + \dfrac{1}{L_n}} \quad \text{H} \qquad (29\text{-}11)$$

Several resistances in a parallel circuit may be combined into a single equivalent resistance by applying Equation 29-12.

$$R_T = \cfrac{1}{\dfrac{1}{R_1} + \dfrac{1}{R_2} + \dfrac{1}{R_3} + \cdots + \dfrac{1}{R_n}} \quad \Omega \qquad (29\text{-}12)$$

EXAMPLE 29-19

Compute the equivalent component for each of the following parallel circuit conditions.

(a) Three resistors in parallel:

$$R_1 = 12 \ \Omega \quad R_2 = 6.2 \ \Omega \quad R_3 = 8.2 \ \Omega$$

(b) Three capacitors in parallel:

$$C_1 = 10 \ \mu\text{F} \quad C_2 = 25 \ \mu\text{F} \quad C_3 = 5 \ \mu\text{F}$$

(c) Three inductors in parallel:

$$L_1 = 500 \ \text{mH} \quad L_2 = 80 \ \text{mH} \quad L_3 = 210 \ \text{mH}$$

Solution (a) Use Equation 29-12 to find the equivalent resistance:

$$R_T = \cfrac{1}{\dfrac{1}{R_1} + \dfrac{1}{R_2} + \dfrac{1}{R_3}}$$

$$= \frac{1}{\dfrac{1}{12} + \dfrac{1}{6.2} + \dfrac{1}{8.2}}$$

$$\therefore \quad R_T = 2.73 \ \Omega$$

(b) Use Equation 29-10 to find the equivalent capacitance:

$$C_T = C_1 + C_2 + C_3$$
$$= 10 + 25 + 5$$
$$\therefore \quad C_T = 40 \ \mu F$$

(c) Use Equation 29-11 to find the equivalent inductance:

$$L_T = \frac{1}{\dfrac{1}{L_1} + \dfrac{1}{L_2} + \dfrac{1}{L_3}}$$

$$= \frac{1}{\dfrac{1}{500} + \dfrac{1}{80} + \dfrac{1}{210}}$$

$$\therefore \quad L_T = 51.9 \ \text{mH}$$

Computing Branch Admittance

The admittance of each branch of a parallel circuit is found by first computing the impedance of each branch in polar form, and then taking the reciprocal of the impedance. This procedure is demonstrated in the following example.

**EXAMPLE
29-20**

Compute the admittance of each branch of the circuit of Figure 29-12.

Solution   Determine the impedance of each branch in polar form:

$$Z_1 = 100 + j200 \ \Omega \ \boxed{\rightarrow P} \ 224\underline{/63.4°} \ \Omega$$
$$Z_2 = 47 - j84 \ \Omega \ \boxed{\rightarrow P} \ 96.3\underline{/-60.8°} \ \Omega$$

Solve for branch admittances. Use Equation 29-6:

$$Y = 1/Z$$

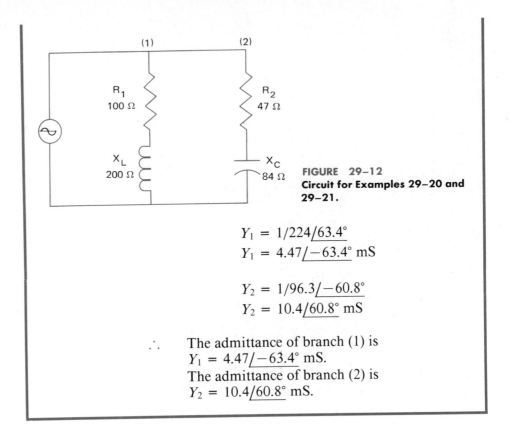

FIGURE 29–12
Circuit for Examples 29–20 and 29–21.

$$Y_1 = 1/224 \underline{/63.4°}$$
$$Y_1 = 4.47 \underline{/-63.4°} \text{ mS}$$

$$Y_2 = 1/96.3 \underline{/-60.8°}$$
$$Y_2 = 10.4 \underline{/60.8°} \text{ mS}$$

∴   The admittance of branch (1) is
$Y_1 = 4.47 \underline{/-63.4°}$ mS.
The admittance of branch (2) is
$Y_2 = 10.4 \underline{/60.8°}$ mS.

Computing Circuit Admittance and Circuit Impedance

Once the branch admittances are known, they are added to obtain the admittance of the entire circuit. Thus,

$$Y_T = Y_1 + Y_2 + Y_3 + \cdots + Y_n \quad \text{S} \qquad (29\text{-}13)$$

The impedance of the entire circuit may be computed by taking the reciprocal of the admittance. Thus,

$$Z_T = 1/Y_T \quad \Omega \qquad (29\text{-}14)$$

EXAMPLE 29-21

Use the information of Example 29-20 to compute the admittance and the impedance of the circuit of Figure 29-12.

Solution Express the branch admittances in rectangular form and add:

$$Y_1 = 4.47\underline{/-63.4°} \text{ mS} \boxed{\rightarrow\text{R}} \quad 2.00 - j4.00 \text{ mS}$$
$$Y_2 = 10.4\underline{/60.8°} \text{ mS} \boxed{\rightarrow\text{R}} \quad \underline{5.07 + j9.08 \text{ mS}}$$
$$Y_T = 7.07 + j5.08 \text{ mS}$$

Convert Y_T to polar form:

$$7.07 + j5.08 \text{ mS} \boxed{\rightarrow\text{P}} 8.71\underline{/35.7°} \text{ mS}$$
$$\therefore \quad Y_T = 8.71\underline{/35.7°} \text{ mS}$$

Compute Z_T. Use Equation 29-14:

$$Z_T = 1/(8.71 \times 10^{-3} \underline{/35.7°})$$
$$\therefore \quad Z_T = 115\underline{/-35.7°} \text{ } \Omega$$

EXAMPLE 29-22

Compute the admittance and the impedance of the circuit in Figure 29-13.

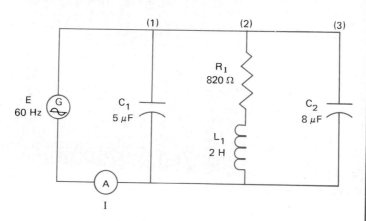

FIGURE 29–13
Circuit for Example 29–22.

Solution Compute the equivalent component for C_1 and C_2:

$$C_T = C_1 + C_2$$
$$C_T = 13 \text{ } \mu\text{F}$$

Compute the reactance of L_1 and C_T:

$$\omega = 2\pi60 = 377 \text{ rad/s}$$
$$X_{L_1} = 377(2) = 754 \ \Omega$$
$$X_{C_T} = 1/(377 \times 13 \times 10^{-6}) = 204 \ \Omega$$

Determine the impedance of each branch in polar form:

$$Z_{1,3} = 0 - j204 \ \Omega \boxed{\rightarrow\text{P}} \ 204\underline{/-90°}$$
$$Z_2 = 820 + j754 \ \Omega \boxed{\rightarrow\text{P}} \ 1.11\underline{/42.6°} \ \text{k}\Omega$$

Determine the branch admittances:

$$Y_{1,3} = 1/Z_{1,3} = 1/204\underline{/-90°}$$
$$Y_{1,3} = 4.90\underline{/90°} \ \text{mS}$$

$$Y_2 = 1/Z_2 = 1/1110\underline{/42.6°}$$
$$Y_2 = 0.901\underline{/-42.6°} \ \text{mS}$$

Express the branch admittances in rectangular form and add:

$$Y_{1,3} = 4.90\underline{/90°} \ \text{mS} \boxed{\rightarrow\text{R}} \quad \ \ 0 \quad \ + j4.9 \ \text{mS}$$
$$Y_2 = 0.901\underline{/-42.6°} \ \text{mS} \boxed{\rightarrow\text{R}} \ \underline{0.663 - j0.610 \ \text{mS}}$$
$$Y_T = 0.663 + j4.29 \ \text{mS}$$
$$Y_T = 0.663 + j4.29 \ \text{mS} \boxed{\rightarrow\text{P}} \ 4.34\underline{/81.2°} \ \text{mS}$$

$\therefore$ Compute Z_T:

$$Z_T = 1/Y_T = 1/(4.34 \times 10^{-3}\underline{/81.2°})$$
$$\therefore \quad Z_T = 230\underline{/-81.2°} \ \Omega$$

Parallel Equivalent Circuit

The rectangular form of the circuit admittance of an ac parallel circuit represents the components of the *parallel equivalent circuit*. That is, $Y_T = G_{eq} + jB_{eq}$. The parallel equivalent circuit is formed by taking the reciprocal of:

- G_{eq} to get the equivalent parallel resistance R_P
- B_{eq} in polar form to get the equivalent parallel reactance X_P.

The value of the reactive component may then be computed if the circuit frequency is known.

EXAMPLE 29-23

Compute the parallel equivalent circuit of Example 29-22 (Figure 29-13) for a frequency of 60 Hz.

Solution Use the total admittance $Y_T = 0.663 + j4.29$ mS, and express Y_T in terms of G_{eq} and B_{eq}:

$$G_{eq} = 0.663 \text{ mS}$$
$$B_{eq} = 4.29 \text{ mS}$$

Compute the equivalent parallel resistance:

$$R_P = 1/G_{eq} = 1/0.663 \times 10^{-3} \underline{/0°}$$
$$R_P = 1.51 \text{ k}\Omega$$

Compute the equivalent parallel reactance:

$$X_P = 1/B_{eq} = 1/(4.29 \times 10^{-3} \underline{/90°})$$
$$X_P = 233 \underline{/-90°} \ \Omega \text{ (capacitive reactance)}$$

Compute the size of the capacitor:

$$X_C = X_{eq} = 1/(\omega C)$$
$$C = 1/(\omega X_C)$$
$$\omega = 377 \text{ rad/s}$$
$$C = 1/(377 \times 233)$$
$$C = 11.4 \ \mu\text{F}$$

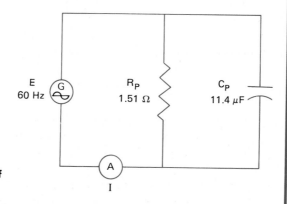

FIGURE 29-14
Parallel equivalent circuit of Figure 29-13.

∴ The parallel equivalent circuit is made up of 1.51 kΩ of resistance in parallel with 11.4 µF of capacitance when the frequency is 60 Hz. Figure 29-14 is the schematic diagram of the parallel equivalent circuit of Figure 29-13.

EXAMPLE 29-24

Determine the components of the parallel equivalent circuit of Figure 29-15 for a frequency of 400 Hz if $E = 10$ V$/0°$ and $I = 602/{-27°}$ mA.

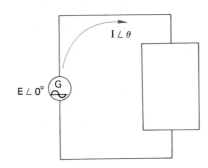

FIGURE 29–15
The box represents an unknown circuit.

Solution Compute the circuit impedance:

$$Z = E/I = 10/\underline{0°} /(602 \times 10^{-3} /\underline{-27°})$$
$$Z = 16.6/\underline{27°} \ \Omega$$

Compute the circuit admittance:

$$Y = 1/16.6/\underline{27°}$$
$$Y = 60.2/\underline{-27°} \ \text{mS} \ \boxed{\rightarrow\text{R}} \ 53.6 - j27.3 \ \text{mS}$$

Solve for R_P and X_P:

$$R_P = 1/(53.6 \times 10^{-3})$$
$$R_P = 18.7 \ \Omega$$
$$X_P = 1/(27.3 \times 10^{-3} /\underline{-90°})$$
$$X_P = 36.6/\underline{90°} \ \Omega \ \text{(inductive reactance)}$$

Compute L:

$$X_L = 2\pi f L$$
$$L = X_L/(2\pi f)$$
$$L = 36.6/(2\pi 400)$$
$$L = 14.6 \text{ mH}$$

∴   The parallel equivalent circuit is a resistance of 18.7 Ω in parallel with an inductance of 14.6 mH when the frequency is 400 Hz.

EXERCISE 29-4

In each of the following, determine the equivalent component. Assume that the components are connected in parallel.

1. $C_1 = 247$ pF, $C_2 = 88$ pF   2. $R_1 = 82$ kΩ, $R_2 = 1.0$ MΩ
3. $L_1 = 50$ μH, $L_2 = 180$ μH   4. $L_1 = 28$ mH, $L_2 = 79$ mH
5. $R_1 = 390$ Ω, $R_2 = 150$ Ω   6. $C_1 = 0.68$ μF, $C_2 = 1200$ pF

Determine the branch admittance for each branch of a two-branch parallel circuit, given the following information.

7. Branch (1) $R = 75$ Ω $X_L = 130$ Ω
   (2) $R = 72$ Ω $X_C = 143$ Ω
8. Branch (1) $R = 560$ Ω $X_C = 318$ Ω
   (2) $R = 2.0$ kΩ
9. Branch (1) $R = 840$ Ω $X_C = 600$ Ω
   (2) $R = 420$ Ω
10. Branch (1) $R = 1.6$ kΩ
    (2) $R = 2.5$ kΩ

In each of the following, determine the circuit admittance and the circuit impedance for a two-branch parallel circuit given the impedance of each branch.

11. Branch (1) $291 + j83$ Ω
    (2) $0 - j153$ Ω
12. Branch (1) $143 + j308$ Ω
    (2) 203 Ω

13. Branch (1) $8.3 + j4.1$ kΩ
    (2) $0 + j3.6$ kΩ

14. Branch (1) $0 - j12$ Ω
    (2) $0 + j48$ Ω

15. Branch (1) $392 - j170$ Ω
    (2) $94 + j568$ Ω

Compute the components of the equivalent circuit contained in the box of Figure 29-15, given the following information. Assume that the components are connected in parallel.

16. $E = 22.0\underline{/0°}$ V $\quad I = 2.2\underline{/72°}$ A $\qquad f = 60$ Hz

17. $E = 80\underline{/0°}$ V $\quad I = 380\underline{/-11°}$ mA $\quad f = 5$ kHz

18. $E = 33\underline{/0°}$ V $\quad I = 92\underline{/31°}$ mA $\qquad f = 400$ Hz

19. $E = 325\underline{/0°}$ V $\quad I = 0.96\underline{/7°}$ A $\qquad f = 60$ Hz

20. $E = 12\underline{/0°}$ V $\quad I = 495\underline{/-65°}$ μA $\quad f = 180$ kHz

Solve the following.

21. Two impedances are connected in parallel: $Z_1 = 210\underline{/81°}$ Ω and $Z_2 = 750\underline{/51°}$ Ω. $E = 100\underline{/0°}$ V.
    (a) Determine the admittance of each branch.
    (b) Determine the circuit admittance.
    (c) Determine the circuit impedance.
    (d) Determine the circuit current.
    (e) Construct the circuit current, source voltage, and phasor diagram. Use E as the reference phasor.
    (f) Construct the circuit admittance diagram.

22. Two impedances are connected in parallel: $Z_1 = 920 - j260$ Ω, $Z_2 = 382\underline{/-56°}$ Ω, and $E = 60\underline{/0°}$ V. Determine:
    (a) The circuit admittance
    (b) The circuit impedance
    (c) The circuit current
    (d) The parallel equivalent circuit

SELECTED TERMS

admittance The reciprocal of impedance; the vector quantity used to describe the ease of the passage of alternating current; expressed in units of siemens.

impedance The opposition to the passage of alternating current through an electrical circuit; expressed in units of ohms.

susceptance The reciprocal of inductive or capacitive reactance; measured in units of siemens.

30

SINUSOIDAL ALTERNATING CURRENT

An ac generator produces a time-varying voltage. When a load is placed across an ac source, the resulting current is also time varying. In this chapter we study how the varying amplitudes of the voltage and the current are related to the sine wave. We also study how power is determined from a sinusoidal voltage and current.

30-1 TIME AND DISPLACEMENT

If the ac wave produced by a generator has a constant frequency, then a rotating vector representing an alternating current or voltage will also have a constant frequency. When a voltage vector is rotated from the reference axis at a constant frequency, the angle generated will be a function of time. Thus,

$$\alpha = \omega t \quad \text{rad} \qquad (30\text{-}1)$$

where α = angular displacement from the reference axis in radians
ω = angular velocity in radians per second
t = time in seconds

From Figure 30-1, an equation that relates the instantaneous voltage (e) to the maximum voltage (E_{mx}) and displacement angle (α) may be derived.

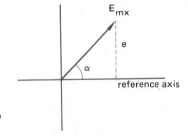

FIGURE 30–1
The instantaneous voltage e is a direct function of the angular displacement α from the reference axis.

Applying the concepts of trigonometry, we write a sine function as $\sin(\alpha) = e/E_{mx}$. Solving for e results in Equation 30-2:

$$e = E_{mx} \sin(\alpha) \quad \text{V} \qquad (30\text{-}2)$$

Similarly,

$$i = I_{mx} \sin(\alpha) \quad \text{A} \qquad (30\text{-}3)$$

Substitute $\alpha = \omega t$:

$$e = E_{mx} \sin (\omega t) \qquad \text{V} \qquad (30\text{-}4)$$
$$i = I_{mx} \sin (\omega t) \qquad \text{A} \qquad (30\text{-}5)$$

where e = instantaneous voltage at time t
$\quad i$ = instantaneous current at time t
$\quad E_{mx}$ = maximum amplitude of the voltage wave in volts (V)
$\quad I_{mx}$ = maximum amplitude of the current wave in amperes (A)
$\quad \omega$ = angular velocity of the generator producing the voltage wave
$\qquad$ in radians per second (rad/s)
$\quad t$ = time in seconds (s)

Figure 30-2 relates the phasor diagram of voltage to the periodic wave of voltage. As shown in Figure 30-2, the phasor diagram indicates only the condition for a particular time (t_1) or angular displacement (ωt_1).

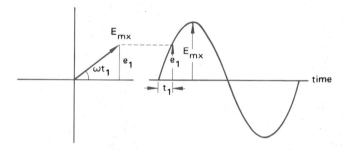

FIGURE 30-2
The phasor diagram displays the conditions of the periodic wave at time t_1.

EXAMPLE 30-1 Determine the instantaneous voltage (e) of a 60-Hz ac source that has a maximum amplitude (E_{mx}) of 165 V for the following times.

(a) $t_1 = 3$ ms $\qquad$ (b) $t_2 = 11$ ms
(c) $t_3 = 18$ ms $\qquad$ (d) $t_4 = 25$ ms

Solution Use Equation 30-4. Remember that $\omega = 2\pi f$.

(a)
$$e = E_{mx} \sin (\omega t)$$
$$\boxed{\times}$$
$$\boxed{\text{RAD}} \;\; \boxed{\text{SIN}}$$
$$e = 165 \sin (2\pi 60 \times 3 \times 10^{-3})$$
$$e = 165 \sin (1.13^r)$$

$$\boxed{\times} \quad e = 165 \,(0.905)$$
$$\therefore \quad e = 149 \text{ V at } t = 3 \text{ ms}$$

(b) $\qquad e = 165 \sin (2\pi60 \times 11 \times 10^{-3})$

Use *chain calculations.*

$$\therefore \quad e = -139 \text{ V at } t = 11 \text{ ms}$$

(c) $\qquad e = 165 \sin (2\pi60 \times 18 \times 10^{-3})$
$$\therefore \quad e = 79.5 \text{ V at } t = 18 \text{ ms}$$

(d) $\qquad e = 165 \sin (2\pi60 \times 25 \times 10^{-3})$
$$\therefore \quad e = 0 \text{ V at } t = 25 \text{ ms}$$

EXAMPLE 30-2 Compute the angular displacement from the reference axis in radians and in degrees for each of the given times of Example 30-1.

Solution Use Equation 30-1:

$$\omega = 2\pi60 = 377 \text{ rad/s}$$

(a) $t_1 = 3$ ms

$$\alpha = \omega t$$
$$\alpha = 377(3 \times 10^{-3})$$
$$\therefore \quad \alpha = 1.13^r = 64.8°$$

(b) $t_2 = 11$ ms

$$\alpha = 377(11 \times 10^{-3})$$
$$\therefore \quad \alpha = 4.15^r = 238°$$

(c) $t_3 = 18$ ms

$$\alpha = 377(18 \times 10^{-3})$$
$$\therefore \quad \alpha = 6.79^r = 389°$$

(d) $t_4 = 25$ ms

$$\alpha = 377(25 \times 10^{-3})$$
$$\therefore \quad \alpha = 9.43^r = 540°$$

EXAMPLE 30-3　Write the instantaneous equation in terms of time for a current of 400 Hz that has a maximum current (I_{mx}) of 22 A.

Solution　Use Equation 30-5. Compute ω for a frequency of 400 Hz:

$$\omega = 2\pi 400$$
$$\omega = 2.51 \text{ krad/s}$$

Substitute $I_{mx} = 22$ A and $\omega = 2.51$ krad/s:

$$\therefore \quad i = 22 \sin (2.51 \times 10^3 \, t) \quad \text{A}$$

EXAMPLE 30-4　Compute the instantaneous value for the following phasors.

(a) $E_{mx} = 100$ V, $\underline{/\theta} = 140°$
(b) $I_{mx} = 40$ mA, $\underline{/\theta} = 0.85^r$

Solution　(a) Use Equation 30-2:

$$e = E_{mx} \sin (\alpha)$$
$$= 100 \sin (140°)$$
$$\therefore \quad e = 64.3 \text{ V}$$

(b) Use Equation 30-3:

$$i = I_{mx} \sin (\alpha)$$
$$= 40 \times 10^{-3} \sin (0.85^r)$$
$$\therefore \quad i = 30.1 \text{ mA}$$

EXAMPLE 30-5   The circuit current is $i = 50 \times 10^{-3} \sin (500t)$ A.

(a) Determine the instantaneous current when $t = 6$ ms.

(b) Draw the phasor diagram for the current.

Solution   (a) Substitute $t = 6$ ms into $i = 50 \times 10^{-3} \sin (500t)$:

$$i = 50 \times 10^{-3} \sin (500 \times 6 \times 10^{-3})$$
$$i = 7.06 \text{ mA} \quad \text{when } t = 6 \text{ ms}$$

(b) Translate $t = 6$ ms into angular displacement in degrees:

$$\alpha = \omega t$$
$$= 500(6 \times 10^{-3})$$
$$\alpha = 3.0^{\text{r}} = 172°$$

∴   The current phasor is $50\underline{/172°}$ mA, as shown in Figure 30-3.

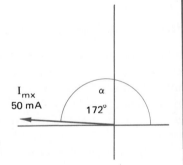

I_{mx}
50 mA

α

172°

FIGURE 30-3
Current phasor diagram for
Example 30-5.

From Example 30-5 it may be seen that the instantaneous equation, $i = 50 \times 10^{-3} \sin (500t)$ A, of the sinusoidal current wave can be translated into the current phasor $50\underline{/172°}$ mA when $t = 6$ ms. Thus,

$$i = 50 \times 10^{-3} \sin (172°) \text{ A} \Rightarrow 50 \times 10^{-3} \underline{/172°} \text{ A}$$

EXERCISE 30-1
With the given information, determine the instantaneous quantity (voltage e or current i) for each of the following.

1. $E_{mx} = 18$ V $f = 100$ Hz $t = 2.3$ ms
2. $I_{mx} = 2$ A $\alpha = 2.3^r$
3. $I_{mx} = 710$ mA $f = 62$ kHz $t = 13.7$ μs
4. $E_{mx} = 120$ V $\alpha = 212°$
5. $I_{mx} = 12.6$ A $\alpha = 4.1^r$
6. $I_{mx} = 4.3$ mA $f = 710$ kHz $t = 500$ ns
7. $E_{mx} = 82$ V $\alpha = 3.1^r$
8. $E_{mx} = 12.6$ V $f = 400$ Hz $t = 1.5$ ms
9. $E_{mx} = 440$ V $\alpha = -36°$
10. $I_{mx} = 1.43$ A $f = 60$ Hz $t = 2.7$ ms

For each of the following instantaneous equations, draw a phasor diagram that expresses the angular displacement (α) in degrees. All angles are in radians.

11. $i = 10 \sin (4.50)$ 12. $i = 0.37 \sin (3.77)$
13. $e = 24 \sin (1.61)$ 14. $i = 15 \times 10^{-3} \sin (5.4)$
15. $e = 100 \sin (2.15)$ 16. $e = 36 \sin (0.53)$

30-2 POWER AND POWER FACTOR

Power in a Resistive Load

When an ac voltage of the general form $e = E_{mx} \sin (\omega t)$ is applied to a resistive load, a circuit current results in the form $i = I_{mx} \sin (\omega t)$, which is in phase with the voltage and the phase angle $\theta = 0$. If the two waves represented by the instantaneous equations are plotted and the instantaneous amplitudes of the waves are multiplied, a third wave representing the instantaneous power ($p = ei$ VA) will result.

Figure 30-4 shows the instantaneous power resulting from multiplying e

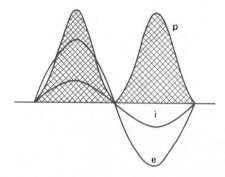

FIGURE 30–4
Instantaneous power in a resistive load is always positive.

and i. Mathematically, an expression for *average power* may be derived from instantaneous power.

$$p = ei$$
$$(1) \quad e = E_{mx} \sin (\omega t)$$
$$(2) \quad i = I_{mx} \sin (\omega t)$$

Substitute (1) and (2) into $p = ei$:

$$p = [E_{mx} \sin (\omega t)][I_{mx} \sin (\omega t)]$$
$$p = E_{mx}I_{mx} \sin^2 (\omega t)$$

However, $\sin^2 (\omega t) = \frac{1}{2}[1 - \cos (2 \omega t)]$:

$$p = \frac{E_{mx}I_{mx}}{2} [1 - \cos (2 \omega t)]$$
$$p = \frac{E_{mx}I_{mx}}{2} - \frac{E_{mx}I_{mx}}{2} \cos (2 \omega t)$$

Notice that the second term is a cosine wave having a frequency twice that of the voltage or current wave. Since the average value of a cosine wave for one cycle is zero, the second term does not contribute to the average. Only the first term remains. This term is called the *average power* and is expressed in a formula as

$$P = \frac{E_{mx}I_{mx}}{2} \quad \text{W} \tag{30-6}$$

Effective Values

Factor the right member of Equation 30-6 using the common denominator $\sqrt{2}$. Thus,

$$P = \frac{E_{mx}}{\sqrt{2}} \cdot \frac{I_{mx}}{\sqrt{2}}$$

Replace $1/\sqrt{2}$ with 0.707:

$$P = (0.707E_{mx})(0.707I_{mx})$$

The factor $0.707E_{mx}$ is given the symbol E, which stands for the *effective voltage;* that is,

$$E = 0.707E_{mx} \quad \text{V} \tag{30-7}$$

And the factor $0.707I_{mx}$ is given the symbol I, which stands for the *effective current;* that is,

$$I = 0.707I_{mx} \quad A \tag{30-8}$$

Therefore,

$$P = EI \quad W \tag{30-9}$$

The average power (also called *true power*) dissipated by a resistance in an ac circuit is computed by a formula using the same letters as the formula for power in a dc circuit.

The product of the effective voltage (E) and the effective current (I) results in the same power (watts) as the product of a dc voltage (E) and a dc current (I) of the same magnitudes.

The effective voltage and current are also called root-mean-square (rms) values. This name comes from the mathematical procedure used to derive the effective value. Table 30-1 is a summary of ac voltage and current notation. From the table it may be noted that no subscript notation is used with effective quantities. Example 30-6 demonstrates converting from effective to maximum voltage.

TABLE 30-1

AC VOLTAGE AND CURRENT NOTATION

AC Quantity	Notation	Example	Visual Representation	Comments
Peak-to-peak	E_{pp}, V_{pp}, or I_{pp}	$E_{pp} = 20$ V	Peak-to-peak 20 V	Twice maximum
Maximum or peak	E_{mx} or E_p I_{mx} or I_p	$E_p = 10$ V	Peak 10 V	One half peak to peak
Effective or rms	E or I	$E = 7.07$ V	rms 7.07 V	No subscript 0.707 × peak

EXAMPLE 30-6 Determine the maximum voltage of a 117-V line.
Solution Remember that 117 V means 117 V effective. Use Equation 30-7:

$$E = 0.707E_{mx}$$

Solve for E_{mx}:

$$E_{mx} = E/0.707$$
$$= 117/0.707$$
$$\therefore \quad E_{mx} = 165 \text{ V}$$

Power in a Reactive Load

When an ac voltage is applied to a reactive load, the voltage is 90° *out of phase* with the current. As with the case of a resistive load, when the instantaneous amplitudes of the voltage and current are multiplied, the resultant wave represents the instantaneous power of the reactor, as shown in Figure 30-5. When the power wave is averaged, the result is zero, which leads to the conclusion that *reactive loads do not dissipate power.*

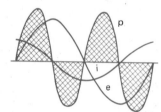

FIGURE 30-5
Instantaneous power in a reactive load is both positive and negative.

Power Factor

Because only the resistive portion of an ac circuit dissipates power, we are interested in knowing what part of the impedance is resistive. The ratio of the circuit resistance to the magnitude of the circuit impedance is called the *power factor.* Thus,

$$\text{Power factor} = R/|Z| \tag{30-10}$$

When the impedance of an ac circuit is resistive ($Z = R$), the power factor is 1. When the impedance is reactive ($Z = jX$), the power factor is zero. The power factor is a positive, unitless number that can be expressed as a percentage or a decimal fraction.

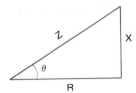

FIGURE 30-6
The power factor of a circuit is cos (θ), which equal R/Z.

The power factor may be related to the phase angle through the impedance diagram. Looking at Figure 30-6, we may recognize the ratio of $R/|Z|$ as the cosine function of θ. Thus,

$$\text{Power factor} = \cos{(\theta)} = R/|Z| \qquad (30\text{-}11)$$

EXAMPLE 30-7   Determine the power factor for each of the following circuits:

    (a) $Z = 650\underline{/-34°}\ \Omega$
    (b) $Z = 125 + j90\ \Omega$
    (c) $|Z| = 143\ \Omega, \quad R = 18\ \Omega$

Solution   (a) $\theta = -34°$.

$$\cos{(\theta)} = \cos{(-34°)}$$
$$\therefore \quad \text{Power factor} = 0.829$$

(b) Convert $125 + j90\ \Omega$ to polar form:

$$125 + j90 \boxed{\rightarrow P}\ 154\underline{/35.8°}\ \Omega$$
$$\cos{(\theta)} = \cos{(35.8°)}$$
$$\therefore \quad \text{Power factor} = 0.811$$

(c) Substitute $R = 18$ and $|Z| = 143$ into:

$$\text{Power factor} = R/|Z|$$
$$= 18/143$$
$$\therefore \quad \text{Power factor} = 0.126$$

The word *leading* or the word *lagging* is often used with power factor. The relationship of the current through the load to the voltage as the reference phasor determines *lead* or *lag*. Since current leads voltage in a capacitive load, a *capacitive circuit has a leading power factor*. Since current lags voltage in an inductive load, an *inductive circuit has a lagging power factor*.

The generalized equations for computing the power dissipation in an ac circuit include the power factor as shown in the following:

$$P = |E||I| \cos (\theta) \quad W \qquad (30\text{-}12)$$

$$P = |I|^2 |Z| \cos (\theta) \quad W \qquad (30\text{-}13)$$

$$P = \frac{|E|^2}{|Z|} \cos (\theta) \quad W \qquad (30\text{-}14)$$

EXAMPLE 30-8 Determine the average power dissipated in an ac circuit having an impedance of $Z = 750\underline{/-55°}$ Ω and a source voltage of $e = 165 \sin (377t)$ V.

Solution Solve for E in $E = 0.707 E_{mx}$:

$$E = 0.707(165)$$
$$E = 117 \text{ V}$$

Substitute into Equation 30-14:

$$P = \frac{|E|^2}{|Z|} \cos (\theta)$$

$$= \frac{117^2}{750} \cos (55°)$$

$$\therefore \quad P = 10.5 \text{ W}$$

EXAMPLE 30-9 A certain ac circuit has a leading power factor of 0.62, an impedance of 310 Ω, and a circuit current $i = 2.3 \sin (754t)$ A.

(a) Determine the power dissipation.
(b) Determine the source voltage E.
(c) Determine the series equivalent components.
(d) Draw the voltage current phasor with the current as the reference phasor.

Solution (a)

$$P = |I|^2 |Z| \cos (\theta)$$
$$= [2.3(0.707)]^2 (310)(0.62)$$
$$\therefore \quad P = 508 \text{ W}$$

(b)

$$E = IZ$$
$$|E| = (2.3)(0.707)(310)$$
$$\therefore \quad |E| = 504 \text{ V}$$

(c)

$$\angle|\theta| = \text{Arccos (power factor)}$$
$$\angle|\theta| = \text{Arccos } (0.62) = 51.7°$$
$$\angle\theta = -51.7°$$

Observation Because current is the reference phasor and the load is capacitive (leading power factor), the voltage lags. Thus, $-51.7°$.

$$310\underline{/-51.7°}\;\boxed{\rightarrow\text{R}}\;192 - j243 \; \Omega$$

Compute the capacitance: ($\omega = 754$ rad/s)

$$C = 1/(\omega X_C)$$
$$= 1/[754(243)]$$
$$C = 5.5 \;\mu\text{F}$$

$\therefore$ The equivalent series circuit is a resistance of 192 Ω and a capacitance of 5.5 μF at an $\omega = 754$ rad/s.

(d) Remember, in a capacitive circuit the current leads the voltage and voltage lags current. See Figure 30-7.

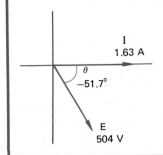

I
1.63 A

θ
$-51.7°$

E
504 V

FIGURE 30–7
Phasor diagram for Example 30–9.
Notice $i = 0.707 I_{mx}$.

Convert each of the following to its effective value.

1. $E_{mx} = 14.3$ V
2. $I_{mx} = 7.4$ A
3. $E_{mx} = 17.2$ V
4. $I_{mx} = 10.3$ mA
5. $E_{mx} = 180$ V
6. $E_{mx} = 4.3$ kV
7. $I_{mx} = 15.7$ A
8. $E_{mx} = 62$ V
9. $I_{mx} = 8.1$ mA

Convert each of the following to its maximum value.

10. $E = 110$ V
11. $I = 1.32$ A
12. $I = 81$ mA
13. $I = 18$ mA
14. $E = 2.3$ kV
15. $E = 15$ V
16. $I = 76.2$ μA
17. $E = 44$ V
18. $I = 604$ μA

Compute the average power in watts given the following phasor diagrams.

19.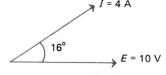

20.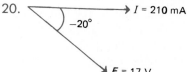

21.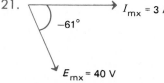

22.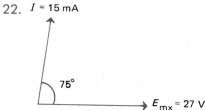

Determine the power factor for each of the following circuit conditions and state if it is leading or lagging.

23. $Z = 12 - j8$ Ω
24. $Z = 42\underline{/-81°}$ Ω
25. $R = 33$ Ω, $|Z| = 89$ Ω
26. $Z = 218 + j143$ Ω

Solve the following.

27. Determine the power dissipated in an ac circuit having a peak source voltage of $E_p = 18$ V and an impedance of $Z = 216\underline{/42°}$ Ω.

28. Determine the power dissipated in a capacitive load having an impedance of $Z = 717\underline{/-19°}$ Ω and an instantaneous circuit current of $i = 1.25 \sin(377t)$.

29. Determine the equivalent series circuit for an ac circuit having a power dissipation of $P = 30$ W and a lagging power factor of 0.5. The circuit current is $i = 132 \times 10^{-3} \sin(5027t)$ A.

30. Determine the components of the equivalent series circuit for an ac circuit having a power dissipation of $P = 10$ W and a leading power factor of 0.72. The source voltage is $e = 14.3 \sin(377t)$ V.

30-3 INSTANTANEOUS EQUATIONS AND THE *EI* PHASOR DIAGRAM

The graph of the instantaneous equation of voltage and current gives a continuous *picture* of the interrelation between the circuit current and the source voltage. The phasor diagram of current and voltage gives a picture of the circuit conditions for a selected instant of time. Given sufficient information, each may be developed from the other.

The phasor diagram is constructed with one of the quantities acting as the reference phasor and the other phasor located θ radians or degrees from the reference phasor. Thus, if current was the reference phasor, it would be located at $I\underline{/0}$ and the voltage would be located at $E\underline{/\theta}$. We see that the reference phasor assumes 0 as its direction, while the other phasor assumes the phase angle θ as its direction.

When writing the instantaneous equations, one equation is written as the reference equation, while the other contains the phase angle. Either equation may be selected as the reference. The following equations are the generalized equations for the instantaneous voltage and current.

$$e = E_{mx} \sin(\omega t \pm \theta) \quad V \qquad (30\text{-}15)$$
$$i = I_{mx} \sin(\omega t \pm \theta) \quad A \qquad (30\text{-}16)$$

where ωt is in radians and θ is the phase angle in radians or degrees.

Special care must be taken when evaluating the sine of $(\omega t \pm \theta)$. Both ωt and θ must be in the same units before adding and evaluating.

EXAMPLE 30-10

Use the phasor diagram of Figure 30-8 to write the instantaneous equations for voltage and current at a frequency of 60 Hz ($\omega = 377$ rad/s).

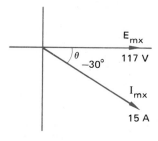

FIGURE 30–8
Phasor diagram for
Example 30–10.

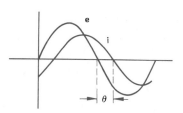

FIGURE 30–9
The current, i, lags the voltage,
e, by ∠θ, the phase angle.

Solution   Because voltage is the reference phasor, the instantaneous voltage equation will have a zero phase angle:

$$e = E_{mx} \sin (\omega t \pm \theta)$$
$$\therefore \quad e = 117 \sin (377t) \text{ V}$$

The current equation will indicate the phase angle:

$$i = I_{mx} \sin (\omega t \pm \theta)$$
$$\theta = -30° \left(\frac{\pi}{180°}\right) = -0.524^r$$
$$\therefore \quad i = 15 \sin (377t - 0.524) \text{ A}$$

Figure 30-9 shows the phase relationship between the instantaneous current and the instantaneous voltage waves.

EXAMPLE
30-11

Construct a phasor diagram to represent the conditions of the following equation when $t = 4$ ms:

$$e = 10 \sin (200t^r + 20°) \text{ V}$$
$$i = 2 \sin (200t) \text{ A}$$

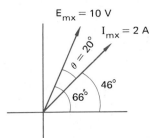

$E_{mx} = 10$ V

$I_{mx} = 2$ A

$\theta = 20°$

$46°$

$66°$

FIGURE 30–10
Phasor diagram for Example 30–11.

Solution   Compute the position in degrees of each phasor at $t = 4$ ms:

$$\alpha = \omega t \text{ (rad)}$$
$$= 200(4 \times 10^{-3})$$
$$\alpha = 0.80^r = 46°$$

$\therefore$ E_{mx} is located at 66° (46° + 20°) and I_{mx} is located at 46°. See Figure 30-10.

EXAMPLE 30-12

Use the phasor diagram of Figure 30-11 to write the instantaneous equations for voltage and current at a frequency of 400 Hz.

Solution   Compute ω using $\omega = 2\pi f$:

$$\omega = 2\pi(400)$$
$$\omega = 2.51 \text{ krad/s}$$

Convert E to E_{mx}:

$$E = 0.707E_{mx}$$
$$E_{mx} = E/0.707$$
$$= 24/0.707$$
$$E_{mx} = 34 \text{ V}$$

Convert I to I_{mx}:

$$I_{mx} = I/0.707$$
$$= 3/0.707$$
$$I_{mx} = 4.2 \text{ A}$$

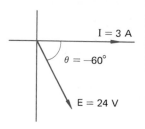

FIGURE 30-11
Phasor diagram for Example 30-12.

The current phasor is the reference, so the instantaneous equation for current will have a zero phase angle:

$$i = I_{mx} \sin (\omega t \pm \theta) \text{ A}$$
$$\therefore \quad i = 4.2 \sin (2.51 \times 10^3 \, t) \text{ A}$$

$$e = E_{mx} \sin (\omega t - \theta) \text{ V}$$
$$\therefore \quad e = 34 \sin (2.51 \times 10^3 t^r - 60°) \text{ V}$$

EXAMPLE 30-13

The voltage ($E = 36$) in an ac circuit leads the current ($I = 0.31$ A) by 27°. Determine the instantaneous value of current when the voltage has completed 2 radians of its cycle.

Solution Because the instantaneous current *tracks* with the voltage, the current has also completed 2 radians of its cycle. Assume voltage is the reference phasor. Write the instantaneous equation for current:

$$i = I_{mx} \sin (\omega t - \theta)$$

Substitute $\omega t = 2^r$, $\theta = 27°$, and $I_{mx} = 0.31/0.707$ A:

$$i = (0.31/0.707) \sin (2^r - 27°)$$

Express 27° in radians:

$$i = 0.439 \sin (2 - 0.47)$$
$$= 0.439 \sin (1.53)$$
$$\therefore \quad i = 0.439 \text{ A}$$

Write the instantaneous equations for voltage and current from the given phasor diagram and angular velocity in each of the following.

1.

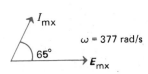

$\omega = 900$ rad/s

$10°$

2.

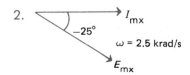

$-25°$

$\omega = 2.5$ krad/s

3.

$\omega = 377$ rad/s

$65°$

4.

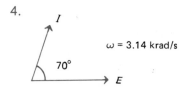

$\omega = 3.14$ krad/s

$70°$

Express the phase angle in degrees.

5. $e = 18 \sin (200t^r)$ V
   $i = 2 \sin (200t^r - 38°)$ A

6. $e = 100 \sin (600t^r + 0.4^r)$ V
   $i = 7 \sin (600t^r)$ A

7. $e = 190 \sin (377t^r - 0.7^r)$ V
   $i = 0.62 \sin (377t^r)$ A

8. $e = 143 \sin (700t^r)$ V
   $i = 3 \sin (700t^r - 81°)$ A

Solve the following.

9. A 400-Hz, 2.5-V test oscillator delivers 30 mA to a reactive load. The voltage leads the current by 48°.
   (a) Write the instantaneous equations for voltage and current.
   (b) Determine the value of the instantaneous current at $t = 50$ μs.
   (c) Determine the average power delivered to the load.

10. A 60-Hz, 117-V generator delivers 20 A to a reactive load. The current lags the voltage by 1.1^r.
    (a) Write the instantaneous equations for voltage and current.
    (b) Determine the value of instantaneous current at $t = 4$ ms.
    (c) Determine the average power delivered to the load.

SELECTED TERMS

average power   The power (in watts) dissipated by the resistive component of the impedance in an ac circuit. Sometimes referred to as true power.

power factor   The ratio of resistance to impedance; the cosine of the circuit phase angle.

31

ADDITIONAL TRIGONOMETRIC AND EXPONENTIAL FUNCTIONS

This and the following chapter are designed to help you prepare for a course in calculus. The functions and concepts introduced in this chapter will be encountered again in the following chapter and in most calculus textbooks. As in calculus textbooks, all angles are assumed to be in radians unless specified to be in degrees.

31-1 AUXILIARY TRIGONOMETRIC FUNCTIONS

In this section, three additional trigonometric functions are introduced. These functions are not as common in electronics as the sine, cosine, and tangent functions. However, they do occur in a study of calculus and so we introduce them at this time. These functions are the secant, cosecant, and cotangent.

The first function is the *secant,* which is equal to the reciprocal of the cosine. Thus, for any angle θ, sec $(\theta) = 1/\cos(\theta)$. Many calculators do not have the secant function, but they do have a reciprocal key. The way to calculate the secant function is to calculate the cosine, and then calculate the reciprocal.

EXAMPLE 31-1   Use your calculator to find sec (45°).
  Solution   Set your calculator to work in degrees and enter 45:

$$\boxed{\text{COS}}\,45 \Rightarrow 0.707$$

Form the reciprocal:

$$\boxed{1/x} \quad 1.414$$

$$\therefore \quad \text{sec } (45°) = 1.414$$

The second function is the *cosecant,* which is equal to the reciprocal of the sine. Thus, for any angle θ, csc $(\theta) = 1/\sin(\theta)$. The cosecant function may be calculated as the reciprocal of the sine.

EXAMPLE 31-2   Use your calculator to find csc (30°).
  Solution   Set your calculator to work in degrees and enter 30:

$$\boxed{\text{SIN}}\,30 \Rightarrow 0.500$$

Form the reciprocal:

$$\boxed{1/x} \quad 2.0$$
$$\therefore \quad \text{csc } (30°) = 2.0$$

The third function is the *cotangent,* which is equal to the reciprocal of the tangent. Thus, for any angle θ, cot $(\theta) = 1/\tan (\theta)$. Note, some authors use ctn ( ) to indicate the cotangent. Again, if your calculator does not have this function, use the tangent and the reciprocal to calculate it.

EXAMPLE 31-3   Use your calculator to find cot (1.42).

Solution   Set your calculator to work in radians and enter 1.42:

$$\boxed{\text{TAN}}\,1.42 \Rightarrow 6.58$$

Form the reciprocal:

$$\boxed{1/x} \quad 0.152$$

$$\therefore \quad \cot (1.42) = 0.152$$

The definitions of the trigonometric functions are presented in Table 31-1. This table refers to Figure 31-1.

TABLE 31-1

DEFINITIONS OF TRIGONOMETRIC FUNCTIONS

Function	Abbreviation	Definition	Relation to Other Functions
Sine (θ)	sin (θ)	y/z	$1/\csc (\theta)$
Cosine (θ)	cos (θ)	x/z	$1/\sec (\theta)$
Tangent (θ)	tan (θ)	y/x	sin (θ)/cos (θ)
Secant (θ)	sec (θ)	z/x	$1/\cos (\theta)$
Cosecant (θ)	csc (θ)	z/y	$1/\sin (\theta)$
Cotangent (θ)	cot (θ)	x/y	$1/\tan (\theta)$

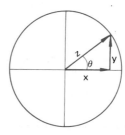

FIGURE   31–1

Circle used to define the auxiliary trigonometric functions.

EXERCISE 31-1

Calculator Drill
Evaluate the following functions.

1. sec (29°)	2. sec (50°)	3. sec (−30°)
4. sec (150°)	5. sec (2.3)	6. sec (−1.05)
7. sec (0.56)	8. sec (6.28)	9. csc (82°)
10. csc (98°)	11. csc (−25°)	12. csc (−135°)
13. csc (1.4)	14. csc (6.0)	15. csc (−2.0)
16. csc (4.0)	17. cot (12.5°)	18. cot (37.5°)
19. cot (75°)	20. cot (115°)	21. cot (−0.32)
22. cot (−1.0)	23. cot (−2.0)	24. cot (1.05)

31-2 GRAPHS OF THE AUXILIARY TRIGONOMETRIC FUNCTIONS

The secant, cosecant, and cotangent functions are periodic functions, as are their reciprocals. From the previous section we know the definitions of the functions and how to calculate their values. To have a better idea of how a function behaves, it is useful to graph the function.

EXAMPLE 31-4   Create a plot of the secant function for angles from −1.57 to 7.85 radians. On the same graph, plot the cosine function.

Solution   Build a table of θ, cos (θ), and sec (θ) for θ varying from −1.57 to 7.85:

SAMPLE FUNCTION TABLE

θ	cos (θ)	sec (θ)
−1.57	0.0008	1260.0
−1.5	0.071	14.1
−1.4	0.170	5.88
−1.3	0.267	3.74
−1.2	0.362	2.76
−1.0	0.540	1.85

See Figure 31-2 for the graphs.

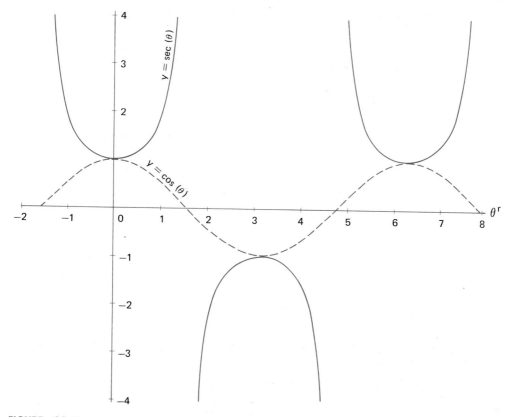

FIGURE 31-2

Graph of secant and cosine functions for Example 31–4. Notice that the secant goes off the graph wherever the cosine is "close" to zero.

EXERCISE 31-2

1. Form a graph of the sine and cosecant functions for angles from 0 to 6.28 radians.
2. Form a graph of the tangent and cotangent functions for angles from −1.57 to 7.85 radians.

31-3 TRIGONOMETRIC IDENTITIES

A trigonometric identity is an equation that is true for all angles. For example, the definition of the secant function sec $(\theta) = 1/\cos(\theta)$ is an identity. Many identities can be derived from the definitions of the circular functions. We write these identities using the special symbol $\equiv$ for identically equal.

CHAPTER 31 ADDITIONAL TRIGONOMETRIC AND EXPONENTIAL FUNCTIONS

EXAMPLE 31-5 Show that $\tan(\theta) \equiv \sin(\theta)/\cos(\theta)$.

Solution Substitute the ratio definitions for the circular functions.

Left member:

$$\tan(\theta) = y/x$$

Right member:

$$
\begin{aligned}
\sin(\theta)/\cos(\theta) &= (y/z)/(x/z) \\
&= (y/z)(z/x) \\
&= y/x
\end{aligned}
$$

This is the same value as for the tangent (independent of θ).

$$\therefore \quad \tan(\theta) \equiv \sin(\theta)/\cos(\theta)$$

EXAMPLE 31-6 Check $\cos^2(\theta) + \sin^2(\theta) \equiv 1$. Remember $\cos^2(\theta)$ means $\cos(\theta)\cos(\theta)$.

Solution Substitute the ratio definitions. Left member:

$$
\begin{aligned}
\cos^2(\theta) + \sin^2(\theta) &= (x/z)^2 + (y/z)^2 \\
&= x^2/z^2 + y^2/z^2 \\
&= \frac{x^2 + y^2}{z^2}
\end{aligned}
$$

Apply the Pythagorean theorem ($x^2 + y^2 = z^2$):

$$\cos^2(\theta) + \sin^2(\theta) = z^2/z^2 = 1$$

This is the same value as the right member:

$$\therefore \quad \cos^2(\theta) + \sin^2(\theta) \equiv 1$$

EXERCISE 31-3

Check the following identities.

1. $\cot(\theta) \equiv \cos(\theta)/\sin(\theta)$
2. $\tan(\theta) \equiv \sin(\theta)\sec(\theta)$
3. $\csc(\theta) \equiv \cot(\theta)\sec(\theta)$
4. $\sec^2(\theta) \equiv 1 + \tan^2(\theta)$

5. $\tan(\theta) \equiv \sec(\theta)/\csc(\theta)$ 6. $\cot(\theta) \equiv \csc(\theta)/\sec(\theta)$

7. $\tan(\theta)\cot(\theta) \equiv 1$ 8. $\sin(\theta)\csc(\theta) \equiv 1$

9. $\tan(\theta) + \cot(\theta) \equiv \sec(\theta)\csc(\theta)$

10. $\cos(\theta)\sec(\theta) \equiv 1$

11. $\sin(\theta)\cos(\theta)\sec(\theta)\csc(\theta) \equiv 1$

12. $1 - 2\sin^2(\theta) \equiv 2\cos^2(\theta) - 1$

13. $\cos(\theta) + \tan(\theta)\sin(\theta) \equiv \sec(\theta)$

14. $\cos^4(\theta) - \sin^4(\theta) \equiv \cos^2(\theta) - \sin^2(\theta)$

15. $\cot(\theta) + \tan(\theta) \equiv \cot(\theta)\sec^2(\theta)$

16. $\csc^2(\theta)\tan^2(\theta) \equiv 1 + \tan^2(\theta)$

17. $\dfrac{1}{1 - \sin(\theta)} + \dfrac{1}{1 + \sin(\theta)} \equiv 2\sec^2(\theta)$

18. $\dfrac{\tan(\theta) - \cot(\theta)}{\tan(\theta) + \cot(\theta)} \equiv 2\sin^2(\theta) - 1$

19. $\dfrac{1 - \cos(\theta)}{\sin(\theta)} \equiv \dfrac{\sin(\theta)}{1 + \cos(\theta)}$

20. $\dfrac{\cos(\theta) - \sin(\theta)}{\cos(\theta) + \sin(\theta)} \equiv \dfrac{\cot(\theta) - 1}{\cot(\theta) + 1}$

31-4 HYPERBOLIC FUNCTIONS

These interesting functions occur in the study of the interaction of electric and magnetic fields. They also occur in power transmission applications. These functions can be identified as ratios in geometric figures known as hyperbolas. However, we will present them in terms of the exponential function.

The first hyperbolic function that we introduce is the *hyperbolic sine function,* which is defined by $\sinh(x) = (e^x - e^{-x})/2$. The symbol sinh ( ) is read "hyperbolic sine" and is used to indicate the function. Check your calculator and your owner's guide to see if your calculator will calculate this function for you. Look for a key like $\boxed{\text{HYP}}$ used with the sine function key $\boxed{\text{SIN}}$ to calculate the hyperbolic sine. If your calculator does not have this ability, you can use the definition and the exponential function to calculate the hyperbolic sine. Remember $e^{-x} = 1/e^x$.

EXAMPLE 31-7   Use the definition $\sinh(x) = (e^x - e^{-x})/2$ to calculate $\sinh(1.5)$.

Solution   Use your calculator; enter 1.5:

$$\boxed{e^x} \qquad e^{1.5} = 4.48$$

Store this value; then change its sign and form its reciprocal:

$$\boxed{\text{CHS}} \;\; \boxed{1/x} \qquad -e^{-1.5} = -0.223$$

Form the sum of this and the previous value:

$$\boxed{+} \qquad e^{1.5} - e^{-1.5} = 4.26$$

Divide by 2:

$$\boxed{\div} \qquad \frac{e^{1.5} - e^{-1.5}}{2} = 2.13$$

$$\therefore \qquad \sinh(1.5) = 2.13$$

The second hyperbolic function is the *hyperbolic cosine function,* which is defined by cosh $(x) = (e^x + e^{-x})/2$. The symbol cosh ( ) is read "hyperbolic cosine" and is used to indicate the function. The hyperbolic cosine is calculated in a manner similar to the hyperbolic sine.

EXAMPLE 31-8   Use the definition cosh $(x) = (e^x + e^{-x})/2$ to calculate the cosh (0.11).

Solution   Using your calculator, enter 0.11 and exponentiate:

$$\boxed{e^x} \qquad e^{0.11} = 1.12$$

Store this value and form its reciprocal:

$$\boxed{1/x} \qquad e^{-0.11} = 0.896$$

Form the sum of this and the previous value:

$$\boxed{+} \qquad e^{0.11} + e^{-0.11} = 2.01$$

Divide by 2:

$$\boxed{\div} \qquad \frac{e^{0.11} + e^{-0.11}}{2} = 1.01$$

$$\therefore \qquad \cosh(0.11) = 1.01$$

EXAMPLE 31-9 Evaluate cosh (-0.21).

  Solution Use the hyperbolic function key; enter -0.21.

$$\boxed{\text{HYP}}\ \boxed{\text{COS}}\qquad 1.02$$
$$\therefore\qquad \cosh(-0.21) = 1.02$$

The third hyperbolic function is the *hyperbolic tangent function,* which is defined by $\tanh(x) = (e^x - e^{-x})/(e^x + e^{-x})$. The symbol tanh ( ) is read "hyperbolic tangent" and is used to indicate the function.

EXAMPLE 31-10

Use the definition $\tanh(x) = (e^x - e^{-x})/(e^x + e^{-x})$ to calculate $\tanh(0.5)$.

  Solution Using your calculator, enter 0.5 and exponentiate:

$$\boxed{e^x}\qquad e^{0.5} = 1.65$$

Store this value, change the sign, and form the reciprocal:

$$\boxed{\text{CHS}}\ \boxed{1/x}\qquad -e^{-0.5} = -0.61$$

Form the sum of this and the previous result:

$$\boxed{+}\qquad e^{0.5} - e^{-0.5} = 1.04$$

Save this result as the numerator. To form the denominator recall $e^{0.5}$:

$$\boxed{\text{RCL}}\qquad e^{0.5} = 1.65$$

Form the reciprocal:

$$\boxed{1/x}\qquad e^{-0.5} = 0.61$$

Sum this and the previous result:

$$\boxed{+}\qquad e^{0.5} + e^{-0.5} = 2.26$$

Divide into the numerator:

$$\boxed{\div} \quad (e^{0.5} - e^{-0.5})/(e^{0.5} + e^{-0.5}) = 0.462$$
$$\therefore \quad \tanh(0.5) = 0.462$$

EXAMPLE 31-11   Use the hyperbolic key to calculate tanh (1.0).

Solution   Using your calculator, enter 1.0:

$$\boxed{\text{HYP}} \quad \boxed{\text{TAN}} \quad 0.762$$
$$\therefore \quad \tanh(1.0) = 0.762$$

EXERCISE 31-4

Calculator Drill

1. Find the hyperbolic sine of the following numbers:
   (a) -1.0 (b) -0.5 (c) 0.0
   (d) 0.6 (e) 1.2 (f) 2.0
2. Find the hyperbolic cosine of the numbers in problem 1.
3. Find the hyperbolic tangent of the numbers in problem 1.

31-5 GRAPHING THE HYPERBOLIC FUNCTIONS

The hyperbolic functions are not periodic functions. This becomes clear when we look at graphs of the functions. In the previous section we developed techniques for using a calculator to evaluate the functions. We can use those techniques to build tables and graphs of the functions.

EXAMPLE 31-12   Graph $y = \tanh(x)$ for values of x between -3 and 3.

Solution   Build a table of function values; then sketch the graph:

x	0.0	0.1	0.2	0.5	0.75	1.0
tanh (x)	0.0	0.100	0.197	0.462	0.635	0.762

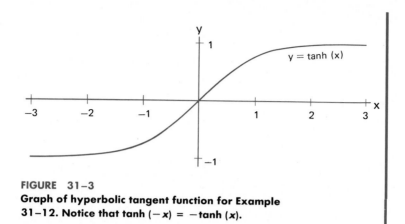

FIGURE 31-3
Graph of hyperbolic tangent function for Example 31–12. Notice that tanh $(-x) = -$tanh (x).

EXERCISE 31-5

Graph the following functions for x between -2.3 and 2.3.

1. $y = e^x/2$ 2. $y = e^{-x}/2$
3. $y = \cosh(x)$ 4. $y = \sinh(x)$

31-6 HYPERBOLIC IDENTITIES

A hyperbolic identity is an equation involving the hyperbolic functions that is true for all values of the argument. Again, we will use the symbol $\equiv$ for identically equal.

EXAMPLE 31-13

Check $\cosh^2(x) - \sinh^2(x) \equiv 1$.

Solution Substitute the exponential definition for $\cosh(x)$ and $\sinh(x)$ in the left member:

$$\cosh^2(x) - \sinh^2(x) = \left(\frac{e^x + e^{-x}}{2}\right)^2 - \left(\frac{e^x - e^{-x}}{2}\right)^2$$

$$= \frac{e^{2x} + 2 + e^{-2x}}{4} - \frac{e^{2x} - 2 + e^{-2x}}{4}$$

$$= 4/4$$

$$= 1$$

This is the value of the right member and it is independent of the value of x.

$$\therefore \quad \cosh^2(x) - \sinh^2(x) \equiv 1$$

EXERCISE 31-6

Check the following identities.

1. $\sinh(x) + \cosh(x) \equiv e^x$
2. $\tanh(-x) \equiv -\tanh(x)$
3. $\tanh(x) \equiv \sinh(x)/\cosh(x)$
4. $\cosh(x) \equiv \cosh(-x)$
5. $\sinh(x) \equiv -\sinh(-x)$
6. $\sinh(2x) \equiv 2\sinh(x)\cosh(x)$
7. $\cosh(2x) \equiv \cosh^2(x) + \sinh^2(x)$
8. $\sinh(2x) + \cosh(2x) \equiv [\sinh(x) + \cosh(x)]^2$
9. $\tanh^2(x) + 1/\cosh^2(x) \equiv 1$
10. $1/\tanh^2(x) - 1/\sinh^2(x) \equiv 1$

31-7 INVERSE HYPERBOLIC FUNCTIONS

If your calculator has the hyperbolic function key, check your owner's guide for the procedure to calculate the inverse hyperbolic functions. The keystroke sequence should look like $\boxed{\text{INV}}$ $\boxed{\text{HYP}}$ $\boxed{\text{SIN}}$, $\boxed{f^{-1}}$ $\boxed{\text{HYP}}$ $\boxed{\text{SIN}}$, or $\boxed{\text{HYP}^{-1}}$ $\boxed{\text{SIN}}$.

For those calculators without the hyperbolic functions, we can use some of the identities of the previous section to develop formulas that can be used to evaluate the inverse hyperbolic functions.

EXAMPLE 31-14

Develop a formula for the inverse hyperbolic sine. For $S = \sinh(x)$, find x.

Solution We know from the preceding section that $e^x = \sinh(x) + \cosh(x)$. If we can express $\cosh(x)$ in terms of $\sinh(x)$, we can use natural logarithms to find x. We also know that $\cosh^2(x) - \sinh^2(x) = 1$.

Solve for cosh (x):

$$\cosh^2(x) = \sinh^2(x) + 1$$
$$\cosh(x) = \sqrt{\sinh^2(x) + 1}$$

Substitute for cosh (x) in $e^x = \sinh(x) + \cosh(x)$:

$$e^x = \sinh(x) + \sqrt{\sinh^2(x) + 1}$$

Substitute S for sinh (x):

$$e^x = S + \sqrt{S^2 + 1}$$

Apply natural logarithms to both members:

$$x = \ln(S + \sqrt{S^2 + 1})$$

$$\therefore \quad \sinh^{-1}(S) = \ln(S + \sqrt{S^2 + 1})$$

**EXAMPLE
31-15**

Solution

Check the formula of Example 31-14 for $x = -2$.
Use your calculator to find sinh (-2) and store the result:

$$\boxed{\text{HYP}}\;\boxed{\text{SIN}}\quad S = \sinh(-2) = -3.6269$$

Use the formula with $S = -3.6269$. Square the result:

$$\boxed{x^2}\quad S^2 = 13.15$$

Add 1:

$$\boxed{+}\quad S^2 + 1 = 14.15$$
$$\boxed{\sqrt{x}}\quad \sqrt{S^2 + 1} = 3.76$$

Recall and add:

$$\boxed{+}\quad S + \sqrt{S^2 + 1} = 0.135$$
$$\boxed{\ln x}\quad \ln(S + \sqrt{S^2 + 1}) = -2.00$$

$$\therefore \quad \text{The formula checks for this value.}$$

A similar formula for the inverse hyperbolic cosine is given in Equation 31-1.

$$\cosh^{-1}(C) = \ln(C + \sqrt{C^2 - 1}) \qquad (31\text{-}1)$$

Because $\cosh(-x) = \cosh(x)$, there are two possible solutions for $\cosh^{-1}$. Equation 31-1 gives the positive solution, which is called the *principal* solution.

EXAMPLE 31-16

Check Equation 31-1 for cosh (1.75).

Solution Use your calculator to find cosh (1.75) and store the result:

$$\boxed{\text{HYP}} \ \boxed{\text{COS}} \qquad C = \cosh(1.75) = 2.9642$$

Use Equation 31-1 with $C = 2.9642$:

$$\boxed{x^2} \qquad C^2 = 8.79$$

Subtract 1:

$$\boxed{-} \qquad C^2 - 1 = 7.79$$
$$\boxed{\sqrt{x}} \qquad \sqrt{C^2 - 1} = 2.79$$

Recall and add:

$$\boxed{+} \qquad C + \sqrt{C^2 - 1} = 5.75$$
$$\boxed{\ln x} \qquad \ln(C + \sqrt{C^2 - 1}) = 1.75$$

$\therefore$ The formula checks for this value.

The formula for the inverse hyperbolic tangent is given by Equation 31-2.

$$\tanh^{-1}(T) = \tfrac{1}{2}\ln[(1 + T)/(1 - T)] \qquad (31\text{-}2)$$

EXAMPLE 31-17

Check Equation 31-2 for tanh (−1.00).

Solution Use your calculator to find tanh (−1.00) and store the result:

HYP TAN $T = \tanh(-1.00) = -0.7616$

Use Equation 31-2 with $T = -0.7616$. Start by adding 1 to T:

$$\boxed{+} \quad 1 + T = 0.238$$

Store and form $(1 - T)$:

$$\boxed{-} \quad 1 - T = 1.76$$

Recall and divide:

$$\boxed{\div} \quad (1 + T)/(1 - T) = 0.135$$
$$\boxed{\ln x} \quad \ln[(1 + T)/(1 - T)] = -2.00$$
$$\tfrac{1}{2}\ln[(1 + T)/(1 - T)] = -1.00$$

$\therefore$ Equation 31-2 checks for $\tanh(-1.00)$.

EXERCISE 31-7
Evaluate the following inverse functions.

1. $\sinh^{-1}(0.8223)$
2. $\sinh^{-1}(10.018)$
3. $\sinh^{-1}(-6.0502)$
4. $\sinh^{-1}(-1.1752)$
5. $\cosh^{-1}(1.5431)$
6. $\cosh^{-1}(1.1276)$
7. $\cosh^{-1}(2.1509)$
8. $\cosh^{-1}(27.308)$
9. $\tanh^{-1}(0.5005)$
10. $\tanh^{-1}(0.9051)$
11. $\tanh^{-1}(-0.9640)$
12. $\tanh^{-1}(0.2449)$

SELECTED TERMS

cosecant   The reciprocal of the sine function: $\csc(x) = 1/\sin(x)$.
cotangent   The reciprocal of the tangent function: $\cot(x) = 1/\tan(x)$.
identity   An equation that is true for all values of the variable.
secant   The reciprocal of the cosine function: $\sec(x) = 1/\cos(x)$.

32

MATHEMATICAL ANALYSIS

This chapter examines several properties of functions and their behaviors. It is not a complete discussion of analysis; it is only an introduction to the types of questions addressed in a course on calculus.

32-1 DOMAIN AND RANGE

This section presents two useful properties of functions, the domain and the range of the function. The domain is associated with the values of the independent variable; the range is associated with the values of the dependent variable or function.

The *domain* of a function is defined as the *list* of values of the independent variable for which the function is defined. For example, the sine function is defined for all values of the angle while the Arc sine function is only defined for values from -1 to $+1$. Thus, the domain of $Y = \sin (\theta)$ equals all values

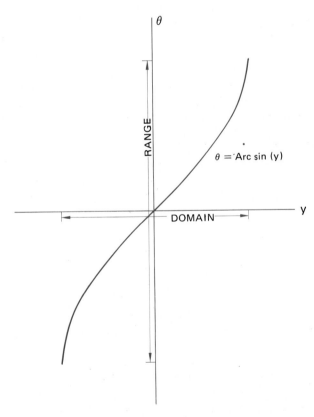

FIGURE 32–1
The domain and range of the Arc sine function. The domain equals $-1 \leq Y \leq 1$, and the range equals $-1.57^r \leq \theta \leq 1.57^r$.

of θ, sometimes written $-\infty < \theta < \infty$. The domain of $\theta = \text{Arcsin}\,(Y)$ equals $-1 \leqslant Y \leqslant 1$.

The *range* of a function is defined as the *list* of values that the function can assume over the entire domain. For example, the sine function takes on all values from -1 to 1, while the Arc sine function takes on the values from -1.57^r to 1.57^r. Thus, the range of $Y = \sin\,(\theta)$ equals $-1 \leqslant Y \leqslant 1$. The range of $\theta = \text{Arcsin}\,(Y)$ equals $-1.57^r \leqslant \theta \leqslant 1.57^r$. Figure 32-1 illustrates the concepts of range and domain for the Arc sine function.

EXERCISE 32-1

Determine the domain and range of the following functions.

1. $Y = \cos\,(\theta)$ 2. $\theta = \text{Arccos}\,(Y)$
3. $Y = \sinh\,(X)$ 4. $Y = \cosh\,(X)$
5. $Y = \tanh\,(X)$ 6. $Y = \ln\,(R)$
7. $Z = e^u$ 8. $r = \sqrt{S}$
9. $t = 2s$ 10. $t = |2s|$

32-2 DISCONTINUITIES

In the preceding section we looked at concepts of domain and range, which are related to how much of the independent axis and the dependent axis are used to graph the whole function. In this section, we look at the graph itself to see whether there are any "breaks" in it. These breaks are called *discontinuities.*

If there are no breaks or discontinuities in a function, it is said to be continuous. The graph of $s = 3r - 1$ shown in Figure 32-2 is for only a portion of the domain of the function. For any other portion, the graph would

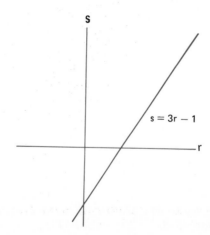

FIGURE 32-2
Graph of a continuous function.

look the same. The function $s = 3r - 1$ is a continuous function because there are no breaks in the graph.

There are three types of discontinuities. The first type is called a *removable discontinuity,* because if the function is redefined at a single point, it becomes a continuous function. The function $y = (x^2 - 1)/(x - 1)$ is not defined when x is equal to 1. Examination of the graph in Figure 32-3 shows that there is a break in the graph at $x = 1$. Furthermore, to fill the break would require setting the function to 2 at this point. Thus, the function

$$Y = \begin{cases} (x^2 - 1)/(x - 1) & x \neq 1 \\ 2, & x = 1 \end{cases}$$

is a continuous function. The discontinuity has been *removed.*

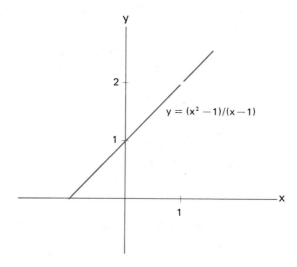

FIGURE   32–3
Graph of a removable discontinuity.

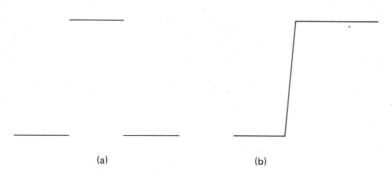

(a)   (b)

FIGURE   32–4
Oscilloscope trace of a square wave: (a) step discontinuity; (b) ramp for faster sweep time.

The second type of discontinuity is called a *step discontinuity* because of the appearance of the graph. The trace of a square wave with a rise time of 20 ns on an oscilloscope set at 1 μs per division will look like a function with step discontinuities, as shown in Figure 32-4(a). A faster sweep will produce a continuous trace with ramps as in Figure 32-4(b). A step discontinuity can be replaced with a very steep ramp without totally destroying the function.

EXAMPLE 32-1   The greatest integer function, $G(x)$, is defined as the greatest integer less than or equal to x. Replace each step discontinuity with a ramp with a slope of 10.

Solution   Create a table of function values for $-0.5 \leqslant x \leqslant 2.5$ and plot the function:

x	-0.5	-0.1	0.0	0.5	1.0	1.5	1.99	2.0	2.5
G(x)	-1.0	-1.0	0.0	0.0	1.0	1.0	1.0	2.0	2.0

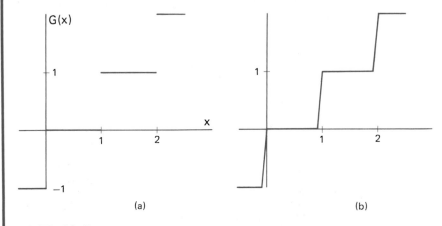

(a)   (b)

FIGURE 32-5
Graphs for Example 32-1: (a) greatest integer function showing step discontinuities; (b) modified greatest integer function with step discontinuities removed.

The points are plotted in Figure 32-5(a). At each integer value of x, $G(x)$ has a step increase of 1. To replace the step with a ramp of slope 10, use the definition of slope and solve for Δx:

$$\text{slope} = \Delta G/\Delta x$$
$$10 = 1/\Delta x$$
$$\Delta x = 0.1$$

> Redraw the function as in Figure 32-5(b). Begin a ramp at 0.1 ahead of each integer value of x. End the ramp when x is an integer at the proper value of the function.

The third type of discontinuity is called an *essential discontinuity*. It is termed *essential* because it cannot be removed or glossed over, as in the case of removable and step discontinuities. The function $s = 1/r$ is plotted in Figure 32-6. This function has an essential discontinuity at $r = 0$.

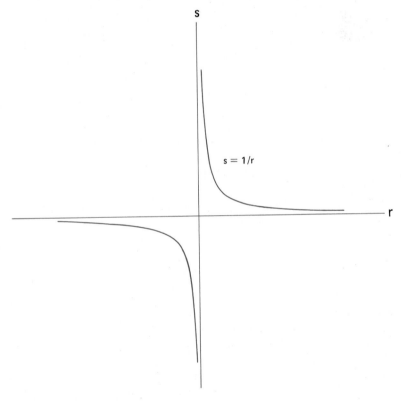

FIGURE 32-6
Graph of $s = 1/r$ showing an essential discontinuity.

EXERCISE 32-2
Determine whether the following eight functions are continuous or discontinuous, and, if discontinuous, the type of discontinuity.

1. $y = \cos(\theta)$ 2. $S = \tanh(r)$
3. $Z = \tan(\omega t)$ 4. $u = (v^2 + v - 2)/(v + 2)$
5. $V = e^t$ 6. $y = |x|$
7. $w = \sec(\phi)$

8. $y = \text{sign}(x) = \begin{cases} 1 \text{ if } x > 0 \\ 0 \text{ if } x = 0 \\ -1 \text{ if } x < 0 \end{cases}$

9. Define the remainder function $r(x)$ in terms of the greatest integer function $G(x)$ by $r(x) = x - G(x)$. Plot the function for $0 \leqslant x \leqslant 3$. Describe its behavior.

32-3 FUNCTIONS OF LARGE NUMBERS

A concept that underlies much of calculus is the concept of a *limit*. How does a function behave as the independent variable approaches a specific value? In this section, we look at the behavior of several functions as the independent variable is allowed to become *large without bound* or to approach *infinity*.

To aid in our discussions, we introduce several symbols. The first symbol, ∞, is read as "infinity" and is used to mean "larger than any number you can think of." The second symbol, $\lim\limits_{x \to \infty} f(x)$, is read as "limit of $f(x)$ as x approaches infinity."

Now let's consider the behavior of several familiar functions as the independent variable is allowed to approach infinity.

EXAMPLE 32-2   Discuss the behavior of $f(x) = x$ as x approaches infinity.

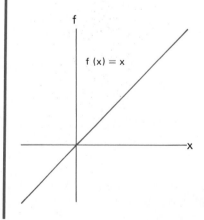

FIGURE 32–7
Graph for Example 32–2.

Solution Sketch a graph of $f(x)$; see Figure 32-7.
The function is always equal to x; so as x grows, so does the function.

$$\therefore \quad \lim_{x \to \infty} x = \infty$$

EXAMPLE 32-3 Discuss the behavior of $S(r) = 1/r$ as r approaches infinity.

Solution Sketch a graph of $S(r)$; see Figure 32-6.
As r increases beyond zero, $1/r$ becomes a smaller and smaller positive number. When $r \gg 1$, $0 < 1/r \ll 1$.

$$\therefore \quad \lim_{r \to \infty} 1/r = 0$$

EXAMPLE 32-4 Find $\lim_{\theta \to \infty} \sin(\theta)$.

Solution Sketch a graph of $\sin(\theta)$; see Figure 32-8.

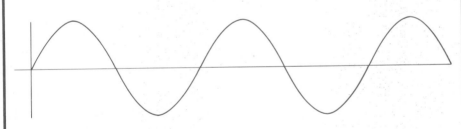

FIGURE 32–8
Sketch of sine function for Example 32–4.

The sine function is a periodic function, and as θ increases, $\sin(\theta)$ keeps repeating its oscillation between 1 and -1.

$$\therefore \quad \lim_{\theta \to \infty} \sin(\theta) \text{ does not exist.}$$

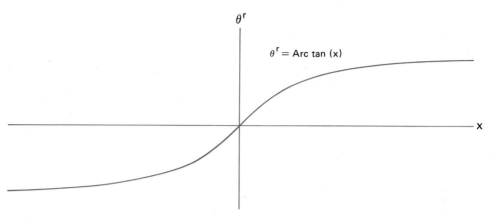

FIGURE 32–9
Sketch of Arctan (x) for Example 32–5.

EXAMPLE 32-5   Find $\lim\limits_{x \to \infty}$ Arctan (x) in radians.

Solution   Sketch a graph of Arctan (x); see Figure 32-9.
The Arc tangent function is a continuous function that increases with increasing values of x. Consider the following table:

x	1	10	100	1000	10 000
Arctan (X)	0.79	1.47	1.56	1.57	1.57

$$\therefore \quad \lim_{x \to \infty} \text{Arctan}\ (x) \approx 1.57$$

There are three possible answers to the question of what $\lim\limits_{x \to \infty} f(x)$ is. The limit may be a finite number such as zero or 1.57. The limit may be infinite, either $+\infty$ or $-\infty$. Finally, the limit may not exist because the function keeps oscillating and the amplitude of oscillation is not *tending to zero*.

EXERCISE 32-3
Find the following limits.

1. $\lim\limits_{\theta \to \infty} \cos (\theta)$

2. $\lim\limits_{x \to \infty} e^x$

3. $\lim\limits_{t \to \infty} e^{-t}$

4. $\lim\limits_{Z \to \infty} \sinh (Z)$

5. $\lim\limits_{Z \to \infty} \cosh (Z)$

6. $\lim\limits_{t \to \infty} \tanh (t)$

7. $\lim\limits_{x \to \infty} \sqrt{x}$

8. $\lim\limits_{w \to \infty} \ln (w)$

9. $\lim\limits_{x \to \infty} (2x + 3)/x$

10. $\lim\limits_{t \to \infty} e^{1/t}$

32-4 ASYMPTOTES

Some complicated functions behave like much simpler functions for large values of the independent variables. When this happens, the simpler function is referred to as an *asymptote* for the more complicated function. Strictly, an asymptote is a straight line, but we will use this more general definition.

EXAMPLE 32-6 Examine the behavior of tanh (x) as x increases.

Solution Sketch the graph of $y = \tanh (x)$; see Figure 32-10.

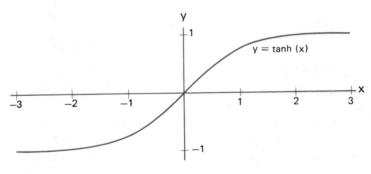

FIGURE 32–10
Sketch of $y = \tanh (x)$ for Example 32–6.

Notice that for $x > 2.5$, tanh $(x) \approx 1.0$. Also for $x < -2.5$, tanh $x \approx -1.0$.

∴ $y = 1$ and $y = -1$ are asymptotes for $y = \tanh (x)$.

Check Examine the definition of the hyperbolic tangent to see if these results make sense:

$$\tanh (x) = \frac{e^x - e^{-x}}{e^x + e^{-x}}$$

When $x > 2.3$, $e^x > 10$, and $e^{-x} < 0.1$; so $e^x \gg e^{-x}$ and the definition can be simplified:

$$\tanh(x) \approx e^x/e^x$$
$$\therefore \quad \tanh(x) \approx 1 \text{ for } x > 2.3$$

When $x < -2.3$, $e^x < 0.1$, and $e^{-x} > 10$; so $e^x \ll e^{-x}$ and:

$$\tanh(x) \approx -e^{-x}/e^{-x}$$
$$\therefore \quad \tanh(x) \approx -1 \text{ for } x < -2.3$$

EXAMPLE 32-7   Examine $s = \sqrt{t^2 + 4}$ for asymptotic behavior.
Solution   When does $t^2 + 4$ look like t^2?

$$\text{When } t^2 \gg 4, t > 7, \text{ or } t < -7.$$
$$s \approx \sqrt{t^2} = |t| \text{ for } |t| > 7$$

Plot the two functions $s = \sqrt{t^2 + 4}$ and $s = |t|$ on the same graph; see Figure 32-11.

$$\therefore \quad s = |t| \text{ is an asymptote.}$$

EXERCISE 32-4
Examine the following functions for asymptotic behavior.

1. $y = e^{-x}$ 2. $y = 1/x^2$
3. $s = 1 - e^{1/t}$ 4. $u = \cosh(V) - \sinh(V)$
5. $u = \sqrt{v^2 - 1}$ 6. $V = 2e^{2/u^2}$
7. $R = \dfrac{2r}{r + 2}$ 8. $X_C = 1/(\omega C); C = 1\ \mu\text{F}$

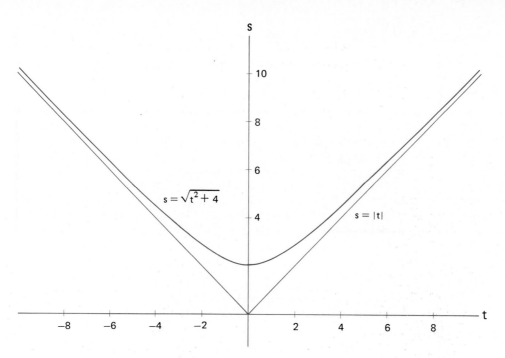

$s = \sqrt{t^2 + 4}$

$s = |t|$

FIGURE 32–11
Graph for Example 32–7.

SELECTED TERMS

discontinuity   A break in the graph of a function.
domain   The list of all values of the independent variable for which the function is defined.
range   The list of all values that a function assumes over its domain.

33

COMPUTER NUMBER SYSTEMS

More and more computers are coming into use. As an electronic technician you may use a computer in your work, or you may be employed to work on computers. Either way you need knowledge of the number systems in common use with computers. The topics presented here have been selected to help you gain this knowledge.

33-1 DECIMAL NUMBER SYSTEM

The number system we are familiar with is the decimal system. There are three other number systems that are important in electronics because they are used with digital computers. Before we look at these other systems, we need to review the decimal system.

The decimal system uses ten symbols called *digits* to form numbers. The digits are 0, 1, 2, 3, 4, 5, 6, 7, 8, and 9. The number of digits in a system is called the *radix* or *base*. Thus, the radix of the decimal system is 10. Notice that the radix is not one of the digits. Besides the ten digits, the decimal system includes a minus sign ($-$) and a decimal point (.) to indicate negative numbers and fractions.

The reason that the decimal system needs only ten digits is that the system assigns different values to a digit depending on its position relative to the decimal point. Thus, the 9 takes on three different values in 93, 19, and 1.09. The concept of positional value is summarized in Table 33-1.

The leading or leftmost digit is called the *most significant digit,* or *MSD,* because it corresponds to the highest power of ten and contributes most to the value of the number. The trailing or rightmost digit is called the *least significant digit,* or *LSD,* because it corresponds to the lowest power of ten and contributes least to the value of the number.

EXAMPLE 33-1   Find the value of 5, 3, and 1 in 543.21.

Solution   Determine the positions of 5, 3, and 1 and then use Table 33-1:

5 is the third digit left of the decimal point

∴   Its value is 500.

3 is the first digit left of the decimal point

∴   Its value is 3.

> 1 is the second digit to the right of the
> decimal point
>
> ∴   Its value is 0.01.

Decimal numbers can extend farther left and right than shown in Table 33-1. Because of this, we need a rule for the general case.

RULE 33-1. VALUE OF DIGITS IN DECIMAL NUMBERS

To determine the value of a digit in a decimal number:

1. Determine its position relative to the decimal point.
2. For a digit n places to the *left* of the decimal point, the value is given by

$$value = digit \times 10^{n-1}$$

3. For a digit m places to the *right* of the decimal point, the value is given by

$$value = digit \times 10^{-m}$$

EXERCISE 33-1

1. Find the value of 1 in the following numbers.
   (a) 5107.26   (b) 21 760.5   (c) 50.7126
   (d) 6075.12   (e) 72.0165   (f) 157.602
2. Find the value of 2 in the numbers in problem 1.
3. Find the value of 5 in the numbers in problem 1.
4. Find the value of 6 in the numbers in problem 1.
5. Find the value of 7 in the numbers in problem 1.
6. Find the most significant digit (MSD) of each number in problem 1.
7. Find the least significant digit (LSD) of each number in problem 1.

33-2 THREE ADDITIONAL NUMBER SYSTEMS

Three additional number systems are becoming more important in electronics. These systems are associated with computers, digital logic trainers, and all digital devices. These number systems are the *binary, octal,* and *hexadecimal* systems.

716

TABLE 33-1

POSITIONAL VALUE AND THE DECIMAL NUMBER SYSTEM

Position		To Left of Decimal Point								To Right of Decimal Point		
	5	4	3	2	1		1	2	3	4	5
Position value	10 000	1000	100	10	1		0.1	0.01	0.001	0.0001	0.000 01
Power of ten	10^4	10^3	10^2	10^1	10^0		10^{-1}	10^{-2}	10^{-3}	10^{-4}	10^{-5}

The binary system has only two digits, called bits: 0 and 1. These can represent the states of a two-state device. Examples are (1) a switch is on or off, or (2) a diode is conducting or not. Because of this relationship, all computers work in binary *on the device level.*

The radix of the binary system is 2 because there are two digits. Like the decimal system, the binary system uses positional value. However, the period used to separate the integer and fractional parts of a number is called a *binary point,* not a decimal point.

Since we are working with four different systems, and the period looks and functions the same in each system, we will give it a new name that will be the same in each system. From now on, we will refer to it as the *radix point.*

The *octal* system has eight digits: 0, 1, 2, 3, 4, 5, 6, and 7. The radix of the octal system is 8. The octal system is often used to display binary information from computers.

The third system is the *hexadecimal* system, which has 16 digits: 0, 1, 2, 3, 4, 5, 6, 7, 8, 9, A, B, C, D, E, and F. The radix of the hexadecimal system is 16. The hexadecimal system is also used to display binary information. Table 33-2 lists the first 21 integers in binary, octal, decimal, and hexadecimal.

Notice that the digits in the binary and octal systems are familiar digits, but the hexadecimal system uses six symbols we normally use for letters, not digits. It is sometimes convenient to refer to these digits by their phonetic

TABLE 33-2

COMPARISON OF FOUR NUMBER SYSTEMS			
Binary System	Octal System	Decimal System	Hexadecimal System
0	0	0	0
1	1	1	1
10	2	2	2
11	3	3	3
100	4	4	4
101	5	5	5
110	6	6	6
111	7	7	7
1000	10	8	8
1001	11	9	9
1010	12	10	A
1011	13	11	B
1100	14	12	C
1101	15	13	D
1110	16	14	E
1111	17	15	F
10 000	20	16	10
10 001	21	17	11
10 010	22	18	12
10 011	23	19	13
10 100	24	20	14

alphabet names: alpha, bravo, charley, delta, echo, and fox. Take the time now to memorize the decimal values for these special hexadecimal digits given in Table 33-2.

To distinguish the system in which a number is written, we will add the radix as a subscript to the number. Whenever we write the radix, we will write it as a decimal number. For example, 100 in hexadecimal is written 100_{16}.

EXAMPLE 33-2 Indicate that 101 and 110 are binary numbers.

Solution The radix of the binary system is 2. Append a subscript of 2 to each number:

$$\therefore \quad 101_2 \text{ and } 110_2$$

All four systems, binary, octal, decimal, and hexadecimal, use positional values. Whole numbers without a written radix point have an implied radix point to the right of the LSD. Some positional values are illustrated in Table 33-3. We can use these facts to find the decimal value of any digit in any number, as outlined in Rule 33-2.

TABLE 33-3

POSITIONAL VALUES
$1111_2 = 1 \times 2^3 + 1 \times 2^2 + 1 \times 2^1 + 1 \times 2^0$
$1001_8 = 1 \times 8^3 + 0 \times 8^2 + 0 \times 8^1 + 1 \times 8^0$
$1100_{10} = 1 \times 10^3 + 1 \times 10^2 + 0 \times 10^1 + 0 \times 10^0$
$1010_{16} = 1 \times 16^3 + 0 \times 16^2 + 1 \times 16^1 + 0 \times 16^0$

RULE 33-2. DECIMAL VALUE OF DIGITS IN NUMBERS

To determine the value of a digit in a number:

1. Determine its position relative to the radix point.
2. For a digit n places to the left of the radix point, the value is given by

$$\text{Value} = \text{digit} \times R^{n-1}$$

   where R is the radix.
3. For a digit m places to the right of the radix point, the value is given by

$$\text{Value} = \text{digit} \times R^{-m}$$

EXAMPLE 33-3   Find the decimal value of 7 in 70_8.
     Solution   Use Rule 33-2.
     Step 1:   7 is two places left of the implied radix point.
     Step 2:

$$\text{Value} = 7 \times 8^{(2-1)}$$
$$= 7 \times 8$$
$$\therefore \quad \text{Value} = 56_{10}$$

Table 33-4 lists the decimal values for some powers of 2, which will aid in applying Rule 33-2. Similarly, Table 33-5 lists the decimal values for some powers of 8, while Table 33-6 lists the decimal values for some powers of 16. Notice that the decimal values in Table 33-5 and 33-6 are contained in Table 33-4. Since 8 and 16 are integer powers of 2, every integer power of 8 and every integer power of 16 can be expressed as an integer power of 2.

TABLE 33-4

POWERS OF 2

2^n	n	2^{-n}
1	0	1
2	1	0.5
4	2	0.25
8	3	0.125
16	4	0.0625
32	5	0.03125
64	6	0.015625
128	7	0.0078125
256	8	0.00390625
512	9	0.001953125
1024	10	0.0009765625
2048	11	0.00048828125
4096	12	0.000244140625

TABLE 33-5

POWERS OF 8

8^n	n	8^{-n}
1	0	1
8	1	0.125
64	2	0.015625
512	3	0.001953125
4096	4	0.000244140625

TABLE 33-6

POWERS OF 16		
16^n	**n**	**16^{-n}**
1	0	1
16	1	0.0625
256	2	0.00390625
4096	3	0.000244140625

EXAMPLE 33-4 Find the decimal value of B in $1B40_{16}$.

Solution Use Rule 33-2.

Step 1: B is three places left of the implied radix point.

Step 2:

$$\text{Value} = B \times 16^{(3-1)}$$
$$= B \times 16^2$$

Convert the hexadecimal digit B to decimal by referring to Table 33-2. Thus, $B \Rightarrow 11$:

$$\text{Value} = 11 \times 16^2$$

Refer to Table 33-6 to convert 16^2 to 256.

$$\therefore \quad \text{Value} = 11 \times 256 = 2816_{10}$$

EXAMPLE 33-5 Find the decimal value of 1 in 0.01_2.

Solution Use Rule 33-2.

Step 1: 1 is two places to the right of the radix point.

Step 2:

$$\text{Value} = 1 \times 2^{-2}$$

Refer to Table 33-4 to convert 2^{-2} to 0.25.

$$\therefore \quad \text{Value} = 0.25_{10}$$

EXERCISE 33-2

Find the decimal values of the indicated digits.

1. All the 1s in 101.01_2
2. All the 1s in $11\ 010.1_2$
3. The 3 and the 5 in $730\ 51_8$
4. The 1s in 100.1_8

5. The 1s in 100.1_{16} 6. The 7 and the A in $47F.A_{16}$
7. The 4 and the 7 in 7620.04_8 8. The 2 and the 6 in 420.65_8
9. The C and the 9 in $C90_{16}$ 10. The 5 and the E in $50.E_{16}$

33-3 CONVERTING NUMBERS TO THE DECIMAL SYSTEM

In the preceding section we learned to calculate the decimal value of any digit in a binary, octal, or hexadecimal number. It is an easy extension to calculate the decimal value of a binary, octal, or hexadecimal number. The procedure is stated in Rule 33-3.

RULE 33-3. CALCULATING THE DECIMAL VALUE OF A NUMBER

To calculate the decimal value of a number:

1. Calculate the decimal value of each digit.
2. The decimal value of the number is the sum of the values of the digits.

EXAMPLE 33-6   Find the decimal value of $10\ 001_2$.
   Solution   Use Rule 33-3; begin with the LSD:
   Step 1:   The position of the LSD is one place to the left of the implied radix point:

$$\text{Value} = 1 \times 2^0$$
$$= 1_{10}$$

The next three digits are each zero. Zeros only act as place keepers; their value is always zero. The position of the MSD is five places to the left of the implied radix point:

$$\text{Value} = 1 \times 2^4$$
$$= 16_{10}$$

   Step 2:

$$10\ 001_2 = 16_{10} + 0_{10} + 0_{10} + 0_{10} + 1_{10}$$
$$\therefore \quad 10\ 001_2 = 17_{10}$$

EXAMPLE 33-7   Find the decimal value of 2001_8.

   Solution   Use Rule 33-3; begin with the LSD.

   Step 1:   The LSD is one place to the left of the implied radix point:

$$\text{Value} = 1 \times 8^0$$
$$= 1_{10}$$

The MSD is four places to the left of the implied radix point:

$$\text{Value} = 2 \times 8^3$$
$$= 1024_{10}$$

   Step 2:

$$2001_8 = 1024_{10} + 0_{10} + 0_{10} + 1_{10}$$
$$\therefore \quad 2001_8 = 1025_{10}$$

EXAMPLE 33-8   Find the decimal value of 2001_{16}.

   Solution   Use Rule 33-3; begin with the LSD.

   Step 1:   The LSD is one place to the left of the implied radix point:

$$\text{Value} = 1 \times 16^0$$
$$= 1_{10}$$

The MSD is four places to the left of the implied radix point:

$$\text{Value} = 2 \times 16^3$$
$$= 8192_{10}$$

   Step 2:

$$2001_{16} = 8192_{10} + 0_{10} + 0_{10} + 1_{10}$$
$$\therefore \quad 2001_{16} = 8193_{10}$$

EXAMPLE 33-9   Find the decimal value of $C3.0D2_{16}$.

   Solution   Use Rule 33-3; begin with the LSD.

Step 1:   The 2 is three places to the right of the radix point:

$$Value = 2 \times 16^{-3}$$
$$= 0.000\ 488\ 281\ 25_{10}$$

The D is two places to the right of the radix point:

$$Value = 13 \times 16^{-2}$$
$$= 0.050\ 781\ 25_{10}$$

The 3 is one place to the left of the radix point:

$$Value = 3_{10}$$

The C is two places to the left of the radix point:

$$Value = 12 \times 16$$
$$= 192_{10}$$

Step 2:

$$C3.0D2_{16} = 192_{10} + 3_{10} + 0_{10} + 0.050\ 781\ 25_{10}$$
$$+ 0.000\ 488\ 281\ 25_{10}$$
$$\therefore \quad C3.0D2_{16} \approx 195.051\ 27_{10}$$

EXERCISE 33-3

Find the decimal value of the following numbers.

1. 110_2	2. 1010_2	3. $110\ 000_2$
4. $11\ 111_2$	5. 0.11_2	6. 1.01_2
7. 10.11_2	8. 1.011_2	9. 11.0101_2
10. $0.111\ 11_2$	11. 110_8	12. 1010_8
13. 270_8	14. $64\ 000_8$	15. 0.11_8
16. 1.01_8	17. 76.2_8	18. 14.7_8
19. 5.34_8	20. 0.76_8	21. 110_{16}
22. 1010_{16}	23. 270_{16}	24. $ABCD_{16}$
25. 0.11_{16}	26. 1.01_{16}	27. 76.2_{16}
28. $0.B1_{16}$	29. $F0.CC_{16}$	30. $E67.D_{16}$

33-4 CONVERTING DECIMAL NUMBERS TO OTHER SYSTEMS

To convert an octal number to decimal, we use multiplication by powers of eight. To convert a whole number from decimal to octal, we reverse the procedure and use division by eight.

EXAMPLE 33-10

Convert 1024_8 to decimal and back to octal.

Solution Use Rule 33-2 to indicate the value of each digit:

$$1024_8 = 1 \times 8^3 + 0 \times 8^2 + 2 \times 8^1 + 4$$
$$= 512 + 0 + 16 + 4$$
$$\therefore \quad 1024_8 = 532_{10}$$

Reverse the order of the terms, and factor the right member of the first line of the solution:

$$4 + 8^1 \times 2 + 8^2 \times 0 + 8^3 \times 1$$
$$= 4 + 8\,[2 + 8(0 + 8 \times 1)]$$

This factored form shows that the conversion can be considered repeated multiplication by 8. To reverse the process, divide 532 by 8. The remainder is the LSD.

$$
\begin{array}{r}
8\,\underline{|\,532} \\
8\,|\quad 66 \text{ R } 4 \quad \text{LSD} \\
8\,|\quad\;\; 8 \text{ R } 2 \\
8\,|\quad\;\; 1 \text{ R } 0 \\
\quad\;\; 0 \text{ R } 1 \quad \text{MSD} \longrightarrow 1\,024_8
\end{array}
$$

$$\therefore \quad 532_{10} = 1024_8$$

The process used in Example 33-10 can be formalized as a rule for converting whole numbers from decimal to any other system.

RULE 33-4. CONVERTING DECIMAL WHOLE NUMBERS

To convert a whole number from decimal to binary, octal, or hexadecimal, use repeated division by the new radix. The remainder upon the first division is the LSD; the remainder of the last division is the MSD.

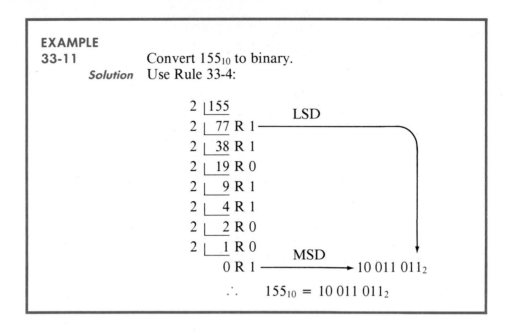

EXAMPLE
33-11

Convert 155_{10} to binary.

Solution Use Rule 33-4:

$$
\begin{array}{rl}
2\ \lfloor 155 \\
2\ \lfloor\ 77\ R\ 1 & \text{LSD} \\
2\ \lfloor\ 38\ R\ 1 \\
2\ \lfloor\ 19\ R\ 0 \\
2\ \lfloor\ \ 9\ R\ 1 \\
2\ \lfloor\ \ 4\ R\ 1 \\
2\ \lfloor\ \ 2\ R\ 0 \\
2\ \lfloor\ \ 1\ R\ 0 & \text{MSD} \\
0\ R\ 1 & \longrightarrow 10\ 011\ 011_2
\end{array}
$$

$$\therefore \quad 155_{10} = 10\ 011\ 011_2$$

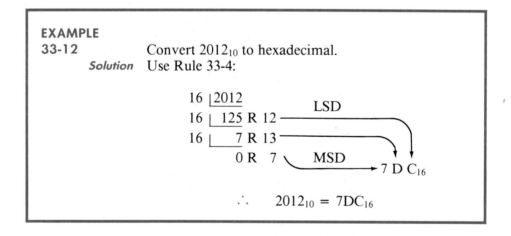

EXAMPLE
33-12

Convert 2012_{10} to hexadecimal.

Solution Use Rule 33-4:

$$
\begin{array}{rl}
16\ \lfloor 2012 \\
16\ \lfloor\ 125\ R\ 12 & \text{LSD} \\
16\ \lfloor\ \ \ 7\ R\ 13 \\
0\ R\ \ \ 7 & \text{MSD} \\
& \longrightarrow 7\ D\ C_{16}
\end{array}
$$

$$\therefore \quad 2012_{10} = 7DC_{16}$$

The process of converting a binary fraction to decimal involves division by 2. To convert a decimal fraction to binary, reverse the process and multiply by 2. The integer result of the first multiplication is the MSD of the binary fraction. The fractional part of the result is again multiplied by 2 to obtain the next digit. The process is terminated when the fractional part is zero or when as many binary digits as desired have been found.

**EXAMPLE
33-13**

Solution

Convert $0.110\ 11_2$ to decimal and back to binary.

$$0.110\ 11_2 = 1 \times 2^{-1} + 1 \times 2^{-2} + 0 \times 2^{-3}$$
$$+ 1 \times 2^{-4} + 1 \times 2^{-5}$$
$$= 0.5 + 0.25 + 0 + 0.0625 + 0.03125$$

$$\therefore \quad 0.110\ 11_2 = 0.843\ 75_{10}$$

Use repeated multiplication by 2 to convert $0.843\ 75_{10}$ to binary:

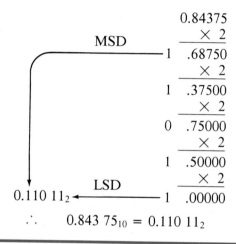

$$\therefore \quad 0.843\ 75_{10} = 0.110\ 11_2$$

The process illustrated in Example 33-13 is summarized in the following rule:

RULE 33-5. CONVERTING DECIMAL FRACTIONS

To convert a decimal fraction to binary, octal, or hexadecimal, multiply the fraction by the new radix. The integer portion of the product is the MSD. Repeat the process with the fractional remainder for the next digit. The process is terminated when the remaining fraction is zero or when all the desired digits have been determined.

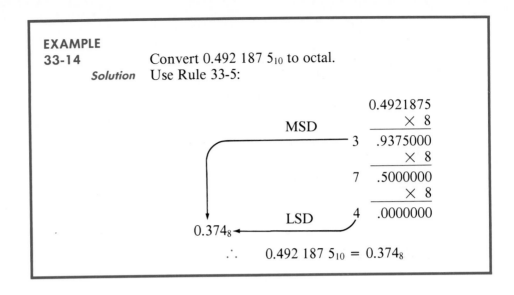

EXAMPLE 33-14

Convert $0.492\ 187\ 5_{10}$ to octal.

Solution   Use Rule 33-5:

$$\therefore \quad 0.492\ 187\ 5_{10} = 0.374_8$$

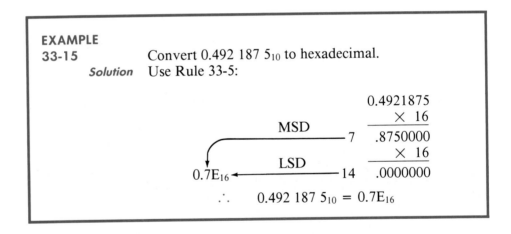

EXAMPLE 33-15

Convert $0.492\ 187\ 5_{10}$ to hexadecimal.

Solution   Use Rule 33-5:

$$\therefore \quad 0.492\ 187\ 5_{10} = 0.7E_{16}$$

In Examples 33-14 and 33-15, notice that the octal and hexadecimal representations require fewer digits than the decimal representation. Some other decimal fractions cannot be represented exactly in binary, octal, or hexadecimal.

**EXAMPLE
33-16**   Convert 0.1_{10} to octal to nine places.

     Solution   Use Rule 33-5:

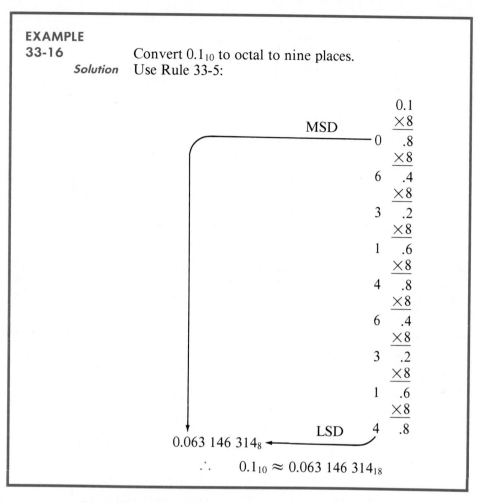

$$0.1_{10} \approx 0.063\ 146\ 314_{18}$$

To convert a decimal number to binary, octal, or hexadecimal, first convert the integer part and then convert the fractional part.

EXERCISE 33-4

1. Convert the following decimal numbers to binary, octal, and hexadecimal:
   (a) 19_{10} (b) 22_{10} (c) 15_{10} (d) 36_{10}
   (e) 48_{10} (f) 39_{10} (g) 64_{10} (h) 100_{10}
2. Convert the following decimal numbers to octal and hexadecimal:
   (a) 164_{10} (b) 212_{10} (c) 408_{10} (d) 512_{10}
   (e) 575_{10} (f) 650_{10} (g) 749_{10} (h) 800_{10}

3. Convert the following decimal fractions to binary, octal, and hexadecimal (to five significant digits):
   - (a) 0.5_{10}
   - (b) 0.25_{10}
   - (c) 0.75_{10}
   - (d) 0.125_{10}
   - (e) 0.0625_{10}
   - (f) $0.156\ 25_{10}$
   - (g) 0.156_{10}
   - (h) 0.995_{10}

4. Convert 3.5_{10} to binary.

5. Convert 24.375_{10} to octal.

6. Convert 28.4375_{10} to hexadecimal.

33-5 CONVERTING BETWEEN BINARY, OCTAL, AND HEXADECIMAL

The octal and hexadecimal systems were chosen for the ease of conversion to and from binary. We begin with octal conversion. To convert an octal number to binary, replace each octal digit with the appropriate set of three binary digits from Table 33-7.

EXAMPLE 33-17   Convert the octal number 25.24_8 to binary.

Solution   Use Table 33-7 to convert each octal digit:

$$25.24_8 = 010\ 101.010\ 100_2$$

Delete leading and trailing zeros:

$$10\ 101.0101_2$$
$$\therefore \quad 25.24_8 = 10\ 101.0101_2$$

To convert a binary number to octal, start at the radix point and collect the binary digits into groups of three. If necessary, add *leading zeros* to the

TABLE 33-7

OCTAL BINARY CONVERSION			
Octal	*Binary*	*Octal*	*Binary*
0	000	4	100
1	001	5	101
2	010	6	110
3	011	7	111

integer part and *trailing zeros* to the fractional part. Use Table 33-7 to convert each set of three binary digits to octal digits.

EXAMPLE
33-18 Convert the binary number $10\ 110.1001_2$ to octal.
  Solution Group the binary digits into sets of three; start counting at the radix point:

$$10\ 100.1001_2 = 10\ 110.100\ 1_2$$

Add extra leading and trailing zeros to complete the sets of three:

$$10\ 110.100\ 1_2 = 010\ 110.100\ 100_2$$

Use Table 33-7 to convert to octal:

$$\therefore \qquad 010\ 110.100\ 100_2 = 26.44_8$$

The conversion of hexadecimal to binary is performed in a manner similar to that of octal to binary conversion. Table 33-8 is used to replace each hexadecimal digit with four binary digits.

EXAMPLE
33-19 Convert the hexadecimal number $A3.0F_{16}$ to binary.
  Solution Use Table 33-8 to convert each digit:

$$A3.0F = 1010\ 0011.0000\ 1111$$
$$\therefore \qquad A3.0F_{16} = 10\ 100\ 011.000\ 011\ 11_2$$

TABLE 33-8

HEXADECIMAL BINARY CONVERSION

Hexadecimal	Binary	Hexadecimal	Binary
0	0000	8	1000
1	0001	9	1001
2	0010	A	1010
3	0011	B	1011
4	0100	C	1100
5	0101	D	1101
6	0110	E	1110
7	0111	F	1111

The conversion of binary to hexadecimal is performed like binary to octal, except the binary digits are grouped in "sets of four" instead of three. Table 33-8 is used to convert the sets of four binary digits to hexadecimal digits.

EXAMPLE 33-20

Convert the binary number $101\ 100\ 101.100\ 111_2$ to hexadecimal.

Solution Group the binary digits into sets of four; begin at the radix point:

$$101\ 100\ 101.100\ 111_2 = 1\ 0110\ 0101.1001\ 11_2$$

Add leading and trailing zeros to fill out the sets of four:

$$1\ 0110\ 0101.1001\ 11_2 = 0001\ 0110\ 0101.1001\ 1100_2$$

Use Table 33-8 to convert the sets of four to hexadecimal digits:

$$\therefore \quad 0001\ 0110\ 0101.1001\ 1100_2 = 165.9C_{16}$$

A simple way to convert between octal and hexadecimal is to convert first to binary, and then from binary to the desired system.

EXAMPLE 33-21

Convert the octal number 157_8 to hexadecimal.

Solution First convert to binary:

$$157_8 = 001\ 101\ 111_2$$

Regroup in sets of four:

$$157_8 = 0110\ 1111_2$$

Convert to hexadecimal:

$$\therefore \quad 157_8 = 6F_{16}$$

EXERCISE 33-5
Convert the following octal numbers to binary and hexadecimal.

    1. 124_8 2. 75_8 3. 2001_8 4. 25.76_8

Convert the following hexadecimal numbers to binary and octal.

    5. 124_{16} 6. $F2D2_{16}$ 7. AB_{16} 8. $A.71_{16}$

Convert the following binary numbers to octal and hexadecimal.

    9. 1011.1_2 10. $11\ 100.01_2$

    11. $101\ 010\ 101_2$ 12. $111\ 110\ 110\ 101_2$

33-6 BINARY ADDITION AND SUBTRACTION

Binary addition and subtraction are easy to learn for those who know decimal addition and subtraction. There are only four combinations of digits to learn. The following rule governs binary addition.

RULE 33-6. BINARY ADDITION

$$0 + 0 = 0$$
$$0 + 1 = 1$$
$$1 + 0 = 1$$
$$1 + 1 = 0 \text{ plus a carry (or } 10_2)$$

EXAMPLE 33-22

Add the two binary numbers $10\ 110_2$ and $11\ 010_2$.

Solution   Use Rule 33-6 and show carries:

$$
\begin{array}{r}
1111 \leftarrow \text{carry} \\
10110 \\
+\ \ 11010 \\
\hline
110000_2
\end{array}
$$

Binary subtraction is the inverse of addition, and is just as easy to learn. The following rule governs binary subtraction.

RULE 33-7. BINARY SUBTRACTION

$$1 - 0 = 1$$
$$1 - 1 = 0$$
$$0 - 0 = 0$$
$$0 - 1 = 1 \text{ and a borrow } (10_2 - 1 = 1)$$

EXAMPLE 33-23   Subtract 1011_2 from $11\ 101_2$.

Solution   Use Rule 33-7:

$$
\begin{array}{r}
{}^{0}{}_{1} \leftarrow \text{borrow} \\
11\cancel{1}01 \\
-\ \ 1011 \\
\hline
10010_2
\end{array}
$$

Rules 33-6 and 33-7 are summarized in an addition table, Table 33-9. To add two digits, find the column with the first digit at the top, and find the row with the second digit on the left end; the answer is at the point of intersection of this row and column.

To use Table 33-9 for subtraction, find the column with the digit to be subtracted at the top, and find the row with the digit (or 10_2) to be subtracted from in this column; the answer is at the left end of this row.

TABLE 33-9

BINARY ADDITION TABLE

+	0	1
0	0	1
1	1	10

EXERCISE 33-6
Perform the indicated binary arithmetic.

1. $\begin{array}{r} 101 \\ +\ 101 \\ \end{array}$ 2. $\begin{array}{r} 1010 \\ +\ 1100 \\ \end{array}$

3. $\begin{array}{r} 1111 \\ +\ 1011 \\ \end{array}$ 4. $\begin{array}{r} 11 \\ +\ 11 \\ \end{array}$

5. $\begin{array}{r} 10111 \\ +\ \ \ 111 \\ \end{array}$ 6. $\begin{array}{r} 10101 \\ +\ 1110 \\ \end{array}$

7.	101 110 + 11	8.	10 100 + 11
9.	100 − 10	10.	101 − 11
11.	1111 − 111	12.	1011 − 110
13.	1000 − 1	14.	1001 − 11
15.	1010 − 101	16.	10101 − 1110

33-7 OCTAL ADDITION AND SUBTRACTION

Frequently, numbers in octal are used for troubleshooting. Skill in adding and subtracting octal numbers is necessary to use these numbers. Table 33-10 is an octal addition table. It is used the same way that Table 33-9 is used for binary addition and subtraction. As in decimal arithmetic, the path to proficiency is practice.

**EXAMPLE
33-24**

Add the following pairs of octal numbers.

(a) 2_8 and 4_8 (b) 6_8 and 7_8

Solution   Use Table 33-10.

(a) Look across the top row to find the 2; it is in the fourth column. Look down the left column to find the 4; it is in the sixth row. Look at the point where this row and column intersect; the answer is 6.

$$\therefore \quad 2_2 + 4_8 = 6_8$$

(b) Look across the top row to find the 6; it is in the eighth column. Look down the left column to find the 7; it is in the ninth row. Look at the point where the eighth column and the ninth row intersect; the answer is 15.

$$\therefore \quad 6_8 + 7_8 = 15_8$$

Observation When two digits sum to less than 8 in decimal addition, the answer is the same in octal addition. When two digits sum to more than 7 in decimal addition, the answer in octal is two more. The leading "1" in these entries in Table 33-10 indicates a carry.

**EXAMPLE
33-25** Add $30\ 477_8$ and $26\ 605_8$.

Solution Use Table 33-10:

$$
\begin{array}{r}
1\,1\,1 \quad \leftarrow\text{carry}\\
30477\\
+26605\\
\hline
57304_8
\end{array}
$$

**EXAMPLE
33-26** Subtract:

(a) 3_8 from 7_8
(b) 6_8 from 13_8.

Solution Use Table 33-10.

(a) Find the column headed by 3, the digit to be subtracted. Look down the column to find 7, the digit to be subtracted from. Look at the digit at the left end of this row, the answer, 4.

$$\therefore \quad 7_8 - 3_8 = 4_8$$

(b) Find the column headed by 6, the digit to be subtracted. Look down the column to find 13, the number to be subtracted from. Look at the digit at the left end of this row, the answer, 5.

$$\therefore \quad 13_8 - 6_8 = 5_8$$

TABLE 33-10

OCTAL ADDITION TABLE

+	0	1	2	3	4	5	6	7
0	0	1	2	3	4	5	6	7
1	1	2	3	4	5	6	7	10
2	2	3	4	5	6	7	10	11
3	3	4	5	6	7	10	11	12
4	4	5	6	7	10	11	12	13
5	5	6	7	10	11	12	13	14
6	6	7	10	11	12	13	14	15
7	7	10	11	12	13	14	15	16

EXAMPLE 33-27

Subtract 237_8 from 1052_8.

Solution Use Table 33-10:

$$
\begin{array}{r}
4_1 \leftarrow \text{borrow} \\
10\cancel{5}2 \\
-\ \ 237 \\
\hline
613_8
\end{array}
$$

EXERCISE 33-7

Perform the indicated octal arithmetic.

1. $3 + 5$	2. $2 + 1$	3. $6 + 1$
4. $4 + 3$	5. $3 + 2$	6. $7 + 6$
7. $5 + 4$	8. $6 + 3$	9. $7 + 5$
10. $2 + 7$	11. $32 + 17$	12. $124 + 73$

13.   5605
   + 2170

14.   3651
   + 2026

15.   50770
   + 7426

16.   35407
   + 6777

17. $7 - 4$

18. $12 - 7$

19. $5 - 3$	20. $13 - 5$	21. $34 - 7$
22. $63 - 15$	23. $135 - 66$	24. $215 - 120$

25.   7365
   − 1024

26.   5057
   − 4011

27.   70000
   − 5321

28.   25042
   − 6043

29.   10000
   −   1

30.   65432
   −56343

33-8 HEXADECIMAL ADDITION AND SUBTRACTION

A growing number of computers are using hexadecimal for *troubleshooting* numbers. Just as we took a quick look at octal addition and subtraction, we now look at hexadecimal addition and subtraction using Table 33-11.

As long as two digits add up to less than 10 in decimal, they have the same sum in hexadecimal. Be careful, remember A1 is a number—not a steak sauce!

EXAMPLE 33-28

Perform the following addition in hexadecimal.

(a) $3 + 6$ (b) $7 + 7$ (c) $6 + D$

Solution   Use Table 33-11. Look up the first number at the top of the table, the second at the left. The answer is where the row and column cross.

(a) $\begin{array}{r} 3 \\ + 6 \\ \hline 9_{16} \end{array}$ (b) $\begin{array}{r} 7 \\ + 7 \\ \hline E_{16} \end{array}$ (c) $\begin{array}{r} 6 \\ + D \\ \hline 13_{16} \end{array}$

EXAMPLE 33-29

Find the sum of $24F_{16}$ and $1AAA_{16}$ in hexadecimal.

Solution   Use Table 33-11.

$$\begin{array}{r} 1 \leftarrow \text{carry} \\ 1\,AAA \\ +\ \ \ 24\,F \\ \hline 1\,CF9_{16} \end{array}$$

$\therefore \quad 1\,AAA_{16} + \ 24F_{16} = 1CF9_{16}$

Subtraction is the reverse of addition. With the aid of Table 33-11, subtraction in hexadecimal is not too difficult.

EXAMPLE 33-30

Perform the following subtraction in hexadecimal.

(a) $9 - 6$ (b) $7 - 7$ (c) $F - A$ (d) $19 - B$

Solution Use Table 33-11. Look up the digit to be subtracted at the top of the table, look down the column for the second number. The digit at the left of this row is the answer.

(a) $\begin{array}{r} 9 \\ -6 \\ \hline 3_{16} \end{array}$ (b) $\begin{array}{r} 7 \\ -7 \\ \hline 0_{16} \end{array}$ (c) $\begin{array}{r} F \\ -A \\ \hline 5_{16} \end{array}$ (d) $\begin{array}{r} 19 \\ -B \\ \hline E_{16} \end{array}$

EXAMPLE 33-31

Subtract $2ABC_{16}$ from 3000_{16}.

Solution Use Table 33-11.

$$\begin{array}{r} {}_{2}\,F\,F\,{}_{1}\leftarrow\text{borrow} \\ 3\cancel{0}\cancel{0}\cancel{0} \\ -2ABC \\ \hline 5\,4\,4_{16} \end{array}$$

EXERCISE 33-8

Perform the indicated hexadecimal arithmetic.

1. $7 + 7$	2. $8 + 3$	3. $3 + A$
4. $6 + 4$	5. $1 + 7$	6. $9 + 5$
7. $12 + 57$	8. $35 + 73$	9. $A + B$
10. $F + 6$	11. $67 + 67$	12. $179 + 4F$

13. $\begin{array}{r} 785 \\ + BA1 \end{array}$ 14. $\begin{array}{r} 96A \\ + C0D \end{array}$ 15. $\begin{array}{r} 2001 \\ + FFF \end{array}$

16. $\begin{array}{r} D2C3 \\ + 98A \end{array}$ 17. $A - 1$ 18. $F - 4$

19. $9 - 6$ 20. $F - 5$ 21. $D - A$

22. $E - B$ 23. $12 - F$ 24. $A - 6$

TABLE 33-11

HEXADECIMAL ADDITION TABLE

+	0	1	2	3	4	5	6	7	8	9	A	B	C	D	E	F
0	0	1	2	3	4	5	6	7	8	9	A	B	C	D	E	F
1	1	2	3	4	5	6	7	8	9	A	B	C	D	E	F	10
2	2	3	4	5	6	7	8	9	A	B	C	D	E	F	10	11
3	3	4	5	6	7	8	9	A	B	C	D	E	F	10	11	12
4	4	5	6	7	8	9	A	B	C	D	E	F	10	11	12	13
5	5	6	7	8	9	A	B	C	D	E	F	10	11	12	13	14
6	6	7	8	9	A	B	C	D	E	F	10	11	12	13	14	15
7	7	8	9	A	B	C	D	E	F	10	11	12	13	14	15	16
8	8	9	A	B	C	D	E	F	10	11	12	13	14	15	16	17
9	9	A	B	C	D	E	F	10	11	12	13	14	15	16	17	18
A	A	B	C	D	E	F	10	11	12	13	14	15	16	17	18	19
B	B	C	D	E	F	10	11	12	13	14	15	16	17	18	19	1A
C	C	D	E	F	10	11	12	13	14	15	16	17	18	19	1A	1B
D	D	E	F	10	11	12	13	14	15	16	17	18	19	1A	1B	1C
E	E	F	10	11	12	13	14	15	16	17	18	19	1A	1B	1C	1D
F	F	10	11	12	13	14	15	16	17	18	19	1A	1B	1C	1D	1E

25. B − 4	26. 15 − C	27. 100
		− AA

28. F88	29. 2677	30. 4A00
− A51	− 1F0A	− 35 51

33-9 COMPLEMENTS

When two numbers sum to a specified number, they are called *complements*. In trigonometry, two angles that sum to a right angle are called complements. Computers use complements to perform certain arithmetic functions. There are two systems of complements in use in computers today. The first system is based on the largest digit. The second system is based on the radix.

Largest-Digit Complements

Each number system has a largest digit. The largest digit of the binary system is 1; the largest digit of the octal system is 7. Largest-digit complements (also called *base minus one complements*) for the binary system are based on 1 and are called *one's complements*. The largest-digit complements for the octal system are based on 7 and are called *seven's complements*. The largest digit and the corresponding name for complements are presented in Table 33-12 for the various number systems.

TABLE 33-12

LARGEST-DIGIT COMPLEMENTS

System	Largest Digit	Largest-Digit Complements
Binary	1	One's complements
Octal	7	Seven's complements
Decimal	9	Nine's complements
Hexadecimal	F	F's complements

In the rest of this chapter, we will work eight-digit numbers. While microcomputers commonly use 8-bit groupings to form a "word," there is nothing special about eight. The following concepts would work with words of any fixed length.

To find the *eight*-digit one's complement of a binary number, simply subtract the number from the *eight*-digit binary number of all ones (1111 1111).

**EXAMPLE
33-32**   Form the eight-digit one's complements of:

(a) $101\ 0110_2$ (b) $110\ 0000_2$

Solution   (a) $\begin{array}{r} 1111\ 1111 \\ -\ \ \ 101\ 0110 \\ \hline 1010\ 1001_2 \end{array}$ (b) $\begin{array}{r} 1111\ 1111 \\ -\ \ \ 110\ 0000 \\ \hline 1001\ 1111_2 \end{array}$

Observation   To form the one's complement of a binary number, replace each 1 with a 0 and each 0 with a 1.

To find the *eight*-digit seven's complement of an octal number, simply subtract the number from the *eight*-digit octal number of all sevens (7777 7777).

**EXAMPLE
33-33**   Form the eight-digit seven's complements of:

(a) $545\ 6700_8$ (b) $6321\ 0754_8$

Solution   (a) $\begin{array}{r} 7777\ 7777 \\ -\ \ \ 545\ 6700 \\ \hline 7232\ 1077_8 \end{array}$ (b) $\begin{array}{r} 7777\ 7777 \\ -\ 6321\ 0754 \\ \hline 1456\ 7023_8 \end{array}$

To form the *eight*-digit nine's complement of a decimal number, subtract the number from the *eight*-digit decimal number of all nines (9999 9999).

**EXAMPLE
33-34**   Find the eight-digit nine's complements of:

(a) $3791\ 0684_{10}$ (b) $5327\ 9926_{10}$

Solution   (a) $\begin{array}{r} 9999\ 9999 \\ -\ 3791\ 0684 \\ \hline 6208\ 9315_{10} \end{array}$ (b) $\begin{array}{r} 9999\ 9999 \\ -\ 5327\ 9926 \\ \hline 4672\ 0073_{10} \end{array}$

To form the *eight*-digit F's complement of a hexadecimal number, subtract the number from the *eight*-digit hexadecimal number of all Fs (FFFF FFFF).

EXAMPLE 33-35

Find the eight-digit F's complements of:

(a) $3791\ 0684_{16}$ (b) $AB25\ CDEF_{16}$

Solution (a) $\begin{array}{r} \text{FFFF FFFF} \\ -\ 3791\ 0684 \\ \hline \text{C86E F97B}_{16} \end{array}$ (b) $\begin{array}{r} \text{FFFF FFFF} \\ -\ \text{AB25 CDEF} \\ \hline 54\text{DA}\ 32\ 10_{16} \end{array}$

Table 33-13 contains the F's complements for all the hexadecimal digits. Take time to look over the table to see the patterns. It will aid you in forming complements of hexadecimal numbers.

TABLE 33-13

F'S COMPLEMENTS

Digit	Complement	Digit	Complement
0	F	8	7
1	E	9	6
2	D	A	5
3	C	B	4
4	B	C	3
5	A	D	2
6	9	E	1
7	8	F	0

EXERCISE 33-9

Find the eight-digit one's complements of the following binary numbers.

1. $111\ 1111_2$
2. $100\ 0110_2$
3. $110\ 0110_2$
4. $101\ 0101_2$
5. $111\ 0001_2$
6. $11\ 0011_2$

Find the eight-digit seven's complements of the following octal numbers.

7. $6420\ 1357_8$
8. $777\ 7777_8$
9. $111\ 1111_8$
10. $3534\ 6721_8$
11. $2451\ 7000_8$
12. $14\ 2167_8$

Find the eight-digit nine's complements of the following decimal numbers.

    13. $1746\ 2001_{10}$ 14. $111\ 1111_{10}$ 15. $9253\ 8679_{10}$

    16. $777\ 7777_{10}$ 17. $9999\ 9999_{10}$ 18. $3796\ 8421_{10}$

Find the eight-digit F's complements of the following hexadecimal numbers.

    19. $111\ 1111_{16}$ 20. $777\ 7777_{16}$ 21. $1FAB\ 9824_{16}$

    22. $C6A2\ ED57_{16}$ 23. $3056\ 7CDE_{16}$ 24. $BF98\ 0134_{16}$

True Complements

The second type of complements (called *true complements*) are named for the radix of each system. Thus, the true complements for the binary system are called *two's complements*. The eight-digit true complement of a number is formed by subtracting the number from 1 0000 0000. This number is 1 larger than the number used to find the largest-digit complement. Thus, the two's complement of a binary number is equal to the one's complement plus 1.

    In like manner, the true complements for octal numbers are called *eight's complements* and are equal to seven's complements plus 1. The true complements for decimal numbers are called *ten's complements* and are equal to nine's complements plus 1. The true complements for hexadecimal numbers are called *sixteen's complements* and are equal to F's complements plus 1.

**EXAMPLE
33-36**

Find the eight-digit two's complement of the binary number $101\ 0101_2$.

Solution   (1) Subtract from 1 0000 0000:

$$
\begin{array}{r}
1\ 0000\ 0000 \\
-\ \ \ \ \ \ 101\ 0101 \\
\hline
1010\ 1011_2 \text{ (two's complement)}
\end{array}
$$

(2) Calculate the two's complement from the one's complement:

$$
\begin{array}{r}
1111\ 1111 \\
-\ \ \ 101\ 0101 \\
\hline
1010\ 1010 \text{ (one's complement)} \\
+\ \ \ \ \ \ \ \ \ \ \ \ 1 \text{ (add 1)} \\
\hline
1010\ 1011_2 \text{ (two's complement)}
\end{array}
$$

$\therefore$ Two's complements can be calculated in two ways.

EXAMPLE 33-37

Find the eight-digit eight's complement of the octal number $3721\ 5640_8$.

Solution Calculate the seven's complement and add 1:

$$
\begin{array}{r}
7777\ 7777 \\
-\ 3721\ 5640 \\
\hline
4056\ 2137 \quad \text{(seven's complement)} \\
+1 \quad \text{(add 1)} \\
\hline
4056\ 2140_8 \quad \text{(eight's complement)}
\end{array}
$$

EXAMPLE 33-38

Find the eight-digit ten's complement of the decimal number $3721\ 5640_{10}$.

Solution Calculate the nine's complement and add 1:

$$
\begin{array}{r}
9999\ 9999 \\
-\ 3721\ 5640 \\
\hline
6278\ 4359 \quad \text{(nine's complement)} \\
+1 \quad \text{(add 1)} \\
\hline
6278\ 4360_{10} \quad \text{(ten's complement)}
\end{array}
$$

EXAMPLE 33-39

Find the eight-digit sixteen's complement of the hexadecimal number $A375\ B1EF_{16}$.

Solution Calculate the F's complement and add 1:

$$
\begin{array}{r}
FFFF\ FFFF \\
-\ A375\ B1EF \\
\hline
5C8A\ 4E10 \quad \text{(F's complement)} \\
+1 \quad \text{(add 1)} \\
\hline
5C8A\ 4E11_{16} \quad \text{(sixteen's complement)}
\end{array}
$$

Find the eight-digit two's complements of the following binary numers:

1. $111\ 1111_2$ 2. $1100\ 1110_2$ 3. $101\ 0111_2$
4. $100\ 1100_2$ 5. $110\ 1100_2$ 6. $111\ 0001_2$

Find the eight-digit eight's complements of the following octal numbers:

7. $2651\ 0777_8$ 8. $72\ 1654_8$ 9. $1743\ 1111_8$
10. $1000\ 0000_8$ 11. $111\ 1111_8$ 12. $3421\ 5720_8$

Find the eight-digit ten's complements of the following decimal numbers:

13. $2561\ 0777_{10}$ 14. $94\ 3876_{10}$ 15. $1000\ 0000_{10}$
16. $4716\ 9582_{10}$ 17. $3490\ 1627_{10}$ 18. $2815\ 9376_{10}$

Find the eight-digit sixteen's complements of the following hexadecimal numbers:

19. $71AE\ 642F_{16}$ 20. $1000\ 0000_{16}$ 21. $3D8B\ C905_{16}$
22. $FA\ 9EDC_{16}$ 23. $2561\ 0777_{16}$ 24. $111\ 1111_{16}$

33-10 BINARY ARITHMETIC WITH COMPLEMENTS

To work with signed numbers, we had to learn several rules for adding and subtracting that depend on the signs of the numbers and their relative sizes. For example, to add two numbers with opposite signs, subtract the smaller from the larger and retain the sign of the larger number. All the rules we learned must be "taught" to a computer's arithmetic unit. This is done through the design of the logic circuits. A common technique used by computer manufacturers is to use complements to represent negative numbers.

Most computers work with only fixed-length words. In such a system there is only a finite number of possible binary numbers. Suppose that we want to be able to count up to 127. If we will settle for the positive integers 1 through 127, we can use a "word" of seven binary digits (called *bits*). Now, if we want to include negative integers, we need an extra bit for the sign (0 for positive and 1 for negative).

For negative numbers, we could simply turn the sign bit "on." Instead, computer manufacturers have chosen to complement the number. Figure 33-1 shows the bit patterns of 53_{10} and -53_{10} in an 8-bit word.

The advantage of using complements to store negative numbers lies in

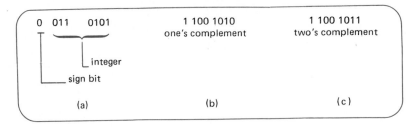

| 0 011 0101 | 1 100 1010 | 1 100 1011 |
| one's complement | two's complement |

integer
sign bit

(a)   (b)   (c)

FIGURE 33–1
Bit patterns in computer words: (a) word containing 53_{10}; (b) word
containing -53_{10} in one's complement; (c) word containing -53_{10} in
two's complement.

simplifying the computer's arithmetic circuits. Addition of signed numbers, when negative numbers are stored as complements, is a simple extension of binary addition. Subtraction is performed by complementing the number to be subtracted and adding. Some computers have subtractive adders; that is, they only know how to subtract, not how to add.

Addition using one's complements is governed by the following rule:

RULE 33-8. ADDING NUMBERS IN ONE'S COMPLEMENT

To add two numbers in one's complement:

1. Use binary addition.
2. Check for a carry beyond the most significant digit (MSD):
   (a) If there is no carry beyond the MSD, the result of step 1 is the answer.
   (b) If there is a carry beyond the MSD, add 1 to the result obtained in step 1.

EXAMPLE 33-40

Use one's complement to add 75_{10} and -16_{10}. Leave a space between the sign bit and the integer.

Solution Convert to eight-digit binary and use Rule 33-8.

$$75_{10} = 0\ 100\ 1011_2$$
$$-16_{10} = -0\ 001\ 0000_2$$

Form one's complement:

$$-16_{10} = 1\ 110\ 1111_2$$

Step 1:   Add:

$$0\ 100\ 1011$$
$$+\ 1\ 110\ 1111$$
$$\overline{1\ \ \ 0\ 011\ 1010}$$

Step 2:   Yes, there is a carry beyond the eighth digit, so add 1 to the result obtained in step 1:

$$0\ 011\ 1010$$
$$+\ \ \ \ \ \ \ \ \ \ \ 1$$
$$\overline{0\ 011\ 1011_2}$$

Check   Convert the answer to decimal; it should be 59:

$$0\ 011\ 1011_2 = 1 + 2 + 8 + 16 + 32$$
$$0\ 011\ 1011_2 = 59_{10}$$

Observation   The process of checking for a carry beyond the most significant digit and conditionally adding 1 to the result is called *end-around carry*.

RULE 33-9. SUBTRACTING NUMBERS IN ONE'S COMPLEMENT

To subtract binary numbers using one's complements:

1. Form the one's complement of the number to be subtracted.
2. Add, using one's complement arithmetic.

EXAMPLE 33-41

Subtract $0\ 101\ 1000_2$ from $0\ 011\ 0111_2$ using one's complements. Express the answer as a signed binary number.

Solution   Use Rule 33-9.

Step 1:   Form the one's complement:

$$1\ 111\ 1111$$
$$-\ 0\ 101\ 1000$$
$$\overline{1\ 010\ 0111}$$

Step 2: Add:

$$0\ 011\ 0111$$
$$+\ 1\ 010\ 0111$$
$$\overline{1\ 101\ 1110}$$

Observation   The sign bit (MSD) is 1; this indicates a negative result, which is in one's complement. To interpret the result, complement the number.

$$1\ 111\ 1111$$
$$-\ 1\ 101\ 1110$$
$$\overline{0\ 010\ 0001}$$

Delete the leading zeros and prefix the answer with a minus sign.

$$-10\ 0001_2$$

$$\therefore \qquad 0\ 011\ 0111_2 - 0\ 101\ 1000_2 = -10\ 0001_2$$

The rule for two's complement addition is slightly different than the rule for one's complement.

RULE 33-10. ADDING NUMBERS IN TWO'S COMPLEMENT

To add two numbers in two's complement:

1. Use binary addition.
2. Any carry beyond the most significant digit (MSD) is discarded *(end-off carry)*.

EXAMPLE 33-42

Use two's complements to add 75_{10} and -16_{10}.

Solution   Convert to eight-digit binary numbers:

$$75_{10} = 0\ 100\ 1011_2$$
$$-\ 16_{10} = -0\ 001\ 0000_2$$

Form two's complement:

$$-16_{10} = 1\ 111\ 0000_2$$

Apply Rule 33-10:

Step 1: Add:

$$
\begin{array}{r}
0\ 100\ 1011 \\
+\ 1\ 111\ 0000 \\
\hline
1\quad 0\ 011\ 0000
\end{array}
$$

Step 2: End-off carry:

$$0\ 011\ 1011_2 = 59_{10}$$

Observation   The sign bit is off (zero), so the result is positive.

The procedure for subtraction with two's complement numbers is similar to subtraction with one's complement numbers.

RULE 33-11. SUBTRACTING NUMBERS IN TWO'S COMPLEMENT

To subtract binary numbers using two's complements:

1. Form the two's complement of the number to be subtracted.
2. Add, using two's complement arithmetic.

EXAMPLE
33-43
Subtract $0\ 101\ 1000_2$ from $0\ 011\ 0111_2$ using two's complements. Express the answer as a signed binary number.

Solution   Use Rule 33-11.

Step 1:   Convert to two's complement form:

$$
\begin{array}{l}
1\ 111\ 1111 \\
-\ 0\ 101\ 1000 \\
\hline
1\ 010\ 0111\ \text{(one's complement)} \\
+1\ \text{(add 1)} \\
\hline
1\ 010\ 1000\ \text{(two's complement)}
\end{array}
$$

Step 2:   Add, using two's complement arithmetic.

$$
\begin{array}{r}
0\ 011\ 0111 \\
+\ 1\ 010\ 1000 \\
\hline
1\ 101\ 1111_2
\end{array}
$$

$$1\ 111\ 1111$$
$$-\ 1\ 101\ 1111$$
$$0\ 010\ 0000\ \text{(one's complement)}$$
$$1\ \text{(add 1)}$$
$$0\ 010\ 0001$$

Delete the leading zeros and prefix the answer with a minus sign.

$$-10\ 0001_2$$

$$\therefore \quad 0\ 011\ 0111_2 - 0\ 101\ 1000_2 = -10\ 0001_2$$

EXAMPLE 33-44

Subtract $1\ 011\ 0110_2$ from $1\ 110\ 1100_2$ using two's complements.

Solution   Use Rule 33-11.

Step 1:   Form two's complement:

$$1\ 111\ 1111$$
$$-\ 1\ 011\ 0110$$
$$0\ 100\ 1001\ \text{(one's complement)}$$
$$1\ \text{(add 1)}$$
$$0\ 100\ 1010\ \text{(two's complement)}$$

Step 2:   Add, using two's complement arithmetic:

$$1\ 110\ 1100$$
$$+\ 0\ 100\ 1010$$
$$1\ \ 0\ 011\ 0110$$

End-off carry:

$$0\ 011\ 0110_2$$

Observation   The sign bit is off; therefore, the result is positive.

Use one's complement to perform the following eight-digit binary arithmetic (if the sign bit is already on, the number is in one's complement).

1.   0 110 1010
   +0 000 1111

2.   0 011 1111
   +0 011 1111

3.   1 100 1010
   +0 100 1010

4.   0 011 0110
   +1 011 0111

5.   0 101 1101
   −0 110 0110

6.   0 011 1011
   −0 001 0100

7.   1 100 1000
   −1 010 0110

8.   1 011 1011
   −0 110 0101

9.   0 011 0101
   −1 100 1010

10.   1 100 0000
   +1 100 0000

11–20. Repeat problems 1–10 using two's complements. (If the sign bit is already on, the number is in two's complement form.)

33-11 REVIEW

This section will provide a review of the previous material by asking you to solve arithmetic problems for numbers with mixed bases. The proper procedure is to convert the numbers to a common base, evaluate the answer and convert the answer to the desired base. This procedure will be illustrated in the following examples.

EXAMPLE 33-45

Add 150_{10} to 333_8; express the answer as a hexadecimal number.

Solution   Convert 150_{10} to octal:

$$150_{10} = 222_8$$

Perform octal addition:

$$222_8 + 333_8 = 555_8$$

Convert to binary:

$$555_8 = 101\ 101\ 101_2$$

Convert to hexadecimal:

$$0001\ 0110\ 1101_2 = 16D_{16}$$

$$\therefore \quad 150_{10} + 333_8 = 16D_{16}$$

EXAMPLE 33-46

Add 256_8 and -127_{16}; express the answer as a decimal number.

Solution   Convert both numbers to binary:

$$256_8 = 010\ 101\ 110_2$$
$$-127_{16} = -0001\ 0010\ 0111_2$$

Convert to 10-digit one's complements and add:

$$\begin{array}{r} 0\ 0\ 1010\ 1110 \\ +\ 1\ 0\ 1101\ 1000 \\ \hline 1\ 1\ 1000\ 0110_2 \end{array}$$

The sign bit is on, so the answer is negative. Complement to obtain:

$$-111\ 1001_2$$

Convert to decimal:

$$-(64 + 32 + 16 + 8 + 1) = -121_{10}$$

$$\therefore \quad 256_8 - 127_{16} = -121_{10}$$

EXERCISE 33-12

Express the answers to the following problems as decimal numbers.

1. $\begin{array}{r} 35_8 \\ +67_8 \end{array}$ 2. $\begin{array}{r} 57_8 \\ -29_8 \end{array}$ 3. $\begin{array}{r} 43_8 \\ -72_8 \end{array}$

4. $\begin{array}{r} 146_8 \\ +\ 35_8 \end{array}$ 5. $\begin{array}{r} 1A_{16} \\ +\ F_{16} \end{array}$ 6. $\begin{array}{r} 29_{16} \\ +100_{16} \end{array}$

7. $\quad 40_{16}$ $\quad -22_{16}$	8. $\quad 99_{16}$ $\quad -8E_{16}$	9. $\quad 10_{16}$ $\quad +\ 4_8$
10. $\quad 29_{16}$ $\quad +21_8$	11. $\quad 306_8$ $\quad -\ FA_{16}$	12. $\quad 100_{16}$ $\quad -\ 77_8$
13. $\quad 1F9_{16}$ $\quad -776_8$	14. $\quad 587_8$ $\quad -\ AB_{16}$	15. $\quad 11001_2$ $\quad +\quad 43_8$
16. $\qquad 37_8$ $\quad +110101_2$	17. $\qquad D6_{16}$ $\quad -1100\ 1\ 1_2$	18. 10101110_2 $\qquad 8E_{16}$
19. $\quad 100_8$ $\quad -\ 25_{10}$	20. $\quad 263_{10}$ $\quad -342_8$	21. $\quad 79_{10}$ $\quad -1E_{16}$
22. $\quad 308_{16}$ $\quad -517_{10}$	23. $\quad 1A1B_{16}$ $\quad +\ 315_{10}$	24. $\quad F3C_{16}$ $\quad +429_{10}$

SELECTED TERMS

binary   The number system based on the number 2.

complements   A method used to store negative numbers and to perform subtraction in computers.

hexadecimal   The number system based on the number 16.

octal   The number system based on the number 8.

radix   The base of a number system: thus, 2 is the radix of the binary system.

34

MATHEMATICS OF COMPUTER LOGIC

The ideas, rules, and theorems of Boolean algebra (in conjunction with other techniques and methods) are used to solve logic problems. Some of the logic problems solved with Boolean concepts have only two possible values for the logical variables. These *two-state* types of logic functions occur in the switching circuitry of digital computers. The output of these logic circuits is either on or off.

Since digital computer circuits have only two states (on or off, conducting or nonconducting, high or low), it is natural that the binary number system be used to describe these two conditions. The *on* state is indicated by a one (1) while the *off* state is indicated by a zero (0).

34-1 INTRODUCTORY CONCEPTS

When evaluating a logic expression (also referred to as a Boolean expression), the variables and constants of the expression may be assigned only one of two possible values, either a 1 or a 0. A 1 in a logic expression indicates the presence of a certain voltage level (e.g., 5 V) at the input or output of the computer's switching circuitry. A 0 in a logic expression indicates the presence of a voltage level (e.g., 0 V) that is different from that represented by a 1. When analyzing a digital logic system, it is simpler to think in terms of 1s and 0s rather than in terms of voltage levels. Because of the discrete nature of the voltage levels at the inputs and outputs of the logic circuitry, it has become a common practice to let 1 stand for the presence of a specified voltage level and 0 the absence of that level. Thus, a logic 0 might represent a voltage level of 0 V, while a logic 1 might represent a voltage level of 5 V.

In the course of this chapter, we will use switch symbols to represent the *two-state* logic conditions. An open switch will be indicated by a 0, while a closed switch will be indicated by a 1. Figure 34-1 shows a single-pole single-throw (SPST) switch. The condition of the switch is represented by the logic variable A. When the switch is open $A = 0$; when the switch is closed $A = 1$.

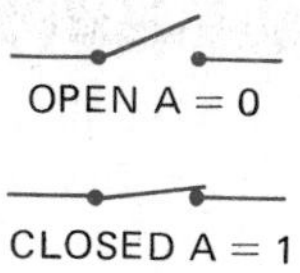

FIGURE 34–1
The condition of the switch is represented by the logic variable A.

OPEN $A = 0$

CLOSED $A = 1$

Logic Expressions

Consider the series circuit of Figure 34-2. The presence or absence of light depends on the condition of the switch. We will let A represent the condition of the switch; closed = 1, open = 0. The condition of the light will be represented by f_o (standing for output function); light = 1, dark = 0. The table

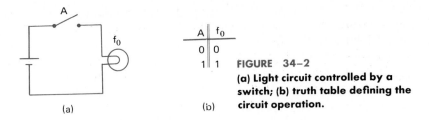

(a)

A	f_0
0	0
1	1

(b)

FIGURE 34–2
(a) Light circuit controlled by a switch; (b) truth table defining the circuit operation.

next to the circuit, Figure 34-2(b), defines the operation of the circuit. This type of table is called a *truth table.*

The lamp will not produce light until the switch is closed. A logical expression representing the direct relationship that exists between the switch and the lamp is

$$f_o = A$$

We will use uppercase letters to represent logical conditions and f_o to represent the output of a logical equation.

Logic Operators

Only three basic logic operations are used to design all the logic circuits of a digital computer. These logic operations are:

- The *inversion operator,* commonly called the NOT operator. The NOT operator is indicated by an overbar ($^-$) as in $\overline{A}$; read "NOT A."
- The *disjunction operator,* commonly called the OR operator. The OR operator is indicated by a plus sign ($+$) as in A + B; read "A OR B."
- The *conjunction operator,* commonly called the AND operator. The AND operator is indicated by any of the usual signs of multiplication ($\cdot$, $\times$, etc.), as in A $\cdot$ B or AB; read "A AND B."

EXAMPLE 34-1   Read each of the following logical expressions.

(a) $A + B + \overline{C}$ (b) $\overline{A}B$ (c) $\overline{A} + BC$

Solution   Use the words OR, AND, and NOT:

(a) $A + B + \overline{C}$ is read A OR B OR NOT C.

(b) $\overline{A}B$ is read NOT A AND B.

(c) $\overline{A} + BC$ is read NOT A OR B AND C.

Using the words of the logical operators, NOT, OR, and AND, write out the following logical expressions.

1. $\overline{B}$
2. $A + B$
3. AB
4. $AB + C$
5. $C + AB$
6. ABC
7. $A\overline{B}C$
8. $A + \overline{B}$
9. $\overline{A} + \overline{B}$
10. $\overline{A}B + C\overline{D}$
11. $A + BC + A\overline{B}$
12. $\overline{A}\overline{B} + \overline{C}\overline{D} + E$

34-2 INVERSION OPERATOR (NOT)

We will begin our study of the NOT operator by studying the behavior of the switch pictured in Figure 34-3(a). The switch is a two-state device and, as such, can be in only one state (position) at a time. If setting the logic variable A to 1 represents the switch in the closed position, then the open position (not closed) is indicated by setting the logic variable NOT A to 1 ($\overline{A} = 1$).

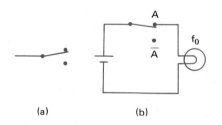

A	$\overline{A}$	f_0
0	1	0
1	0	1

(a) (b) (c)

FIGURE 34-3
(a) Schematic symbol of a single-pole, double-throw switch; (b) circuit used for Example 34-2; (c) truth table for the circuit of part (b).

EXAMPLE 34-2   If the closed switch of Figure 34-3(b) is indicated by $A = 1$, then what is the condition of the lamp (light or dark) when $\overline{A} = 1$?

Solution   Since $\overline{A} = 1$ indicates the switch is in the $\overline{A}$ position, then the switch is open and $A = 0$. The lamp is dark.

Observation   The truth table for the circuit conditions of Figure 34-3(b) is noted in Figure 34-3(c). In this truth table, $f_o = 1$ means "lamp lit" and $f_o = 0$ means "lamp dark."

From the truth table of Figure 34-3(c), we learn that the lamp is lit when $A = 1$. Another way of putting this statement would be, the lamp is lit when $\overline{A} = 0$. Furthermore, the lamp is dark when $A = 0$; or, putting this statement another way, the lamp is dark when $\overline{A} = 1$. Thus, A and $\overline{A}$ are the inverse of one another.

Inverter

The design of digital circuits requires the inversion of logic levels. The digital device used for this operation is called the *inverter,* which performs the NOT operation. The logic diagram symbol for the inverter is shown in Figure 34-4.

A	$\overline{A}$
0	1
1	0

(b)

A INPUT ⊳o $\overline{A}$ OUTPUT

(a)

FIGURE 34–4
(a) Inverter symbol used to indicate the NOT operation; (b) inverter truth table.

When talking about the NOT operator in relation to logic levels of 1 and 0, the term *complement* is used. For example, $\overline{1} = 0$ and $\overline{0} = 1$. The complement or NOT of 1 is 0, and the complement or NOT of 0 is 1.

The recomplement of a complemented variable or logic level results in the original uncomplemented variable or logic level. Thus, 1 complemented is 0, and 0 recomplemented is 1. Similarly, A inverted is $\overline{A}$, and $\overline{A}$ reinverted is $\overline{\overline{A}}$ or A. Figure 34-5 illustrates this concept.

The words NOT, *complement,* and *inverse* indicate that the opposite condition of the logic variable or logic level is indicated. Thus $\overline{1} = 0$ and $\overline{0} = 1$; NOT $A = \overline{A}$ and NOT $\overline{A} = A$.

(a)   (b)

FIGURE 34–5
(a) The output logic level of an inverter is the complement of the input logic level. (b) Logic variables are NOTed by an inverter.

⌐EXERCISE 34-2

Solve the following.

1. $\overline{1} = $ ___?___   2. $\overline{0} = $ ___?___   3. $\overline{\overline{0}} = $ ___?___

4. $\overline{\overline{1}} = $ ___?___   5. NOT $A = $ ___?___   6. NOT $\overline{A} = $ ___?___

7. 1 ▷o— ?   8. 0 ▷o— ?   9. $\overline{B}$ ▷o— ?

10. ? ▷o— A 11. ? ▷o—▷o—▷o— 0

34-3 CONJUNCTION OPERATOR (AND)

The conjunction operator (commonly called the AND operator) depends upon two or more events happening at the same time. When events occur in *conjunction,* a desired function will result. For example, an automobile will start if certain conditions are present in conjunction with one another. Let f_o = car starts, B = charged battery, G = gasoline in tank, K = key in ignition, S = shift lever in park. Then

$$f_o = B \text{ AND } G \text{ AND } K \text{ AND } S$$
$$f_o = B \cdot G \cdot K \cdot S$$
$$f_o = BGKS$$

That is, the car will start "if the battery is charged *and* the tank has gasoline *and* the ignition is unlocked with the key *and* the shift lever is in park."

The AND operator may be visualized as a circuit with two or more series switches. In Figure 34-6(a), we see that the lamp will light when both switches are closed at the same time. From the truth table of Figure 34-6(b), we also see that both switches must be closed before the lamp will light. When $A = 1$ (closed) and $B = 1$ (closed), then $f_o = 1$ (lit). Notice that when either one or both of the switches are open, the lamp is dark. Thus:

$$f_o = 0 \text{ (dark)} \quad \text{when} \quad A = 0 \text{ (open) and} \quad B = 1 \text{ (closed)}$$
$$f_o = 0 \text{ (dark)} \quad \text{when} \quad A = 1 \text{ (closed) and } B = 0 \text{ (open)}$$
$$f_o = 0 \text{ (dark)} \quad \text{when} \quad A = 0 \text{ (open) and} \quad B = 0 \text{ (open)}$$

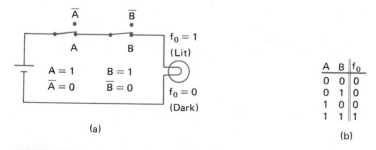

(a)

A	B	f_0
0	0	0
0	1	0
1	0	0
1	1	1

(b)

FIGURE 34-6

(a) AND **operation may be thought of as two or more series switches; (b) truth table for the two-variable** AND **circuit of part (a).**

In the AND operation, all input logic levels must be 1 if the logic level of the output function is to be 1.

AND Gate

The operation of digital computers uses the conjunction operation to combine several logic levels at the same time. The digital device used for this operation is called the AND gate. The logic diagram symbol for the AND gate is shown in Figure 34-7.

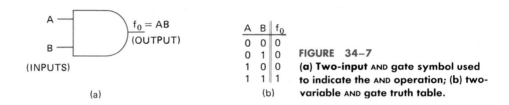

(a)

A	B	f_0
0	0	0
0	1	0
1	0	0
1	1	1

(b)

FIGURE 34-7
(a) Two-input AND gate symbol used to indicate the AND operation; (b) two-variable AND gate truth table.

For a two-input AND gate, there are 2^2 or 4 possible combinations of the variables because each variable has 2 possible values. The general rule is

$$\text{Number of possible combinations} = 2^n \qquad (34\text{-}1)$$

Where n is the number of input variables.

EXAMPLE 34-3   (a) Construct the logic diagram for the logic expression $f_o = ABC$.

(b) With the aid of Equation 34-1, determine the number of combinations needed to form the truth table.

(c) Write the truth table for this three-input AND gate.

Solution   (a) Figure 34-8(a) pictures a three-input AND gate with the inputs noted with the logical variables A, B, and C.

(b) The number of combinations needed to describe a three-variable gate is 2^3 or 8 combinations. These eight combinations are written as the binary equivalent of the decimal numbers 0 through 7.

(c) The truth table is formed as pictured in Figure 38-8(b). Here you see the eight possible combi-

---

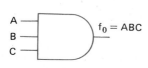

	A	B	C	$f_0 = ABC$
0	0	0	0	0
1	0	0	1	0
2	0	1	0	0
3	0	1	1	0
4	1	0	0	0
5	1	0	1	0
6	1	1	0	0
7	1	1	1	1

(a)   (b)

FIGURE 34–8

(a) Three-input AND gate for Example 34–3; (b) truth table for the three-variable AND gate.

nations for the input variables (A, B, and C) noted by the binary numbers 000 (0_{10}) through 111 (7_{10}). The output AND function (f_o) is determined for each of the listed input combinations by applying the concepts of the AND operation. Thus:

The combination of line 0 yields an output function of zero:

$$0 \text{ AND } 0 \text{ AND } 0 \Rightarrow 0$$

The combination of line 4 yields an output function of zero:

$$1 \text{ AND } 0 \text{ AND } 0 \Rightarrow 0$$

The combination of line 7 yields an output function of one:

$$1 \text{ AND } 1 \text{ AND } 1 \Rightarrow 1$$

In summary, the output function of an AND gate has a logic level of 1 only when all the input logic levels are 1 at the same time. If one or more input logic levels of the AND gate are 0, then the output function of the AND gate is 0. In actual logic circuits, the timing of the logic levels is very important. Figure 34-9 emphasizes this concept. Here we see the important role that timing plays in the operation of the AND gate. Between times T_1 and T_2 of Figure 34-9(b), we see that both input A and B are *high* (logic level 1) resulting in the output also going *high* (logic level 1). This same condition occurs between time T_5 and T_6. The remainder of the time intervals result in the output function being *low* (logic level 0), because either or both inputs A and B are *low* (logic level 0).

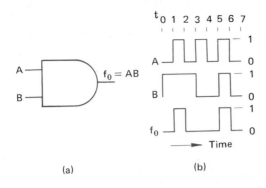

$f_0 = AB$

(a)

FIGURE 34-9
The output logic level (f_o) of the AND gate (a) is 1 only when the input logic levels of *A* and *B* are 1 at the same time as shown in part (b).

(b)

EXAMPLE 34-4

(a) Using the logic diagram symbols for NOT and AND, draw the logic diagram for $f_o = \overline{A}B$.

(b) Write the truth table for the two-variable logic expression.

Solution

(a) Figure 34-10 pictures the implementation of the logic function $f_o = \overline{A}B$. Note the use of the inverter to produce NOT *A*.

(b) The truth table has 2^2 or 4 possible conditions. In constructing the truth table, an additional column is added for the $\overline{A}$ combination. Figure 34-10(b) shows the completed truth table.

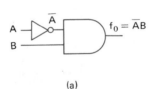

	A	B	$\overline{A}$	$f_0 = \overline{A}B$
0	0	0	1	0
1	0	1	1	1
2	1	0	0	0
3	1	1	0	0

(a)   (b)

FIGURE 34-10
(a) Logic diagram for the function $f_o = \overline{A}B$; (b) truth table for the logic expression $f_o = \overline{A}B$.

Observation

The method for forming the truth table of Figure 34-10(b) proceeds from left to right. For example, the combination of line 1 results in 0 1 for columns (*A, B*) and 1 1 for columns (*B, $\overline{A}$*). 1 AND 1 for $\overline{A}$ AND *B* results in 1 for f_o.

EXERCISE 34-3

Determine the output logic level of f_o in each of the following expressions.

1. $f_o = ABC$ when $A = 1$, $B = 1$, and $C = 1$.
2. $f_o = A\overline{B}C$ when $A = 1$, $B = 0$, and $C = 1$.
3. $f_o = ABC$ when $\overline{A} = 0$, $B = 1$, and $\overline{C} = 1$.
4. $f_o = \overline{A}\,\overline{B}C$ when $A = 0$, $B = 0$, and $C = 1$.
5. $f_o = \overline{A}BC$ when $A = 1$, $B = 1$, and $C = 1$.
6. $f_o = AB\overline{C}$ when $\overline{A} = 0$, $B = 0$, and $C = 0$.
7. $f_o = \overline{A}\,\overline{B}\,\overline{C}$ when $\overline{A} = 1$, $\overline{B} = 1$, and $C = 1$.
8. $f_o = \overline{A}B\overline{C}$ when $A = 1$, $B = 0$, and $C = 1$.

Using the logic diagram symbols for NOT and AND, draw the logic diagram for each of the following expressions.

9. $f_o = AB$ 10. $f_o = ABC$ 11. $f_o = ABCD$
12. $f_o = A\overline{B}$ 13. $f_o = \overline{A}B$ 14. $f_o = AB\overline{C}$

Complete the truth table for each of the following logic diagrams.

15.

	A B	$f_o = AB$
0	0 0	0
1		
2		
3		

16.

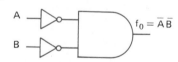

	A B	$\overline{A}$	$\overline{B}$	$f_o = \overline{A}\,\overline{B}$
0	0 0	1	1	
1				
2				
3				

17. For the logic expression $f_o = \overline{A}BC$:
   (a) Determine the number of logic combinations needed to construct the truth table.
   (b) Construct the truth table.
   (c) Draw the logic diagram.

18. For the logic expression $f_o = \overline{A}B\overline{C}$:
   (a) Determine the number of logic combinations needed to construct the truth table.

L (b) Construct the truth table.
(c) Draw the logic diagram.

34-4 DISJUNCTION OPERATOR (OR)

The disjunction operator (commonly called the OR operator) allows for a choice between two or more events occurring at the same time. For example, an automobile with an automatic transmission may not be started with the shift lever in drive or if the battery is dead. Let f_o = car won't start, D = shift lever in drive, and B = dead battery. Then

$$f_o = D \text{ OR } B$$
$$f_o = D + B$$

That is, "the car will not start if the shift lever is in drive *or* the battery is dead."

The OR operator may be visualized as a circuit with two or more parallel switches. In Figure 34-11(a), we see that the lamp will light when either of the switches is closed. From the truth table of Figure 34-11(b), we learn that the OR operation includes the possibility of both variables being 1. This type of OR operator is called an *inclusive* OR because of its inclusive nature. We will use the term OR to mean *inclusive* OR. From the truth table, you may notice that when either or both of the switches are closed the lamp is lit, and the lamp is dark only when both switches are open. Thus

$$f_o = 0 \text{ (dark)} \quad \text{when } A = 0 \text{ (open) and } B = 0 \text{ (open)}$$
$$f_o = 1 \text{ (lit)} \quad \text{when } A = 0 \text{ (open) and } B = 1 \text{ (closed)}$$
$$f_o = 1 \text{ (lit)} \quad \text{when } A = 1 \text{ (closed) and } B = 0 \text{ (open)}$$
$$f_o = 1 \text{ (lit)} \quad \text{when } A = 1 \text{ (closed) and } B = 1 \text{ (closed)}$$

In the OR operation, if any of the input logic levels is 1, then the output logic level of the output function will be 1.

OR Gate

The operation of digital computers uses the disjunction operation to produce a high output (logic level 1) when any one of several inputs is high (logic level 1). The digital device used for this operation is called the OR gate. The logic diagram symbol for the OR gate is shown in Figure 34-12. Like the AND gate, the OR gate has 2^n possible combinations needed to describe its operation (where n is the number of inputs).

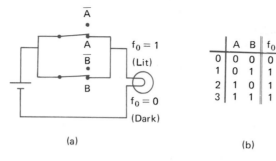

	A	B	f_0
0	0	0	0
1	0	1	1
2	1	0	1
3	1	1	1

(a)   (b)

FIGURE 34–11
(a) OR operation may be thought of as two or more parallel switches; **(b)** truth table for the two-variable OR circuit of part **(a)**.

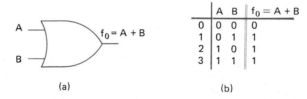

	A	B	$f_0 = A + B$
0	0	0	0
1	0	1	1
2	1	0	1
3	1	1	1

(a)   (b)

FIGURE 34–12
(a) Two-input OR gate symbol used to indicate the OR operation; **(b)** two-variable OR gate truth table.

EXAMPLE 34-5 (a) Construct the logic diagram for the logic expression $f_0 = \overline{A} + B + \overline{C}$.

(b) With the aid of Equation 34-1, determine the number of combinations needed to form the truth table.

(c) Write the truth table for this three-input OR gate.

Solution (a) Figure 34-13(a) shows a three-input OR gate with inverters in the A and C inputs to produce $\overline{A}$ and $\overline{C}$.

(b) The number of combinations needed to describe a three-variable gate is 2^3 or 8 combinations.

(c) The truth table is formed as pictured in Figure 34-13(b). Here you see the eight possible combinations for the input variables (A, B, and C) noted in the binary numbers 000 through 111. In

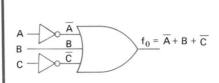

	A	B	C	$\overline{A}$	B	$\overline{C}$	$f_0 = \overline{A} + B + \overline{C}$
0	0	0	0	1	0	1	1
1	0	0	1	1	0	0	1
2	0	1	0	1	1	1	1
3	0	1	1	1	1	0	1
4	1	0	0	0	0	1	1
5	1	0	1	0	0	0	0
6	1	1	0	0	1	1	1
7	1	1	1	0	1	0	1

(a) (b)

FIGURE 34–13
(a) Three-input OR gate with inverts for Example 34–5;
(b) truth table for the three-variable logic expression
$f_o = A + B + C$.

the next column, you see the $\overline{A}$, B, and $\overline{C}$ logic levels listed. Note the combination of line 5, where $\overline{A} = 0$, $B = 0$, and $\overline{C} = 0$ is the only combination that produces an output function of 0 ($f_o = 0$ OR 0 OR 0 = 0).

In summary, the output function of an OR gate has a logic level of 0 only when all the input logic levels are 0 at the same time. If one or more input logic levels of the OR gate is 1, then the output function of the OR gate is 1.

EXERCISE 34-4
Determine the output logic level of f_o in each of the following expressions.

1. $f_o = A + B + C$ when $A = 1$, $B = 1$, and $C = 1$.
2. $f_o = A + \overline{B} + C$ when $\overline{A} = 1$, $B = 1$, and $C = 0$.
3. $f_o = A + B + C$ when $\overline{A} = 0$, $B = 1$, and $\overline{C} = 1$.
4. $f_o = \overline{A} + \overline{B} + C$ when $A = 0$, $B = 0$, and $C = 1$.
5. $f_o = \overline{A} + B + C$ when $A = 1$, $B = 1$, and $C = 1$.
6. $f_o = A + \overline{B} + \overline{C}$ when $\overline{A} = 1$, $B = 1$, and $C = 1$.
7. $f_o = \overline{A} + \overline{B} + \overline{C}$ when $\overline{A} = 1$, $\overline{B} = 1$, and $C = 1$.
8. $f_o = \overline{A} + B + \overline{C}$ when $A = 1$, $B = 0$, and $C = 1$.

Using the logic diagram symbols for NOT and OR, draw the logic diagram for each of the following expressions.

9. $f_o = A + B$
10. $f_o = A + B + C$
11. $f_o = A + B + C + D$
12. $f_o = A + \overline{B}$

13. $f_o = \overline{A} + \overline{B}$ 14. $f_o = A + B + \overline{C}$

Complete the truth table for each of the following logic diagrams.

15.

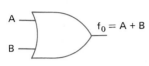

	A	B	$f_o = A + B$
0	0	0	0
1			
2			
3			

16.

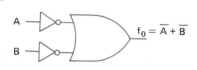

	A	B	$\overline{A}$	$\overline{B}$	$f_o = \overline{A} + \overline{B}$
0	0	0	1	1	
1					
2					
3					

17. For the logic expression $f_o = \overline{A} + B + C$:
   (a) Determine the number of logic combinations needed to construct the truth table.
   (b) Construct the truth table.
   (c) Draw the logic diagram.

18. For the logic expression $f_o = B + \overline{C} + \overline{D}$:
   (a) Determine the number of logic combinations needed to construct the truth table.
   (b) Construct the truth table.
   (c) Draw the logic diagram.

34-5 APPLICATION OF LOGIC CONCEPTS

In this section, you will have an opportunity to expand your knowledge of the logic operators as well as the use of truth tables. As you have already learned, logic diagrams are used to symbolize logic expressions. No matter how complex the logic expression may be, it can be symbolized with the three logic operations. By using the AND, OR, and NOT logic symbols independently or collectively, any logic expression may be symbolized. The following examples will demonstrate further use of these *building blocks* in the construction of logic diagrams and truth tables.

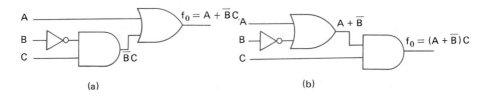

FIGURE 34–14
(a) Logic diagram for the expression $f_o = A + \overline{B}C$ for Example 34–6. (b)
Logic diagram for the expression $f_o = (A + \overline{B})C$. This expression
demonstrates the use of parentheses to indicate the desired order of
operation.

EXAMPLE 34-6   Develop the logic diagram for the logic expression
$f_o = A + \overline{B}C$.

Solution   Figure 34-14(a) is the logic diagram for the logic
expression $f_o = A + \overline{B}C$.

Observation   When working with logic expressions, the AND operation has precedence over the OR operation. Thus, in
the expression $A + \overline{B}C$, NOT B AND C is ORed with A,
as noted in Figure 34-14(a). The defined order of operation may be superseded by the parentheses. In the
expression $(A + \overline{B})C$, A OR NOT B is ANDed with C.
Figure 34-14(b) is the logic diagram for this logic
expression.

In the preceding example, you saw the importance of a defined *order* of
operation for logic expressions. The next example demonstrates the use of
truth tables to determine if two logic expressions are equivalent.

EXAMPLE 34-7   Determine if the logic expressions represented by
the logic diagrams of Figure 34-14(a) and (b) are
equivalent.

Solution   Form the equation $A + \overline{B}C = (A + \overline{B})C$. Construct
a truth table for each member of the equation using 2^3
or 8 lines to represent the possible combinations. Figure 34-15(a) is the truth table of the left member, $A +$
$\overline{B}C$. Figure 34-15(b) is the truth table of the right

---

	A	B	C	$\bar{B}$	$\bar{B}C$	$f_0 = A + \bar{B}C$
0	0	0	0	1	0	0
1	0	0	1	1	1	1
2	0	1	0	0	0	0
3	0	1	1	0	0	0
4	1	0	0	1	0	1
5	1	0	1	1	1	1
6	1	1	0	0	0	1
7	1	1	1	0	0	1

(a)

	A	B	C	$\bar{B}$	$A+\bar{B}$	$f_0 = (A + \bar{B})C$
0	0	0	0	1	1	0
1	0	0	1	1	1	1
2	0	1	0	0	0	0
3	0	1	1	0	0	0
4	1	0	0	1	1	0
5	1	0	1	1	1	1
6	1	1	0	0	1	0
7	1	1	1	0	1	1

(b)

FIGURE 34–15
Truth tables for Example 34–7.

member, $(A + \bar{B})C$. To determine equality, the output function of each member is compared for each of the 8 possible combinations. Make a line-by-line comparison between the output functions of the two truth tables. Notice that lines 4 and 6 do not match. This leads us to conclude that these two logic expressions and their corresponding logic diagrams are not equivalent.

$$\therefore \quad A + \bar{B}C \neq (A + \bar{B})C$$

Observation In Figure 34-15(a), notice how the term $\bar{B}C$ (NOT B AND C) was evaluated as an intermediate step by AND-ing the combinations listed in the C and $\bar{B}$ columns. The use of this intermediate step reduces the possibility of error in the formation of the output function. The output function is formed by ORing the combination of columns A and $\bar{B}C$.

In the preceding example, you saw how two logic diagrams may be examined for equality by comparing the output functions in the corresponding truth tables. This technique is used in the next example to verify that two logic expressions are equivalent.

EXAMPLE 34-8 Determine if $(A + B)(B + C) = AC + B$ is a true statement.

Solution Construct a truth table for each member of the equation using 2^3 or 8 lines for each. Figure 34-16(a) is the

	A	B	C	A+B	B+C	$f_0 = (A+B)(B+C)$
0	0	0	0	0	0	0
1	0	0	1	0	1	0
2	0	1	0	1	1	1
3	0	1	1	1	1	1
4	1	0	0	1	0	0
5	1	0	1	1	1	1
6	1	1	0	1	1	1
7	1	1	1	1	1	1

(a)

	A	B	C	AC	$f_0 = AC + B$
0	0	0	0	0	0
1	0	0	1	0	0
2	0	1	0	0	1
3	0	1	1	0	1
4	1	0	0	0	0
5	1	0	1	1	1
6	1	1	0	0	1
7	1	1	1	1	1

(b)

FIGURE 34–16
Truth tables for Example 34–8.

truth table for the left member, while Figure 34-16(b) is the truth table for the right member. Making a line-by-line comparison between the output functions of the two truth tables leads us to conclude that these two logic expressions and their corresponding logic diagrams are equivalent.

$$\therefore \quad (A + B)(B + C) = AC + B$$

From the preceding example, we learn that the same output may result from several different logic expressions. Although each expression may produce the desired output, only one expression is the simplest and least expensive to implement. Figure 34-17(a) and (b) are the two equivalent logic diagrams of the previous example. The one on the right is simpler and less costly to implement. The next example provides you with a step-by-step development of the truth table for a logic diagram.

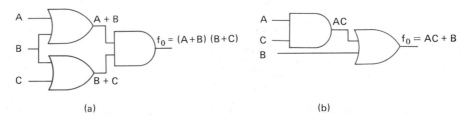

FIGURE 34–17
Logic diagrams for the equivalent expressions used in Example 34–8.
Part (b) is simpler than part (a) because it has fewer logic gages.

EXAMPLE 34-9 Construct the truth table for the logic diagram of Figure 34-18.

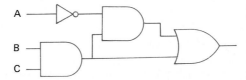

A

B

C

FIGURE 34–18
Logic diagram for Example 34–9.

Solution   Because of the three inputs, select three variables (A, B, and C) to represent them. Use 2^3 or 8 lines in the truth table. Start the truth table by listing the 8 combinations in the left side of the truth table, as in Figure 34-19(a). Next, form a column for the NOT operation; label this column $\overline{A}$. Then, form a column for the AND operation, BC, as indicated in Figure 34-19(b). Finally, form columns for the other AND operation, $\overline{A}BC$, and the output of the OR gate, $BC + \overline{A}BC$, as indicated in Figure 34-19(c).

Observation   The logic levels of the output (f_o) are identical to those of column BC. This means that $BC + \overline{A}BC = BC$ and the logic diagram of Figure 34-18 may be simplified to a single AND gate with two inputs, B and C.

LINE	A	B	C
0	0	0	0
1	0	0	1
2	0	1	0
3	0	1	1
4	1	0	0
5	1	0	1
6	1	1	0
7	1	1	1

(a)

LINE	A	B	C	$\overline{A}$	BC
0	0	0	0	1	0
1	0	0	1	1	0
2	0	1	0	1	0
3	0	1	1	1	1
4	1	0	0	0	0
5	1	0	1	0	0
6	1	1	0	0	0
7	1	1	1	0	1

(b)

LINE	A	B	C	$\overline{A}$	BC	$\overline{A}$BC	$f_0 = BC + \overline{A}BC$
0	0	0	0	1	0	0	0
1	0	0	1	1	0	0	0
2	0	1	0	1	0	0	0
3	0	1	1	1	1	1	1
4	1	0	0	0	0	0	0
5	1	0	1	0	0	0	0
6	1	1	0	0	0	0	0
7	1	1	1	0	1	0	1

(c)

FIGURE 34–19
Truth table developed for the logic diagram of Figure 34–18.

EXERCISE 34-5
Draw the logic diagram for each of the following logic expressions.

1. $\overline{A}B + C$ 2. $AC + AB$ 3. $\overline{A} + ABC$

4. $\overline{A}B + \overline{A}C$ 5. $A(B + C)$ 6. $AB(A + C)$

Construct a truth table for the following logic equations.

7. $A\overline{C} + B$

8. $AB\overline{C} + \overline{A}B + C$

9. $A(C + B\overline{C})$

10. $C(\overline{B} + \overline{A})$

Using truth tables, determine which of the following equations is true.

11. $A(A + B) = A$

12. $A + \overline{A}B = A + B$

13. $A(B + C) = AB + AC$

14. $(A + B)(A + C) = A + BC$

15. $AC + \overline{A}\,\overline{C} = A\overline{C} + \overline{A}C$

16. $A\overline{B} + C(A + \overline{B}) = AC$

17. Construct the truth table for the logic diagram of Figure 34-20.

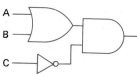

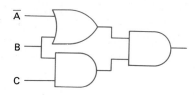

FIGURE 34–20
Logic diagram for Exercise 34–5, problem 17.

FIGURE 34–21
Logic diagram for Exercise 34–5, problem 18.

18. Construct the truth table for the logic diagram of Figure 34-21.

34-6 INTRODUCTION TO KARNAUGH MAPS

One goal for the logic circuit designer is to design circuits with as few parts as possible, because a reduction in parts will result in an increase in circuit reliability and a reduction in circuit cost. The *Karnaugh map,* or *K-map* as it will be called, is a tool for simplifying logic circuits. It is similar in concept to the truth table, but by its organization it allows the designer to simplify a circuit through pattern recognition.

Recall that a truth table contains a line for each combination of values for the input variables. Thus, a truth table for two variables contains four lines, and a truth table for three variables contains eight lines. Furthermore, each line is identified by the decimal number equal in value to the binary number representing the values of the input variables. Similarly, a K-map contains a square for each combination of values for the input variables.

LINE	A B	f_0
0	0 0	
1	0 1	
2	1 0	
3	1 1	

(a)

(b)

FIGURE 34–22
Truth table and Karnaugh map for the two-input variables *A* and *B*.

Thus, a K-map for two variables contains four squares, as shown in Figure 34-22(b), and a K-map for three variables contains eight squares, as shown in Figure 34-23(b). Furthermore, each square in a K-map is numbered. These numbers correspond to the line or combination numbers in the truth table. The usefulness of a K-map comes from the organization of the data. This means that the numbering system of the squares for the K-maps in Figures 34-22(b), 34-23(b), and 34-24 must be carefully followed.

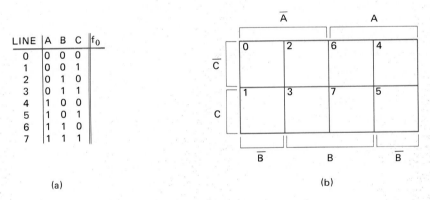

LINE	A B C	f_0
0	0 0 0	
1	0 0 1	
2	0 1 0	
3	0 1 1	
4	1 0 0	
5	1 0 1	
6	1 1 0	
7	1 1 1	

(a)

(b)

FIGURE 34–23
Truth table and Karnaugh map for the three-input variables *A*, *B*, and *C*.

Examine the K-map for the two variables *A* and *B* in Figure 34-22(b). The four small squares are organized as a larger square with two rows and two columns. One column is for *A*, which is labeled; the other column is for $\overline{A}$, which is usually not labeled. Similarly, one row is for *B*, which is labeled; the other row if for $\overline{B}$, which is usually not labeled. The number in the upper left corner of each square is the same as the combination number for the corresponding line in the truth table. Since the upper left square represents the term $\overline{A}\,\overline{B}$, the number in the corner of the upper left square is zero. Similarly, the

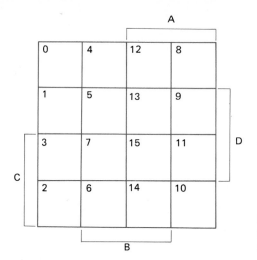

A

0	4	12	8
1	5	13	9
3	7	15	11
2	6	14	10

C

D

B

FIGURE 34–24
Pattern for four variable Karnaugh maps.

first line of the truth table in Figure 34-22(a) is for combination 0_{10} or 00_2; that is, $\overline{A}$ and $\overline{B}$.

The K-map for the three variables A, B, and C of Figure 34-23(b) is slightly more complicated than the two-variable K-map. Two columns are for A and two for $\overline{A}$. In like manner there are two columns for B and two for $\overline{B}$. Notice how the two columns for B are split between A and $\overline{A}$. Finally, one row is for C and the other row is for $\overline{C}$. Since the upper left square represents the term $\overline{A}\,\overline{B}\,\overline{C}$, the number in the corner of the upper left square is zero.

The use of a K-map for logic circuit simplification is part of a systematic process. First, build a truth table from the logic diagram. Second, map the truth table by entering a 1 in each square of the K-map corresponding to a combination having a 1 for the output function in the truth table. (All the other squares are left blank.) Third, group adjacent squares that contain 1s to simplify the logic expression. Fourth, read the simplified logic expression for the circuit from the K-map. Fifth, create a simplified logic diagram from the simplified logic expression. The preceding concepts will be explored in the following examples.

EXAMPLE 34-10

Create a truth table and a K-map for $f_o = AB + \overline{A}B$ and simplify.

Solution Since there are two variables, the truth table will contain four lines and the K-map will contain four squares. See Figure 34-25 for the truth table and the K-map.

Notice that the two 1s in the K-map are adjacent.

LINE	A	B	AB	$\overline{A}B$	f_0
0	0	0	0	0	0
1	0	1	0	1	1
2	1	0	0	0	0
3	1	1	1	0	1

(a)

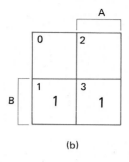

(b)

FIGURE 34–25
Truth table and Karnaugh map for Example 34–10,
$f_o = AB + \overline{A}B.$

Because both 1s are in the B row, they can be represented by a single term in the single variable B. Thus, $AB + \overline{A}B$ simplifies to B.

$$\therefore \qquad f_o = B$$

EXAMPLE 34-11

Create a truth table and a K-map for $f_o = AB + AC + \overline{A}B\overline{C}$ and simplify.

Solution   Since there are three variables, the truth table will contain eight lines and the K-map will contain eight squares. See Figure 34-26 for the truth table and K-map.

Notice that the three-variable term, $\overline{A}B\overline{C}$, is represented by one square (number 2) while each two-variable term occupies two squares (6, 7) and (5, 7). Furthermore, a single square (number 7) can be part of two different terms (AB and AC).

Simplify by grouping the four 1s into pairs: (2, 6) and (5, 7).

Read the map:

Both squares of the pair (2, 6) are in column B and row $\overline{C}$; thus (2, 6) represents $B\overline{C}$.

Both squares of the pair (5, 7) are in column A and row C, thus (5, 7) represents AC.

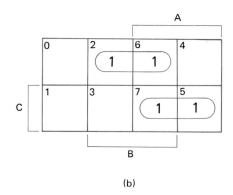

LINE	A	B	C	AB	AC	$\overline{A}B\overline{C}$	f_0
0	0	0	0	0	0	0	0
1	0	0	1	0	0	0	0
2	0	1	0	0	0	1	1
3	0	1	1	0	0	0	0
4	1	0	0	0	0	0	0
5	1	0	1	0	1	0	1
6	1	1	0	1	0	0	1
7	1	1	1	1	1	0	1

(a)

(b)

FIGURE 34–26
Truth table and Karnaugh map for Example 34–11,
$f_o = AB + AC + \overline{A}B\overline{C}.$

Therefore, the simplified expression is $AC + B\overline{C}$.

$$AB + AC + \overline{A}B\overline{C} = B\overline{C} + AC$$

$$\therefore \quad f_o = B\overline{C} + AC$$

Once the K-map is completed for an expression, the next step is to check for recognizable patterns. An isolated 1 represents a term containing all the variables. Two adjacent 1s represent a term containing all but one variable. The two 1s can be side by side or one on top of the other. However, diagonally positioned 1s are not significant and do not result in a reduction of variables. Four adjacent 1s represent a term containing all but two variables. The four 1s can be in a row or a column or a square. For our purpose of defining adjacent squares in K-maps, the top row is considered to be adjacent to the bottom row; and the left column is considered to be adjacent to the right column.

EXAMPLE 34-12

Determine a minimal expression for the K-map in Figure 34-27.

Solution The K-map is for the two variables U and V. The two 1s are not adjacent; they are isolated. Therefore, there are two terms, each containing both variables.

$$\therefore \quad f_o = U\overline{V} + \overline{U}V$$

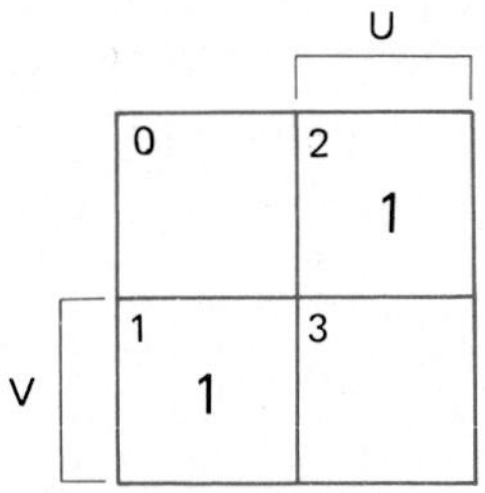

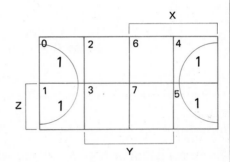

FIGURE 34–27
K-map for Example 34–12.

EXAMPLE 34-13

Determine a minimal expression for the K-map in Figure 34-28.

FIGURE 34–28
K-map for Example 34–13.

Solution   The K-map is for the three variables X, Y, and Z. The four 1s are adjacent; therefore, they represent one term in one variable.

$$\therefore \quad f_o = \overline{Y}$$

EXAMPLE 34-14

Determines a minimal expression for the K-map in Figure 34-29.

Solution   The K-map is for the four variables A, B, C, and D. Square 15 can be combined with 11 or 14. Since square 11 only can be combined with 15, we combine 11 and 15 into one term in three variables. The

---

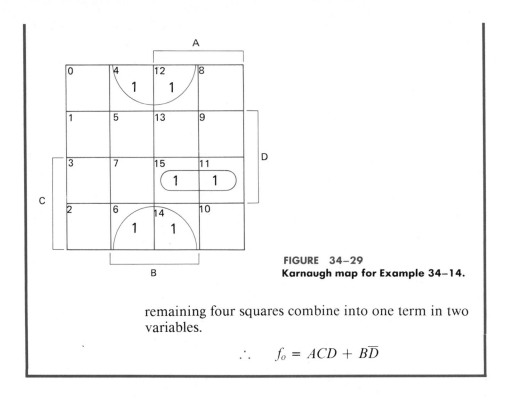

FIGURE 34–29
Karnaugh map for Example 34–14.

remaining four squares combine into one term in two variables.

$$\therefore \quad f_o = ACD + B\overline{D}$$

Summary

The Karnaugh map was introduced in this section as a tool for simplifying logic circuits. Its relation to the truth table has been explored. Its usefulness resides in the ability of the user to recognize patterns. Table 34-1 and Guidelines 34-1 are designed to help in the pattern recognition process.

Table 34-1 gives the relationship between the number of adjacent entries in a K-map and the number of variables in the corresponding term for the three practical sized K-maps. The relationship in the table holds both for mapping an expression and for reading a K-map.

TABLE 34-1

OCCUPANCY RATES FOR KARNAUGH MAPS			
	Number of Occupied Squares		
Number of Variables in Term	**Two-Variable Map**	**Three-Variable Map**	**Four-Variable Map**
1	2	4	8
2	1	2	4
3		1	2
4			1

Guidelines 34-1 are useful in interpreting the patterns found within K-maps. Use the guidelines while working the exercise set at the end of this section.

EXERCISE 34-6
Create a Karnaugh map for each of the following expressions:

1. AB
2. $\overline{A}B$
3. $J\overline{K} + K$
4. $M\overline{N} + \overline{M}$
5. $\overline{P}QR + P\overline{R}$
6. $UV + UW + VW$

Use Karnaugh maps to simplify the following expressions:

7. $S\overline{T} + \overline{S}\,\overline{T}$ 8. $\overline{X}Y + XY$ 9. $H + \overline{G}H$

10. $JK + JK\overline{L}$ 11. $UV + UW + U\overline{V}\,\overline{W}$ 12. $\overline{P}\overline{Q} + PR + QR$

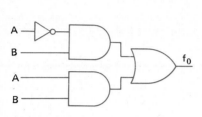

FIGURE 34–30
Logic diagram for Exercise 34–6, problem 13.

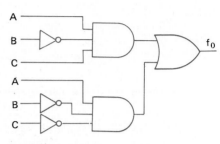

FIGURE 34–31
Logic diagram for Exercise 34–6, problem 14.

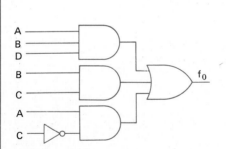

FIGURE 34–32
Logic diagram for Exercise 34–6, problem 15.

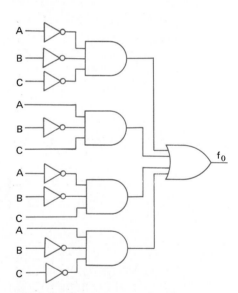

FIGURE 34–33
Logic diagram for Exercise 34–6, problem 16.

13. Create a K-map for the logic diagram in Figure 34-30.
14. Create a K-map for the logic diagram in Figure 34-31.
15. Simplify the logic diagram in Figure 34-32.
16. Simplify the logic diagram in Figure 34-33.

34-7 DeMORGAN'S THEOREM

In Section 34-2 the idea of the NOT operator was introduced along with its effect on a variable. In this section, DeMorgan's theorem is introduced to define the effect of the NOT operator on a logic expression. We will begin with the examination of the following two special cases.

EXAMPLE 34-15

Use a K-map to find another expression for $\overline{U \cdot V}$.

Solution The expression $\overline{U \cdot V}$ is read NOT the product U AND V.

Form a truth table as in Figure 34-34(a).

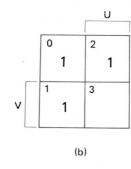

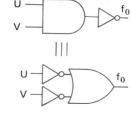

(a)   (b)   (c)

FIGURE 34–34
Illustration of DeMorgan's theorem: $\overline{UV} = \overline{U} + \overline{V}$.

Create a K-map as in Figure 34-34(b).
Read the function from the K-map:

$$f_o = \overline{U} + \overline{V}$$

$\therefore \overline{U \cdot V} = \overline{U} + \overline{V}$, as shown in Figure 34-34(c).

EXAMPLE 34-16

Use a K-map to find another expression for $\overline{Y + Z}$.

Solution The expression $\overline{Y + Z}$ is read NOT the sum Y OR Z.
Form a truth table as in Figure 34-35(a).
Create a K-map as in Figure 34-35(b).

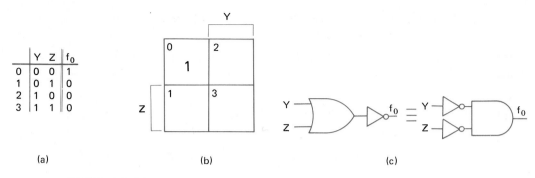

(a)   (b)   (c)

FIGURE 34–35
Illustration of DeMorgan's theorem: $\overline{Y} + \overline{Z} = \overline{YZ}$.

Read the function from the K-map:

$$f_o = \overline{Y} \cdot \overline{Z}$$

$$\therefore \quad \overline{Y + Z} = \overline{Y} \cdot \overline{Z}, \text{ as shown in Figure 34-35(c)}.$$

The results of the preceding two examples are summarized as DeMorgan's theorem in the following rule.

RULE 34-1. DeMORGAN'S THEOREM

To invert a logic expression, invert each variable and exchange all AND and OR operators. Use signs of grouping to preserve the original order of operation. Remember, for any variable X, $\overline{\overline{X}} = X$. Thus, the inverse of

$$f_o = (A \cdot \overline{B}) + C$$

is given by $f_o = (\overline{A} + B) \cdot \overline{C}$

**EXAMPLE
34-17**   Invert $J \cdot \overline{K} + \overline{J} \cdot K \cdot L + \overline{L}$.

Solution   Use DeMorgan's Theorem:

$$J \cdot \overline{K} + \overline{J} \cdot K \cdot L + \overline{L}$$

$$(\overline{J} + K) \cdot (J + \overline{K} + \overline{L}) \cdot L$$

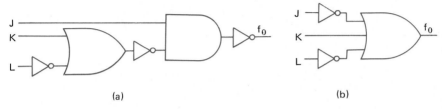

The parentheses are to retain the original order of operation.

EXAMPLE
34-18
Redraw the logic diagram in Figure 34-36(a) with the inverters on the input side of the gates.

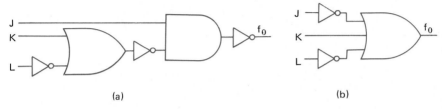

FIGURE 34–36
Logic diagrams for Examples 34–18.

Solution Determine the output function for the logic diagram:

$$f_o = \overline{J(\overline{K + \overline{L}})}$$

Let $M = (K + \overline{L})$ and apply DeMorgan's theorem:

$$f_o = \overline{J\overline{M}}$$
$$f_o = \overline{J} + M$$
$$\therefore \quad f_o = \overline{J} + K + \overline{L}$$

See Figure 34-36(b) for the new logic diagram.

As we saw in Section 34-2, inverting a variable twice results in the original variable. The same holds true for expressions; that is, $f_o = $ NOT NOT f_o. This provides the key to the technique for simplifying circuits using OR, logic as in the following example.

EXAMPLE
34-19
Simplify the logic diagram in Figure 34-37.

Solution Determine the logic expression for the output function:

$$f_o = (\overline{A} + B + C)(\overline{A} + B + D)(\overline{B} + C + \overline{D})$$
$$(\overline{A} + \overline{C} + \overline{D})(A + \overline{B} + \overline{D})$$

Invert the function:

$$\overline{f}_o = \overline{(\overline{A} + B + C)(\overline{A} + B + D)(\overline{B} + C + \overline{D})}$$
$$\cdot \overline{(\overline{A} + \overline{C} + \overline{D})(A + \overline{B} + \overline{D})}$$
$$\overline{f}_o = A \cdot \overline{B} \cdot \overline{C} + A \cdot \overline{B} \cdot \overline{D} + B \cdot \overline{C} \cdot D$$
$$+ A \cdot C \cdot D + \overline{A} \cdot B \cdot D$$

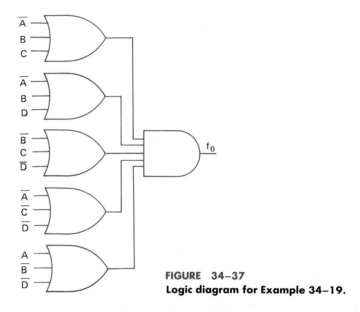

FIGURE 34–37
Logic diagram for Example 34–19.

Form the truth table for $\overline{f}_o$ as in Figure 34-38.
Create the K-map for $\overline{f}_o$ as in Figure 34-39(a).
Read a minimal form from the K-map:

$$\overline{f}_o = A \cdot \overline{B} + B \cdot D$$

Invert $\overline{f}_o$ for a minimal form for f_o:

$$\overline{\overline{f}}_o = \overline{A \cdot \overline{B} + B \cdot D}$$
$$f_o = (\overline{A} + B)(\overline{B} + \overline{D})$$

See Figure 34-39(b) for a simplified logic diagram.

LINE	A	B	C	D	$AB\bar{C}$	$A\bar{B}D$	$B\bar{C}D$	$\bar{A}CD$	$\bar{A}BD$	f_0
0	0	0	0	0	0	0	0	0	0	0
1	0	0	0	1	0	0	0	0	0	0
2	0	0	1	0	0	0	0	0	0	0
3	0	0	1	1	0	0	0	0	0	0
4	0	1	0	0	0	0	0	0	0	0
5	0	1	0	1	0	0	1	0	1	1
6	0	1	1	0	0	0	0	0	0	0
7	0	1	1	1	0	0	0	0	1	1
8	1	0	0	0	1	1	0	0	0	1
9	1	0	0	1	1	0	0	0	0	1
10	1	0	1	0	0	1	0	0	0	1
11	1	0	1	1	0	0	0	1	0	1
12	1	1	0	0	0	0	0	0	0	0
13	1	1	0	1	0	0	1	0	0	1
14	1	1	1	0	0	0	0	0	0	0
15	1	1	1	1	0	0	0	1	0	1

FIGURE 34–38
Truth table for Example 34–19.

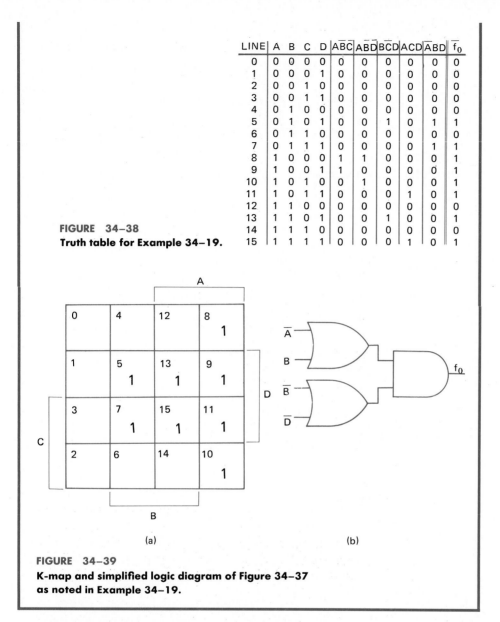

FIGURE 34–39
K-map and simplified logic diagram of Figure 34–37 as noted in Example 34–19.

The methods for simplifying logic diagrams of this and the previous section are summarized in the following procedure.

PROCEDURE 34-1. SIMPLIFYING LOGIC DIAGRAMS

1. Determine the output function from the logic diagram.
2. If there are more OR gates than AND gates, invert the function.

3. Form the truth table for the (inverted) function.
4. Form the corresponding Karnaugh map.
5. Read a minimal (inverted) function from the map.
6. If necessary, reinvert to obtain the minimal function.
7. Draw the logic diagram for the minimal function.

EXERCISE 34-7

Invert the following expressions.

1. $A \cdot B$ 2. $C \cdot \overline{D}$

3. $\overline{A} + B$ 4. $A + \overline{B} \cdot C$

5. $(U + \overline{V} + W)\overline{X}Y$ 6. $A(\overline{B} + C)\overline{D}$

7. $(A + \overline{B})(\overline{C} + B)(\overline{A} + D)$ 8. $XY + \overline{X}\overline{Z} + YZ$

Perform the indicated inversions for the following.

9. $\overline{X + \overline{Y} \cdot Z}$ 10. $\overline{U(\overline{V + W})}$

11. $\overline{J + \overline{KL}}$ 12. $\overline{A(B + \overline{CD})}$

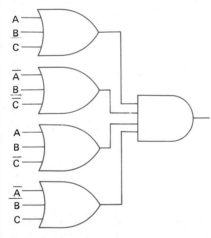

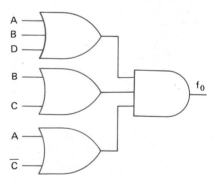

FIGURE 34–40
Logic diagram for Exercise 34–7, problem 13.

FIGURE 34–41
Logic diagram for Exercise 34–7, problem 14.

13. Simplify the logic diagram in Figure 34-40.
14. Simplify the logic diagram in Figure 34-41.

34-8 BOOLEAN THEOREMS

This section is a formal approach to the Boolean algebra for logic (two-state) variables. It provides an alternative approach for logic circuit simplification to the methods in the preceding sections.

Within the two rules given in this section, there is marked symmetry between the role of the AND operator and the OR operator. This symmetry is the basis of the concept of *duality*. The operator AND is the *dual* of the operator OR. The value 0 is the *dual* of the value 1. A variable, A, is the dual of its inverse, $\overline{A}$. Many of the properties listed in Rule 34-2 occur as pairs; that is, as a statement and its *dual*.

RULE 34-2. PROPERTIES OF BOOLEAN ALGEBRA FOR LOGIC VARIABLES

There are only two possible values: 0 and 1.

  1. If $A \neq 0$, then $A = 1$.   2. If $A \neq 1$, then $A = 0$.

The system is closed under conjunction, disjunction, and inversion.

  3. $0 \cdot 0 = 0$ 4. $1 + 1 = 1$
  5. $0 + 0 = 0$ 6. $1 \cdot 1 = 1$
  7. $1 \cdot 0 = 0$ 8. $1 + 0 = 1$
  9. $\overline{0} = 1$ 10. $\overline{1} = 0$

Simple results are independent of the order of operation.

 11. $AB = BA$ 12. $A + B = B + A$
 13. $A(BC) = (AB)C$ 14. $A + (B + C) = (A + B) + C$

However, the results of mixed operations are order dependent.

 15. $A(B + C) = AB + AC$ 16. $A + (BC) = (A + B)(A + C)$

In Boolean algebra, a second NOT counteracts the first NOT.

 17. $\overline{\overline{A}} = A$

The preceding properties can be used directly in the process of logic expression simplification. However, it is easier to use them to develop a set of theorems for logic expression simplification.

EXAMPLE 34-20

Use Rule 34-2 to derive $A \cdot 0 = 0$.

Solution Since A is a logic variable, then $A = 0$ or $A = 1$.
If $A = 0$, then $A \cdot 0 = 0 \cdot 0 = 0$, by Rule 34-2.3.
If $A = 1$, then $A \cdot 0 = 1 \cdot 0 = 0$, by Rule 34-2.7.

$$\therefore \quad A \cdot 0 = 0$$

EXAMPLE 34-21

Use DeMorgan's theorem to derive $A + 1 = 1$.

Solution We just derived $A \cdot 0 = 0$ for any logic variable including $\overline{A}$. Thus

$$\overline{A} \cdot 0 = 0$$

Apply DeMorgan's theorem to both sides:

$$\overline{\overline{A} \cdot 0} = \overline{0}$$
$$\therefore \quad A + 1 = 1$$

The various theorems for simplifying logic expressions contained in the following rule can be derived from the properties listed in Rule 34-2 in a manner similar to that used in the preceding two examples.

RULE 34-3. SOME THEOREMS FOR SIMPLIFYING LOGIC EXPRESSIONS

1. $A \cdot 0 = 0$
2. $A + 1 = 1$
3. $A \cdot 1 = A$
4. $A + 0 = A$
5. $AA = A$
6. $A + A = A$
7. $A\overline{A} = 0$
8. $A + \overline{A} = 1$
9. $(A + B)A = A$
10. $AB + A = A$

In Rule 34-3, notice that theorem 1 is the dual of theorem 2; that is, by applying DeMorgan's theorem to theorem 1 we can derive theorem 2. In the same way, theorem 3 is the dual of theorem 4.

**EXAMPLE
34-22**

Show that $A(B + \overline{B}) = A$.

Solution   Use Rule 34-3.8:

$$B + \overline{B} = 1$$
$$A(B + \overline{B}) = A \cdot 1$$

Apply Rule 34-3.3

$$A \cdot 1 = A$$
$$\therefore \quad A(B + \overline{B}) = A$$

EXERCISE 34-8

Use Rules 34-1 and 34-2 to derive the following relations.

1. $A \cdot 1 = A$
2. $A + 0 = A$
3. $AA = A$
4. $A + A = A$
5. $A\overline{A} = 0$
6. $A + \overline{A} = 1$
7. $(A + B)A = A$
8. $AB + A = A$
9. $A + (B\overline{B}) = A$

Use the previous results to simplify the following expressions.

10. $A(1 + C) + B$
11. $\overline{X}\overline{Y} + \overline{X}Y + XY$
12. $(UV + VW)(UW + VW)$
13. $D(DE + E)$
14. $JKL + JK + L$

34-9 APPLICATIONS

The goal of this section is to show how the concepts of Boolean algebra can be used in practical situations. To achieve this goal, we need the additional concept of *do-not-care* for logic circuit specification. In some applications, the output function for certain input combinations will not be specified because the output will not be used or because the input will not occur. The concept of *do-not-care* will be explored in the following example.

**EXAMPLE
34-23**

You have acquired a tuner amplifier, a tape deck, and a turntable. You want to integrate them into a system.

TABLE 34-2

OPERATION SPECIFICATION FOR STEREO SYSTEM				
Mode	*Tuner Amplifier Selector Switch*	*Turntable*	*Tape Deck*	*Auxiliary Input*
Radio	AM or FM	Off	Off	Do-not-care
Tape from radio	AM or FM	Off	On	Do-not-care
Play a record	Auxiliary	On	Off	Turntable
Tape a record	Auxiliary	On	On	Turntable
Play a tape	Auxiliary	Off	On	Tape deck

Unfortunately, the amplifier has only one pair of auxiliary input jacks. The problem is to design a switching circuit to automatically connect the turntable or the tape deck to the amplifier.

Solution   Begin by specifying the desired system operation as in Table 34-2.

Let A represent the selection of auxiliary input.

Let T represent that the turntable is on.

Let D represent that the tape deck is on.

Let f_o represent the connecting switch in Figure 34-42(a); $f_o = 1$ when the turntable is connected.

Create a truth table for the operation of the connecting switch in terms of A, T, and D. Enter a d for the output

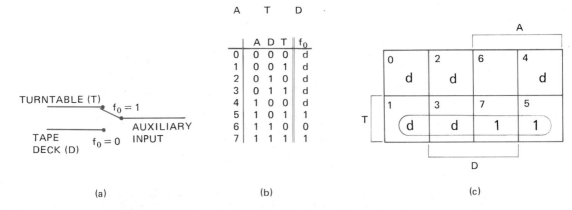

(a)   (b)   (c)

FIGURE 34–42
(a) Selection switch, (b) truth table, and (c) K-map for Example 34–23.

CHAPTER 34   *MATHEMATICS OF COMPUTER LOGIC*

function for each combination for which you *do-not-care* how the auxiliary input is configured. See Figure 34-42(b).

To map this system specification, create a three-variable K-map. Place a 1 in each square for $f_o = 1$. Place a *d* in each square for $f_o = $ *do-not-care*. See Figure 34-42(c). Read the map by grouping all the 1s and as many *d*'s as convenient to make a maximal group (a set of two, four, or eight squares). This will give a minimal expression for f_o. Thus

$$f_o = T$$

Therefore, use a 117-V ac relay with a contact configuration as in Figure 34-43 to perform the switching function. When the turntable is off, the relay is off, and the tape deck is connected to the auxiliary input.

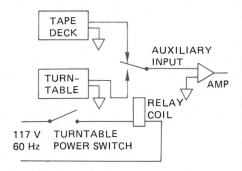

FIGURE 34–43
Relay contact configuration for Example 34–23.

EXERCISE 34-9

1. A certain brand of microwave oven requires that the main power switch be on, the door be latched, and the timer set to a time other than zero before the oven will operate. Draw a logic diagram for the operation of the oven.

2. There are two common mistakes made when a driver gets out of a car: (1) leaving the keys in the ignition and (2) leaving the lights on. Draw the logic diagram for a warning circuit to sound a buzzer when either mistake is made.

3. A novel cigaret lighter is being sold that uses a light beam to light the lighter. The lighter has a cutout on its side that has been fitted with a light source and a light sensor. Opening the lid of the lighter activates

the light source; interrupting the light source causes the lighter to light. Draw a logic diagram for the operation of the lighter.

4. An automatic parking lot gate is shown in Figure 34-44. The ticket vendor is to extend a ticket when a car drives over sensor A and the driver presses button B. Draw a logic diagram to control the extending of a ticket.

5. Once the extended ticket ($C = 1$) of problem 4 is taken ($C = 0$) from dispenser C, then the arm is raised ($E = 1$) to let the car through. Draw a logic diagram to control the raising of the arm in Figure 34-44.

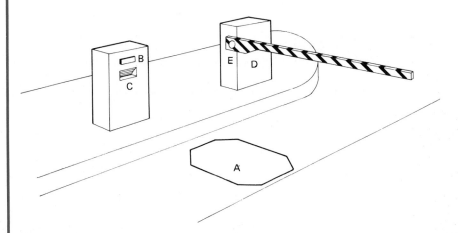

FIGURE 34–44
Entrance to a controlled parking lot. Sensor _A_ detects the presence of a car. Pushing button _B_ extends the ticket. Sensor _C_ detects that the ticket is taken. Sensor _D_ detects the presence of the car in the gate area. Sensor _E_ detects gate position.

6. The raised arm of problem 5 is lowered ($E = 0$) without hitting the car. Draw a logic diagram to control the lowering of the arm. The lowering of the arm causes the system to be reset for the next car.

SELECTED TERMS

Boolean algebra The formal concepts and theorems used to calculate with two-valued variables and functions.

Karnaugh map A graphical presentation of a truth table used to simplify logic circuits.

truth table An exhaustive technique used to list the input and output states for a Boolean or logic expression.

APPENDIX A

REFERENCE
TABLES

SYMBOLS

CONSTANTS

GREEK ALPHABET

SELECTED ABBREVIATIONS

AMERICAN WIRE GAUGE FOR SOLID, ANNEALED COPPER CONDUCTORS AT 20°C

SYMBOLS

$\times$	Multiply by, AND	$\geq$	Greater than or equal to
$\div$	Divide by	$\leq$	Less than or equal to
$+$	Positive, add, OR	$\therefore$	Therefore
$-$	Negative, subtract	$\angle$	Angle
$\pm$	Plus or minus	$\lvert a \rvert$	Absolute value of a
$=$	Equals	Δ	Interval
$\equiv$	Identically equal to	$\%$	Percent
$\approx$	Appoximately equal to	$\sqrt{}$	Radical sign
$\neq$	Not equal	$\llcorner$	Right angle
$>$	Greater than	∞	Undefined, infinity
$\gg$	Much greater than	$\Rightarrow$	Yields
$<$	Less than	$\rho \underline{/\theta}$	Polar coordinates
$\ll$	Much less than	(x, y)	Rectangular coordinates

CONSTANTS

e	2.718 281 828	$\pi/2$	1.570 796 327
π	3.141 592 654	$\sqrt{2}/2$	0.707 106 781
$\ln(10)$	2.302 585 093	j	$\sqrt{-1}$

GREEK ALPHABET

Name	Capital	Lowercase	Name	Capital	Lowercase
Alpha	A	α	Nu	N	ν
Beta	B	β	Xi	Ξ	ξ
Gamma	Γ	γ	Omicron	O	o
Delta	Δ	δ	Pi	Π	π
Epsilon	E	ϵ	Rho	P	ρ
Zeta	Z	ζ	Sigma	Σ	σ
Eta	H	η	Tau	T	τ
Theta	Θ	θ	Upsilon	Υ	υ
Iota	I	ι	Phi	Φ	ϕ
Kappa	K	κ	Chi	X	χ
Lambda	Λ	λ	Psi	Ψ	ψ
Mu	M	μ	Omega	Ω	ω

SELECTED ABBREVIATIONS

alternating current	ac	kilowatt-hour	kWh
ampere	A	least significant digit	LSD
antilogarithm (common)	antilog	limit	lim
antilogarithm (natural)	antiln	logarithm (common)	log
audio frequency	AF	logarithm (natural)	ln
centimeter	cm	maximum	mx
clockwise	cw	mega (1×10^6)	M
cosecant	csc	meter	m
cosine	cos	micro (1×10^{-6})	μ
cotangent	cot	mil (0.001 in.)	mil
coulomb	C	milli (1×10^{-3})	m
counterclockwise	ccw	minute	min
cycles per second (hertz is preferred unit)	c/s	most significant digit	MSD
		nano (1×10^{-9})	n
decibel	dB	ohm	Ω
degrees Celsius	°C	peak-to-peak	p p
degrees Fahrenheit	°F	pico (1×10^{-12})	p
direct current	dc	radian	rad
electromotive force	emf	radio frequency	RF
farad	F	revolution	rev
foot, feet	ft	revolutions per minute	rev/min
giga (1×10^9)	G	root mean square	rms
gram	g	secant	sec
henry	H	second	s
hertz	Hz	siemens	S
horsepower	hp	sine	sin
hour	h	tangent	tan
inch	in	volt	V
International System of Units	SI	voltampere	VA
		watt	W
kelvin	K	watthour	Wh
kilo (1×10^3)	k		

AMERICAN WIRE GAUGE (AWG) FOR SOLID ANNEALED COPPER CONDUCTORS AT 20°C

Gauge No.	SI Metric Units		English Units	
	Dia. (mm)	Ω/km	Dia. (mils)*	Ω/1000 ft
0000	11.68	0.1608	460.0	0.0490
000	10.40	0.2028	409.6	0.0618
00	9.266	0.2557	364.8	0.0779
0	8.252	0.3224	324.9	0.0983
1	7.348	0.4066	289.3	0.1239
2	6.543	0.5127	257.6	0.1563
3	5.827	0.6465	229.4	0.1970
4	5.189	0.8152	204.3	0.2485
5	4.620	1.028	181.9	0.3133
6	4.115	1.296	162.0	0.3951
7	3.665	1.634	144.3	0.4982
8	3.264	2.061	128.5	0.6282
9	2.906	2.599	114.4	0.7921
10	2.588	3.277	101.9	0.9989
11	2.305	4.132	90.74	1.260
12	2.053	5.211	80.81	1.588
13	1.828	6.571	71.96	2.003
14	1.628	8.285	64.08	2.525
15	1.450	10.45	57.07	3.184
16	1.291	13.17	50.82	4.016
17	1.150	16.61	45.26	5.064
18	1.024	20.95	40.30	6.385
19	0.912	26.42	35.89	8.051
20	0.812	33.31	31.96	10.15

*1 mil = 0.001 in.

	SI Metric Units		English Units	
Gauge No.	Dia. (mm)	Ω/km	Dia. (mils)*	Ω/1000 ft
21	0.723	42.00	28.45	12.80
22	0.644	52.96	25.35	16.14
23	0.573	66.79	22.57	20.36
24	0.511	84.21	20.10	25.67
25	0.455	106.2	17.90	32.37
26	0.405	133.9	15.94	40.81
27	0.361	168.9	14.20	51.47
28	0.321	212.9	12.64	64.90
29	0.286	268.5	11.26	81.83
30	0.255	338.6	10.03	103.2
31	0.227	426.9	8.928	130.1
32	0.202	538.3	7.950	164.1
33	0.180	678.8	7.080	206.9
34	0.160	856.0	6.305	260.9
35	0.143	1079	5.615	329.0
36	0.127	1361	5.000	414.8
37	0.113	1716	4.453	523.1
38	0.101	2164	3.965	659.6
39	0.090	2729	3.531	831.8
40	0.080	3441	3.145	1049

*1 mil = 0.001 in.

ANSWERS TO SELECTED PROBLEMS

CHAPTER 2

EXERCISE 2-1

1. -7 3. $+20$ 5. -8

Test Score	55	65	90	40	80	50	75	95	30	65	100	45
Difference	-10	0	25	-25	15	-15	10	30	-35	0	35	-20

EXERCISE 2-2

1. (a) $10 + 5$ $10 - 5$ 10×5 $10 \div 5$
   (b) $16 + 8$ $16 - 8$ 16×8 $16 \div 8$
   (c) $20 + 4$ $20 - 4$ 20×4 $20 \div 4$
   (d) $60 + 12$ $60 - 12$ 60×12 $60 \div 12$

3. (a) $10 + 5 = 15$ $10 - 5 = 5$ $10 \times 5 = 50$ $10 \div 5 = 2$
   (b) $16 + 8 = 24$ $16 - 8 = 8$ $16 \times 8 = 128$ $16 \div 8 = 2$
   (c) $20 + 4 = 24$ $20 - 4 = 16$ $20 \times 4 = 80$ $20 \div 4 = 5$
   (d) $60 + 12 = 72$ $60 - 12 = 48$ $60 \times 12 = 720$ $60 \div 12 = 5$

5. $5 - 3$
7. $6 \div 3 = 2$

EXERCISE 2-3

1. true   3. false, 13   5. true
7. true   9. false, -6

Calculator Drill

11. 30   12. -12 13. 11
14. -3 15. 9   16. 6
17. -24 18. -12 19. 14
20. 29   21. -11 22. 5
23. -11 24. 23   25. 1
26. -10 27. 29   28. 5
29. 10   30. 0

EXERCISE 2-4

1. $(6 + 3)$ or (2×2)

Calculator Drill

3. 1	4. 10	5. 2
6. 5	7. 4	8. 14
9. 13	10. 5	11. 77
12. 22	13. 6	14. 17
15. 194	16. 7	17. 2
18. 9	19. 3	20. 7
21. 6	22. 1	

EXERCISE 2-5

1. $3 - 2$	3. $-8 + 2$	5. $2 - 3 - 7$
7. $1 + 2 - 6$	9. $7 + 1 - 4$	

EXERCISE 2-6

1. 3	3. 9	5. 8
7. 7	9. 1	

EXERCISE 2-7

1. 10	3. 5	5. 12
7. -23	9. -52	11. -4
13. -16	15. -13	17. -19
19. $21 - 16 = 5$	21. $18 - 26 = -8$	23. $-12 + 16 = 4$

EXERCISE 2-8

Calculator Drill

1. -20	2. 30	3. 16
4. -4	5. -19	6. 4
7. -36	8. 5	9. -4
10. -13	11. 8	12. 100
13. 48	14. 11	15. -46

EXERCISE 2-9

1. $7 < 10$	3. $23 > 13$	5. $3 = -10 + 13$
7. $2 + 7 > 3$	9. $>$	11. $>$
13. $<$	15. $>$	17. $>$

1. true 3. false 5. true
7. false 9. false

EXERCISE 2-11

1. -6 3. 10 5. -35
7. 42 9. -4 11. -90

Calculator Drill

13. -1240 14. -420 15. 192
16. -918 17. -9384 18. 936
19. 234 20. -2444 21. 1961
22. 1225 23. -1222 24. -518

EXERCISE 2-12

1. 5 3. -3 5. -2
7. 3 9. -4 11. 5

Calculator Drill

13. -3 14. -9 15. 2
16. -4 17. -3 18. -5
19. 3 20. 1 21. -5
22. 5 23. 1 24. 2
25. -2 26. -1 27. 2
28. 2 29. 0 30. 9

EXERCISE 2-13

9. 3^3 11. 8^4 13. 5^3
15. 9^4 17. 7^7 19. 14^2

EXERCISE 2-14

Calculator Drill

1. 32 2. 64 3. 729
4. 125 5. 4096 6. 343
7. 7776 8. 243 9. 1000
10. 65 536 11. 128 12. 59 049

CHAPTER 3

EXERCISE 3-1

1. 75.4×10^7
3. 3.72×10^{-3}
5. 4.01×10^2
7. 5531×10^{-4}
9. 2.49×10^{-3}
11. 82.92×10^2
13. 28.6514×10^4
15. 1788.2×10^{-3}

EXERCISE 3-2

1. 3×10^2
3. 2.8×10^2
5. 5×10^{-1}
7. 3.6×10^{-1}
9. 7.052×10^3
11. 6.3×10^{-3}
13. 1×10^{-3}
15. 5.78×10^5
17. 5.55×10^3
19. 6.8×10^{-1}
21. 1.313×10^{-2}
23. -2.823×10^2
25. -8.25×10^{-3}
27. -4.78×10^1

EXERCISE 3-3

1. 0.0635
3. $0.000\ 382\ 8$
5. $0.009\ 935$
7. 4.62
9. 88.31
11. $-183\ 000$
13. 0.2882
15. $-0.000\ 715$

EXERCISE 3-4

1. 2.73×10^{-4}
3. 6.04×10^{-3}
5. 1.9×10^5
7. 1.313×10^5
9. 4.26
11. 2.07×10^4
13. 7.89×10^{-4}
15. -4.93×10^{-2}

EXERCISE 3-5

1. 7.57×10^2
3. 7.20×10^1
5. 1.41×10^1
7. 1.07×10^{-1}
9. 1.86×10^2
11. 5.93×10^{-2}
13. 8.78×10^{-4}
15. 1.00×10^5

EXERCISE 3-6

1. 10^5
3. 10^{-7}
5. 10^0
7. 10^{-18}
9. 10^6

Calculator Drill

11. 3.53×10^5
12. 5.34×10^4
13. 2.16×10^{-4}
14. 4.19×10^5
15. 1.08×10^5
16. 1.33×10^2
17. 1.98×10^{-3}
13. 1.87×10^{-1}
19. 1.99×10^6

20. 1.11×10^1 21. 8.76×10^5 22. 3.25
23. 2.11×10^1 24. 1.40×10^{-1} 25. 4.32×10^{-3}
26. 3.59×10^{-5} 27. 1.05×10^5 28. 1.01×10^{-4}
29. 3.57×10^{-9} 30. 2.79×10^2

EXERCISE 3-7

1. 10^1 3. 10^1 5. 10^5
7. 5×10^1 9. 3×10^9 11. 2×10^{-7}

Calculator Drill

13. 3.77×10^3 14. 3.88×10^{-5} 15. 8.62×10^{-6}
16. 2.43×10^4 17. 3.85×10^{-8} 18. 1.35
19. 2.38×10^{-2} 20. 2.49×10^{-3} 21. 4.32×10^8

EXERCISE 3-8

1. 10^6 3. 10^8 5. 10^{18}
7. 10^{-14} 9. 10^{-12} 11. 10^{-10}

EXERCISE 3-9

1. 4×10^6 3. 25×10^{-8} 5. 8×10^{12}
7. 64×10^{-12} 9. 81×10^2 11. 49×10^{-10}

EXERCISE 3-10

1. 10^{-4} 3. 10^{-15} 5. 10^8
7. 1 9. 10^{-18} 11. 10^3

Calculator Drill

13. 4.00×10^{18} 14. 8.10×10^{-11} 15. 5.34×10^{11}
16. 2.51×10^7 17. 4.80×10^7 18. 7.31×10^{12}
19. 8.00×10^{15} 20. 2.70×10^{-2} 21. 2.99×10^{-3}
22. 2.03×10^{-13}

EXERCISE 3-11

Calculator Drill

1. 19.29 2. $0.093\ 27$ 3. ± 0.8860
4. 9.996 5. 31.94 6. -8.119

7. 0.093 94   8. ±0.8535   9. −1387
10. 1.448   11. −7.662   12. ±0.6399

EXERCISE 3-12

Calculator Drill

1. 8.00×10^0 2. 5.00×10^0 3. 1.60×10^1
4. 4.00×10^0 5. 4.44×10^0 6. 4.38×10^1
7. 1.58×10^1 8. 7.58×10^0 9. 1.71×10^2
10. -1.36×10^0 11. 1.81×10^2 12. 4.00×10^0

EXERCISE 3-13

Calculator Drill

1. 1.56×10^1 2. 1.08×10^{-1} 3. 7.24×10^0
4. 3.71×10^1 5. 3.15×10^0 6. 8.20×10^3
7. 1.20×10^{-1} 8. 2.70×10^2 9. 4.51×10^3
10. 2.78×10^1 11. 4.57×10^1 12. 6.62×10^1

EXERCISE 3-14

Calculator Drill

1. 2.00×10^{-1} 2. 2.50×10^{-1} 3. 1.67×10^{-1}
4. 1.11×10^{-1} 5. 3.33×10^{-1} 6. 1.08×10^{-2}
7. 1.20×10^1 8. 1.29×10^{-3} 9. 1.99×10^{-3}
10. 7.25×10^2 11. 1.59×10^{-4} 12. -1.09×10^{-3}
13. 2.44×10^1 14. 5.38×10^2 15. 1.41×10^{-4}
16. -2.11×10^{-2}

EXERCISE 3-15

Calculator Drill

1. 3.000   3. 0.2000   5. 16.00
7. 14.18   9. 211.7   11. 1045

EXERCISE 3-16

1. -6.00×10^0 3. 6.24×10^0 5. 1.56×10^1
7. 7.67×10^0 9. -1.38×10^0 11. 4.08×10^0

13. 4.50×10^1 15. 2.03×10^5 17. 1.00×10^0
19. 8.43×10^1

EXERCISE 3-17

1. 3×10^{-1} 3. 4×10^{-1} 5. 20×10^3
7. 70×10^{-3} 9. 6   11. 60
13. 4×10^{-3} 15. 30   17. 200
19. 18

EXERCISE 3-18

1. 1.32×10^3 3. 85.0×10^{-6} 5. 4.70×10^{-3}
7. 8.00×10^{-3} 9. 0.572×10^{-3} or 572×10^{-6}
11. 284   13. 420 000   15. 570 000
17. 94 500 000

EXERCISE 3-19

1. 6.28E5   3. 1.92E4   5. 8.74E6
7. 9.28E6   9. 1.99E–3   11. 9.03E2
13. 9.2   15. 0.000 381   17. 0.2038
19. 448.2   21. 1776   23. 8034

CHAPTER 4

EXERCISE 4-3

1. rational polynomial   3. rational fractional expression
5. irrational   7. rational polynomial
9. irrational   11. irrational
13. rational polynomial   15. rational fractional expression

EXERCISE 4-4

1. 3   3. 1   5. 1
7. 2   9. $\frac{1}{3}$ 11. $\frac{7}{4}$
13. 2   15. r^2 17. $5b$
19. $16d^2$ 21. 7.4ω

EXERCISE 4-5

1. $5a$ 　　3. $12x$ 　　5. $-7c$
7. $-2b$ 　　9. $5h$ 　　11. $10A + 2B$
13. $-3c^2b$ 　　15. $3x^2 + 3x + 2$ 　　17. $-2x^3 + x^2 - 2x$
19. $-5x^2y$

EXERCISE 4-6

1. a 　　3. b 　　5. d
7. c 　　9. b 　　11. d

EXERCISE 4-7

11. $6y + 9$ 　　13. $2x^2$
15. $3x + 2y + 4z$ 　　17. $V^3 + 3V^2 + 5$
19. $-x - y$ 　　21. $3n + 8$
23. $4x^3 - x^2 - 4x$ 　　25. $7m + n$
27. $2x^2 - 5x$ 　　29. $8\omega + 2.6\rho$
31. $3x^3 - 4x^2 - 12x + 8$ 　　33. $-5a^3 + 2a + 5b^2$
35. 0 　　37. $-2\alpha + 15\beta - 9$
39. $-11x + 4y + 3z$

Calculator Drill

41. $5.45x^2 + 0.76x + 18.13$ 　　42. $35.2x^2 - 1.50x - 27.7$
43. $29.8x^2 - 15.2x + 12.3$ 　　44. $40.1x^2 - 19.7x - 28.2$

CHAPTER 5

EXERCISE 5-1

1. x^9 　　3. 1 　　5. b^2
7. $14y$ 　　9. $-35ab$ 　　11. 10^{-2}
13. $-x^4y^3$ 　　15. $16\eta\theta\phi$ 　　17. 16
19. $-8y^{-2}x^2$ 　　21. $6x^2$ 　　23. $-8y^3$

EXERCISE 5-2

1. 27 　　3. $2y - 6$ 　　5. $3\mu - 15$
7. $-3Z + Z^2$ 　　9. $\beta c - 4\beta$ 　　11. $-3x - 3y$

13. $-2a^2 + 4ab$ 15. $-14\theta\omega - 35\theta\phi$ 17. $-y^7 - 4y^6$

19. $3cd^3 + 3c^3d$ 21. $3s^3 + 12s^2$

EXERCISE 5-3

1. $12x + 8y + 20$ 3. $21u - 14v + 7w$

5. $-40x + 5y - 15$ 7. $6\mu - 12\theta + 18$

9. $-2x^2 - 4x + xy$ 11. $12hj - 15h\theta + 3h$

13. $-6a^2b + 9ab^2 - 12abc$ 15. $6x^4 - 21x^3 - 18x^2$

17. $-2a^3b + 2ab^3 - 2abc + 6ab$ 19. $-a^2 - 2b + 3c - 5$

Calculator Drill

21. $7.55y^3 - 24.8xy + 15.7$ 22. $-68.1b^2 + 45.4bc - 105bd$

23. $-125\mu^3 + 225\mu^2 - 306\mu$ 24. $-16.5a^3 - 62.9a^2b^2 + 32.3a^2$

EXERCISE 5-4

1. $a - 8$ 3. $-a + 9c$

5. $-12x^2 - 6x + 4$ 7. 2

9. $3\theta - \beta$ 11. $-2y$

13. $2x + y$ 15. $\mu + 2\pi$

17. $2V^2 - U$ 19. $x^4 - 4x^2 + x - 2$

21. $-5y^2 + 4y + 2$

EXERCISE 5-5

1. $x^3 - 2x^2 - 4$ 3. $-11y - 4y^2$

5. $\theta - 3\eta + 2$ 7. $-14x^2 + 14x + 3$

9. $-15c^2 - 9c - 6ac$ 11. $6\mu^2 - 7\mu + 7$

13. $-2\theta - 2$ 15. $3x^3 + 6x^2 - 18x - 12$

EXERCISE 5-6

1. a^4 3. c^2 5. μ^7

7. 3 9. $-3y^2z^5$ 11. $2c^{-2}$

13. $-8a^2b^4$ 15. $4c^2$

EXERCISE 5-7

1. $3x + 4$ 3. $7 + 4y$ 5. $3U + V$

7. $2b^3 + 4a^3$ 9. $6x + 3y$ 11. $a^3 + a$

13. $13c + 9 + 11c^2$ 15. $x + 2xy + 3y^2$ 17. $6b + 3$
19. $7\mu^2 + 4 + 5\mu^{-2}$

1. $6(a + 2)$ 3. $3(2 + 3b)$
5. $3(m + 1)$ 7. $\omega(u + v)$
9. $7(\phi + \theta)$ 11. $\rho(\rho + 5)$
13. $7y(x + 3)$ 15. $8a^2(a + 2)$
17. $a(x + 3)$ 19. $a^2b^3c(1 + a^2b^2c^2)$
21. $5(\alpha^2 + 3\alpha + 4)$ 23. $ab(1 + ab + a^2b^2)$
25. $8\beta^2(2\pi^2 + \pi + 4)$ 27. $5\eta\alpha(6\eta + 3 - 5\alpha)$
29. $5y^2z(yz + 3 + 2y^2)$ 31. $3y^3x(7y + 6x + 9)$

Calculator Drill

1. 12.0   2. 2.00   3. 5.00
4. 7.00   5. 80.0   6. 54.0
7. −2.00   8. 47.0   9. 6.00
10. 13.0   11. −1.00   12. −3.00
13. 1.00   14. 10.0   15. 20.0
16. 20.0   17. 2.00   18. 13.0
19. −0.284   20. 5.79   21. 7.75
22. −17.5   23. 584   24. 11.8
25. 34.4

CHAPTER 6

1. true   3. false   5. true
7. true   9. true   11. true
13. true   15. false   17. true
19. false

1. $x = 1$ 3. $m = -8$ 5. $x = -1$
7. $\alpha = 9$ 9. $\phi = 4$

1. $x = 2$
3. $x = 10$
5. $y = 8$
7. $\theta = 5$
9. $y = 3$
11. $n = -1$
13. $C = 4$
15. $\tau = -17$
17. $m = -13$
19. $x = 0$
21. $t = 20$

EXERCISE 6-4

1. $y = 4$
3. $\mu = 6$
5. $x = 15$
7. $y = 60$
9. $k = -6$
11. $P = -7$
13. $y = 96$
15. $\beta = -4$
17. $x = -63$
19. $\phi = 14$
21. $r = -2$
23. $y = 52$
25. $x = 4$
27. $k = -4$
29. $y = -3$

EXERCISE 6-5

1. $x = 2$
3. $x = 3$
5. $R = \frac{1}{2}$
7. $\phi = 3$
9. $m = 1$

EXERCISE 6-6

1. $y = 4$
3. $\phi = 1$
5. $R = -3$
7. $E = -12$
9. $\mu = -5$
11. $t = 4$
13. $u = 3$
15. $\eta = 0$

EXERCISE 6-7

1. $\Delta = 2$
3. $x = 2$
5. $y = -15$
7. $\pi = -1$
9. $\theta = 2$
11. $\eta = 5$
13. $y = 2$
15. $z = -\frac{10}{9}$

EXERCISE 6-8

1. $i = 11$
3. $\beta = -1$
5. $x = 3$
7. $R_1 = 5$
9. $W = 12$
11. $L = 18$
13. $R_1 = -8$
15. $\theta = 4$
17. $N = 0$
19. $V_t = 2$
21. $\alpha = -3$
23. $X_c = \frac{1}{3}$

EXERCISE 6-9

1. $\lambda = v/f$
3. $R = E/I$
5. $g_m = \mu/r_p$
7. $f_0 = BW/Q_0$

9. $E_1 = E_{BB} - I_B E_P$

11. $E_L = E_N - R_O I_L$

13. $R = \dfrac{t}{-xC}$

15. $I_2 = \dfrac{I_1 G}{R_1 + R_2}$

17. $\theta_{JA} = \dfrac{T_J - T_A}{P_D}$

19. $E_1 = E_T - I_1 R$

21. $Y_2 = \dfrac{V_1 Y^2}{V_1 Y_1 - i}$

23. $I_T = E_{th}/(R_{th} + R_L)$

25. $R_L = \dfrac{-A_v r_p}{A_v - \mu} = \dfrac{A_v r_p}{\mu - A_v}$

EXERCISE 6-10

1. $8x = 40, x = 5$ 3. $6x - 5 = 25, x = 5$ 5. $16 - x = 9, x = 7$
7. $2(2 + x) = 18, x = 7$ 9. $6x - 2x = 52, x = 13$

EXERCISE 6-11

1. $x = 5$ 3. $x = -1$ 5. $x = -3$
7. 4336   9. $C = 35¢$ 11. $V = 13$ min
13. resistor = 3¢,   15. diode = 170°C,
      capacitor = 15¢   zener = 150°C

CHAPTER 7

EXERCISE 7-1

1. f and l   3. k   5. j
7. h   9. a   11. d

EXERCISE 7-2

1. 0.282 kΩ   3. 176 μH   5. 1.9 MW
7. 1.52 kHz   9. 0.18 MΩ   11. 6.28 m
13. 0.017 J   15. 0.000 902 S   17. 854 W
19. 0.000 047 F   21. 7.83 kV   23. 65.4 mH
25. 152 mS   27. 681 kΩ   29. 20.4 mA
31. 185 μA   33. 1.53 MΩ   35. 1.11 V

EXERCISE 7-3

1. 111 mA
7. 6.80 kΩ
13. 70.6 N

3. 159 Ω
9. 377 rad/s

5. 599°C
11. 806 m

EXERCISE 7-4

1. 17.2×10^3 m
7. 103×10^2 J
13. 1 yd = 0.914 m

3. 20 ms
9. 2.66 kJ

5. 238 cm²
11. 331×10^2 cm/s

CHAPTER 8

EXERCISE 8-1

1. 27.8 mA
7. 9.00 V

3. 150 C
9. 29.3 Ω

5. 12.0 V

EXERCISE 8-2

1. (a) 1.0 kV
   (b) 20 V
   (c) 28.5 V
   (d) 234 V
7. 2.5 Ω

3. 600 Ω

9. 390 Ω

5. 8.0 A

EXERCISE 8-3

1. 23.5 kΩ

3. 2.75 kΩ

5. (a) 273 kΩ
   (b) 806 μA
   (c) *See* Figure A1

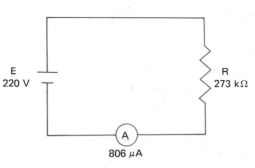

FIGURE A1
Solution to Exercise 8–3, problem 5c.

1. 42 V   3. 37 V   5. $120 = 16 + 40 + 64$

EXERCISE 8-5

1. (a) 14.5 kΩ   3. (a) 2.8 kΩ   5. (a) 20 mA
   (b) 10 mA   (b) 40 mA   (b) 68 V
   (c) 10 mA   (c) 28 V   (c) 1.1 kΩ
   (d) 30 V   (d) 4.5 kΩ

7. (a) 455 μA   9. $R_1 = 25$ kΩ, $R_2 = 25$ kΩ
   (b) 22.3 V $R_4 = 10$ kΩ

CHAPTER 9

EXERCISE 9-1

1. 2.11 W   3. 1.68 MJ   5. 56.2 MJ
7. 160 W   9. 100 V   11. (a) 11 kW
                                          (b) No

EXERCISE 9-2

1. (a) 433×10^2 W   3. 1.5 A
   (b) 43.3 kW
5. 47.1 W   7. 1190 W
9. 0.78   11. (a) 0.809, 0.677
                                       (b) 158 kJ, 320 kJ

EXERCISE 9-3

1. 2.5 kWh   3. $1.09   5. $7.03
7. (a) $50, $30   9. (a) 9.56 kW
   (b) $20   (b) $40.30

CHAPTER 10

EXERCISE 10-1

1. 0   3. −1   5. $\frac{1}{2}$
7. 5   9. 5

1. $2/10 = 12/60$

3. $-3/-39 = 4/52$

5. $-33/77 = 9/-21$

7. $5m/5n = -8m/-8n$

9. $st/t^2 = -2st^2/-2t^3$

11. $-12xy/30x^2 = -6y^3/15xy^2$

13. $-2b^2/(2b^2 - 2ab)$

EXERCISE 10-3

1. $\frac{1}{2}$

3. $\frac{2}{5}$

5. $3b/2$

7. a

9. $2dc/11$

11. x^5

13. $3a^2b^3/2$

15. $(a - 1)$

17. $t/4$

19. $1/(x + 1)$

21. $\theta\beta^2$

EXERCISE 10-4

1. $\frac{1}{2}$

3. $-\frac{2}{5}$

5. $-\frac{3}{20}$

7. -1

9. -6

11. $2(2x + y)/y$

13. $(b - c)/(c + b)$

15. $20b/3a^3$

17. $\pi r/4$

19. $\frac{1}{3}$

EXERCISE 10-5

1. $\frac{3}{2}$

3. $\frac{3}{4}$

5. $21/20$

7. 4

9. $1/y$

11. x/y

13. $-(a - 2) = 2 - a$

15. $(a - b)$

EXERCISE 10-6

1. $\dfrac{\dfrac{8}{3}}{\dfrac{4}{5}}$

3. $\dfrac{\dfrac{-9}{2}}{\dfrac{3}{16}}$

5. $\dfrac{\dfrac{-6x}{7y}}{\dfrac{3x}{14y}}$

7. $\dfrac{\dfrac{R^2}{-t}}{\dfrac{2tR}{t^2}}$

9. $\dfrac{\dfrac{b}{b - 3}}{\dfrac{b^2}{2}}$

11. 10

13. $\frac{3}{4}$

15. $\frac{1}{2}$

17. $\dfrac{x - y}{x + y}$

19. $a/(a - 5)$

EXERCISE 10-7

1. $\dfrac{-1}{4}$

3. $-\dfrac{4}{5}$

5. $\dfrac{y-x}{3a}$

7. $\dfrac{-3y-4}{2x+1}$

9. $\dfrac{-2\pi R}{E+1}$

11. $\dfrac{5\omega+\beta}{6+\beta}$

EXERCISE 10-8

1. $\dfrac{2}{12},\dfrac{8}{12},\dfrac{5}{12}$

3. $\dfrac{-27}{36},\dfrac{12}{36},\dfrac{-20}{36}$

5. $\dfrac{14x}{42},\dfrac{12b}{42},\dfrac{35}{42}$

7. $\dfrac{ay}{xy},\dfrac{3}{xy},\dfrac{ax^2y}{xy}$

9. $\dfrac{x-b}{a-b},\dfrac{x}{a-b}$

EXERCISE 10-9

1. 2

3. -2

5. $10/a$

7. $2/(x+1)$

9. 2

11. $3/y$

EXERCISE 10-10

1. 6

3. 75

5. $2A$

7. $2\pi R$

9. $12\alpha\beta$

11. $8\alpha^2(\beta+1)$

13. $14(7K-1)$

15. $2s(t+1)$

EXERCISE 10-11

1. $\dfrac{2d+c}{d}$

3. $\dfrac{3x+5}{6x}$

5. $\dfrac{y+x}{xy}$

7. $\dfrac{mn+1}{m}$

9. $\dfrac{17ab}{24}$

11. $\dfrac{my+2nx}{xy}$

13. $2t/3$

15. $\dfrac{20a+9b}{6}$

17. $\dfrac{-3x^2+9x-1}{3x}$

19. $\dfrac{6a^2-5ab+6b^2}{12ab}$

21. $-3/(x - 2)$

23. $\dfrac{-11b - 8}{18b(b - 1)}$

25. $\dfrac{-6m^2 - 12m + 60}{m(m + 5)(m - 5)}$

EXERCISE 10-12

1. $\frac{23}{4}$

3. $\frac{7}{3}$

5. $\dfrac{3m - 2}{m}$

7. $\dfrac{9a + 2}{3}$

9. $\dfrac{8x^2 - 7}{2x}$

11. $\dfrac{\omega CL + 3L - z}{L}$

13. $\dfrac{x^2 + 3x + 2}{x + 3}$

15. $\dfrac{x^2 + 10x + 9}{x + 2}$

EXERCISE 10-13

1. $\frac{2}{3}$

3. $\dfrac{ab - 1}{ab + 1}$

5. $\dfrac{1}{2a}$

7. $\dfrac{3a - b}{-2b}$

9. $\dfrac{2x - 5}{x - 2}$

11. $-\dfrac{b}{a}$

13. $\dfrac{3b - 1}{b + 8}$

15. $\dfrac{2(x - 1)(x + 3)}{2x - 7}$

CHAPTER 11

EXERCISE 11-1

1. $x = 2$

3. $a = \frac{3}{2}$

5. $R = -\frac{45}{7}$

7. $m = -1$

9. $V = \frac{16}{21}$

11. $I = -2.5$

13. $x = 30$

15. $a = -10$

17. $x = 10$

19. $I = 3$

21. $x = \frac{39}{7}$

EXERCISE 11-2

1. $x = 4$ 3. $y = 1$ 5. $x = 1$

7. $y = 2$ 9. $y = -3$ 11. $x = 7$

13. $x = 2$ 15. $x = 6$ 17. $y = -2$

19. $y = 2$

EXERCISE 11-3

1. $x = a/b$ 3. $Z = 3a$ 5. $y = \dfrac{2mn}{n + 2m}$

7. $W = \dfrac{b}{c\pi^2 k + 1}$ 9. $s = 3/(D - C)$ 11. $y = \dfrac{m - n}{ab + a}$

13. $x = \dfrac{3k(g - h)}{5k - 3g + 3h}$ 15. $u = \dfrac{2mn}{m + n}$ 17. $x = \dfrac{2b - ab}{b - a}$

19. $V = 3g$

EXERCISE 11-4

1. $F = \dfrac{9}{5} C + 32$ 3. $f = \dfrac{1}{2\pi C X_C}$

5. $w = \dfrac{2kg}{v^2}$ 7. $T_A = T_J - \theta_{JA} P_D$

9. $R_2 = \dfrac{E - C R_1}{C}$ 11. $n = \dfrac{216}{108 - \theta}$

13. $S_2 = \dfrac{\eta F_1 S_1}{A_2}$ 15. $d^2 = \dfrac{q_1 q_2}{4\pi\epsilon_0 F}$

17. $R_2 = \dfrac{R_T R_1}{R_1 - R_T}$ 19. $\alpha = \dfrac{SR_E + SR_\beta - R_1 - R_2}{SR_\beta}$

21. $48.7\ \Omega$ 23. $892\ \Omega$

25. $7.77\ \text{kHz}$ 27. 65.7

29. $403\ \text{k}\Omega$

CHAPTER 12

EXERCISE 12-1

1. $\frac{1}{4}$ 3. $\frac{2}{3}$ 5. $\frac{2}{3}$
7. 4.21/1   9. $\frac{3}{400}$ 11. $\frac{23}{19}$

EXERCISE 12-2

1. -7.00% 3. no   5. 800.000 kHz
7. -5.36% 9. 9.9992 V

EXERCISE 12-3

1. 15 ppm   3. 136 ppm   5. 220 ppm
7. 0.1%   9. 0.01%   11. 0.405%
13. $\pm 0.005\%$

EXERCISE 12-4

1. $x = 4$ 3. $x = 3$ 5. $x = 1.86$
7. true   9. false   11. true

EXERCISE 12-5

1. 1.51 Ω 3. 1.68 Ω 5. 468 m

EXERCISE 12-6

1. 0.967 Ω 3. 130 mΩ 5. (a) 1.26 mm
    (b) #16

EXERCISE 12-7

1. 80.7 mΩ 3. 3.16 m   5. 0.0523×10^{-6} Ωm

EXERCISE 12-8

1. 1.785 Ω 3. 59.4°C

CHAPTER 13

EXERCISE 13-1

1. $V_1 = 0.902$ V 3. $V_2 = 8.33$ V 5. $E = 66.4$ V
7. $\Sigma R = 1.39$ MΩ 9. $R_1 = 870$ Ω

EXERCISE 13-2

1. 300 V 3. 29.9 V 5. 30.1 V
7. 25.5 V 9. 185 V

EXERCISE 13-3

1. 17.9 mS 3. 1.10 mS 5. 161 μS
7. 0.370 mS 9. 0.122 μS 11. 1.79 S

EXERCISE 13-4

1. 18 mS 3. 475 mS 5. 42 μS
7. 20 mS 9. 5.18 mS 11. 2.14 mS
13. 11.4 μS 15. 0.699 μS 17. 3.02 mS
19. 32 μS

EXERCISE 13-5

1. 75.0 Ω 3. 10.3 Ω 5. 1.33 kΩ
7. 44.8 kΩ 9. 9.21 Ω

EXERCISE 13-6

1. 3.00 A 3. 6.40 mA 5. 4.72 mA
7. 0.817 A 9. 20.4 mA

EXERCISE 13-7

1. 1.67 mA 3. 1.20 mA 5. 108 mA
7. $I_1 = 3.33$ A, $I_2 = 1.67$ A 9. $I_3 = 1.67$ A, $I_4 = 3.33$ A

1. $I_N = 167$ mA, $R_N = 150\ \Omega$
3. $E_{Th} = 14.1$ V, $R_{Th} = 47\ \Omega$
5. $E_{Th} = 20$ V, $R_{Th} = 10\ \Omega$
7. $I_N = 221$ mA, $R_N = 30.7\ \Omega$
9. $E_{Th} = 63.9$ V, $R_{Th} = 1.42$ kΩ

CHAPTER 14

EXERCISE 14-1

	Constants	Dependent Variable	Independent Variable
1.	k	F	V, m, r
3.	G	F	d, m_1, m_2
5.	R	V	T, M

EXERCISE 14-2

1. 7

3. 17

5. -13

7. $-5b - 3$

9. $x = g(y)$

11. $z = f(v)$

13. $f(2) = 16, f(-3) = 41$
$\quad f(0.5) = 2.5$

15. $Q(1) = -1.5, Q(0.5) = -1.07$
$\quad Q(5) = 0.833$

EXERCISE 14-3

1. (a) doubled
   (b) one-third
   (c) increase 5 times
   (d) inversely
7. (a) doubled
   (b) multiplied by 0.866

3. (a) one-fourth
   (b) doubled
   (c) not changed

5. (a) half
   (b) doubled
   (c) one-fourth

EXERCISE 14-4

1. $r_p \gg R_0$

3. $G_1 \gg G_2$

5. $K \gg 1$

7. $A'R_f \gg (r_0 + R_f + R_L)$

9. $4k^2 \gg 1$

11. $I \gg I'$

CHAPTER 15

EXERCISE 15-1

5. zero

7. *x*-coordinate

9. II

11. *See* Figure A2.

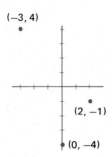

FIGURE A2
Solution to Exercise 15-1, problem 11.

EXERCISE 15-2

1. *See* Figure A3.

3. *See* Figure A4.

5. *See* Figure A5.

7. *See* Figure A6.

9. *See* Figure A7.

11. *See* Figure A8.

13. *See* Figure A9.

15. *See* Figure A10.

17. *See* Figure A11.

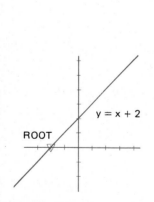

FIGURE A3
Solution to Exercise 15-2, problem 1.

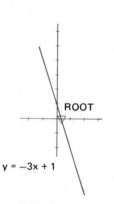

FIGURE A4
Solution to Exercise 15-2, problem 3.

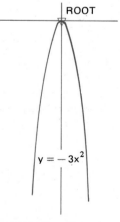

FIGURE A5
Solution to Exercise 15-2, problem 5.

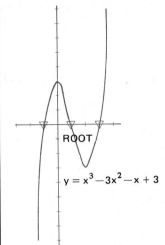

FIGURE A6
Solution to Exercise 15–2, problem 7.

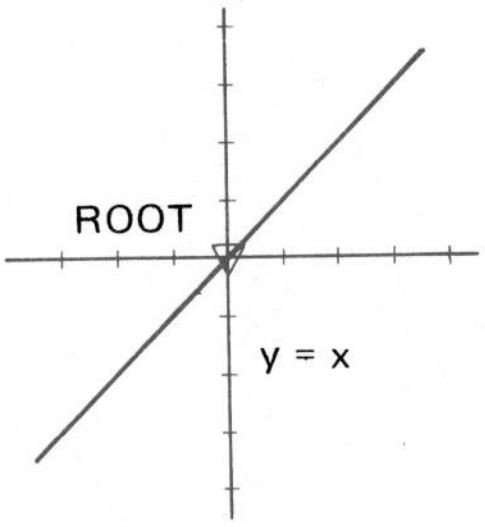

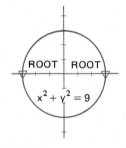

FIGURE A7
Solution to Exercise 15–2, problem 9.

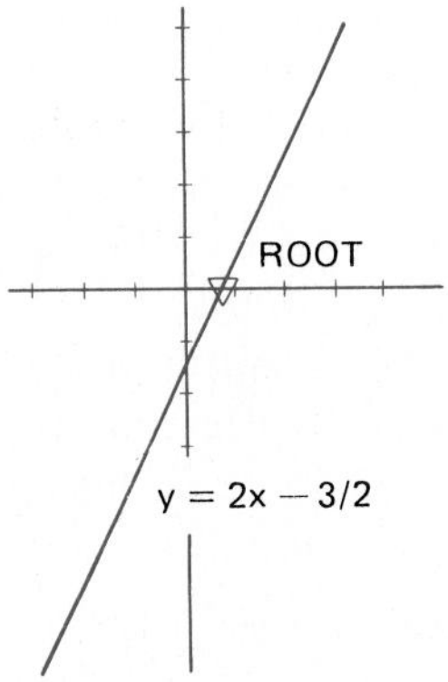

FIGURE A8
Solution to Exercise 15–2, problem 11.

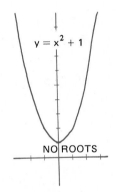

FIGURE A9
Solution to Exercise 15–2, problem 13.

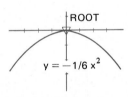

FIGURE A10
Solution to Exercise 15–2, problem 15.

FIGURE A11
Solution to Exercise 15–2, problem 17.

EXERCISE 15-3

1. (a) negative
   (b) -1
   (c) $(0, 5)$

3. (a) negative
   (b) -2
   (c) $(0, -4)$

5. (a) positive
   (b) 4
   (c) $(0, 0)$

7. (a) negative
   (b) $-\frac{2}{3}$
   (c) $(0, 12)$

9. (a) positive
   (b) 1
   (c) $(0, 0)$

11. positive slope, y-intercept at $(0, 3)$, root at $(-3/10, 0)$

13. positive slope, y-intercept at $(0, 0)$, root at $(0, 0)$

15. positive slope, y-intercept at $(0, 0)$, root at $(0, 0)$

17. negative slope, y-intercept at $(0, 6)$, root at $(12, 0)$

19. positive slope, y-intercept at $(0, -2)$, root at $(\frac{2}{5}, 0)$

EXERCISE 15-4

1. $y = -x + 9$

3. $y = -\frac{1}{4}x + 4.5$

5. $y = \frac{2}{5}x - 2$

7. $y = -x + 7$

9. $y = -\frac{6}{11}x + \frac{3}{11}$

11. $y = 5$

EXERCISE 15-5

1. *See* Figure A12.

3. *See* Figure A13.

5. *See* Figure A14.

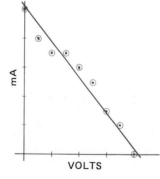

FIGURE A12
Solution to Exercise 15–5, problem 1.

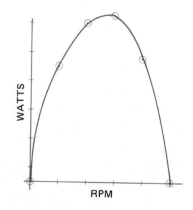

FIGURE A13
Solution to Exercise 15–5, problem 3.

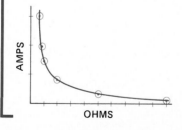

FIGURE A14
Solution to Exercise 15–5, problem 5.

CHAPTER 16

1. intersection at $(1, 1)$
3. intersection at $(-3, -6)$
5. intersection at $(3, 1)$
7. intersection at $(2, 3)$
9. intersection at $(2, -2)$

EXERCISE 16-2

1. $x = 5, y = -1$
3. $x = 2, y = 7$
5. $x = 3, y = 11$
7. $x = 4, y = 8$
9. $x = 0, y = 1$

EXERCISE 16-3

1. $x = 1, y = 1$
3. $x = 4, y = -1$
5. $x = -4, y = \frac{5}{2}$
7. $x = 8, y = 1$
9. $x = 0, y = 1$
11. $x = 1, y = 6$
13. $x = -1, y = 5$

EXERCISE 16-4

1. $E = 1.11 \, E_{av}$
3. $L = X_L/\omega$
5. $w = I^2 L$
7. $Q_0 = X_L/R$
9. $R_L = \dfrac{X_L Q_0' Q_0}{Q_0 - Q_0'}$

EXERCISE 16-5

1. $\Delta = 1 + 2 = 3$
3. $\Delta = 4 - 4 = 0$
5. $x = 3, y = 1$
7. $x = 5, y = 3$
9. $x = 1, y = 8$
11. $x = 3, y = -\frac{2}{3}$
13. $x = 2, y = -2$
15. $x = 10, y = -2$
17. $x = 20, y = 21$
19. $x = 6, y = -10$

EXERCISE 16-6

1. 4
3. 0
5. $x = 1, y = 3, z = 5$
7. $x = 4, y = -1, z = 3$
9. $x = -2, y = 4, z = 1$
11. $I_A = 0.40, I_B = 0.80, I_C = 0.40$

CHAPTER 17

EXERCISE 17-1

1. 0.3 W
3. 0.4 A
5. $-50°C$
7. 9.5 mm
9. 9 mA

EXERCISE 17-2

1. 1 A   3. 125°C/W   5. ≈430 μS

EXERCISE 17-3

1. 18 mA, 1.4 V   3. 75 mA, 2.3 V   5. $I = 0.43$ A, $V_L = 25$ V, $V_R = 35$ V

CHAPTER 18

EXERCISE 18-1

1. $I = 2.22$ mA in a ccw direction

3. $I = 8.89$ mA in a cw direction

5. $I = 0.667$ mA in a ccw direction

7. $I = 1.33$ mA in a cw direction

9. $V_3 = 8$ V

EXERCISE 18-2

1. 1 A down   3. 2 A up   5. 9 A up

7. 6 A down   9. 2 A down

EXERCISE 18-3

1. $I_1 = -5$ A, $I_2 = -5$ A

3. $I_1 = -2.28$ A, $I_2 = -2.76$ A

5. $I_1 = -46.32$ mA, $I_2 = -36.76$ mA

7. $I_1 = 0.7930$ A, $I_2 = -0.9745$ A, $I_3 = -0.9936$ A

9. $I_1 = 2.679$ A, $I_2 = 1.355$ A, $I_3 = 0.9894$ A

11. $I_{R2} = 0.3656$ A to the right

13. $I_{RL} = 25.5$ A; battery is being charged

15. $I_{R3} = 6.14$ A to the right

CHAPTER 19

EXERCISE 19-1

1. 6   3. $15y^2$, 1

5. $x^2 + 2x + 1$ 7. $x^2 + 5x + 6$

9. $x^2 - 8x + 15$

11. $x^2 + 10x + 24$

13. $x^2 - 12x + 20$

15. $x^2 - 3x - 10$

17. $6y^2 - 5y - 21$

19. $90y^2 - 29y + 2$

21. $-6x^2 + 17x - 12$

23. $-18x^2 + 36x - 16$

25. $4x^2 - 16$

27. $x^2 + \frac{3}{4}x + \frac{1}{8}$

29. $y^2 + \frac{2}{5}y - \frac{1}{64}$

EXERCISE 19-2

1. $x^2 - 1$

3. $x^2 - 25$

5. $4x^2 - 1$

7. $25x^2 - 25$

9. $16x^2 - 49$

11. $y^2 - a^2$

13. $4y^2 - c^2$

15. $9y^2 - 4t^2$

17. $4x^2 - \frac{1}{4}$

19. $\frac{4}{25}y^2 - \frac{1}{64}$

EXERCISE 19-3

1. $x^2 + 6x + 9$

3. $x^2 - 4x + 4$

5. $x^2 + 10x + 25$

7. $4x^2 + 8x + 4$

9. $16x^2 - 16x + 4$

11. $4x^2 + 4x + 1$

13. $16x^2 - 24x + 9$

15. $x^2 + \frac{2}{5}x + \frac{1}{25}$

17. $x^2 + x + \frac{1}{4}$

19. $\frac{1}{16}x^2 + \frac{1}{8}x + \frac{1}{16}$

21. $\frac{16}{25}x^2 + \frac{1}{5}x + \frac{1}{64}$

EXERCISE 19-4

1. $(x - 2)(x + 2)$

3. $(2x - 8)(2x + 8)$

5. $(y - c)(y + c)$

7. $(1 - 4c)(1 + 4c)$

9. $(2c - 3y)(2c + 3y)$

11. $(ab - x)(ab + x)$

13. $(\frac{1}{4} - x)(\frac{1}{4} + x)$

15. $(\frac{1}{3} - 5z)(\frac{1}{3} + 5z)$

EXERCISE 19-5

1. $(x - 3)^2$

3. $(x - 2)^2$

5. $(y + 1)^2$

7. $(3y - 5)^2$

9. $(2x - 5c)^2$

11. $(5x - 3y)^2$

EXERCISE 19-6

1. $(3 - x)(a + 2)$

3. $(m - 7)(x - 5)$

5. $(c - a)(x + y)$

7. $(z^2 + 1)(z + 1)$

9. $(b - 3)(b + 3)(4b + 5)$

11. $(x - 1)(x + 1)(b - 1)$

1. $3(a + 1)$
3. $(y^2 + c^2)(y + c)(y - c)$
5. $3(a - 1)(a + 1)$
7. $8y(y - 2)(y + 2)$
9. $5(y + 2)^2$
11. $2(2a + b)^2$
13. $5(y - m - n)(y + m + n)$
15. $(x - 4)(x + 3)$
17. $(x^2 + 1)(x + 1)$
19. $(2y - a - c)(2y - a + c)$

1. $x = 4/(1 - a)$
3. $x = b/(5 - c)$
5. $x = 1$
7. $x = 2$
9. $x = b - 3$
11. $x = b + 4$
13. $E = R/(2I + 3)$
15. $\alpha = \beta/(\beta + 1)$
17. $R = i_1 Z/(\beta i - i_1)$
19. $n = IZ/(V - IR)$
21. $R_1 = \dfrac{R_E(1 - S)}{S(1 - \alpha) - 1}$
23. $R_E = \dfrac{iR_1 R_2 - SR_1 R_2 (1 - \alpha)}{S(R_1 + R_2) - (R_1 + R_2)}$
25. $I_B = \dfrac{V_{CC}}{R_1 + R_E(\beta + 1)}$

CHAPTER 20

1. ± 5
3. ± 2.83
5. ± 7
7. ± 4.24
9. ± 10
11. ± 2
13. ± 0.167
15. ± 3
17. ± 2.28
19. ± 4.24

1. $(x + 5)(x + 2)$
3. $(a + 7)(a + 1)$
5. $(x + 1)(x + 1)$
7. $(a - 4)(a - 1)$
9. $(y - 4)(y - 3)$
11. $(x - 6)(x - 4)$
13. $(a - 9)(a + 2)$
15. $(x + 8)(x - 5)$
17. $(2x + 3)(x + 1)$
19. $(3x - 2)(x + 3)$
21. $(2b + 5)(2b - 3)$
23. $(5a - 7)(a - 2)$
25. $(3x + 2)(2x + 7)$
27. $(9a + 4)(2a - 3)$
29. $(4b - 3)(b - 2)$

1. $y = -2$ and $y = -4$
3. $x = -5$ and $x = -1$
5. $y = 1$ and $y = 5$
7. $x = -1$
9. $y = 3$ and $y = -1$
11. $x = 7$ and $x = -2$
13. $y = \frac{2}{5}$ and $y = 1$
15. $x = -\frac{1}{3}$ and $x = 5$
17. $x = -\frac{5}{3}$ and $x = \frac{3}{4}$
19. $x = \frac{9}{8}$ and $x = -\frac{4}{3}$

EXERCISE 20-4

1. $x = 0.5$ and $x = -3$
3. $x = 0.667$ and $x = -3$
5. $y = 1.5$ and $y = -0.444$
7. $x = -1.67$ and $x = 0.75$
9. $y = 0.966$ and $y = 0.438$

EXERCISE 20-5

1. (a) up
   (b) one real
   (c) (0, 0)
   (d) (0, 0)
3. (a) up
   (b) none
   (c) (0, 2)
   (d) none
5. (a) down
   (b) two real
   (c) (2, 4)
   (d) $x = 0$ and $x = 4$
7. (a) up
   (b) none
   (c) $(-0.833, 0.917)$
   (d) none
9. (a) down
   (b) none
   (c) $(-1, -1)$
   (d) none
11. (a) up
    (b) two real
    (c) $(6, -13)$
    (d) $x = 11.1$ and $x = 0.9$

EXERCISE 20-6

1. $R_1 = 20\ \Omega$, $R_2 = 22\ \Omega$
3. $V = 3$ V
5. $E = 2.45$ V
7. $R_1 = 300\ \Omega$, $R_2 = 600\ \Omega$
9. $R_1 = 38.92\ \Omega$, $R_2 = 28.92\ \Omega$
   $R_3 = 77.83\ \Omega$
11. $200\ \Omega$

CHAPTER 21

EXERCISE 21-1

1. a^8
3. $8b^3$
5. c
7. $-49c^2$
9. y^{4x}
11. $8x^3/(27a^3)$

13. a^{bc} 15. $-2x^4y^3$ 17. y^{a+2}
19. $8x^9y^3$ 21. $25S^4/T^2$

EXERCISE 21-2

1. 1 3. 4 5. 1
7. 1 9. 0.25 11. 9
13. 2 15. 0.5 17. x/y^2
19. $2/(ba^3)$ 21. $3ab^2$ 23. y^3/x^2

25. $\dfrac{y^2}{x^2}$ 27. $\dfrac{1}{27b^6c^3}$

EXERCISE 21-3

Calculator Drill

1. 5 2. 6 3. 12

4. 3 5. 4 6. 2

7. 8 8. 25 9. 4

10. 25.2 11. 5.54 12. 239

13. 0.25 14. 0.296 15. 0.396

16. 0.0442 17. 2.45 18. 125

19. 5.76 20. 9 21. 3.17

23. $\sqrt[3]{a}$ 25. $x^2\sqrt{x}$ 27. $7a\sqrt[3]{a}$

29. $-b\sqrt[3]{7b^2}$ 31. $2^{1/2}$ 33. $-2a^{1/2}$

35. $-5bc^{2/5}d^{1/5}$ 37. $(T+s)^{1/3}$ 39. $-(R-t)^{1/5}$

41. $a^{1/3}$ 43. x 45. $a^{7/12}$

EXERCISE 21-4

1. 7 3. -6^3 5. x^{14}
7. $\sqrt[3]{a}\ \sqrt[3]{b}$ 9. $-\sqrt[4]{c}\ \sqrt[4]{d}$ 11. $\sqrt[7]{4}\ \sqrt[7]{m}\ \sqrt[7]{n}$
13. $\sqrt{2}/\sqrt{3}$ 15. $\sqrt{a}/\sqrt{b}$ 17. $-\sqrt{3x}/\sqrt{5y}$
19. $\sqrt[6]{7}$ 21. $\sqrt[6]{a}$ 23. $-\sqrt[10]{aby}$
25. $3^{2/3}$ 27. $6^{5/2}$ 29. $-a^{7/9}$

830

EXERCISE 21-5

1. $8xy^2 \sqrt{xy}$

3. $3x^3y^2 \sqrt[3]{2x^2y^2}$

5. $-3a^3b^4 \sqrt{2}$

7. $2a^2 \sqrt{3a}/(3b^5)$

9. $\sqrt[3]{4}$

11. $\sqrt{2}$

13. $\sqrt[4]{9a}$

15. $\sqrt{15}/5$

17. $\sqrt{15}/(3x)$

19. $\sqrt[6]{108a^5}/(3a)$

21. $\frac{1}{2}$

23. $\frac{1}{2}$

25. $2 \sqrt{y^2 + 9}$

27. $8a^6b^2c \sqrt{c}$

29. $\dfrac{2x + 3}{2x} \sqrt{2x}$

31. $\dfrac{\sqrt[3]{ay^3} - 24}{2y}$

EXERCISE 21-6

1. $x = 49$

3. $x = 16$

5. $x = 4$

7. $a = 62$

9. $x = 6$

11. $y = 3$

13. $a = 2.25$

15. $a = 8.17$

17. $c = 2.20$

19. $a = 2$

21. $y = 0$

23. $x = 0.75$

25. $y = 48$

27. $g = \dfrac{2S}{t^2}$

29. $V = \dfrac{4\pi r^3}{3}$

CHAPTER 22

EXERCISE 22-1

1. 3

3. 4

5. 2.5

Calculator Drill

7. 0.3010

8. 0.6021

9. 0.9031

10. 1.2041

11. 1.5052

12. 0.3222

13. 0.4914

14. 0.6128

15. 0.7076

16. 0.7853

17. 0.8513

18. 0.7924

19. 0.5328

20. 0.5775

21. 0.7723

22. 0.0212

23. 0.4698

24. 0.9036

25. 0.8820

26. 0.9614

27. 0.6637

1. (a) 0, 0.8451, 0.8451
   (b) 1, 0.8451, 1.8451
   (c) 2, 0.8451, 2.8451
   (d) 3, 0.8451, 3.8451
5. (a) 0, 0.2095, 0.2095
   (b) 1, 0.2095, 1.2095
   (c) 2, 0.2095, 2.2095
   (d) 3, 0.2095, 3.2095
9. (a) $-$ 3, 0.0212, -2.9788
   (b) -2, 0.0212, -1.9788
   (c) -1, 0.0212, -0.9788
   (d) 0, 0.0212, 0.0212

3. (a) -2, 0.8751, -1.1249
   (b) -1, 0.8751, -0.1249
   (c) 0, 0.8751, 0.8751
   (d) 1, 0.8751, 1.8751
7. (a) 0, 0.7443, 0.7443
   (b) 1, 0.7443, 1.7443
   (c) 2, 0.7443, 2.7443
   (d) 3, 0.7443, 3.7443

EXERCISE 22-3

Calculator Drill

1. 2.25
3. 1.44
5. 3.33
7. 8.86×10^{-2}
9. 2.93×10^{1}
11. 2, 2×10^{1}, 2×10^{2}
13. 4, 4×10^{2}, 4×10^{4}
15. 7×10^{-1}, 7×10^{-2}, 7×10^{-3}

2. 9.35
4. 5.06
6. 4.10×10^{-1}
8. 2.48×10^{1}
10. 6.84×10^{2}
12. 3, 3×10^{1}, 3×10^{2}
14. 5, 5×10^{2}, 5×10^{4}

EXERCISE 22-4

1. $1.1761 = 0.4771 + 0.6990$
5. $0.9542 \approx 1.2553 - 0.3010$
9. false
13. 4

3. $1.7745 = 1.2304 + 0.5441$
7. false
11. 28
15. 6

EXERCISE 22-5

1. $0.9542 = 2(0.4771)$
5. $0.9542 \approx (1.9085)/2$
9. 3.28×10^{4}
13. 1.59

3. $2.9248 = 2(1.4624)$
7. $0.4515 = 3(0.3010)/2$
11. 1.87
15. 4.06

EXERCISE 22-6

1. 1.2	3. 0.75	5. −4.1
7. a	9. $-z$	

Calculator Drill

11. 0.2231	12. 1.1506	13. 1.3863
14. 1.6094	15. 1.7918	16. 1.9459
17. −1.8579	18. −3.5405	19. −4.1997
20. 2.9124	21. 6.8320	22. 8.8594
23. 4.9642	24. 3.7121	25. 11.0588
26. 6.9518	27. −0.1851	28. −3.2391
29. 4.3732	30. 7.6014	31. 2.718
32. 1.00	33. 10.0	34. 19.0
35. 9.01	36. 5.10×10^{-3}	37. 2.72×10^{-2}
38. 0.250	39. 1.69	40. 9.22×10^{-44}
41. 0.943	42. 1.12×10^2	43. 8.55×10^{-4}
44. 2.40	45. 9.37×10^{-3}	46. 4.97×10^{-2}
47. 1.67×10^6	48. 1.19×10^{-2}	49. 8.90×10^{-8}
50. 1.75		

EXERCISE 22-7

1. 1.3863	3. −0.7134	5. −0.2357
7. 1.7781	9. 0.7559	11. −0.5376

EXERCISE 22-8

1. $3.9318 = 2.8332 + 1.0986$ 3. $2.7881 = 4.1744 − 1.3863$
5. $5.4161 \approx 2(2.7081)$ 7. $2.0794 \approx (4.1589)/2$
9. 2.33   11. 292
13. 2.65   15. 8.44

EXERCISE 22-9

1. $x = 3.35$ 3. $x = 10.0$
5. $x = 0.465$ 7. $x = 8.48$
9. $5 \log (\sqrt{x}) = \ln (x) + 1.59$
   $5 \log (x)/2 = 2.3026 \log (x) + 1.59$
   $\log (x) = 1.59/(2.5 − 2.3026)$
   $x = 1.13 \times 10^8$

11. $\ln (V) = \ln (17) − 4.5$
    $V = 0.189$

13. $2 \ln (V) - \ln (V) = 3.29$
$$\ln (V) = 3.29$$
$$V = \text{antiln } 3.29$$
$$V = e^{3.29} = 26.84$$

15. $\log (Z) = \dfrac{1}{2.3026}$
$$Z = 10^{1/2.3026} = 2.718$$

17. $P_S/P_N = 10^{N_{dB}/10}$
$$P_N = P_S 10^{-(N_{dB}/10)}$$

19. $N \log (2) = \log (f_2) - \log (f_1)$
$$\log (f_2) = \log f_1 + \log (2^N)$$
$$\log (f_2) = \log (f_1 2^N)$$
$$f_2 = f_1 2^N$$

EXERCISE 22-10

1. $x = 1.15$
7. $x = 1.39$
13. $t = 3.25$
19. $a = -1.95$

3. $y = 1.95$
9. $x = -4.84$
15. $t = 9.23$

5. $s = 0.342$
11. $t = 17.9$
17. $y = 5.00$

EXERCISE 22-11

1. See Figure A15.
7. See Figure A18.

3. See Figure A16.

5. See Figure A17.

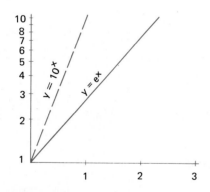

FIGURE A15
Solution to Exercise 22–11, problems 1 and 2.

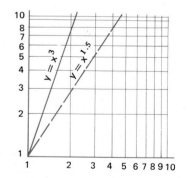

FIGURE A16
Solution to Exercise 22–11, problems 3 and 4.

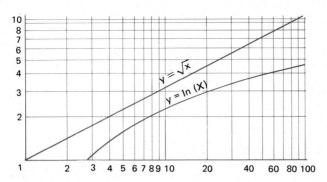

FIGURE A17
Solution to Exercise 22–11, problem 5.

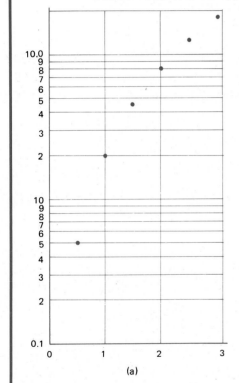

(a)

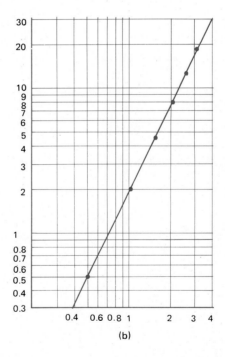

(b)

FIGURE A18
Solution to Exercise 22–11, problems 7a and b.

EXERCISE 22-12

1. 3.95   3. −17 dB   5. 1.3 kΩ
7. 950 μH   9. 1.6 MHz

CHAPTER 23

1. 24.3 dB   3. 0.377 W   5. 46 dB
7. 3.61 V   9. 70.5 dB

EXERCISE 23-2

1. (a) 11.8 dBm   (b) 23 dBm   (c) −17.5 dBm
   (d) 33.4 dBm   (e) −2.15 dBm   (f) 19.4 dBm

3. (a) 50.1 dBmV   (b) − 16.5 dBmV   (c) 71.6 dBmV
   (d) 37.1 dBmV   (e) 58.5 dBmV   (f) −31.4 dBmV

5. 13.4 dBm V   7. 18.8 dBmV   9. 48.9 V

EXERCISE 23-3

1. 8.26 A   3. (a) $i = 216$ mA
                                      (b) $v_L = 6.77$ V
                                      (c) $v_R = 43.2$ V
5. $i = 0.276$ A, $v_R = 15.5$ V   7. $L = 16.6$ H, $i = 875$ mA
   $v_L = 2.5$ V $v_R = 7.0$ V

EXERCISE 23-4

1. 10.9 μA   3. 1.48 V
5. $t = 1.54$ s, $i = 1.8$ mA   7. $i = 165$ μA, $v_R = 16.5$V
   $v_C = 7$ V $v_C = 83.5$ V

CHAPTER 24

EXERCISE 24-1

1. 60 rev   3. 0.8 rev, 288°   5. 90°
7. 317°   9. 446°   11. 25.8°
13. 0.833 rev   15. 2.40 rev   17. 4.71 rev
19. 2.64 rev   21. 2.00 rad   23. 4.71 rad
25. 3.05 rad   27. 2.36 rad

EXERCISE 24-2

1. complementary   3. supplementary   5. complementary
7. neither   9. supplementary   11. 36°
13. 99.3°   15. 90°   17. 0.132^r
19. 0.532^r 21. 32.7°   23. 45°
25. 40.1°   27. 0.501^r

EXERCISE 24-3

1. $\angle\gamma = 101°$, c 3. $\angle\beta = 49°$, a
5. $\angle\gamma = 100°$, c 7. $\angle\gamma = 1.14$ rad, a
9. $\angle\alpha = 1.14$ rad, a 11. $\angle\beta = \pi/4$, a
13. $\angle\gamma = \pi/3$, all same size   15. $\angle\gamma = 65.4°$, c
17. $\angle\beta = 20.9°$, c 19. $\angle\alpha = 90°$, a

EXERCISE 24-4

1. $c = 2.5$ m, $\angle\beta = 53.1°$ 3. $b = 35.6$ m, $\angle\alpha = 24.2°$
5. $c = 90.1$ m, $\angle\alpha = 60.6°$ 7. $a = 3.05$ m, $\angle\beta = 40.4°$
9. $b = 11$ m, $\angle\beta = 10.4°$

EXERCISE 24-5

Calculator Drill

	s	c	t		s	c	t
1.	0.0175,	0.9998,	0.0175	2.	0.4226,	0.9063,	0.4663
3.	0.9816,	0.1908,	5.1446	4.	0.2588,	0.9659,	0.2679
5.	0.9998,	0.0175,	57.290	6.	0.0,	1.0,	0.0
7.	0.7826,	0.6225,	1.2572	8.	0.9449,	0.3272,	2.8878
9.	0.1977,	0.9803,	0.2016	10.	0.2974,	0.9548,	0.3115
11.	0.0009,	1.0000,	0.0009	12.	0.0936,	0.9956,	0.0940
13	0.9893,	0.1461,	6.7720	14.	0.9770,	0.2130,	4.5864
15.	0.7570,	0.6534,	1.1585	16.	0.8980,	0.4399,	2.0413
17.	0.7683,	0.6401,	1.2002	18.	0.9478,	0.3190,	2.9714
19.	0.3633,	0.9317,	0.3899	20.	0.6211,	0.7837,	0.7926
21.	0.9975,	0.0707,	14.101	22.	0.2474,	0.9689,	0.2553
23.	0.7104,	0.7038,	1.0092	24.	0.0170,	0.9999,	0.0170
25.	0.0,	1.0,	0.0	26.	0.4794,	0.8776,	0.5463
27.	0.0998,	0.9950,	0.1003	28.	0.9320,	0.3624,	2.5722

	s	*c*	*t*		*s*	*c*	*t*
29.	0.9927, 0.1205, 8.2381			30.	0.6210, 0.7838, 0.7923		
31.	0.1583, 0.9874, 0.1604			32.	0.3127, 0.9499, 0.3292		
33.	0.9208, 0.3902, 2.3600			34.	0.1987, 0.9801, 0.2027		
35.	1.000, 0.0008, 1255.8			36.	0.9425, 0.3342, 2.8198		
37.	0.8260, 0.5636, 1.4655			38.	0.7229, 0.6909, 1.0463		
39.	0.3476, 0.9376, 0.3707			40.	0.5818, 0.8133, 0.7154		

EXERCISE 24-6

1. $\angle\beta = 70°$, $a = 3.42$ m, $b = 9.40$ m
3. $\angle\beta = 72.5°$, $a = 7.88$ m, $b = 25$ m
5. $\angle\beta = 61.6°$, $b = 36.1$ m, $c = 41.0$ m
7. $\angle\beta = 57.7°$, $b = 4.41$ m, $c = 5.22$ m
9. $\angle\beta = 75.9°$, $a = 3.54$ m, $c = 14.5$ m
11. $\angle\beta = 45°$, $a = 0.70$ km, $c = 0.99$ km
13. $\angle\alpha = 74.7°$, $a = 289$ m, $b = 79.2$ m
15. $\angle\alpha = 39.2°$, $a = 553$ mm, $b = 678$ mm
17. $\angle\alpha = 53.8°$, $b = 43.5$ m, $c = 73.6$ m
19. $\angle\alpha = 14.4°$, $a = 0.275$ km, $c = 1.10$ km

EXERCISE 24-7

Calculator Drill

1. (a) 6.14°   (b) 52.3°   (c) 14.8°
   (d) 34.6°   (e) 30.0°   (f) 23.3°
   (g) 39.9°   (h) 53.7°   (i) 24.3°
   (j) 90.0°   (k) 2.46°   (l) 67.07°

2. (a) 0.107 rad   (b) 0.912 rad   (c) 0.258 rad
   (d) 0.604 rad   (e) 0.524 rad   (f) 0.406 rad
   (g) 0.697 rad   (h) 0.937 rad   (i) 0.424 rad
   (j) 1.57 rad   (k) 0.043 rad   (l) 1.17 rad

3. (a) 83.9°   (b) 37.7°   (c) 75.2°
   (d) 55.4°   (e) 60.0°   (f) 66.7°
   (g) 50.1°   (h) 36.3°   (i) 65.7°
   (j) 0.000°   (k) 87.5°   (l) 22.9°

4. (a) 1.46 rad   (b) 0.658 rad   (c) 1.31 rad
   (d) 0.967 rad   (e) 1.05 rad   (f) 1.16 rad

(g) 0.874 rad   (h) 0.633 rad   (i) 1.15 rad

(j) 0.000 rad   (k) 1.53 rad   (l) 0.40 rad

5. (a) 84.3°   (b) 35.3°   (c) 72.4°

    (d) 23.9°   (e) 57.5°   (f) 45.0°

    (g) 67.0°   (h) 16.2°   (i) 80.6°

    (j) 0.688°   (k) 86.0°   (l) 82.0°

6. (a) 1.47 rad   (b) 0.615 rad   (c) 1.26 rad

    (d) 0.418 rad   (e) 1.00 rad   (f) 0.785 rad

    (g) 1.17 rad   (h) 0.282 rad   (i) 1.41 rad

    (j) 0.012 rad   (k) 1.5 rad   (l) 1.43 rad

EXERCISE 24-8

1. $b = 20$ m, $\angle\alpha = 36.9°$, $\angle\beta = 53.1°$

3. $b = 36.2$ mm, $\angle\alpha = 25.8°$, $\angle\beta = 64.2°$

5. $a = 7.0$ m, $\angle\alpha = 16.3°$, $\angle\beta = 73.7°$

7. $a = 31.3$ km, $\angle\alpha = 54.8°$, $\angle\beta = 35.2°$

9. $c = 21$ mm, $\angle\alpha = 48.8°$, $\angle\beta = 41.2°$

11. $c = 1000$ m, $\angle\alpha = 30.0°$, $\angle\beta = 60.0°$

CHAPTER 25

EXERCISE 25-1

1. 5°   3. −175°   5. 90°

7. 61°   9. 22°   11. −155°

EXERCISE 25-2

1. I   3. III   5. III

7. I   9. I   11. I

Calculator Drill

13. 0.966   14. −0.829   15. −0.866

16. 0.017   17. −0.643   18. 0.829

19. 0.500   20. 0.970   21. 0.259

22. −0.985   23. 0.999   24. 0.707

25. 0.259   26. 0.559   27. −0.500

28. -1.00	29. -0.766	30. 0.559
31. 0.866	32. -0.242	33. 0.966
34. -0.174	35. 0.052	36. -0.707
37. 3.73	38. -1.48	39. 1.73
40. -0.017	41. 0.839	42. 1.48
43. 0.577	44. -4.01	45. 0.268
46. 5.67	47. 19.1	48. -1.00

EXERCISE 25-3

1. *See* Figure A19.   3. *See* Figure A20.   5. *See* Figure A21.
7. *See* Figure A22.

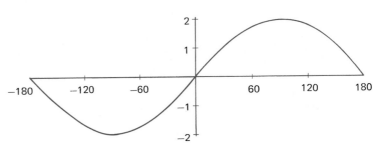

FIGURE A19
Solution to Exercise 25–3, problem 1.

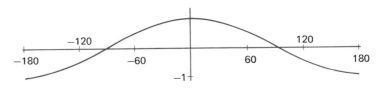

FIGURE A20
Solution to Exercise 25–3, problem 3.

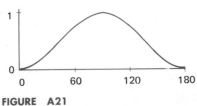

FIGURE A21
Solution to Exercise 25–3, problem 5.

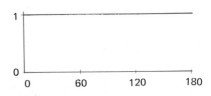

FIGURE A22
Solution to Exercise 25–3, problem 7.

Calculator Drill

1. 45.0°	2. 60.0°	3. 60.0°
4. 30.0°	5. 45.0°	6. 35.0°
7. −30.0°	8. −45.0°	9. −60.0°
10. −20.0°	11. 120°	12. 150°
13. 135°	14. 110°	15. −45.0°
16. −57.0	17. −35°	18. −11.3°
19. −80°	20. −89°	21. 89.9°

EXERCISE 25-5

1. $\angle\alpha = 36.9°$, $\angle\beta = 53.1°$, $\angle\gamma = 90°$
3. $\angle\alpha = 41.3°$, $\angle\beta = 94.1°$, $\angle\gamma = 44.6°$
5. $\angle\alpha = 36.9°$, $\angle\beta = 63.1°$, $c = 57.4$ mm
7. $\angle\beta = 51.3°$, $\angle\gamma = 105.4°$, $a = 40.3$ m
9. $\angle\gamma = 110°$, $b = 31$ m, $c = 58.5$ m
11. $\angle\gamma = 115.7°$, $a = 2.62$ m, $b = 11.1$ m

EXERCISE 25-6

1. 1.5/30°	3. 2/−30°	5. 3.7/−47°
7. 3.9/143°	9. 2/150°	11. 3/−175°
13. 4.5/−83°	15. 4.6/73°	17. 4.2/55°
19. 4.2/88°	21. 4.5/125°	23. 3.5/−100°

EXERCISE 25-7

Calculator Drill

1. 15.0/0°	2. 3.57/−90°	3. 8.70/180°
4. 1.41/45°	5. 10.0/60.1°	6. 7.61/−33.6°
7. 6.57/−50.8°	8. 205/−31.8°	9. 154/97.5°
10. 29.8/154.6°	11. 1.00/−150°	12. 32.7/61.5°
13. 44.1/−27.1°	14. 34.2/−117.3°	15. 5.76/135°
16. 15.2/−7.54°	17. 14.5/157.3°	18. 262/59.2°
19. 3.67/110°	20. 1036/−43°	21. 2.33/−37.2°

22. (0.00, 17)	23. (−16, 0.00)	24. (0.00, −12.7)
25. (64.3, 76.6)	26. (21.7, 12.5)	27. (6.99, −26.1)
28. (53.0, 18.6)	29. (4.64, 4.22)	30. (6.77, −79.0)
31. (0.365, −0.449)	32. (−314, −314)	33. (−68.2, 30.3)
34. (1.12, −12.8)	35. (−11.8, 13.1)	36. (−9.18, − 4.88)
37. (10.1, 19.9)	38. (−6.55, 2.33)	39. (29.8, −7.39)
40. (47.7, −70.0)	41. (18.4, −76.1)	42. (−379, − 352)

CHAPTER 26

EXERCISE 26-1

1. $j3$	3. $j6$	5. $-j5$
7. $j13.5$	9. $-j12$	11. $-j10.4$
13. $j0.935$	15. $-j7.27$	

EXERCISE 26-2

1. $15 + j0$	3. $0 + j1.7$	5. $-5 - j4$
7. $-1 + j0$	9. $17 - j1$	11. complex
13. real	15. real	17. real

EXERCISE 26-3

1. *See* Figure A23.	3. *See* Figure A23.	5. SeeFigure A23.
7. *See* Figure A23.	9. *See* Figure A23.	11. $0 + j0$
13. $-4 - j2$	15. $-1 - j2$	17. $-1 + j0$
19. $2 + j0$	21. $-2 + j4$	

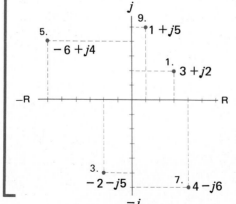

FIGURE   A23
**Solution to Exercise 26–3, problems 1,
3, 5, 7, and 9.**

1. $3\underline{/120°}$ 3. $1\underline{/150°}$ 5. $3\underline{/-90°}$

7. $2\underline{/30°}$ 9. $3\underline{/-150°}$ 11. *See* Figure A24.

13. *See* Figure A24.   15. *See* Figure A24.   17. *See* Figure A24.

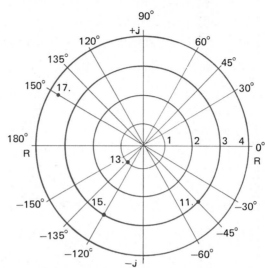

FIGURE A24
Solution to Exercise 26–4, problems 11, 13, 15, and 17.

EXERCISE 26-5

1. $10\underline{/53.1°}$ 3. $4.47\underline{/116.6°}$ 5. $12.4\underline{/-104°}$

7. $6.6\underline{/55°}$ 9. $6.38\underline{/-59.9°}$ 11. $24\underline{/-126°}$

13. $17\underline{/-4.04°}$ 15. $4.21\underline{/-97°}$ 17. $0.931\underline{/76.0°}$

19. $1.34 + j4.82$ 21. $3.78 + j5.89$ 23. $5.87 - j1.25$

25. $-4.10 + j11.3$ 27. $1.09 - j1.68$ 29. $-3.50 - j6.86$

31. $585 + j579$ 33. $-0.132 - j0.137$ 35. $-9.59 - j296$

37. $-125 + j28.8$ 39. $0.0443 + j0.355$

EXERCISE 26-6

1. $8\underline{/0°}$ 3. $10\underline{/0°}$ 5. $10\underline{/-90°}$

7. $15.6\underline{/45°}$ 9. $30\underline{/90°}$ 11. $9.88\underline{/91.7°}$

13. $3.22\underline{/90°}$ 15. $13.0\underline{/1.37^r}$ 17. $10.4\underline{/-0.180^r}$

CHAPTER 27

EXERCISE 27-1

1. $7 + j2$, $7.28\underline{/16°}$
3. $-12 + j1$, $12\underline{/175°}$
5. $15.4 + j8$, $17.4\underline{/27.5°}$
7. $15.5 + j0.33$, $15.5\underline{/1.22°}$
9. $34.7 + j22.4$, $41.3\underline{/32.8°}$
11. $2.66 + j0.94$, $2.82\underline{/19.5°}$
13. $-1.21 + j0.979$, $1.56\underline{/141°}$
15. $-23.8 - j7.45$, $24.9\underline{/-163°}$
17. $58.3 + j39.6$, $70.5\underline{/34.2°}$
19. $-84.08 - j101.4$, $131.8\underline{/-2.26^r}$

EXERCISE 27-2

1. $12\underline{/90°} = 0 + j12$
3. $46.6\underline{/67.2°} = 18.1 + j43.0$
5. $27\underline{/-31.3°} = 23.1 - j14.0$
7. $15\underline{/27°} = 13.4 + j6.81$
9. $12\underline{/-65°} = 5.07 - j10.9$
11. $646\underline{/33°} = 542 + j352$
13. $10.9 \times 10^3\underline{/98°} = -1.52 \times 10^3 + j10.8 \times 10^3$
15. $2.21 \times 10^3\underline{/37°} = 1.76 \times 10^3 + j1.33 \times 10^3$
17. $364\underline{/-3.5°} = 363 - j22.2$
19. $2.82\underline{/-56.9°} = 1.54 - j2.36$
21. $197.1\underline{/3.7^r} = -167.2 - j104.4$
23. $11\,855\underline{/30.7°}$, $10\,194 + j6052$

EXERCISE 27-3

1. $5 + j3$
3. $12.5 + j4$
5. $-9.4 - j10.6$
7. $0.722\underline{/3.2°} = 0.721 + j0.040$
9. $0.354\underline{/81.8°} = 0.05 + j0.35$
11. $3.96\underline{/161.6°} = -3.76 + j1.25$
13. $3\underline{/60°} = 1.5 + j2.6$
15. $3\underline{/-40°} = 2.3 - j1.93$
17. $2\underline{/149°} = -1.71 + j1.03$
19. $0.413\underline{/-90.5°} = -0.004 - j0.413$
21. $3.09\underline{/0.358^r} = 2.89 + j1.08$
23. $0.589\underline{/-4.64^r} = -0.043 + j0.587$

EXERCISE 27-4

1. $9\underline{/20°}$

3. $16\underline{/40°}$

5. $144\underline{/-34°}$

7. $204.5\underline{/112.4°}$

9. $300\underline{/-28°}$

11. $7\underline{/20°}$

13. $5\underline{/-32°}$

15. $1.65\underline{/8.8°}$

17. $4.05\underline{/56°}$

19. $9.05\underline{/-85.6°}$

21. $4.52\underline{/-27.5°}$

23. $0.918\underline{/0.286^r}$

CHAPTER 28

EXERCISE 28-1

1. 17 Hz

3. 5 Hz

5. 35.7 Hz

7. 60 Hz

9. 16.7 ms

11. 83.3 μs

13. 2.4 s

15. 12.3 ms

17. 200 rad/s

19. 1.88 krad/s

21. 37.7 krad/s

23. 314 rad/s

EXERCISE 28-2

1. 10 Ω

3. 107 Ω

5. 35.3 kΩ

7. 164 mΩ

9. 62.5 V

11. 39.6 V

13. 368 μA

15. 1.74 A

EXERCISE 28-3

1. (a) 1.51 kΩ
   (b) 126 kΩ
   (c) 50.3 MΩ

3. (a) 94.2 Ω
   (b) 7.85 kΩ
   (c) 3.14 MΩ

5. (a) 4.52 kΩ
   (b) 377 kΩ
   (c) 151 MΩ

7. (a) 10 Ω
   (b) 1.59 mH

9. (a) 2.5 kΩ
   (b) 398 mH

11. 200 V

13. 22.3 mA

15. 8.34 mA

EXERCISE 28-4

1. (a) 177 Ω
   (b) 2.12 Ω
   (c) 5.31 mΩ

3. (a) 5.31 kΩ
   (b) 63.6 Ω
   (c) 159 mΩ

5. (a) 2.65 Ω
   (b) 31.8 mΩ
   (c) 79.6 $\mu\Omega$

7. (a) 32.3 Ω
   (b) 9.87 μF

9. (a) 3.75 Ω
   (b) 84.9 μF

11. 5.53 V

13. 493 mA

15. 1.79 A

1. $70.7\underline{/45°}$ V $= 50 + j50$ V
3. $5.39\underline{/21.8°}$ V $= 5.0 + j2.0$ V
5. $30\underline{/90°}$ V $= 0 + j30$ V
7. $20\underline{/90°}$ V $= 0 + j20$ V
9. $76.7\underline{/12.0°}$ V $= 75.0 + j16$ V
11. $2.95 + j16.7$ V
13. $6.93 - j4.0$ V
15. $5.58 + j63.76$ V
17. $8.96 - j27.6$ V
19. $27.5 - j47.6$ V
21. $5.83\underline{/-31°}$ V $= 5 - j3$ V
23. $12.8\underline{/38.7°}$ V $= 10 + j8$ V
25. $21\underline{/90°}$ V $= 0 + j21$ V
27. (a) $V_R = 1.0\underline{/0°}$ kV,
   $V_L = 754\underline{/90°}$ V
   (b) $E = 1.25\underline{/37°}$ kV

29. (a) $V_L = 339\underline{/90°}$ V,
   $V_C = 239\underline{/-90°}$ V
   (b) $E = 100\underline{/90°}$ V

1. $2 - j5$ A $= 5.39\underline{/-68.2°}$ A
3. $0.3 - j1.2$ A $= 1.24\underline{/-76°}$ A
5. $0 - j1.6$ A $= 1.6\underline{/-90°}$ A
7. $0 - j6$ A $= 6\underline{/-90°}$ A
9. $0.510 + j0.462$ A $= 0.688\underline{/42.2°}$ A
11. $3.71 + j11.4$ mA
13. $390 - j134$ mA
15. $38.4 + j54.9$ μA
17. $1.6 + j0.126$ A
19. $0.329 + j2.68$ A
21. $3 - j2$ A $= 3.61\underline{/-33.7°}$ A
23. $8 + j7$ A $= 10.6\underline{/41.2°}$ A
25. $3.7 + j1.2$ A $= 3.89\underline{/18°}$ A
27. (a) $I_R = 21.3\underline{/0°}$ mA
   $I_L = 32.7\underline{/-90°}$ mA
   (b) $I = 39\underline{/-57°}$ mA

29. (a) $I_L = 53\underline{/-90°}$ mA
   $I_C = 120\underline{/90°}$ mA
   (b) $I = 67\underline{/90°}$ mA

CHAPTER 29

1. $40\underline{/20°}$ Ω
3. $56.3\underline{/0°}$
5. $17.2\underline{/-8°}$ Ω
7. resistive

9. resistive and inductive

11. resistive and capacitive

13. $10 + j0 \Rightarrow 10\underline{/0°}\ \Omega$

15. $5 - j2 \Rightarrow = 5.39\underline{/-21.8°}\ \Omega$

17. $14\underline{/90°} \Rightarrow 0 + j14\ \Omega$

19. $R = 490\ \Omega,\ X_L = 141\ \Omega$

21. $R = 3.14\ \text{k}\Omega,\ X_C = 1.02\ \text{k}\Omega$

23. $R = 32.4\ \Omega,\ X_L = 76.4\ \Omega$

25. $138.7\underline{/-78.8°}\ \Omega$ and
   $R = 27\ \Omega,\ C = 19.5\ \mu\text{F}$

27. $852.1\underline{/-54.1°}\ \Omega$ and
   $R = 500\ \Omega,\ C = 0.577\ \mu\text{F}$

29. $3.26\underline{/-47.5°}\ \text{k}\Omega$ and
   $R = 2.2\ \text{k}\Omega,\ C = 0.0663\ \mu\text{F}$

EXERCISE 29-2

1. (a) $1880 - j1522 \Rightarrow 2.42\underline{/-39°}\ \text{k}\Omega$
   (b) $-39°$
   (c) 41.3 mA
   (d) $R_{eq} = 1.88\ \text{k}\Omega$ and $C_{eq} = 1.74\ \mu\text{F}$

3. (a) $51 - j291 \Rightarrow 295\underline{/-80°}\ \Omega$
   (b) $-80°$
   (c) 407 mA
   (d) $R_{eq} = 51\ \Omega$ and $C_{eq} = 9.12\ \mu\text{F}$

5. (a) $62 - j79 \Rightarrow 101\underline{/-52°}$
   (b) $-52°$
   (c) 356 mA
   (d) $R_{eq} = 62\ \Omega$ and $C_{eq} = 5.0\ \mu\text{F}$

7. (a) 21.99 kΩ
   (b) $34.8\underline{/39.2°}\ \text{k}\Omega$
   (c) $39.2°$
   (d) $287\ \mu\text{A}$
   (e) $7.75\ \text{V}\underline{/0°}$
   (f) $6.31\ \text{V}\underline{/90°}$
   (g) $7.75 + j6.31 = 10\underline{/39.2°}\ \text{V}$

EXERCISE 29-3

1. $100\underline{/-15°}\ \text{mS}$

3. $1.0\underline{/-82°}\ \text{mS}$

5. $17.9\underline{/19°}\ \text{mS}$

7. $8.33\underline{/90°}\ \text{mS}$

9. $1.1\underline{/5°}\ \text{mS}$

11. $667\underline{/90°}\ \mu\text{S}$

13. $2.4\underline{/-90°}\ \text{mS}$

15. $14.7\underline{/90°}\ \mu\text{S}$

17. $1.96\underline{/90°}\ \mu\text{S}$

19. $1.17\underline{/17°}\ \text{mS}$

21. $8.08\underline{/-61°}\ \mu\text{S}$

23. $79.4\underline{/15°}\ \text{mS}$

EXERCISE 29-4

1. 335 pF

3. 39.1 μH

5. 108 Ω

7. (1) 6.66$\underline{/-60°}$ mS
   (2) 6.25$\underline{/63.2°}$ mS

9. (1) 969$\underline{/35.5°}$ μS
   (2) 2.38$\underline{/0°}$ mS

11. $Y_T = 6.47\underline{/60.5°}$ mS
    $Z_T = 155\underline{/-60.5°}$ Ω

13. $Y_T = 340\underline{/-73.5°}$ μS
    $Z_T = 2.94\underline{/73.5°}$ kΩ

15. $Y_T = 2.55\underline{/-17.9°}$ mS
    $Z_T = 392\underline{/17.9°}$ Ω

17. $R_p = 214$ Ω, $L_p = 35.1$ mH

19. $R_P = 342$ Ω, $C_P = 0.954$ μF

21. (a) $Y_1 = 4.76\underline{/-81°}$ mS
    $Y_2 = 1.33\underline{/-51°}$ mS
    (b) 5.95$\underline{/-74.6°}$ mS
    (c) 168$\underline{/74.6°}$ Ω
    (d) 595$\underline{/-74.6°}$ mA

CHAPTER 30

EXERCISE 30-1

1. 17.9 V

3. −0.576 mA

5. −10.3 A

7. 3.41 V

9. −259 V

11. 258°

13. 92.3°

15. 123°

EXERCISE 30-2

1. 10.1 V

3. 12.2 V

5. 127.3 V

7. 11.1 A

9. 5.73 mA

11. 1.87 A

13. 25.5 mA

15. 21.2 V

17. 62.2 V

19. 38.5 W

21. 29.1 W

23. 0.832 lead

25. 0.371 unknown

27. 557 mW

29. $R_{eq} = 3.45$ kΩ
    $L_{eq} = 1.19$ H at
    $\omega = 5027$ rad/s

EXERCISE 30-3

1. $i = I_{mx} \sin (900t)$ A
   $e = E_{mx} \sin (900t + 10°)$ V

3. $e = E_{mx} \sin (377t)$ V
   $i = I_{mx} \sin (377t + 65°)$ A

5. i lags e by 38°

7. e lags i by 40°

9. (a) $i = 42.4 \times 10^{-3} \sin (2513t)$ A

$\quad\quad e = 3.54 \sin (2513t^r + 48°)$ V

(b) $i = 5.31$ mA   (c) $P = 50.2$ mW

CHAPTER 31

EXERCISE 31-1

Calculator Drill

1. 1.14	2. 1.56	3. 1.15
4. −1.15	5. −1.50	6. 2.01
7. 1.18	8. 1.00	9. 1.01
10. 1.01	11. −2.37	12. −1.41
13. 1.01	14. −3.58	15. −1.10
16. −1.32	17. 4.51	18. 1.30
19. 0.268	20. −0.466	21. −3.02
22. −0.642	23. 0.458	24. 0.574

EXERCISE 31-3

1. Begin with right member

$$\cos (\theta)/\sin (\theta) = (x/z)/(y/z)$$
$$= x/y$$
$$= \cot (\theta)$$

which is the left member

3. Begin with right member

$$\cot (\theta) \sec (\theta) = (x/y)(z/x)$$
$$= z/y$$
$$= \csc (\theta)$$

which is the left member

5. Begin with left member

$$\tan (\theta) = \sin (\theta) \sec (\theta) \text{ (See problem 2)}$$
$$= \sec (\theta)/\csc (\theta)$$

which is the right member

7. Begin with left member

$$\tan(\theta)\cot(\theta) = [\sin(\theta)/\cos(\theta)][\cos(\theta)/\sin(\theta)]$$
$$= 1$$

which is the right member

9. Begin with left member

$$\tan(\theta) + \cot(\theta) = \frac{\sin(\theta)}{\cos(\theta)} + \frac{\cos(\theta)}{\sin(\theta)}$$
$$= \frac{\sin^2(\theta) + \cos^2(\theta)}{\cos(\theta)\sin(\theta)}$$
$$= \frac{1}{\cos(\theta)\sin(\theta)}$$
$$= \sec(\theta)\csc(\theta)$$

which is the right member

11. Begin with the left member

$$\sin(\theta)\cos(\theta)\sec(\theta)\csc(\theta) = \frac{\sin(\theta)\cos(\theta)}{\sin(\theta)\cos(\theta)}$$
$$= 1$$

which is the right member

13. Begin with the right member

$$\sec(\theta) = 1/\cos(\theta)$$
$$= \frac{\sin^2(\theta) + \cos^2(\theta)}{\cos(\theta)}$$
$$= \frac{\sin(\theta)}{\cos(\theta)}\sin(\theta) + \cos(\theta)$$
$$= \tan(\theta)\sin(\theta) + \cos(\theta)$$

which is the left member

15. Begin with the right member

$$\cot(\theta)\sec^2(\theta) = \cot(\theta)(1 + \tan^2(\theta)) \quad \text{(by problem 4)}$$
$$= \cot(\theta) + \cot(\theta)\tan^2(\theta)$$
$$= \cot(\theta) + \tan(\theta) \quad \text{(by problem 7)}$$

which is the left member

17. Begin with the left member

$$\frac{1}{1 - \sin(\theta)} + \frac{1}{1 + \sin(\theta)} = \frac{1 + \sin(\theta)}{1 - \sin^2(\theta)} + \frac{1 - \sin(\theta)}{1 - \sin^2(\theta)}$$

$$= \frac{2}{1 - \sin^2(\theta)}$$

$$= 2/\cos^2(\theta)$$

$$= 2\sec^2(\theta)$$

which is the right member

19. Begin with the right member

$$\frac{\sin(\theta)}{1 + \cos(\theta)} = \frac{\sin(\theta)}{1 + \cos(\theta)} \times \frac{1 - \cos(\theta)}{1 - \cos(\theta)}$$

$$= \frac{\sin(\theta)(1 - \cos(\theta))}{1 - \cos^2(\theta)}$$

$$= \frac{\sin(\theta)(1 - \cos(\theta))}{\sin^2(\theta)}$$

$$= \frac{1 - \cos(\theta)}{\sin(\theta)}$$

which is the left member

EXERCISE 31-4

Calculator Drill

1. (a) -1.175
   (b) -0.521
   (c) 0.000
   (d) 0.637
   (e) 1.509
   (f) 3.627

2. (a) 1.543
   (b) 1.128
   (c) 1.000
   (d) 1.185
   (e) 1.811
   (f) 3.762

3. (a) -0.762
   (b) -0.462
   (c) 0.000
   (d) 0.537
   (e) 0.834
   (f) 0.964

EXERCISE 31-5

1. *See* Figure A25.   3. *See* Figure A26.

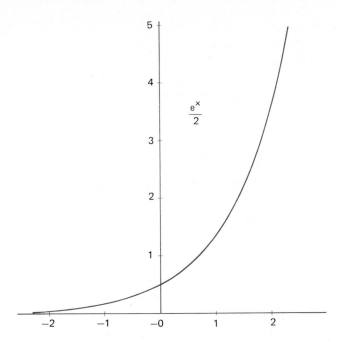

FIGURE   A25
Solution to Exercise 31–5, problem 1.

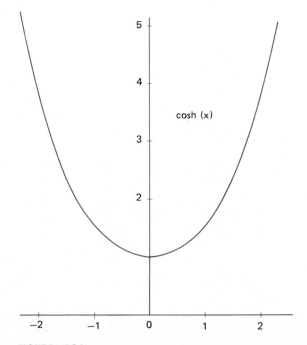

FIGURE   A26
Solution to Exercise 31–5, problem 3.

1. Left member

$$\sinh(x) + \cosh(x) = \frac{e^x - e^x}{2} + \frac{e^x + e^{-x}}{2}$$

$$= e^x$$

which is the right member

3. Begin with right member

$$\sinh(x)/\cosh(x) = \left(\frac{e^x - e^{-x}}{2}\right) \bigg/ \left(\frac{e^x + e^{-x}}{2}\right)$$

$$= (e^x - e^{-x})/(e^x + e^{-x})$$

$$= \tanh(x)$$

which is the left member

5. Begin with right member

$$-\sinh(-x) = -(e^{-x} - e^x)/2$$

$$= (e^x - e^{-x})/2$$

$$= \sinh(x)$$

which is the left member

7. Begin with left member

$$\cosh(2x) = (e^{2x} + e^{-2x})/2$$

Right member

$$\cosh^2(x) + \sinh^2(x) = \left(\frac{e^x + e^{-x}}{2}\right)^2 + \left(\frac{e^x - e^{-x}}{2}\right)^2$$

$$= \frac{e^{2x} + 2 + e^{-2x}}{4} + \frac{e^{2x} - 2 + e^{-2x}}{4}$$

$$= (e^{2x} + e^{-2x})/2$$

9. Left member

$$\tanh^2(x) + 1/\cosh^2(x) = \sinh^2(x)/\cosh^2(x) + 1/\cosh^2(x)$$

$$= (\sinh^2(x) + 1)/\cosh^2(x))$$

$$= \left[\left(\frac{e^x - e^{-x}}{2} \right)^2 + 1 \right] \bigg/ \left(\frac{e^x + e^{-x}}{2} \right)^2$$

$$= \left(\frac{e^{2x} - 2 + e^{-2x}}{4} + \frac{4}{4} \right) \bigg/ \left(\frac{e^{2x} + 2 + e^{-2x}}{4} \right)$$

$$= \left(\frac{e^{2x} + 2 + e^{-2x}}{4} \right) \bigg/ \left(\frac{e^{2x} + 2 + e^{-2x}}{4} \right)$$

$$= 1$$

which is the right member

EXERCISE 31-7

1. 0.75   3. -2.50 5. 1.00
7. 1.40   9. 0.55   11. -2.00

CHAPTER 32

EXERCISE 32-1

1. $-\infty < \theta < \infty$ 3. $-\infty < X < \infty$ 5. $-\infty < X < \infty$
   $-1 \leqslant Y \leqslant 1$ $-\infty < Y < \infty$ $-1 < Y < 1$
7. $-\infty < U < \infty$ 9. $-\infty < s < \infty$
   $0 < Z < \infty$ $-\infty < t < \infty$

EXERCISE 32-2

1. continuous   3. essential discontinuities
5. continuous   7. essential discontinuities
9. Sawtooth curve with a step discontinuity at x equal an integer

EXERCISE 32-3

1. does not exist   3. 0   5. ∞
7. ∞ 9. 2

EXERCISE 32-4

1. $y = 0$ for $1 \ll x$ 3. $s = 0$ for $1 \ll |t|$
5. $u = |v|$ for $1 \ll V$ 7. $R = 2$ for $2 \ll |r|$

854

CHAPTER 33

1. (a) 100
   (b) 1000
   (c) 0.01
   (d) 0.1
   (e) 0.01
   (f) 100

3. (a) 5000
   (b) 0.5
   (c) 50
   (d) 5
   (e) 0.0005
   (f) 50

5. (a) 7
   (b) 700
   (c) 0.7
   (d) 70
   (e) 70
   (f) 7

7. (a) 6
   (b) 5
   (c) 6
   (d) 2
   (e) 5
   (f) 2

┌EXERCISE 33-2

1. 4, 1, 0.25
7. 0.0625, 3584

3. 1536, 40
9. 3072, 144

5. 256, 0.0625

┌EXERCISE 33-3

1. 6
7. 2.75
13. 184
19. 5.4375
25. 0.066 406 25

3. 48
9. 3.3125
15. 0.140 625
21. 272
27. 118.125

5. 0.75
11. 72
17. 62.25
23. 624
29. 240.796 875

┌EXERCISE 33-4

1. (a) 100011_2, 23_8, 13_{16}
   (b) 101110_2, 26_8, 16_{16}
   (c) 1111_2, 17_8, F_{16}
   (d) $100\ 100_2$, 44_8, 24_{16}
   (e) $110\ 000_2$, 60_8, 30_{16}
   (f) $100\ 111_2$, 47_8, 27_{16}
   (g) $1\ 000\ 000_2$, 100_8, 40_{16}
   (h) $1\ 100\ 100_2$, 144_8, 64_{16}

3. (a) 0.1_2, 0.4_8, 0.8_{16}
   (b) 0.01_2, 0.2_8, 0.4_{16}
   (c) 0.11_2, 0.6_8, $0.C_{16}$
   (d) 0.001_2, 0.1_8, 0.2_{16}
   (e) 0.0001_2, 0.04_8, 0.1_{16}
   (f) 0.00101_2, 0.12_8, 0.28_{16}
   (g) $0.001\ 010\ 0_2$, 0.11770_8, $0.27EFA_{16}$
   (h) 1.00000_2, 0.77534_8, $0.FEB85_{16}$

5. 30.3_8

┌EXERCISE 33-5

1. $1\ 010\ 100_2$, 54_{16}
5. $1\ 0010\ 0100_2$, 444_8
9. 13.4_8, B.8

3. $10\ 000\ 000\ 001_2$, 401_{16}
7. $1010\ 1011_2$, 253_8
11. 525_8, 155_{16}

EXERCISE 33-6
All numbers are binary

1. 1010	3. 11 010	5. 11 110
7. 1110	9. 10	11. 1000
13. 111	15. 101	

EXERCISE 33-7
All numbers are octal

1. 10	3. 7	5. 5
7. 11	9. 14	11. 51
13. 7775	15. 60 416	17. 3
19. 2	21. 25	23. 47
25. 6341	27. 62 457	29. 7777

EXERCISE 33-8
All numbers are hexadecimal

1. E	3. D	5. 8
7. 69	9. 15	11. CE
13. 1326	15. 3000	17. 9
19. 3	21. 3	23. 3
25. 7	27. 56	29. 76 D

EXERCISE 33-9

1. $1000\ 0000_2$	3. $1001\ 1001_2$	5. $1000\ 1110_2$
7. $1357\ 6420_8$	9. $7666\ 6666_8$	11. $5326\ 0777_8$
13. $8253\ 7998_{10}$	15. $0746\ 1320_{10}$	17. $0000\ 0000_{10}$
19. $FEEE\ EEEE_{16}$	21. $E054\ 67DB_{16}$	23. $CFA9\ 8321_{16}$

EXERCISE 33-10

1. $1000\ 0001_2$	3. $1010\ 1001_2$	5. $1001\ 0100_2$
7. $5126\ 7001_8$	9. $6034\ 6667_8$	11. $7666\ 6667_8$
13. $7438\ 9223_{10}$	15. $9000\ 0000_{10}$	17. $6509\ 8373_{10}$
19. $8E51\ 9BD1_{16}$	21. $C274\ 36FB_{16}$	23. $DA9E\ F889_{16}$

EXERCISE 33-11
All numbers are binary

1. 0111 1001	3. 0001 0101	5. 1111 0110
7. 0010 0010	9. 0110 1010	11. 0111 1001

13. 0001 0100   15. 1111 0111   17. 0010 0010
19. 0110 1011

EXERCISE 33-12
All numbers are decimal

1. 100   3. −23   5. 41
7. 30   9. 12   11. −53
13. −5   15. 60   17. 163
19. 39   21. 49   23. 6998

CHAPTER 34

EXERCISE 34-1

1. NOT B 3. A AND B 5. C OR A AND B
7. A AND NOT B AND C 9. NOT A OR NOT B
11. A OR B AND C OR A AND NOT B

EXERCISE 34-2

1. 0   3. 0   5. $\overline{A}$
7. 0   9. B 11. 1

EXERCISE 34-3

1. 1   3. 0   5. 0   7. 0

9. *See* Figure A27.   11. *See* Figure A28.
13. *See* Figure A29.

15.

	A	*B*	*AB*
0	0	0	0
1	0	1	0
2	1	0	0
3	1	1	1

FIGURE A27
Solution to Exercise 34–3, problem 9.

FIGURE A28
Solution to Exercise 34–3, problem 11.

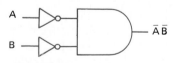

FIGURE A29
Solution to Exercise 34–3, problem 13.

17. (a) 8 (b)

	A	B	C	$\overline{A}$	$\overline{A}BC$
0	0	0	0	1	0
1	0	0	1	1	0
2	0	1	0	1	0
3	0	1	1	1	1
4	1	0	0	0	0
5	1	0	1	0	0
6	1	1	0	0	0
7	1	1	1	0	0

(c) *See* Figure A30.

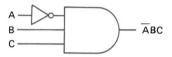

FIGURE A30
Solution to Exercise 34–3, problem 17c.

EXERCISE 34-4

1. 1 3. 1 5. 1 7. 1

9. *See* Figure A31. 11. *See* Figure A32. 13. *See* Figure A33.

15.

	A	B	A + B
0	0	0	0
1	0	1	1
2	1	0	1
3	1	1	1

17. (a) 8

(b)

	A	B	C	$\overline{A}$	$\overline{A} + B + C$
0	0	0	0	1	1
1	0	0	1	1	1
2	0	1	0	1	1
3	0	1	1	1	1
4	1	0	0	0	0
5	1	0	1	0	1
6	1	1	0	0	1
7	1	1	1	0	1

(c) *See* Figure A34.

FIGURE A31
Solution to Exercise 34–4, problem 9.

FIGURE A32
Solution to Exercise 34–4, problem 11.

FIGURE A33
Solution to Exercise 34–4, problem 13.

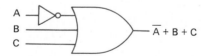

FIGURE A34
Solution to Exercise 34–4, problem 17c.

EXERCISE 34-5

1. *See* Figure A35. 3. *See* Figure A36. 5. *See* Figure A37.

FIGURE A35
Solution to Exercise 34–5, problem 1.

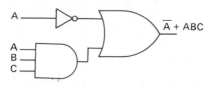

FIGURE A36
Solution to Exercise 34–5, problem 3.

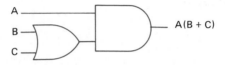

FIGURE A37
Solution to Exercise 34–5, problem 5.

7.

	A	B	C	$\bar{C}$	$A\bar{C}$	$A\bar{C} + B$
0	0	0	0	1	0	0
1	0	0	1	0	0	0
2	0	1	0	1	0	1
3	0	1	1	0	0	1
4	1	0	0	1	1	1
5	1	0	1	0	0	0
6	1	1	0	1	1	1
7	1	1	1	0	0	1

9.

	A	B	C	$\bar{C}$	$B\bar{C}$	$C + B\bar{C}$	$A(C = B\bar{C})$
0	0	0	0	1	0	0	0
1	0	0	1	0	0	1	0
2	0	1	0	1	1	1	0
3	0	1	1	0	0	1	0
4	1	0	0	1	0	0	0
5	1	0	1	0	0	1	1
6	1	1	0	1	1	1	1
7	1	1	1	0	0	1	1

11. True   13. True   15. False

17.

	A	B	C	$A + B$	$\bar{C}$	$(A + B)\bar{C}$
0	0	0	0	0	1	0
1	0	0	1	0	0	0
2	0	1	0	1	1	1
3	0	1	1	1	0	0
4	1	0	0	1	1	1
5	1	0	1	1	0	0
6	1	1	0	1	1	1
7	1	1	1	1	0	0

EXERCISE 34-6

1. *See* Figure A38.   3. *See* Figure A39.   5. *See* Figure A40.

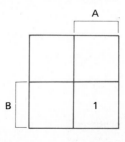

FIGURE   A38
Solution to Exercise 34–6, problem 1.

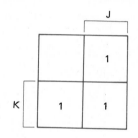

FIGURE A39
Solution to Exercise 34–6, problem 3.

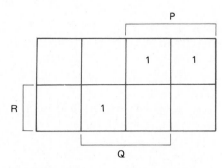

FIGURE A40
Solution to Exercise 34–6, problem 5.

7. $\overline{T}$ 9. H 11. U

13. *See* Figure A41.   15. *See* Figure A42.

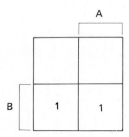

FIGURE A41
Solution to Exercise 34–6, problem 13.

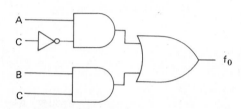

FIGURE A42
Solution to Exercise 34–6, problem 15.

EXERCISE 34-7

1. $\overline{A} + \overline{B}$

3. $A\overline{B}$

5. $\overline{U}V\overline{W} + X + \overline{Y}$

7. $\overline{A}B + C\overline{B} + A\overline{D}$

9. $\overline{X}YZ$

11. $\overline{J}KL$

13. *See* Figure A43.

FIGURE   A43

B ———— f_0 **Solution to Exercise 34–7, problem 13.**

EXERCISE 34-8

1. $A = 0$ or $A = 1$
        If $A = 0$, then $A \cdot 1 = 0 \cdot 1 = 0 = A$
        If $A = 1$, then $A \cdot 1 = 1 \cdot 1 = 1 = A$
   $\therefore$ $A \cdot 1 = A$

3. $A = 0$ or $A = 1$
        If $A = 0$, then $AA = 0 \cdot 0 = 0 = A$
        If $A = 1$, then $AA = 1 \cdot 1 = 1 = A$
   $\therefore$ $AA = A$

5. $A = 0$ or $A = 1$
        If $A = 0$, then $A\overline{A} = 0 \cdot 1 = 0$
        If $A = 1$, then $A\overline{A} = 1 \cdot 0 = 0$
   $\therefore$ $A\overline{A} = 0$

7. $B = 0$ or $B = 1$
        For $B = 0$:
        $(A + B)A = (A + 0)A = AA = A$
        For $B = 1$:
        $(A + B)A = (A + 1)A = 1 \cdot A = A$
   $\therefore$ $(A + B)A = A$

9. By problem 5, $B\overline{B} = 0$, thus:
   $A + (B\overline{B}) = A + 0 = A$

11. $\overline{XY} + (\overline{X}Y + \overline{X}Y) = XY$
    $\overline{X}(Y + \overline{Y}) + Y(X + \overline{X})$
    $\overline{X} + Y$

13. DE

EXERCISE 34-9

1. *See* Figure A44, where:
   $P = 1$ Main power ON
   $L = 1$ Door latched
   $T = 1$ Timer $\neq 0$
   $f_0 = 1$ Oven ON

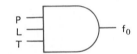

FIGURE   A44
Solution to Exercise 34–9, problem 1.

3. *See* Figure A45, where:
   $L = 1$ Lid UP
   $B = 0$ Light beam OFF
   $f_0 = 1$ Lighter lit

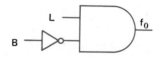

FIGURE   A45
Solution to Exercise 34–9, problem 3.

5. *See* Figure A46, where:
   $C = 0$ Ticket taken
   $D = 0$ No car near gate
   $f_0 = 1$ Gate UP

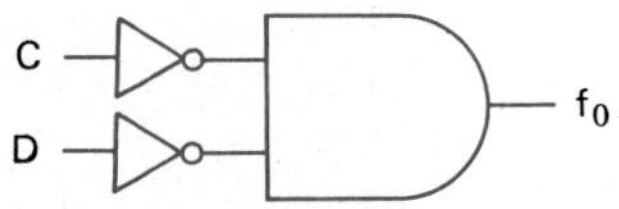

FIGURE   A46
Solution to Exercise 34–9, problem 5.

INDEX

864

$$r_{ac} = \frac{\Delta v_f}{\Delta i_f} \quad \text{(ohms)} \tag{17-4}$$

$$E = V_F + V_{R_L} \tag{17-5}$$

$$E = V_F + IR_L \tag{17-6}$$

$$I = \frac{E - V_F}{R_L} \tag{17-7}$$

$$I = \frac{-1}{R_L} V_F + \frac{E}{R_L} \tag{17-8}$$

$$\Sigma V = \Sigma E \tag{18-1}$$

$$V_1 + V_2 + V_3 = E_1 + E_2 + E_3 \tag{18-2}$$

$$IR_1 + IR_2 + IR_3 = E_1 + E_2 + E_3 \tag{18-3}$$

$$(a + b)^2 = a^2 + 2ab + b^2 \tag{19-1}$$

$$(c - d)^2 = c^2 - 2cd + d^2 \tag{19-2}$$

$$ax^2 + bx + c = 0 \tag{20-1}$$

$$x = -\frac{b}{2a} \pm \sqrt{\left(\frac{b}{2a}\right)^2 - \frac{c}{a}}$$

$$\text{or} \quad x = \frac{-b}{2a} \pm \sqrt{\left(\frac{-b}{2a}\right)^2 - \frac{c}{a}} \tag{20-2}$$

$$x = -\frac{b}{2a} \pm \sqrt{\frac{b^2}{4a^2} - \frac{4ac}{4a^2}}$$

$$= -\frac{b}{2a} \pm \sqrt{\frac{b^2 - 4ac}{4a^2}}$$

$$= -\frac{b}{2a} \pm \frac{\sqrt{b^2 - 4ac}}{2a}$$

$$x = \frac{-b \pm \sqrt{b^2 - 4ac}}{2a} \tag{20-3}$$

$$y = ax^2 + bx + c \tag{20-4}$$

$$x_0 = -\frac{b}{2a} \tag{20-5}$$

$$y_0 = ax_0^2 + bx_0 + c \tag{20-6}$$

$$N_{dB} = 10 \log \left(\frac{P_{out}}{P_{in}}\right) dB \tag{23-1}$$

$$N_{dB} = 10 \log \left(\frac{E_{out}^2/R_{out}}{E_{in}^2/R_{out}}\right)$$

$$= 10 \log \left[\left(\frac{E_{out}}{E_{in}}\right)^2\right]$$

$$= 2(10) \log \left(\frac{E_{out}}{E_{in}}\right)$$

$$N_{dB} = 20 \log \left(\frac{E_{out}}{E_{in}}\right) dB \tag{23-2}$$

$$N_{dBm} = 10 \log \left(\frac{P}{1 \text{ mW}}\right) dBm \tag{23-3}$$

$$N_{dBmV} = 20 \log \left(\frac{V}{1 \text{ mV}}\right) dBmV \tag{23-4}$$

$$i = \frac{E}{R} \left[1 - e^{-t(L/R)}\right] \tag{23-5}$$

$$v_L = Ee^{-t/(L/R)} \tag{23-6}$$

$$v_R = E[1 - e^{-t/(L/R)}] \tag{23-7}$$

$$i = \frac{E}{R} e^{-t/RC} \tag{23-8}$$

$$v_C = E(1 - e^{-t/RC}) \tag{23-9}$$

$$v_R = Ee^{-t/RC} \tag{23-10}$$

$$y = Y_{mx}(1 - e^{-t/\tau}) \tag{23-11}$$

$$y = Y_{mx}e^{-t/\tau} \tag{23-12}$$

$$T = 1/f \text{ s} \tag{28-1}$$